Contents

KU-757-918

GATEWAY COLLEGE
LE2 7BY
LIBRARY
ACC. No. 17707
Dewey No. 598

Preface

In July 1978 a joint committee of the Royal Society and the Institute of Biology produced a report on the teaching of Human Biology. In their discussion of syllabus content the Committee stressed the necessity of 'a clear committment to practical work for pupils' and emphasised that the practical work should 'illustrate the investigatory quality of science as well as providing the basic laboratory skills and habits'. We have endeavoured to incorporate these desirable features into this book, together with the factual content currently required by all Human Biology syllabuses at this level and certain extra topics suggested by the joint committee. These latter include botanical topics such as photosynthesis and soil science as well as the life of Man from birth to senility, health and disease, homeostasis and genetics. Broader aspects of the everyday life of Man, from his clothing to his education, have also been covered, each with the necessary factual material supported by an enquiry approach.

Furthermore, the committee emphasised that examinations must reflect this new approach by rewarding candidates skilled or experienced in practical activities and investigation. We consider that the investigative approach, the questions asked by and for the child, together with the emphasis on understanding in this book will fulfill these requirements. In addition, the chapters dealing with the socially important topics of ecology, population, birth control, pollution and disease are designed to help young persons to appreciate the problems of the world in which they live and hopefully to contribute towards the judgements they will need to make in the future.

The book covers material required by all existing examination boards in the United Kingdom and overseas.

Introductory Chapter

I 1.00 SI Units

Historical note

The idea of units based on the decimal system was conceived by Simon Stevin (1548–1620). In the early days of the French Academie des Sciences (founded 1666) decimal units were also considered, but the development and adoption of the metric system followed the French Revolution.

Talleyrand, the statesman, was advised by scientists to establish a system of weights and measures based on the metre as the unit of length and the gram as a unit of mass. The metre was intended to be one ten-millionth part of the distance from the North Pole to the equator at sea level, and the gram was to be the mass of 1 cubic centimetre of water at 0°C.

In 1873, the British Association for the Advancement of Science selected the centimetre and the gram as basic units of length and mass for physical purposes. The second was adopted as the base-unit for time and thus gave rise to the *centimetre-gram second* (c.g.s.) system. At the beginning of the twentieth century, practical measurements in this system were replaced by larger units, the metre and the kilogram. These, combined with the second as the unit of time, gave the *metre-kilogram-second* (MKS) system.

Certain other base units of temperature, electric current and light intensity have been added since, and in 1960 the comprehensive system was named the *International System of Units*, abbreviated to SI units in all languages.

I 1.10 Important rules for use of SI units

There are three categories of SI units:

a) Base units
There are six base units.

Quantity	SI unit	Symbol
Length	metre	m
Mass	kilogram	kg
Time	second	s
Electric current	ampere	A
Thermodynamic temperature	kelvin	K
Luminous intensity	candela	cd

Table I.1 Base Units

Quantity	SI unit	Symbol	Expressed in terms of SI units or derived units
Force	newton	N	$1N = 1\ kg\ m/s^2$
Work, energy quantity of heat	joule	J	1J = 1 Nm
Power	watt	W	1W = 1 J/s
Quantity of electricity	coulomb	C	1C = 1 As
Electric potential	volt	V	1V = 1 W/A
Electric resistance	ohm	Ω	1Ω = 1 V/A

Table I.2 Derived units

b) Derived units
These are stated in terms of base units, and for some of these derived units, special names and symbols exist. The useful ones in biology are listed in table 1.2.

c) Supplementary units
This third type of SI unit is not likely to be used in biology at this level. It includes the units for plane and solid angles.

I 1.11 Decimal multiples and sub-multiples are given below:

Multiplication factor for the unit	Prefix	Symbol
10^{12}	tera	T
10^{9}	giga	G
10^{6}	mega	M
10^{3}	kilo	k
10^{2}	hecto	h
10	deca	da
10^{-1}	deci	d
10^{-2}	centi	c
10^{-3}	milli	m
10^{-6}	micro	μ
10^{-9}	nano	n
10^{-12}	pico	p
10^{-15}	femto	f
10^{-18}	atto	a

Table I.3 Decimal multiples and sub-multiples

Thus:

$$1\ cm^3 = (10^{-2}\ m)^3 = 10^{-6}\ m^3$$

1 centimetre (cm) = 1 x 10^{-2} m
1 nanometre (nm) = 1 x 10^{-9} m

It should be noted that the kilogram is somewhat out of place in the above table since it is a basic SI unit. It is sometimes more convenient in the school laboratory to work in terms of grams and cubic centimetres, but where possible the basic units should be used.

I 1.12 Special units with names not included in the SI system

Certain specially named units in biology are in common use but do not form part of the SI system. These should be progressively abandoned.

The following can continue to be used:
minute (min) = 60 s; hour (h) = 60 min = 3 600 s; day (d) = 24 h.

Quality	Unit name and symbol	Conversion factor
Length	Ångström (Å)	10^{-10} metre = 0.1 nm
Length	micron (μm)	10^{-6} metre = 10^{-3} mm
Volume	litre (l)	10^{-3} metre^3 = 1 dm^3
Mass	tonne (t)	10^3 kilogram = 10^3 kg
Heat energy	calorie (cal)	4.1855 joule = 4.1855 J

Table 1.4 Special Units

I 1.13 Conventions agreed for writing SI units

a) Units should be written out in full or using the agreed symbols

b) The letter 's' is never used to indicate a plural form, thus 10 kg and 10 cm^3, **not** 10 kgs or 10 cms^3.

c) A full stop is **not** written after symbols except where the symbol may occur at the end of a sentence.

d) Capital initial letters are never used for units except in the case of certain units named after famous scientists e.g. N (Newton) and W (Watt). Names of units written in full do not have a capital letter, e.g. metres, kilogram and second. Even those named after scientists do not have a capital when written, thus newton and watt.

e) When symbols are combined as a quotient e.g. metre per second, they can be written as m/s or ms^{-1}. The use of the *solidus* (stroke) must be restricted to once only. In acceleration it must be m/s^2 and not m/s/s.

f) The use of the raised decimal point 2·1 is **not** correct. The internationally accepted decimal sign is placed *level with the feet* of the numerals, e.g. 2.1. The comma is no longer used in large numbers to divide them into groups of three digits. A space should be left instead so the figure appears as 345 423 123 and **not** 345,423,123.

g) The mole replaced the gram-molecule, gram atom etc. It is not based upon the kilogram, which is the SI unit, but on the gram. The mole is the amount of substance which contains as many entities as there are atoms in 12 g of carbon 12. The number of entities in a mole is 6.022 169 x 10^{23} and is called the *Avogadro constant.* Molar means *per mole.*

I 1.14 Values of imperial units in terms of SI units

Length	1 yd = 0.914 4 m 1 ft = 304.8 mm 1 mile = 1.609 344 km	Area	1 in² = 645.16 mm² 1 ft² = 0.092 903 m² 1 yd² = 0.836 127 m²
Volume	1 in³ = 16 387.1 mm³ 1 ft³ = 0.028 311 68 m³	Mass	1 lb = 0.453 592 37 kg
Density	1 lb/in³ = 2.767 99 x 10⁴ kg/m³ 1 lb/ft³ = 16.018 5 kg/m³	Power	1 horse power = 745.700 W

Table I.5 Imperial units in terms of SI Units

I 2.00 Use of the hand lens and the microscope

The biologist should only accept the evidence of his senses when investigating organisms. His eyes are the most useful, and they can be helped by instruments which produce a magnified image of the object under investigation.

I 2.10 The hand lens

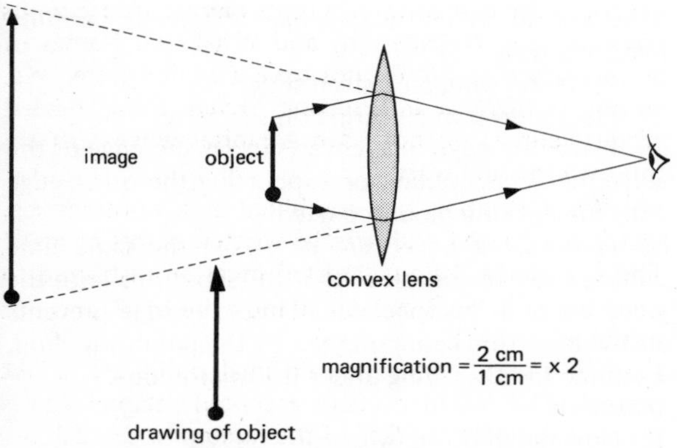

fig I.1 The hand lens and magnification

A hand lens is a convex lens generally mounted in a frame. In order for it to be used as a magnifier, the lens should be placed a short distance from the eye, and the object under investigation brought towards the lens until the enlarged image can be seen. The lens is marked with its magnifying power, such as x8 or x10, which is an indication of how much larger the image has been made compared with the object. If a drawing is to be made with the help of the lens, then the magnification of the drawing in relation to the size of the object must be calculated. The magnification of the lens bears no relation to the size of the drawing.

$$\text{Drawing magnification} = \frac{\text{linear dimension of the drawing}}{\text{linear dimension of the object}}$$

I 2.20 The microscope

The structure of the microscope is essentially that of an instrument using two convex lenses to obtain a greatly enlarged image of a very small object. Examine a microscope and note its parts, as shown in fig I.2.

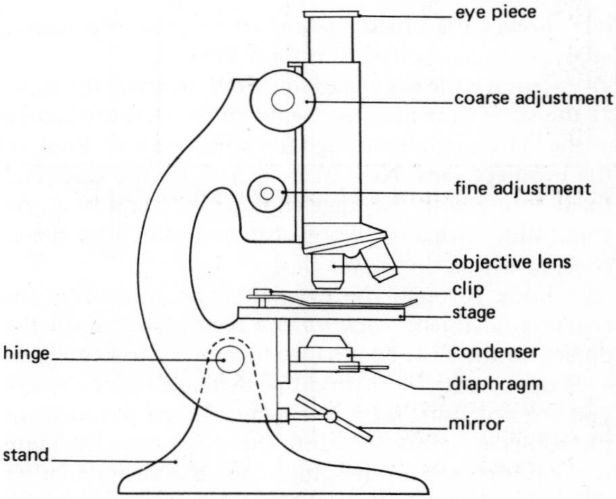

fig I.2 The structure of the microscope

I 2.21 Handling the microscope

A microscope is a very expensive instrument and should be *handled with great care* in order to maintain its precision. Pay particular attention to the following rules when you use the microscope:

i) Always *use two hands* to carry the instrument. Hold the microscope with one hand and place the other under the base.

ii) When putting the microscope on the bench or table, place it *down carefully* in order that the delicate mechanism is not jarred.

iii) *Clean the lenses* by wiping them with lens paper. Never touch the lenses with the fingers or use a coarse cloth for cleaning purposes. The lenses should never be wetted.
iv) Keep the stage of the microscope *dry and clean*. Wipe it immediately if it becomes wet.
v) *Do not tilt* the microscope when using a wet preparation on the slide.
vi) In order to protect the objective lens always cover the object on the slide with a cover *slip*.
vii) Always move the lens *up* when focusing, to avoid breaking the slide.

I 2.22 Using the microscope

i) The instrument should be placed on the bench or table with the arm towards you and the stage away from you. Sit behind the microscope in a comfortable position.
ii) Light must be made to shine through the object: this could be light from a window or an electric lamp on the bench. The mirror under the stage is used to reflect the light. Swing the mirror so that the flat surface is uppermost.
iii) Rack up the sub-stage condenser until it is within 5 mm of the stage.
iv) Lower the objective lens by means of the coarse adjustment to about 5 mm from the stage.
v) Remove the eyepiece lens and look down the tube, at the same time moving the mirror so that the source of the light is visible through the objective lens. Replace the eyepiece lens. Now place a slide on the stage and fix it firmly with the stage clips. This should be a prepared slide with a small organism mounted in gum and covered with a thin cover slip.
vi) Look through the eyepiece lens and, using the coarse adjustment, *rack up the tube slowly* until the object on the slide comes into focus. Try to keep both eyes open, but if you find this difficult cover the eye you are not using with one hand. Adjust the focus of the sub-stage condenser until the window or the lamp is super-imposed on the slide. Then put the condenser just out of focus, so that the window or lamp disappears. The lighting should now be optimum.
vii) The object can be focused clearly at different levels, according to the thickness of the specimen on the slide, by using the fine adjustment screw.

I 2.23 Investigating the image

Procedure

Cut out a letter ‘p’ from a newspaper, place it in the middle of a clean slide and cover with a cover slip. Put the slide on the stage of the microscope and focus as described above.

Questions

1 Describe the appearance of the newspaper. How is it different from that seen by the naked eye?
2 Does the letter look larger?
3 Is the letter still the same? Which of the following shapes is similar to the image of the letter ‘p’?

p b q d

4 How would you describe the image of the letter ‘p’ as seen under the microscope?

I 2.24 Making a temporary slide mount

Procedure

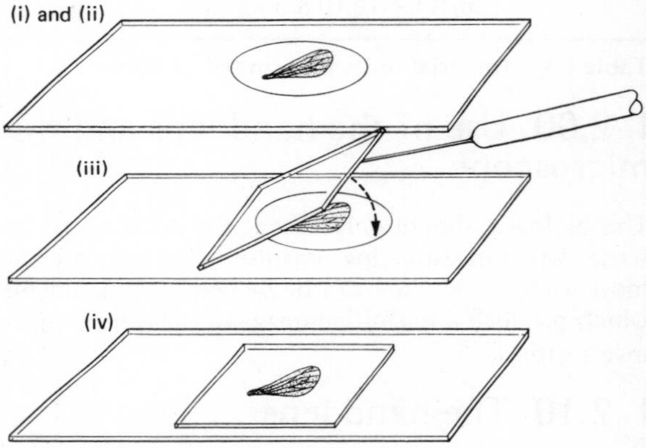

fig I.3 A temporary mount of a fly’s wing

Take a dead fly and with a pair of forceps carefully remove a wing.
i) Place the wing on a clean slide.
ii) Place a drop of glycerine on the wing.
iii) Place one edge of a cover slip on the edge of the glycerine drop, holding or supporting the other edge with forceps or a mounted needle.
iv) *Lower the cover slip gently* at the same time pulling away the needle so that the slip drops onto the glycerine and the specimen. This technique prevents air bubbles from being trapped by the mounting fluid. Examine the fly’s wing under the microscope.

Questions

1 How much of the wing can you see?
2 How does its appearance compare with what can be seen with a hand lens?
3 Focus carefully on the small bars in the wing. Now move the slide to the right whilst observing down the microscope. In which direction does the image move?
4 Now push the slide away from you. Which way does the image move now?

I 2.25 Using the high power of the microscope

Procedure

i) If the microscope is not already adjusted for light and low power work, proceed as in section I 2.22.
ii) Position the object on the slide *in the centre of the field* of view. This is important for tiny structures since the field of view is smaller under high power.

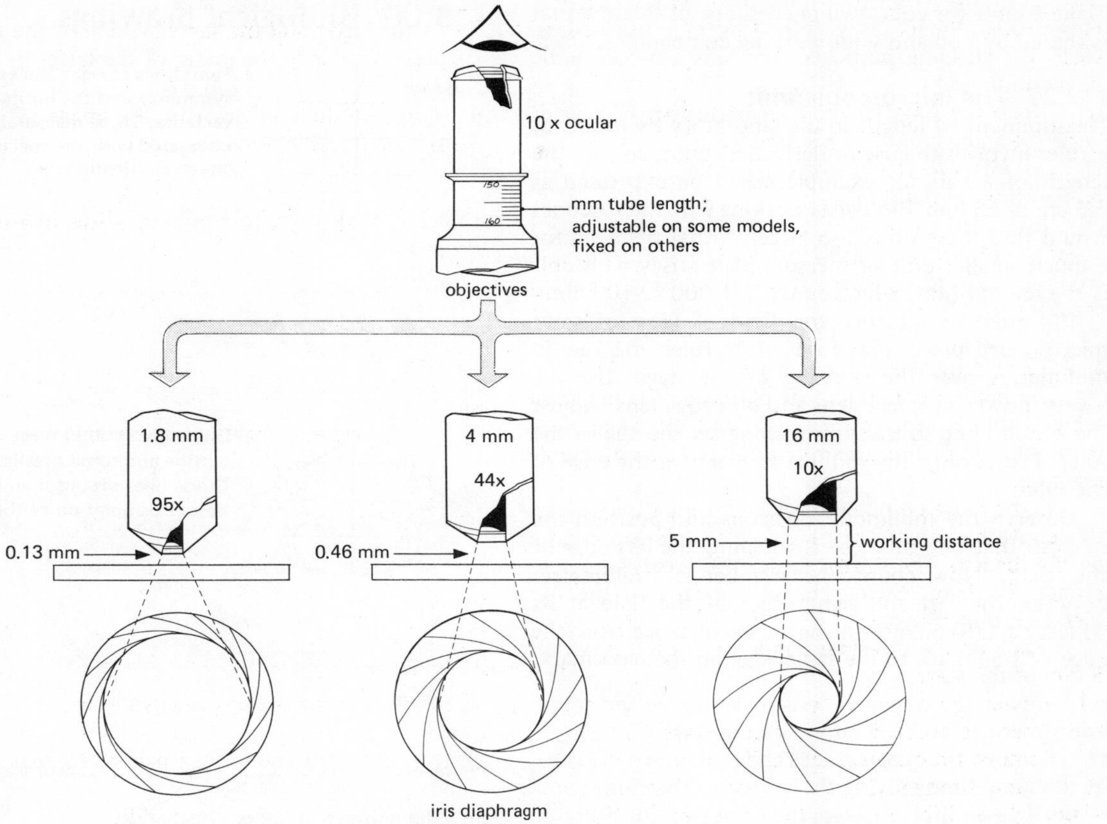

fig I.4 The relationship between the working distance of the objective lens and adjustment of the diaphragm

iii) Bring the high power objective lens into position by swinging the revolving nosepiece or screwing the appropriate lens into the tube.
iv) Look at the stage from the side and lower the tube and *objective lens* until it is *almost touching the slide.* It is of help when performing this operation to watch the reflection of the objective lens in the glass slide. Aim to make the image and lens almost meet.
v) Look into the microscope through the eyepiece lens and *slowly* raise the tube by means of the *fine adjustment screws* until the image comes into focus. Move the slide so that the image is in the centre of the field of view (remember your investigation of this aspect in section I 2.24).
vi) The diaphragm can now be adjusted to allow the light to give the best viewing conditions. Fig 1.4 shows the relationship between objective lens and diaphragm that gives the best results.

I 2.26 Magnification and the microscope

The microscope makes things look much bigger than a hand lens and this magnification can be up to 500 times or more on your microscope. Your microscope may have two or three objective lenses and one or two eyepiece lenses. The top of the eyepiece lens will have its particular magnification written on it, such as x7 or x10. Sometimes these figures may be marked on the sides of the lenses. The objective lenses are also marked in this way, and the low power objective that you used in section I 2.22 will probably have x10 (or 16 mm) as its magnification. There is a high power objective lens which may have x44 or (4 mm) as its magnification.

The total magnification using any combination of these lenses is obtained simply by multiplying the magnification of the eyepiece lens by that of the objective lens.
Thus:

Objective lens	Eyepiece lens	Magnification of the microscope
x 10	x 7	x 70
x 44	x 7	x 308
x 10	x 10	x 100
x 44	x 10	x 440

Table I.6 Magnification

Make a table for your own microscope or the one that is shared by you and your working companions.

I 2.27 The microscopic unit

Measurement of length in the laboratory by means of a ruler involves the use of derived SI units, so that the length of a hair for example could be expressed as 6.5 cm or 65 mm. Biologists working with microscopes would find these units too large, so that the need for a much smaller unit of measurement arises. This unit is the *micron* (μm) which equals 1/1 000 or 10^{-3} mm.

In order to measure the field of view of your microscope, place the edge of a ruler marked in millimetres over the opening of the stage. Use the lowest power eyepiece lens and objective lens. Adjust the bench lamp so that it is shining on the top of the ruler. Focus onto the millimetre marks at the edge of the ruler.

Observe the millimetre divisions and position the ruler so that one of these is touching the left edge of the field. Now count the number of millimetres between the left and right edges of the field at its widest part. Remember 1 mm is the distance from the edge of one mark to the same edge on the next mark.

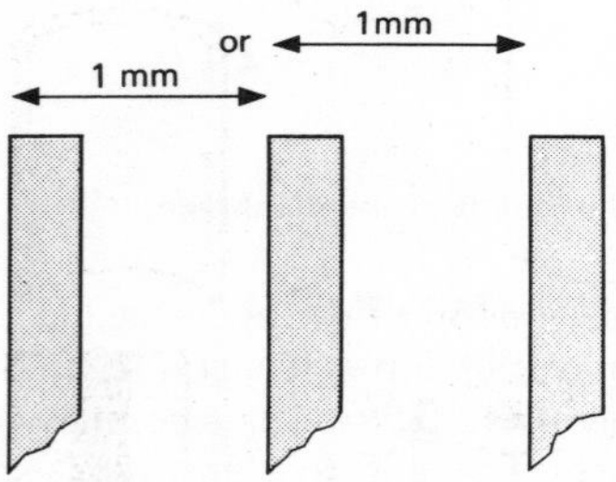

fig I.5 View of marks of a ruler under the microscope

Questions

1 What is the approximate diameter of the low power field?

2 Convert this figure to microscopic units.

3 Repeat this exercise and complete the following table:

I 3.00 Biological drawings

a)

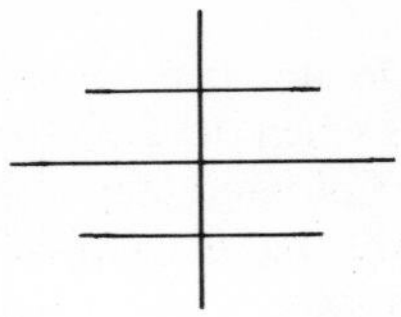

Faint lines at right angles give the symmetry and the limits of the vertebra. These measurements compared with the specimen give the magnification.

b)

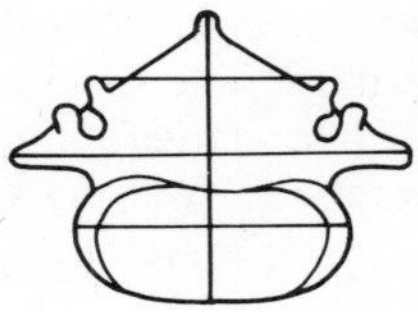

Guide lines still in place with the outline and some details filled in. Guide lines should then be erased before completion of the drawing.

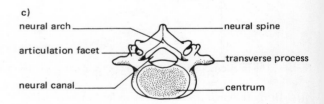

c)

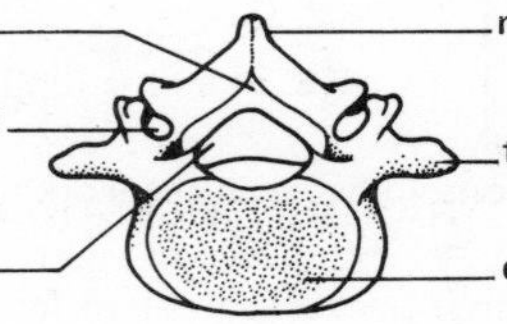

fig I.6 Three stages in drawing the posterior view of a lumbar vertebra
(a) Preliminary limits and symmetry
(b) An intermediate stage of drawing
(c) A completed drawing and labels

The human biologist must include as part of his work the accurate and careful study of living and dead animals. Observations made on the structure and behaviour of animals may include dissection or making sections in order to display the internal parts. These observations are recorded in a *drawing* as a permanent record of what has been seen.

Lens combinations	Eyepiece lens	Objective lens	Magnification	Field of view (μ)
1				
2				
3				
etc				

The drawing made in this way is an important part of the training of a biologist, and there are a number of points to remember.

i) Draw in *lead pencil* only. Do not use coloured pencils, coloured crayons or pens when drawing from specimens during practical work. Coloured crayons or pens may be useful when producing diagrams for theory notes or examination questions.

ii) Keep the *pencil sharp* and use an HB grade. This is hard enough for all thicknesses of line.

iii) All drawings should be as *large as possible* within the space available. Never take less than half a page, and if possible use a whole page. Leave enough space at either side for labels or annotations, and leave additional space beneath the drawing for a full title.

iv) Observe carefully and *draw first a faint outline* determining the overall magnification. The specimen should be measured in at least two dimensions to ensure that it is drawn accurately in the correct proportions. Never draw less than x 2 for a small specimen such as a vertebra.

v) Fill in the details of your drawing using *firm lines* and *simple outlines. Do not use shading.* Decide where to draw your lines and then do so without lifting the pencil from the paper until the line is completed. A bold single stroke is required, not a succession of half-hearted scratchings. Observe carefully and put all your observations into your drawing.

vi) Take great care with labels and label lines. The latter should *never cross* each other. *Print* the labels *in pencil* so that you can rub out mistakes. The label line should be exactly at the centre of the structure labelled. Labels must be correct e.g. never use a plural for a single structure: facet, not facets.

vii) Drawings should have a *full title below* to show:

a) the *name of the organism,*

b) the *position* and *type* of section (or dissection) and

c) its *magnification.*

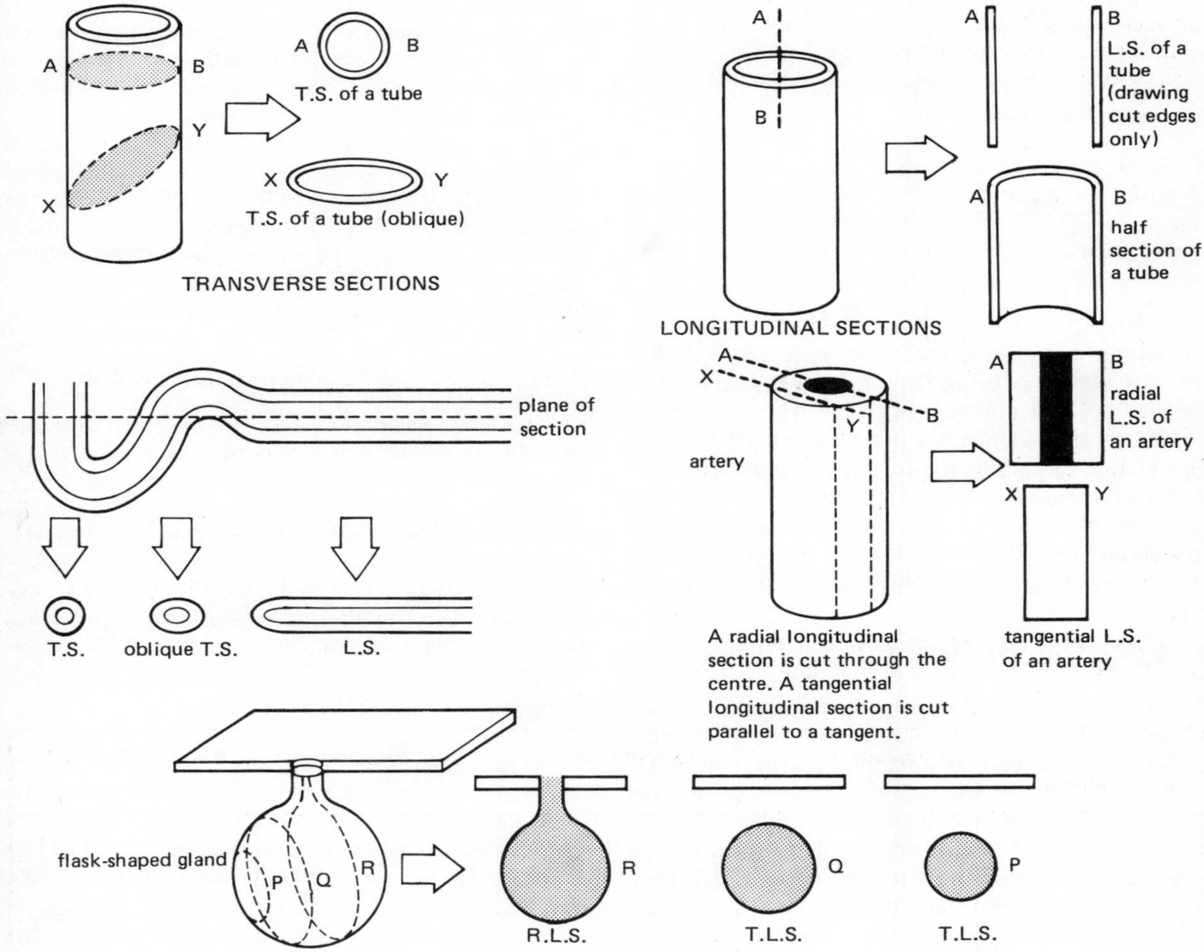

fig I.7 Types of section

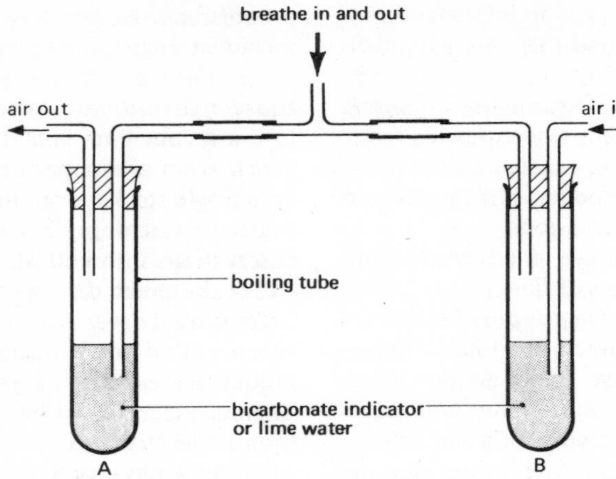

fig I.8 The experimental apparatus for bubbling inhaled and exhaled air through bicarbonate indicator solution

viii) Where a large number of small structures are present in a specimen there is no need to repeat all of them in the drawing. This often leads to careless rapid repetition and consequently a poor drawing. Draw with great care from 5 to 10 structures to show their arrangement, e.g. oocytes in an ovary, epithelial cells in a section through the gut.

I 3.10 Types of section

Almost any microtome section through an organ of the body will cut through tubular structures, such as blood vessels, in different planes. Some of the more usual variations are shown in fig I.7.

I 3.20 Apparatus

Apparatus should always be *drawn in section* as though the cut edge were being viewed. Thus any tube which permits the flow of liquids or gases will be shown open.

Fig I.8 shows apparatus designed to investigate the production of carbon dioxide and its presence in exhaled air. Notice that there is a free flow of air through the tube into the first test-tube containing bicarbonate indicator, then into the second test-tube and its bicarbonate indicator and finally out into the air.

I 4.00 Scientific method and records

Science is built upon *facts* which are *agreed by trained observers.* No scientist, however, is satisfied with simply collecting facts until they fit into a pattern. Biology, which may involve the study of a very complicated living organism or a habitat, necessitates only that patterns relate to small parts of systems of the organisms or habitat. Once the pattern has been developed, the next stage is the *prediction of new observations* based on the pattern. If these predictions prove accurate then a *law,* which is a statement of this pattern or order, can be put forward.

A *hypothesis* is a suggested explanation of certain observed phenomena. It can only be tested by *experimentation.* Very often the hypothesis is incorrect, but it is still useful if it leads to experiments which produce a more acceptable hypothesis. The majority of hypotheses in scientific experimentation prove to be incorrect and so they are discarded.

An experiment is usually performed *to test a hypothesis,* but it may equally produce unexpected facts which give rise to further hypotheses. Thus the scientific enquiry is sent in a new direction. Experiments are *investigations* carried out *under carefully controlled conditions* so that the various factors operating are identifiable. Generally only *one variable factor is observed.* All experiments need to be planned very carefully so that they do in fact investigate the variable with which the hypothesis is concerned.

In order to ensure that conclusions drawn from the experiment are valid, it is essential to carry out at the same time *a control experiment* in which the factor being investigated is absent. For example, in an experiment to investigate the action of saliva on starch, the control would have distilled water in place of saliva. Therefore, when starch is converted to sugar in the experiment, we can be sure that this is caused by the saliva and not by any other factor.

Any experiment can be considered under the following headings.

1 Hypothesis

This may be presented as a question which the experimenter is trying to answer, e.g.: 'Does saliva convert starch to sugar?' Alternatively the *object or aim* of the experiment could be stated e.g.: 'To investigate the relation between height and age in school children between the ages of 10 and 16'. The aim should be stated as clearly as possible in order to give an accurate idea of the scope of the experiment.

2 Method

This describes the means used to investigate the hypothesis. It should be a full description of the apparatus, the chemicals used, the living organisms, the measurements made, and any other information that is necessary for any other experimenter to repeat the experiment. **No experiment is valid unless its methods and results can be repeated by another scientist.** The method description should include *drawings* of experimental *apparatus,* together with the *control* apparatus. If apparatus is not used then details of that test group and the control group, measurements and any other relevant facts are required.

3 Results

The results are always expressed as a *series of measurements* or *observations,* and it is important that laboratory notebooks should include full information. It may be that there are minor results that do not fit in the general pattern and such results should not be overlooked. The degree of accuracy of the measurement should always be stated. Results are often best displayed in tabular form. *Histograms or graphs* may be a final part of the presentation.

4 Discussion

The final part of the experiment is the discussion of the results and how they affect the hypothesis. Great care must be taken when generalising from them. It may be that *no* final conclusions can be drawn but *further hypotheses suggested* as a result of this particular experiment must be tested. Thus the discussion may form a basis for further experimentation to test these hypotheses and not until a whole series of experiments have been performed can conclusions be drawn.

In the chapters which follow, all practical experimental work should be written up in this way. Where questions are asked about experiments, the *answers should be included in the discussion.* The method need not be written out completely since the procedure is clearly described in the text of each chapter. It can save time to refer simply to the page and section of the book where the method is set down. The results and discussions should always be complete because this is your own work. Below is an example of the recording of an experiment by a fourteen-year-old schoolboy, whose writing and presentation are of quite a high standard.

Experiment No. 00

Hypothesis

We have discovered that saliva changes starch into sugar. Does variation in temperature have any effect on this reaction?

Method

A water bath was set up (a treacle tin 2/3 full of water over a bunsen) and the temperature of the water was brought to 20°C. The bunsen burner flame was then turned down and the water temperature was checked from time to time to make sure that it stayed constant. Saliva was prepared by spitting into a beaker and diluting with an equal volume of distilled water. 1 cm^3 of the diluted saliva was placed in a test tube and 1 cm^3 of a 1% starch suspension was placed in a second test tube. Both tubes were then placed in the water bath and then left there for five minutes in order for their contents to reach 20°C.

Meanwhile, a white tile was prepared with twelve spots of iodine solution (in potassium iodide) arranged in three rows of four.

The test tube with starch suspension was then shaken and its contents poured into the test tube with saliva. The starch and saliva were mixed with a glass rod and a drop of the mixture was placed, using the rod, onto one of the iodine spots on the tile. The colour of the spot was recorded, the test tube was returned to the water bath and the glass rod was washed. The operation was carried out three further times at thirty second intervals.

The entire experiment was repeated at 40°C and again at 60°C.

Apparatus

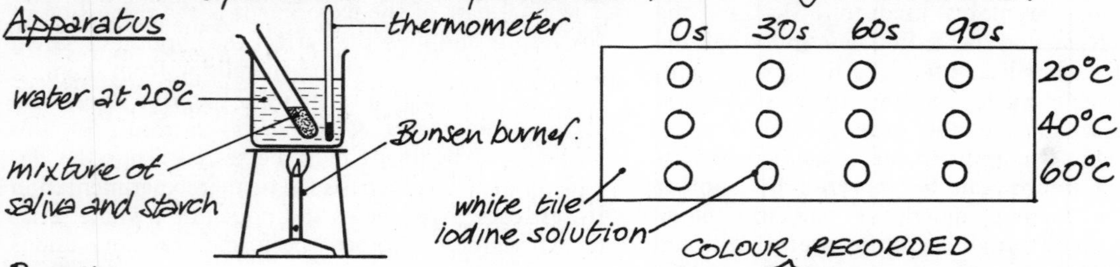

Results.

Temperature	COLOUR RECORDED 0 seconds	30s	60s	90s.
20°C	blue-black	blue-black	blue	yellow-brown
40°C	blue-black	yellow-brown	yellow-brown	yellow-brown
60°C	blue-black	blue-black	blue	yellow-brown

Discussion

A blue-black colour shows the presence of starch. When the colour is yellow-brown (the original iodine colour) it means that the starch has been changed to sugar.

The class results were compared and they all showed the pattern seen in the results table. It can be seen that quickest disappearance of starch occurs at 40°C

40°C is the nearest of the three temperatures to the normal human body tempera-ture (37°C) and therefore it is not surprising that saliva works best at this temperature. We should check that the starch has in fact been changed to sugar by testing with Benedict's or Fehling's solution.

I 4.10 Representing data obtained from experiments

An experiment needs to be *repeated several times* to ensure that results obtained are consistent, but when working in class there is seldom time for this to be done. The same result can be arrived at when the experiment is performed by *several groups* of pupils in a class. The results of the class can be tabulated on the blackboard and used for analysis to test the hypothesis.

Suppose that the heights of all the pupils in the class are measured, and they are to be recorded and analysed to find out how they vary. The results would first be set out as follows:

Pupil	Height (cm)	Pupil	Height (cm)
1	132	19	146
2	140	20	128
3	135	21	145
4	129	22	136
5	133	23	135
6	127	24	142
7	134	25	143
8	139	26	121
9	141	27	148
10	130	28	134
11	136	29	127
12	123	30	138
13	139	31	136
14	132	32	130
15	137	33	137
16	131	34	129
17	133	35	144
18	125	36	126

Table I.7 Height range of 36 students

Height (cm)	Frequency f	Total in each group
120–122	1	1
123–125	11	2
126–128	1111	4
129–131	~~1111~~	5
132–134	~~1111~~ 1	6
135–137	~~1111~~ 11	7
138–140	~~1111~~	4
141–143	111	3
144–146	111	3
147–149	1	1

Table I.8 The frequency of occurrence of each class of height

The range of heights is from 121 cm to 148 cm which gives 28 groups at 1 cm intervals. This is rather a large number and so for simplicity they should be classified into *ten groups* each having *three heights*, e.g. 120 to 122. The number of groups will depend upon the total number of measurements, but as a general rule there should be between 6 and 20 groups.

From the data in table I.7 we can construct the following table I.8 which shows the number of pupils in each height group. Notice that the fifth mark under frequency is used to cross out the previous four, giving a group of five marks.

We now have the *frequency of occurrence* of each height group. One method of displaying this frequency is by a *histogram* or *bar chart.* The groups are shown along the horizontal axis and the frequency on the vertical axis. Rectangles are constructed whose areas are proportional to the frequency in each group. The bases of the rectangles are equal to the group widths.

A second method is by plotting a *graph.* In this case the data in table I.8 is plotted with the median measurement of each group recorded as a point on the graph.

In this particular example for pupil heights it is probably better to display the results as a histrogram rather than as a graph. If, however, you were plotting the growth of a colony of bacteria by counting the number of cells at stated intervals, then the graph would be the better method. The resulting curve, shown in fig I.11, is called an *exponential curve.*

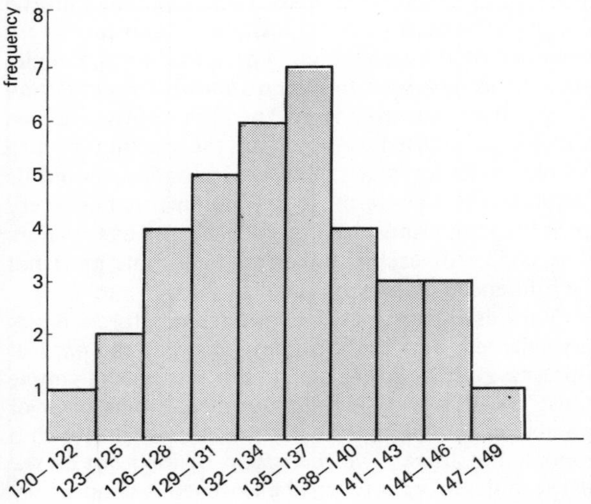

fig I.9 A histogram of heights

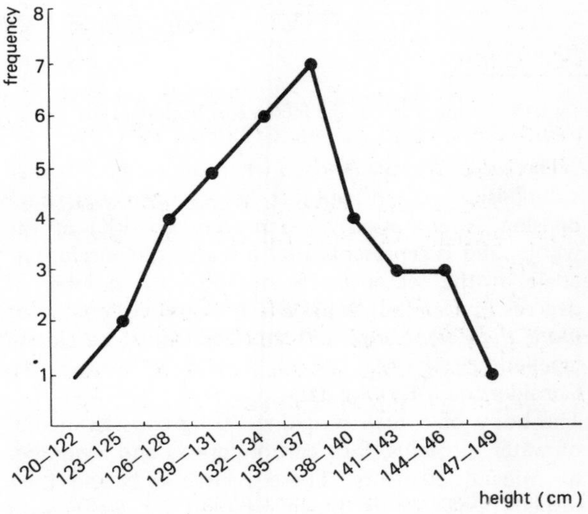

fig I.10 A graph of heights

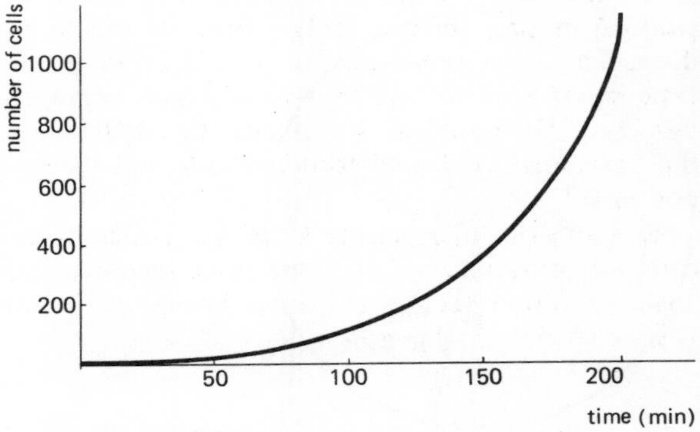

fig I.11 A graph of increase of bacterial cells

This graph also has its limitations, since after about two hours the increase in the number of bacterial cells is so rapid that the scale of the graph is unable to cope with these numbers. A more convenient method in this case would be to use a *logarithmic* scale instead of a *linear* scale for the numbers of bacteria. The exponential growth represented on a logarithmic scale would be a straight line.

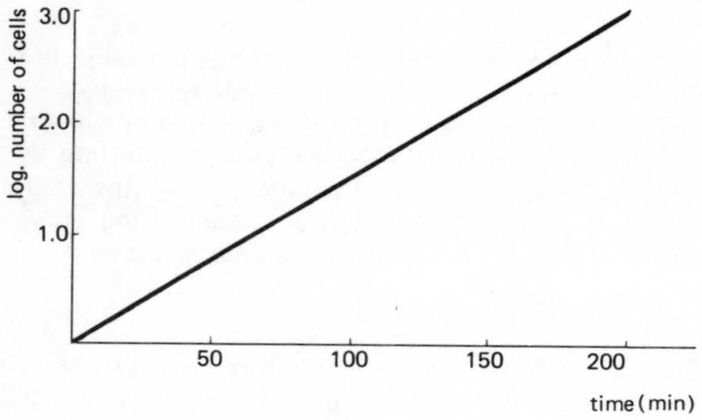

fig I.12 A log graph of the increase of bacterial cells

The *pie chart* can be used when the data contain all sub-divisions of the subject being considered. Each division is expressed by a percentage (P%) of the whole, and is represented by a sector of a circle. The angle of the sector is P% of 360° (the number of degrees in a circle). A pie chart should *not consist of more than eight segments* and each must be clearly labelled.

Consider the following data:

The body of a man weighing 70 kg consists mainly of water, proteins, fats, carbohydrates and a number of mineral elements. These constituents and their respective masses are shown in table I.9, together with the method of calculating their angles for the construction of a pie chart (fig I.13).

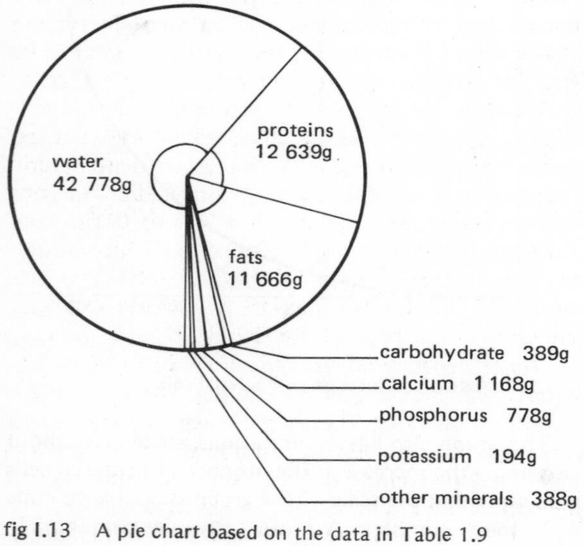

fig I.13 A pie chart based on the data in Table 1.9

Constituent	Mass (g)	Angle of sector to nearest degree (= fraction of total mass x 360)
Water	42 778	$\frac{42\,778}{70\,000} \times 360 = 220°$
Proteins	12 639	$\frac{12\,639}{70\,000} \times 360 = 65°$
Fats	11 666	$\frac{11\,666}{70\,000} \times 360 = 60°$
Carbohydrates	389	$\frac{389}{70\,000} \times 360 = 2°$
Calcium	1 168	$\frac{1\,168}{70\,000} \times 360 = 6°$
Phosphorus	778	$\frac{778}{70\,000} \times 360 = 4°$
Potassium	194	$\frac{194}{70\,000} \times 360 = 1°$
Other minerals	388	$\frac{388}{70\,000} \times 360 = 2°$
TOTAL	70 000	360°

Table I.9 Major constituents of the body of a 70 kg man

I 4.11 Sampling

Whenever quantitative estimates are made of populations of plants or animals, some sampling must be involved because it is impossible to consider every member of the population. The selection of samples must be *as representative as possible of the population* from which they are taken. Then the results obtained can be generalised to apply to the population as a whole. In order that such a generalisation should be accurate the sample must be random; that is, every possible sample must have an equal chance of selection. The choice of each member of the sample must not be influenced by previous choice.

Various factors must be considered before samples are selected, and the most important is the *kind* of information that is required. In this respect the sample must be of one type (*homogeneous*). The class of pupils measured in table I.7 was a *year group* in a school, but there is *no indication* of their *age or sex* in the table. In order to be a homogeneous sample they should be all female (or all male) and all of the same age.

The curve obtained from this sample, shown in fig I.10, is approximately a *bell-shaped* curve, but this

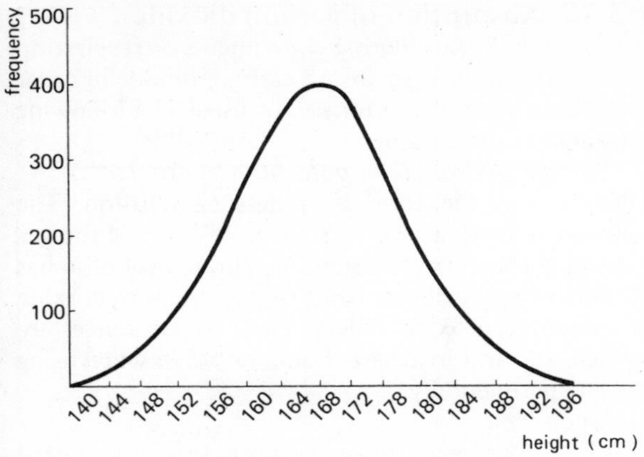

fig I.14 A graph of frequency against height produced as a result of taking a larger sample

would have been a better curve if the sample had been larger, say 100 pupils. Samples of the order of 1 000 or 10 000 are even more valid, and would produce a graph of frequency against height as shown above in fig I.14. The sample taken in table I.7 can be seen therefore to be too small and not reliable on the basis of age or sex. This raises the further point of the reliability of estimates or generalisations based on a few results. As was stated at the beginning of this section, experiments must always be repeated and plenty of results obtained in order to give a *more reliable answer.*

I 5.00 Elementary chemistry

I 5.10 Elements, mixtures and compounds

An *element* is a substance that cannot be broken down into other, different substances. The smallest particle of an element that can normally exist is called an *atom.* For any one element, the atoms are all similar to each other and different from the atoms of other elements. About 100 different elements have been identified, and each has been given a different symbol: for example, carbon (C), chlorine (Cl), iron (Fe), hydrogen (H), iodine (I), oxygen (O), nitrogen (N), sodium (Na), phosphorus (P) and sulphur (S).

Compounds are substances formed by the combination of two or more elements in a way that causes them to change chemically. The smallest particle of a compound is called a *molecule,* and this consists of atoms of two or more different kinds.

The scientific name for common salt is *sodium chloride.* This is a compound consisting of the two elements sodium (a metal) and chlorine (a gas), which have undergone a considerable change on combining. Thus a molecule of sodium chloride (NaCl) consists of one atom of sodium joined to one atom of chlorine.

Water is a compound of the two gases hydrogen and oxygen. Each water molecule consists of two atoms of hydrogen and one of oxygen, and so the formula for water is H_2O.

The term *mixture* is applied when two substances are mixed together but do not change chemically. Thus, if a spoonful of salt and a spoonful of sugar are placed in a beaker and shaken up, they form a mixture and not a compound. Similarly, *air* is a mixture of different gases, principally nitrogen and oxygen (see section 4.01 – content of air).

I 5.20 Water

Water is the most common and in many ways the most important substance to be found in living systems. The human body consists of about 65% water. From where do we obtain this water, and why is it so important?

The obvious answer to this first question is that we obtain it by drinking. This is certainly true, but in addition much of our food contains a high proportion of water: certain vegetables are 85–90% water and many desert animals never drink but obtain all their water in this way.

The answer to the second question is more complex and is related to a number of very special properties which water possesses.

1 Water is a very effective *solvent.* This means that a large number of substances dissolve in it and thus it is a medium in which all the important chemical reactions of the body can take place.

2 Water has a *high specific heat capacity.* In other words, quite a lot of heat energy is required in order to raise the temperature of a given quantity of water by a small amount. This enables organisms to be relatively independent of fluctuations in climatic conditions and also prevents them from overheating as a result of chemical reactions taking place within their bodies.

3 Water is *liquid* at the temperatures found over much of the earth's surface. It does not freeze at too high a temperature or boil at too low a temperature. Thus, as well as ensuring that enzymatic reactions (which operate most effectively at 35–40°C) can take place in a liquid medium, it is suitable for locomotion by swimming animals, for transport of materials within organisms and for the support of those plants that rely upon turgor pressure for rigidity.

4 Water *expands on freezing.* Like many other substances water contracts on cooling, but contraction stops at about 4°C. Thereafter it expands and thus a given volume of water at just above 0°C weighs less than the same volume at 4°C. Thus cold water rises to the top of lakes and oceans in winter and forms ice there first, giving some protection to the organisms

at greater depths. If this were not the case, arctic lakes would freeze solid and be barren of life.

5 Water has a *high surface tension*. This is the tendency of the surface of a liquid to contract to a minimum area. It is an important factor in the water-retaining properties of soil and in the movement of water upwards through the stems of higher plants.

6 Finally, water has a *neutral pH*: it is neither too acid nor too alkaline for many of the reactions taking place within living cells. Linked to this is the fact that it is a major source of hydrogen and oxygen for the body's reactions.

I 5.21 Absorption of water

It may be necessary to keep substances dry or to provide a situation in which there is little or no water in the atmosphere. This is especially necessary in hot, humid climates where delicate instruments may be affected by the growth of fungi under these conditions. The substances that can be used are *calcium chloride* and *silica gel*. A third substance, rather more dangerous but equally effective, is concentrated *sulphuric acid*.

I 5.22 The detection of water

The test reagents used to determine whether or not a colourless liquid is water are *anhydrous copper sulphate* and *cobalt chloride*. Anhydrous copper sulphate is a *white* powder which turns *blue* when wetted. Cobalt chloride is generally used after it has been soaked into a filter paper and dried. When exposed to water or water vapour it changes colour from *blue* to *pink*.

The final tests for the colourless liquid which is suspected of being water are the physical ones to determine whether it boils at 100°C or freezes at 0°C, and whether it has a density of 1 g/cm^3.

Usually, there is not time in the school laboratory to complete these tests and so the chemical confirmation must be sufficient.

I 5.30 Gases in experimental work

I 5.31 Absorption of oxygen

In order to absorb oxygen during experimental work, the liquid used is *potassium pyrogallate* solution. This must be kept under liquid paraffin in order to prevent it absorbing oxygen from the air. It is made by adding *pyrogallol* (125 g) to 65 cm^3 of boiled and cooled tap water and 850 cm^3 of saturated *potassium hydroxide solution*. This should be transferred to a small jar and covered with liquid paraffin.

This liquid will absorb carbon dioxide as well as oxygen but, since it is normally used in an experiment where carbon dioxide has already been extracted, this dual role is not important.

I 5.32 Absorption of carbon dioxide

It is often necessary during experiments on respiration and photosynthesis to absorb carbon dioxide in order to indicate whether it is formed or used. The following substances can be used:

i) *sodium hydroxide* or *potassium hydroxide*

These are in the form of pellets or solution. The solution is used where there is gas flow, and the gas bubbles through the solution. A saturated solution has 50 cm^3 of tap water to 100 g of potassium hydroxide or sodium hydroxide pellets. The two substances are *very caustic* and great care should be taken when using them experimentally.

ii) *soda lime*

This is less dangerous and easy to use. It is often tinted green and when it has reached the limit of absorption of the gas it changes colour from green to grey.

I 5.33 The detection of carbon dioxide

If carbon dioxide is bubbled through *lime water (calcium hydroxide)* it causes very small chalk particles to be set free in suspension and the lime water develops a milky appearance.

A more delicate method of detecting very small changes in carbon dioxide concentration involves using *bicarbonate indicators*. These are indicators of the hydrogen ion (pH) values, which vary according to the acidity of the substance. These indicators are particularly useful in respiration and photosynthesis experiments. A bicarbonate indicator appropriate for these experiments is made up as follows:

Dissolve 0.2 g of *thymol blue* and 0.1 g of *cresol red* in 20 cm^3 of *ethanol*. Weigh out 0.84 g of *sodium hydrogen carbonate* and dissolve it in 900 cm^3 of de-ionised or distilled water. Add the dye solution to the salt solution in a graduated flask and make up to 1 litre with pure water. Glassware must be clear of all traces of dust or dirt.

A working solution of this stock is made by pipetting 25 cm^3 into a 250 cm^3 graduated flask, and making up the volume to 250 cm^3 with distilled water. Before the reagent is used, air from outside the laboratory must be bubbled through by means of a filter pump. This will stabilise the colour at deep-red in the bottle or orange-red in test tubes. The colour indicates the concentration of carbon dioxide in the air at about 0.03% or 300 parts per million. When this bicarbonate indicator solution is placed in any situation where carbon dioxide is added or withdrawn, it will give a colour change as follows:

Less CO_2	←	0.03% CO_2	→	More CO_2
purple		orange-red		yellow
pH 9		8		7

This change is caused by the carbon dioxide forming more, or less, *carbonic acid,* thus decreasing or increasing the pH value.

I 5.40 Acids, bases and pH

Acids are a large group of chemicals which have the property of giving up *hydrogen ions* (H^+). This capacity is responsible for certain chemical reactions by which acids may be identified. For example, when an acid solution comes into contact with a metal, bubbles of hydrogen gas are formed.

Bases are the chemical opposites of acids. They attract hydrogen ions and react with them. The best known base is the *hydroxyl ion* (OH^-) which reacts with hydrogen ions to form water (H_2O). This particular reaction reduces the number of hydroxyl ions and hydrogen ions in a solution so that neutralisation occurs. A solution is acidic if a surplus of hydrogen ions is present and basic if a surplus of hydroxyl ions is present.

Very often in biology it is necessary to indicate how acidic or basic a solution, a soil or some other substance might be. In order to express this, there has been devised a numerical scale, *the pH scale.* This scale runs from one to fourteen.

Numbers 1 to 7 indicate decreasing acidity; pH7 is neutral; and numbers 7 to 14 show increasing basicity or *alkalinity.*

More acid ← Neutral → More basic
pH 1 2 3 4 5 6 7 8 9 10 11 12 13 14

A number of dyes change colour at different points on the pH scale and they can be used as *pH indicators.* The dyes can be absorbed onto strips of paper and these, when dipped into a solution, will change colour according to the pH. By comparing the colour with a reference chart, the pH can be determined.

The most versatile indicator is Universal Indicator, a mixture of several different indicators with a bigger range of colour changes than simple indicators such as litmus. The latter turns red in the presence of acid and blue with an alkali, but Universal Indicator shows the following range of colours in response to pH change:

pH 1 2 3 4 5 6 7 8 9 10 11 12 13 14
red pink yellow green blue indigo violet
acid neutral alkaline

Universal Indicator is used as a solution with a dropper or as an indicator paper which can be dipped in the liquid being tested.

Experiment
To determine the pH of biological substances.
Procedure
i) Obtain several substances for testing, such as vinegar, milk, urine, raw egg, lemon juice, orange juice and blood.
ii) Use small pieces of pH paper to measure the pH of the materials provided. Make a table of your results.
iii) Obtain the following liquids: hydrochloric acid (dilute, say 0.01 M), sodium hydroxide (dilute, say 0.001 M) and distilled or deionised water.
iv) Use small pieces of pH test paper to measure the pH of the three liquids in step (iii).
Questions
1 What is the pH of the two solutions of hydrochloric acid and sodium hydroxide?
2 How does the pH of the distilled water compare with that of the other two liquids?
3 How does the pH of lemon juice compare with that of the solution of hydrochloric acid?
4 Dilute the acid and the alkali with distilled water and test with pH paper. What happens to the pH value in each case?

I 5.50 Osmosis and diffusion

Experiment
What happens if a sugar solution is separated from water by a membrane such as visking tubing?
Procedure
1 Take 10–15 cm of visking tubing and wet it under a tap to make it pliable. Tie a knot tightly in one end of the tubing. Pour into the bag so constructed a 10% sucrose solution, until it is about 3 cm from the top of the bag.
2 Take a capillary tube about 30 cm in length and place it in the bag so that the end dips into the sucrose solution. Tie the open end of the bag tightly around the capillary tube, using a strong thread (take care not to cut the tubing).
3 Fix the capillary tube to a clamp and suspend the bag of sucrose solution in a beaker of water (see fig I.15).
4 Mark the initial level of the sucrose solution in the capillary tube and leave the apparatus to stand. Mark the level of the sucrose solution every ten minutes.
Questions
1 What height did the sucrose solution reach in the capillary tube after 30 minutes?
2 What explanation can you give for the change in height of the liquid in the capillary tube?
3 Is there any evidence to indicate that some form of pressure is acting in the apparatus?

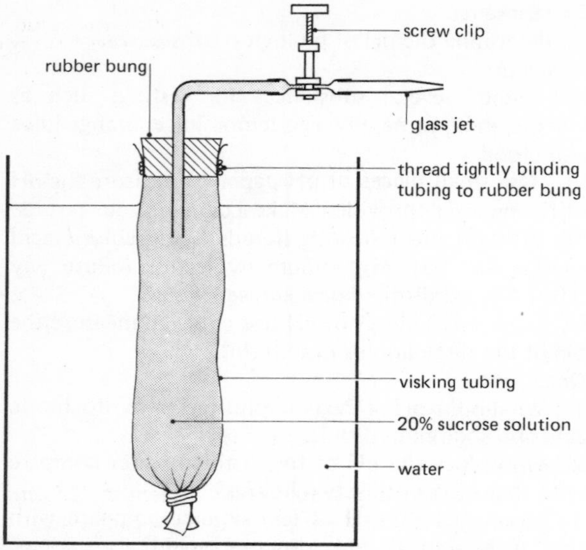

fig I.15 Visking tubing containing a sugar solution, with capillary tube attached, placed in water

Experiment

Is pressure exerted when solutions of different strengths are separated by visking tubing?

Procedure

1 Set up the apparatus as shown in fig I.16. Use visking tubing of wide diameter (3 cm) and having wet it first, tie it tightly into a knot at one end. Tie the other end tightly to a rubber bung by means of strong thread.

2 The rubber bung should have a piece of right-angled glass tubing pushed through it and into the bag. Attached to the other end of the glass tubing is a short length of rubber tubing into which is pushed another piece of glass tubing drawn into a fine jet. The rubber tubing has a screw clip attached.

3 Remove the rubber tubing from the right-angled bend, and, using a pipette, fill the bag with a 20% sucrose solution. Replace the bung and tubing and then place the bag in water in a beaker, leaving it to stand for twenty-four hours.

Questions

4 How does the appearance of the bag after twenty-four hours compare with its state at the beginning of the experiment? How does the bag feel?

5 What explanation can you give for the change in the visking tubing?

6 What happens when you unscrew the clip? Can you give an explanation of the result?

What you have discovered is the phenomenon of *osmosis*. The water surrounding the visking tubing has moved through the tiny pores in the cellulose and increased the quantity of water inside the bag. As a result the liquid has moved up the tube in the first

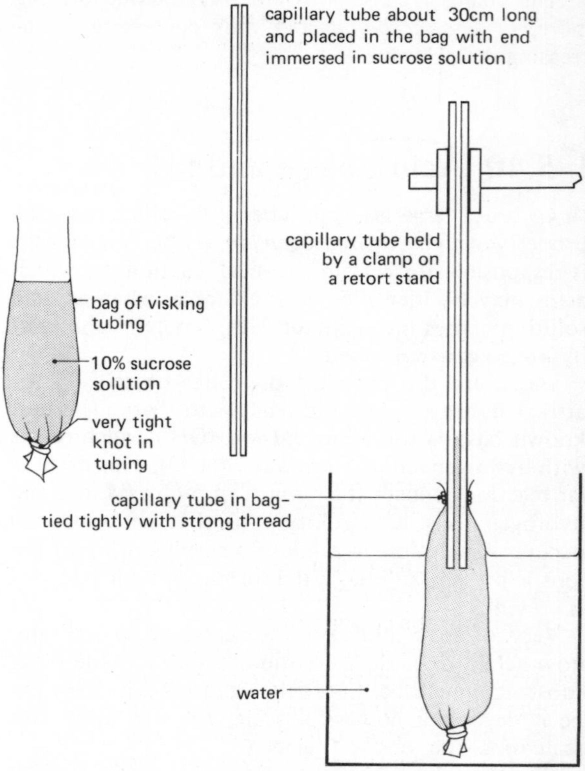

fig I.16 Visking tubing, containing a sugar solution: a closed system with a screw clip, placed in water

experiment, and filled the bag in the second experiment. If a tightly fitting plunger had been present in the bore of the capillary tube, a pressure could have been exerted preventing the rise of the liquid. This pressure would have to be equal to the *osmotic pressure* of the sucrose solution in the bag. The second experiment demonstrates this pressure, which, having built up, causes a jet of liquid to rush out on release of the screw clip. We can define osmosis as the *movement of solvent* (in this case water) through a *partially permeable membrane* from a *dilute solution* (or solvent only) to a *more concentrated solution.* In nature, the solvent is always water, and the passage of water through the membrane continues until the concentration on both sides of the membrane is equal. The simple explanation of this process is shown in fig I.17 which illustrates the partially permeable nature of the visking tubing. The tubing acts as a *molecular sieve* whose pores are so minute that they allow through molecules of water (H_2O), but not the larger molecules of sucrose ($C_{12}H_{22}O_{11}$).

It can be seen in the figure that the molecules of water can pass equally well in either direction through the pores. Since there are more molecules of water

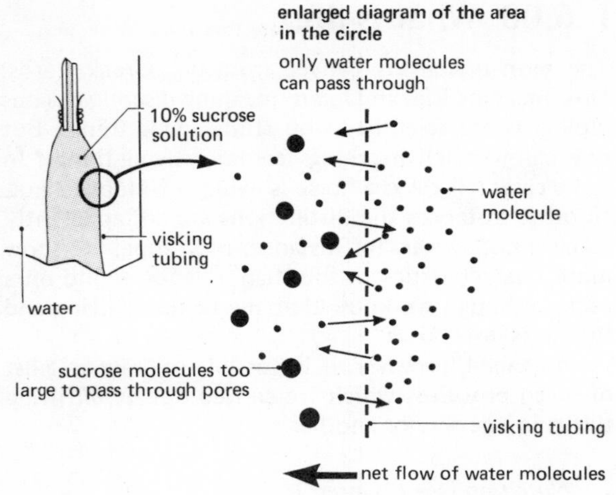

fig I.17 A diagrammatic representation of the pores present in a membrane and the movement of molecules

outside the tubing than there are inside, there is a net movement of water molecules from the water (or weak solution) to the strong solution.

This flow of water can be regarded as a particular example of the phenomenon of *diffusion,* defined as the process by which *molecules of a substance* present in a *region of high concentration* in a *fluid* (liquid or gas) tend to *move into a region of low concentration* until they are evenly *distributed.*

In the two previous experiments, the water molecules surrounding the visking tubing are present in high concentration, while those in the solution of sucrose are present in low concentration, so that the water molecules flow from where they are in a high concentration to where they are in a low concentration.

Experiment
To demonstrate diffusion in a liquid
Procedure
1 Make up a strong solution (20–30%) of copper sulphate – it must be very blue in colour.
2 Introduce 50 cm^3 of the copper sulphate solution into the bottom of a tall jar of distilled water. Use a pipette and let the copper sulphate flow gently so that it forms a distinct layer at the bottom of the jar. The division between water and copper sulphate should show as a clear line.
3 Leave the jar to stand for several days and mark the level reached by the blue colour each day.
Questions
7 When the blue colour has moved throughout the jar how does its colour compare with the colour of the original copper sulphate solution?
8 Can you give an explanation, in terms of diffusion, of any difference in colour?

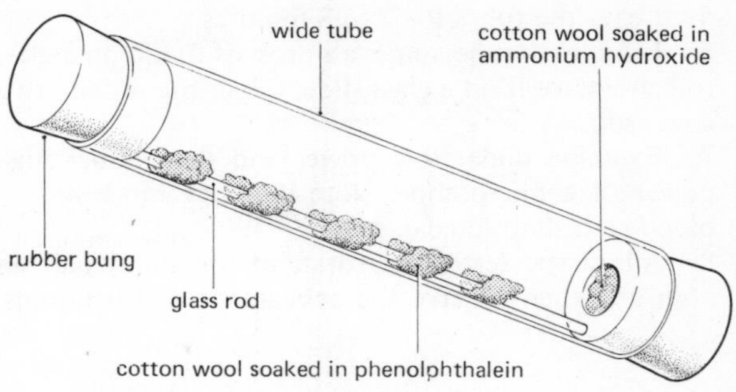

fig I.18 Diffusion in a gas

Experiment
To demonstrate diffusion in a gas
Procedure
1 Take a wide glass tube (50 cm in length; 2.5 cm in diameter), with a rubber bung fitted in each end.
2 Take a 40 cm glass rod, wrap small pieces of cotton wool at each end, and at 10 cm intervals along the rod (5 in all). Each portion of cotton wool must be soaked in phenolphthalein. Alternately, strips of red litmus paper could be used.
3 Place the rod in the wide glass tube, and stopper one end. In the other end place cotton wool soaked in ammonium hydroxide. Immediately stopper this end with the second rubber bung (see fig 2.20).
4 Observe the cotton wool on the glass rod.
Questions
9 What do you observe?
10 How do you know that this is diffusion?
11 What can you say about the comparative speeds of diffusion in liquids and gases?

The process of diffusion is important to all living organisms. Let us now consider the two processes of osmosis and diffusion with reference to living tissues.

Experiment
How do different osmotic conditions affect blood cells?
Procedure
1 Take three test-tubes containing the following:
Tube (i) 1 cm^3 of distilled water,
Tube (ii) 1 cm^3 of salt solution at the same concentration as blood plasma (0.85% sodium chloride solution),
Tube (iii) 1 cm^3 of salt solution at twice the concentration of blood plasma (1.70% sodium chloride solution).
2 Obtain some drops of blood using sterile techniques exactly similar to those used in section 3.20.
3 Allow some drops of blood to fall into each tube. If they fall on the inside wall of the tube, tilt the tube so that the liquid removes the blood. Shake the tubes gently. See fig I.19.
4 Resterilise the little finger.

5 Leave the tubes for 2 to 3 minutes.
6 From each tube, remove a drop of liquid on a glass rod and place it on a glass slide. Cover the drop with a cover slip.
7 Examine under low power and then under high power of a microscope. Note the appearance of any blood cells that you can see.
8 Allow the test-tubes to stand for about half an hour and then observe the appearance of the liquids.

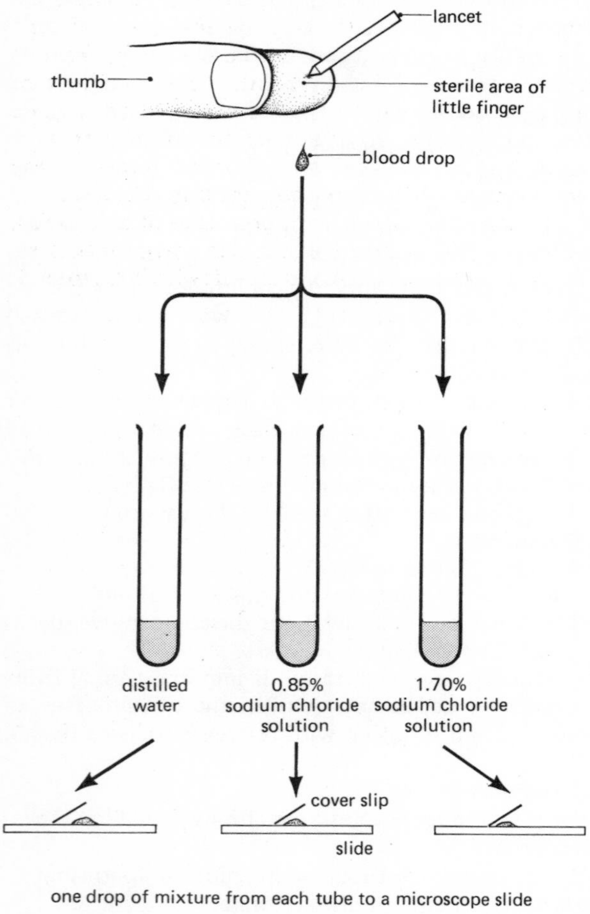

fig I.19 **Blood drops in different salt solutions**

Questions

12 Were blood cells present in tubes (i), (ii) or (iii)?
13 If so describe the appearance of the blood cells, indicating in which tube they were found.
14 Can you give an explanation of your observations?
15 What was the appearance of the liquid in each tube after half an hour?
16 Linking all of these observations together, can you give an explanation of your observations in terms of osmosis?

I 6.00 What is life?

The term *biology* is derived from two Greek words; *bios*, meaning life, and *logos*, meaning discourse. Thus biologists are scientists who study living things. But how can we tell living things from non-living things? In some cases it is easy: a horse is living, a window is not. In other instances the distinctions are not apparently so clear-cut. A car, for instance, may appear to show more characteristics of life than a lichen found on a rock, although we know that the lichen is living and the car is not.

In general, however, it is possible to draw up a list of seven processes which are carried out by *all* living things in one way or another.

1 Nutrition (see Chapter 2)
All living organisms need food. The division of the living world into plant and animal kingdoms is based on the way in which this food is obtained. In most cases plants make their own food by means of a process called *photosynthesis.* This is known as *autotrophic nutrition.* Animals, on the other hand, usually obtain food by eating other organisms, breaking them down with digestive enzymes and absorbing the breakdown products. This is known as *heterotrophic nutrition.*

2 Respiration (see Chapter 4)
Much of the food obtained by either autotrophic or heterotrophic nutrition is used to provide energy for the other vital processes. Energy is released from food when it is broken down, often by combustion or 'burning' in the presence of oxygen. This process is known as *respiration*; it should not be confused with *breathing,* a term which refers specifically in mammals to the means by which the oxygen is brought, via the lungs, into the blood system.

3 Excretion (see Chapter 5)
Just as all life processes require energy, they all result in the production of waste materials. The removal from the body of these waste mateials is known as *excretion.* Living organisms excrete a variety of substances; for example, carbon dioxide and water from the breakdown of sugars and, particularly in animals, chemicals rich in nitrogen from the breakdown of proteins. The removal of undigested food material in the faeces of animals does not qualify as excretion, since this material is not *produced* by the living processes of the animals. Urine, however, contains excreted water and nitrogen.

4 Response (see Chapter 7)
The environments inhabited by living organisms are constantly changing. Both animals and plants have the ability (sometimes termed *irritability*) to respond to these changes and thus ensure, to a greater or lesser degree, that they are not affected adversely by them. For example, the dermal blood vessels of a man may

dilate in response to a rise in the temperature of his surroundings, thus allowing him to lose more heat by radiation and so maintain a constant body temperature: a potted plant which has been grown outdoors may be seen to grow towards a window if it is brought indoors, a response ensuring maximum light for photosynthesis.

5 *Reproduction* (see Chapter 8)

Every living organism has a limited life span. However, all plants and animals have the ability to pass on life, and thus ensure the survival of the species, by producing new individuals with the same general characteristics as themselves. This process is known as reproduction.

6 *Growth* (see Chapter 8)

While a great deal of the food obtained by living organisms through autotrophic and heterotrophic nutrition is respired to produce energy, part is converted into protoplasm. The formation of additional living tissue in this way is termed growth.

7 *Movement* (see Chapter 6)

One of the features that distinguishes animals from plants is the ability of the former to move from place to place: locomotion. This is necessary for animals in order for them to obtain food, whereas plants are able to make their own food by photosynthesis. It should be noted that the cell contents of all living organisms are in a constant state of motion, termed *cyclosis.*

These seven characteristics of living organisms are all possessed by Man and are investigated in some detail throughout the following chapters. However, they are only the observable manifestations of a single all-important property of living material (protoplasm): the ability to extract, convert and use energy from its environment, and thus to maintain and even increase its own energy content. In contrast, dead organic and non-organic material tend to disintegrate as a result of the physical and chemical forces of the environment: their energy content thus decreases.

The maintenance by living organisms of an ordered, self-regulating system without net energy loss is referred to as *homeostasis* and is dealt with in Chapter 5.

1 A brief history of Man

1.00 The position of Man in the animal kingdom

Suppose that you were writing a letter to a boy or girl in another country as part of an English lesson, and that you wanted to describe your surroundings to him or her. One way would be to sit down and make a list of everything that you could see. For example, you could mention every single boy and girl in the classroom by name (John, Mark, Davinda, etc.), then John's shirt, John's desk, Mark's shirt, Mark's desk and so on, and, looking out of the window, you could list an oak tree, a second oak tree, a fir tree and an acacia tree. Of course, this would not be a very sensible way to start your letter, as the lesson would be finished long before your description!

A far better beginning would be: "There are thirty students in the classroom, sitting at desks and wearing the school uniform of white shirts and blue ties. Through the window I can see four large trees." Again, instead of saying "The desks are made of steel, the chairs are made of steel and the tables are made of steel", it is more convenient to say "The furniture is made of steel".

In other words, it is often useful to be able to place individual people and things into groups, so that the members of any group are similar to each other but different from the members of other groups: John, Mark and Davinda are all students; desks, chairs and tables are all furniture; oaks, firs and acacias are all trees. This is called *classification.*

The most useful kinds of classification tell us something important about the things being classified. For example, an aeroplane, a motor car, a housefly and a bedbug can be classified in two alternative ways:

either	*Flying*	*Non-flying*
	aeroplane	motor car
	housefly	bedbug
or	*Living*	*Non-living*
	housefly	aeroplane
	bedbug	motor car

Obviously, the second alternative is the more useful in this case, since it tells us that houseflies and bedbugs have in common with each other and with all other living things (including Man) the vital processes of respiration, growth, reproduction, etc., that were mentioned in the last paragraph of the Introductory Chapter.

Having established that all things are either *living* or *non-living,* what is the next stage in classifying the world in which we live, and where do you and I fit in? The division of the living world into *plant* and *animal* kingdoms is a very old idea that has persisted more or less unchanged to the present day. Plants were recognised as being mainly green, with a fairly limited range of structure and an inability to move from place to place. Today we realise that these features are connected with the ability of all true plants, from a lowly seaweed to a mighty tree, to manufacture their own food from carbon dioxide and water, using the green pigment chlorophyll to trap light energy from the sun. Animals, on the other hand, cannot make their own food and have to eat in order to continue living. They can move around in search of food and have a wide variety of body forms.

There have been several suggestions over the years as to how the animal kingdom might be subdivided. The Ancient Greeks, for example, classified animals as land dwellers, water dwellers and air dwellers, whilst Saint Augustine suggested three other groups, useful, harmful and 'superfluous'! Many early scholars placed Man in a group quite separate from all other animals.

The modern study of animal and plant classification, or *taxonomy* as it is also called, really began in the seventeenth and eighteenth centuries with the work of John Ray in England and Carolus Linnaeus in Sweden, respectively. Ray started by defining the term *species* as a group of animals (or plants) that could interbreed with each other. Thus all domestic dogs belong to the same species and all men belong to the same species, in spite of any differences in appearance. Linnaeus' great contribution was to name large numbers of species and to arrange them according to a system based on *natural* similarities and differences, by which they could easily be identified. Each animal and plant type was given two names, the first being the *genus* name and the second being the *species* name; hence the system was called the *binomial system of nomenclature.* The names were written in Latin (the international language of scholars at the time), the genus or generic name was given a capital letter and both were underlined (in print this appears as slanting, *italic* type). Thus the domestic dog is known as *Canis familiaris* (genus *Canis* and species *familiaris*). A genus may contain a number of closely related species, for example the wolf is *Canis lupus* and the coyote is *Canis latrus.* When differences become more marked, as between wolves and foxes, then another genus is established. The genus *Vulpes* has a number of species including the red fox (*Vulpes fulva*) and the swift fox (*Vulpes velox*). Genera with similar characteristics are grouped into Families, which are then grouped into larger groups – Orders, Classes and Phyla. The Phylum is generally regarded as the largest grouping within the animal kingdom and each member of a phylum is built on the same basic plan. For example, the phylum Arthropoda includes all animals with an exoskeleton and jointed legs.

The wolves and dogs mentioned above would, therefore, be completely classified as follows:

Kingdom	Animalia	
Phylum	Chordata	
Sub-phylum	Vertebrata (Craniata)	
Class	Mammalia	Possess milk glands and diaphragm
Order	Carnivora	Meat-eating, large canine teeth
Family	Canidae	Long slender limbs, four toes on the front feet, non-retractile claws
Genus	*Canis*	Dog-like
Species	*C. familiaris*	(domestic dog)
	C. mesomelas	(black-haired jackal)
	C. lupus	(wolf)

Once a classification scheme has been formalised, a method must be devised whereby other scientists can determine the identity of a particular organism. This is done by constructing an identification key based on the classification. To illustrate the relationship between classification and identification keys, consider how you might classify and make a key for the following sports: badminton, basketball, football, running, tennis and weight-lifting.

The sports can be classified by selecting two different characteristics several times in succession to produce a *dichotomous* key as shown at the foot of the page:

This type of key, as used to identify organisms, is more commonly presented in the following form:

1 Played with racquet . . . . .see 2
  Played without racquet . . . . .see 3

2 Played with ball . . . . .tennis
  Played with shuttlecock . . . . .badminton

3 Team sport . . . . .see 4
  Individual sport . . . . .see 5

4 11 per side . . . . .football
  5 per side . . . . .basketball

5 depends upon speed . . . . .running
  depends upon strength . . . . .weight-lifting

Investigation

To construct dichotomous keys and to identify the following:

1 Modern methods of transport – car, bus, cycle, three-wheeler car, motor-cycle, motor-cycle and sidecar, six-wheeled lorry, etc.

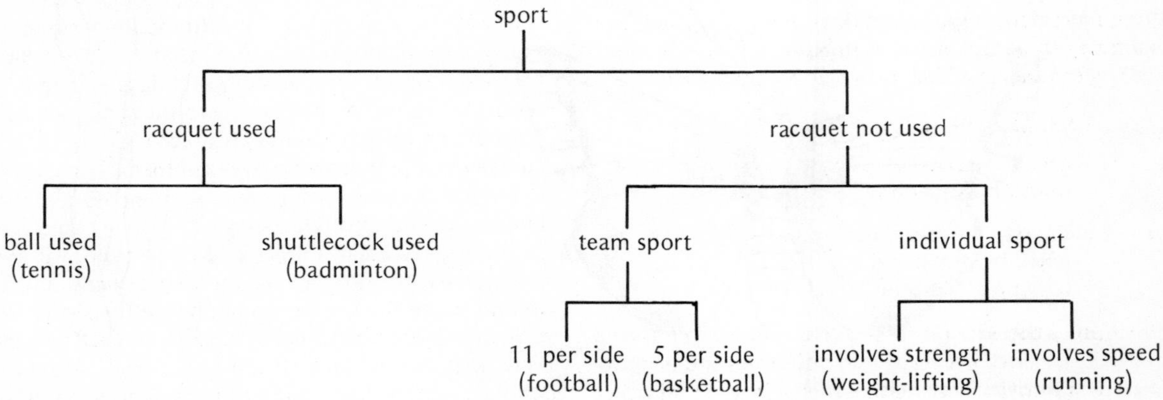

2 Staff cars in the school car-park – colour, make, four seats, two seats, four door, two door, saloon, sports car, etc.

3 Members of a biology class – hair colour, eye colour, male, female, clothing, glasses, etc.

4 Trees and shrubs in the school grounds – leaf size, leaf shape, leaf texture, colour of twigs, presence of leaf stalk, etc.

1.01 Classification of Man

Look carefully at the organisms shown in fig 1.1 and then answer the questions that follow.

Questions

1 Which of the organisms (a) – (e) can manufacture their own food? (see Section I 6.00 and Chapter 2).

2 Which of the organisms (a) – (e) are able to move easily from place to place?

3 Which of the organisms (a) – (e) are composed of cells covered by thick walls made of cellulose?

The three features mentioned above help to distinguish between the two major divisions of the living world, the animal kingdom and the plant kingdom. (a) and (b) are plants; (c), (d) and (e) are animals. There are other, less obvious, differences between animal and plant cells, and the distinction between the two kingdoms is not always as clearcut as in these five examples. Fungi and bacteria, for instance, are usually placed within the plant kingdom, even though they do not photosynthesise. One thing is certain however, Man is an animal.

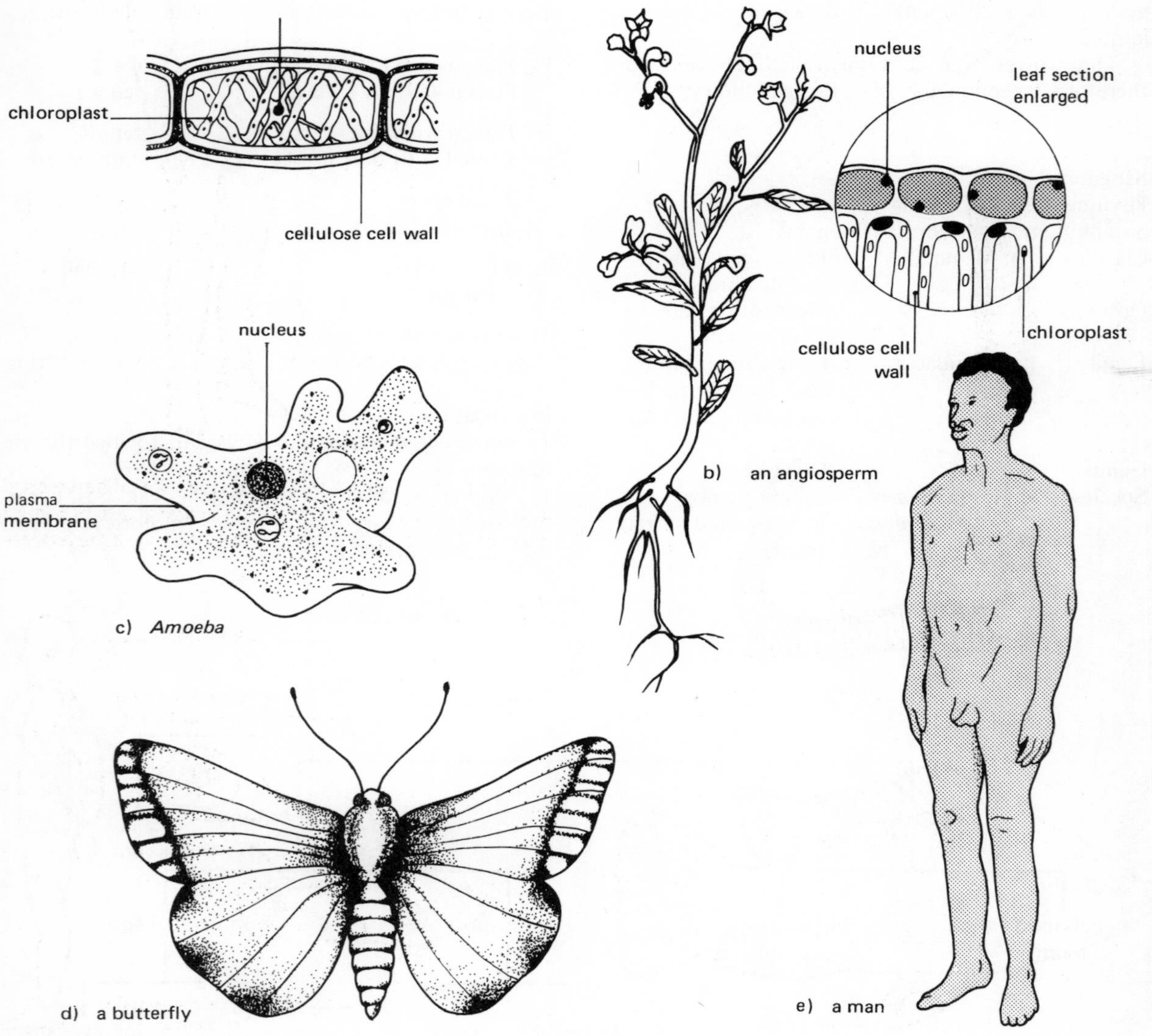

fig 1.1 Five living organisms

1.02 Look carefully at the animals shown in fig 1.2, and then answer the questions that follow.

Questions

1 Which of the animals (a) – (e) have backbones?
2 Which of the animals (a) – (e) have main nerve-cords situated in the dorsal (back) regions of their bodies?
3 Which of the animals (a) – (e) possess gills?

a) *Paramecium*

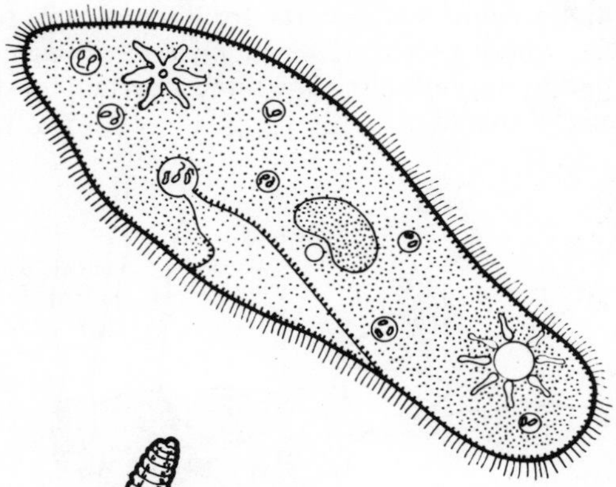

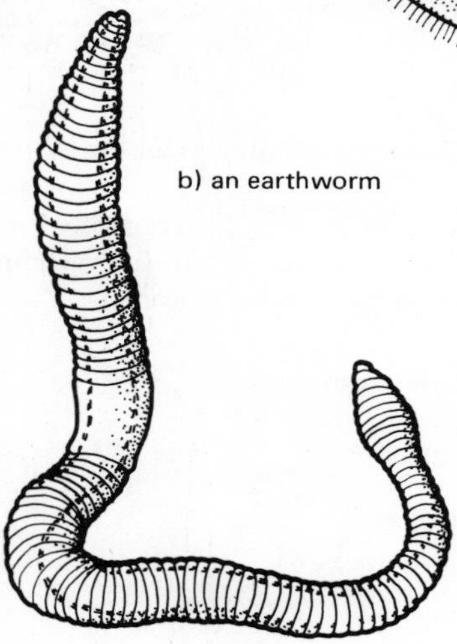

b) an earthworm

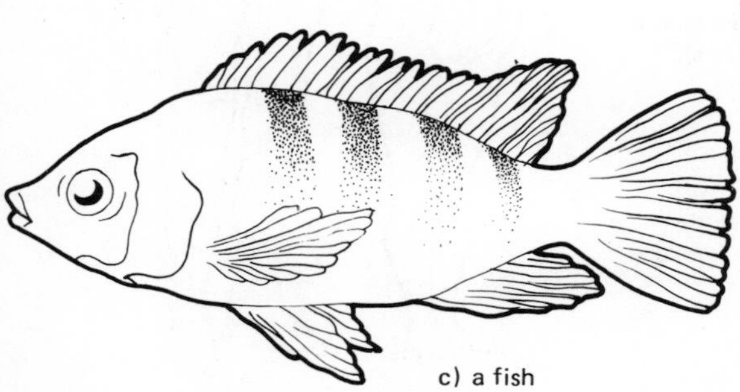

c) a fish

d) an insect

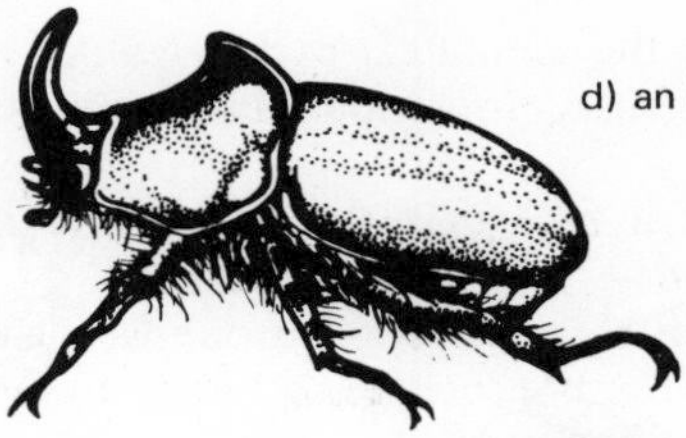

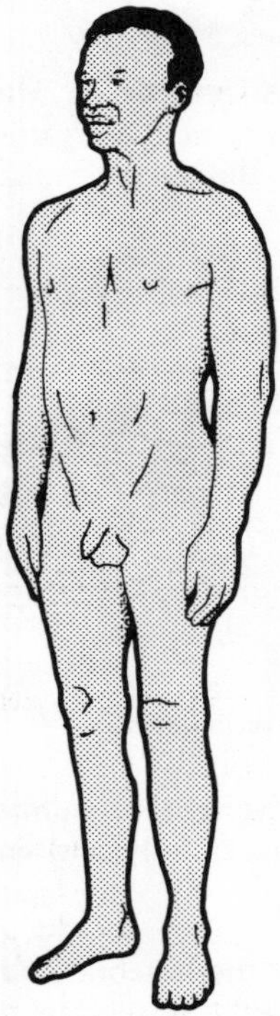

e) a man

fig 1.2 Five animals

These three features – backbone, dorsal nerve-cord and gill slits are characteristic of a large group of higher animals called *vertebrates*. Thus Man is a *vertebrate animal.*

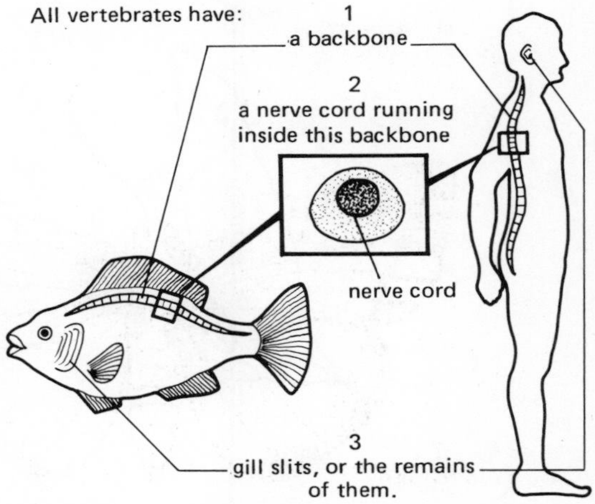
All vertebrates have:
1
a backbone
2
a nerve cord running inside this backbone
nerve cord
3
gill slits, or the remains of them.

fig 1.3 Characteristic features of vertebrates

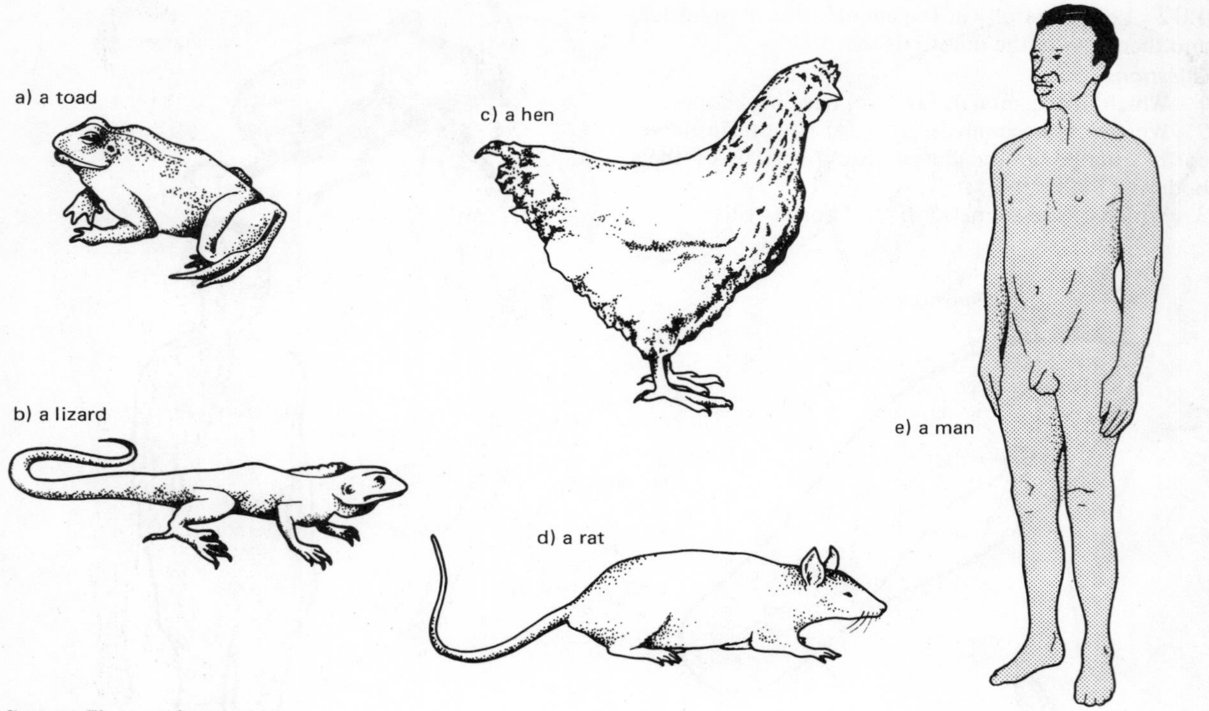

fig 1.4 Five vertebrates

1.03 Five vertebrate animals are shown in fig 1.4. Study them carefully and answer the questions that follow.

Questions

1 Which of the vertebrates (a) – (e) have hairy skins?

2 Which of the vertebrates (a) – (e) give birth to live young that have received nourishment via a placenta during development in the maternal body?

3 Which of the vertebrates (a) – (e) suckle their young?

A more or less hairy skin, a placenta and mammary glands are all features exclusive to the vertebrate class Mammalia. Man is therefore a mammal.

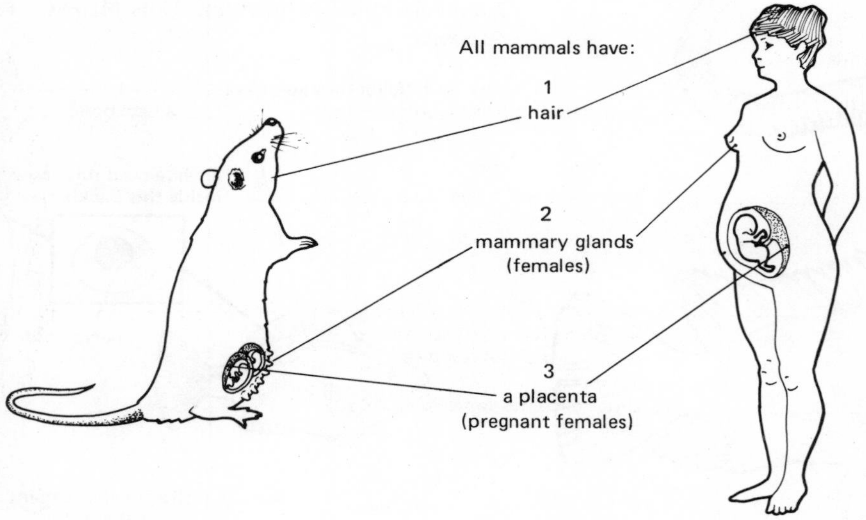

fig 1.5 Characteristic features of mammals

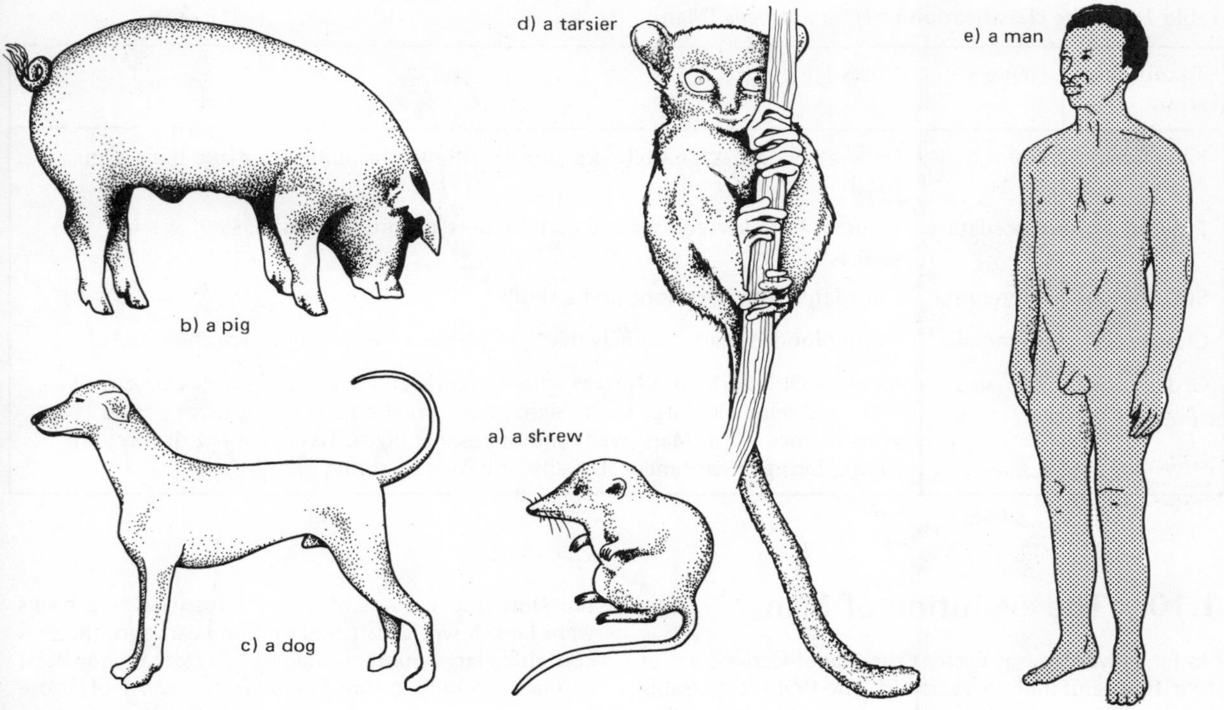

d) a tarsier
e) a man
b) a pig
a) a shrew
c) a dog

fig 1.6 Five mammals

1.04 Fig 1.6 shows five members of the class Mammalia. Look at them carefully and then answer the questions that follow.

Questions

1 In which of the mammals (a) – (e) are the limbs adapted for gripping?

2 Which of the mammals (a) – (e) possess fingernails and toenails instead of claws?

3 In which of the mammals (a) – (e) do you think the sense of sight is more strongly developed than the sense of smell?

Opposable thumbs, fingernails and stereoscopic vision are all characteristic features of the mammalian order Primates, which includes monkeys, apes and Man. Man differs from other living primates in having no tail or cheek pouches and a big toe that cannot be used for grasping.

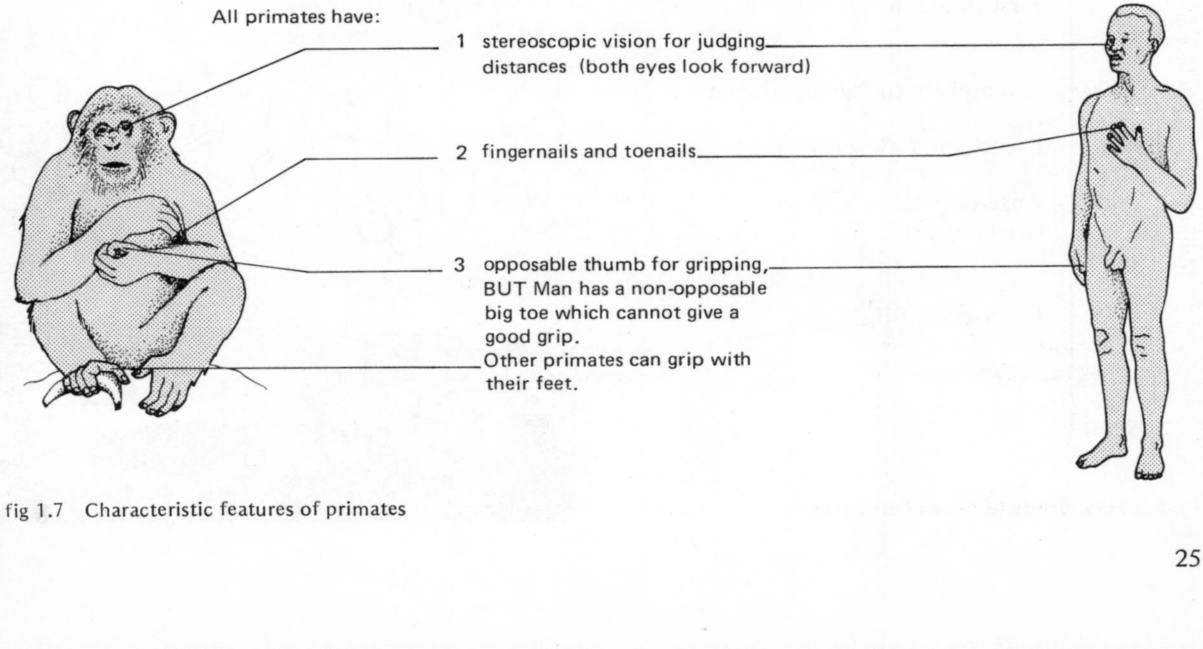

All primates have:
1 stereoscopic vision for judging distances (both eyes look forward)
2 fingernails and toenails
3 opposable thumb for gripping, BUT Man has a non-opposable big toe which cannot give a good grip.
Other primates can grip with their feet.

fig 1.7 Characteristic features of primates

Table 1.1 The classification of *Homo sapiens* (Man)

Taxonomic group	Name	Notes
Kingdom	Animal	Generally compact, mobile organisms, without the ability to make their own food
Phylum	Chordata	Includes both vertebrates and certain more primitive animals, such as sea squirts
Sub-phylum	Vertebrata	Chordates with backbone and a skull
Class	Mammalia	Warm-blooded, hairy animals that give birth to live young which they suckle
Order	Primates	Includes Old World and New World monkeys, and apes. Great development of brain and relatively large skull. Sight good: smell poor. Long growth period (12 years in apes, 17 in Man), nails and opposable digits. Development of forelimb for exploring environment. Usually only one offspring per birth.

1.10 The evolution of Man

As far as we know at present the planet Earth is about four thousand million years old. The first recognisable forms of life, probably simple, bacteria-like organisms, appeared about one thousand million years later. By the time the earliest of today's fossil-bearing rocks were laid down, about 570 million years ago, the seas contained large numbers of invertebrate (backboneless) animals, including forms similar to many of those living now.

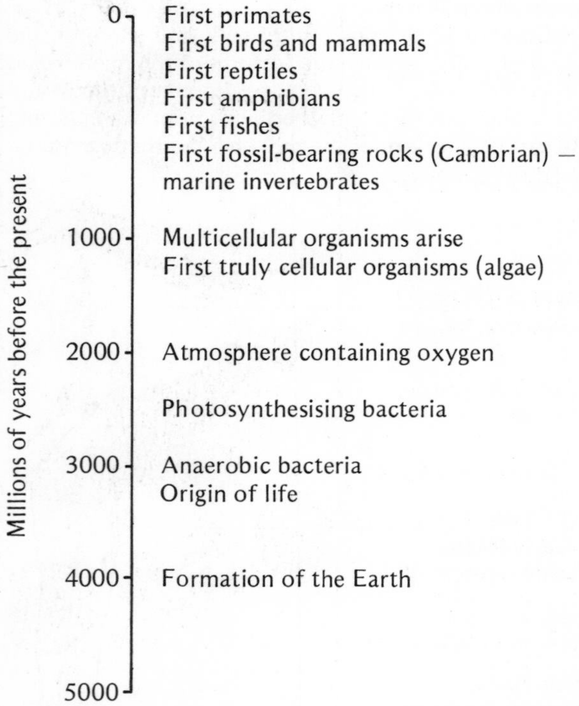

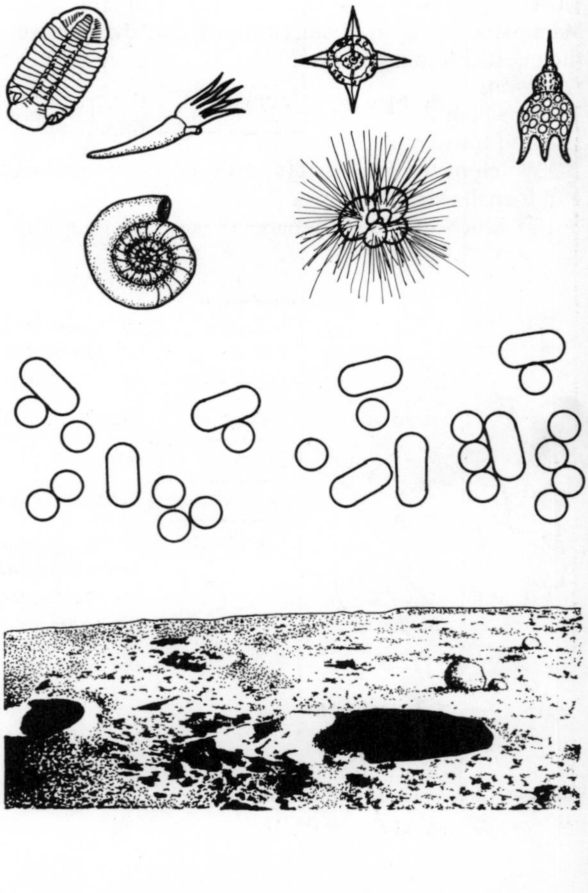

fig 1.8 The history of life on Earth

The first vertebrates appeared 480 million years ago as eel-shaped fish, without proper jaws. After another seventy-five million years, the fishes had evolved to become the dominant animal group, and from them had developed the earliest four-footed creatures, amphibian ancestors of the present-day frogs, toads, newts and salamanders. Over the next fifty million years, some of these amphibians became very large creatures, two metres or more in length, able to move clumsily over land but returning to the water to lay their thousands of small eggs. One group of smaller, thinner amphibians became more independent of water, giving rise to the first reptiles.

Millions of years before present	Era	Period	Major events in evolution	
0	**Quaternary**	Recent Pleistocene	Modern Man	
		Pliocene	"Man-like" and "ape-like" lines separate	
		Miocene	Rise of "modern" mammals. First hominoids	
	Tertiary	Oligocene	Formation of Alps and Himalayas	
		Eocene	Primitive mammals dominant	
		Paleocene	First primates	
100		Cretaceous	Formation of Rocky Mountains and Andes Spread of birds and primitive mammals Extinction of dinosaurs	
	Mesozoic	Jurassic	Dinosaurs dominant. First birds and mammals	
200		Triassic	First dinosaur reptiles	
			Formation of Ural and Appalachian mountains	
		Permian	Rise of reptiles	
300		Carbonif-erous	Amphibians dominant. First reptiles. Great swamp forests giving rise to coal measures	
	Paleozoic	Devonian	Fishes dominant. First amphibians	
400		Silurian	Armoured fishes	
		Ordovician	First vertebrates – jawless fishes	
500		Cambrian	Early arthropods (trilobites) dominant. Many other marine invertebrates, including sponges and molluscs	
			Period of great erosion of rocks	
600	**Precambrian**		Few fossils present in rocks. Evolution of multicellular animals and plants	

fig 1.9 Geological time scale (I)

The reptiles were a very important class, dominating the land for the next two hundred million years. From them arose both the birds and, about 180 million years ago, the earliest mammals. Some of these were small, insect-eating creatures, not unlike the shrews of today. The end of the age of reptiles, 80–100 million years later, coincided with enormous changes in the climate and vegetation of the Earth. The mammals, which had evolved a constant body temperature, were better able to withstand these changes and took over from the reptiles as the most important group of land vertebrates.

About eighty million years ago the earliest primates (the prosimians) appeared. They were small, tree-living animals which lived on insects and/or fruit and probably resembled modern tree-shrews in many respects. The tarsier of Indonesia and the more familiar lemurs, lorises and bush-babies are descended from closely related forms.

By about thirty million years ago, the prosimians had given rise to three other important groups: the Old World monkeys, the New World monkeys and the hominoids. These last-named animals were the ancestors of present-day apes and Man.

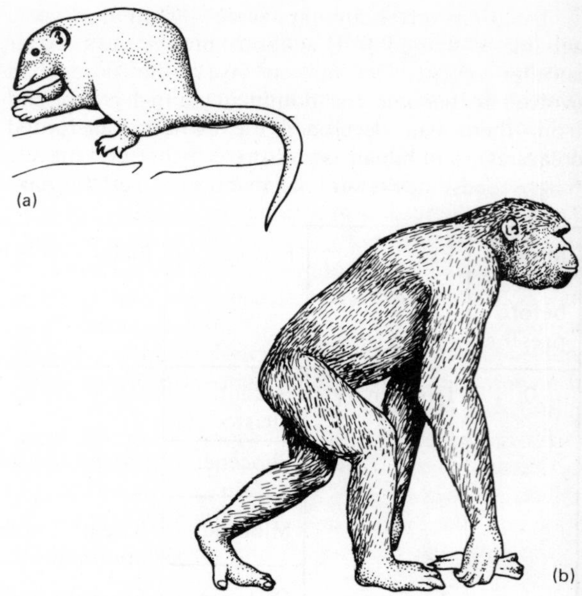

fig 1.10
(a) A tree shrew
(b) An early hominoid

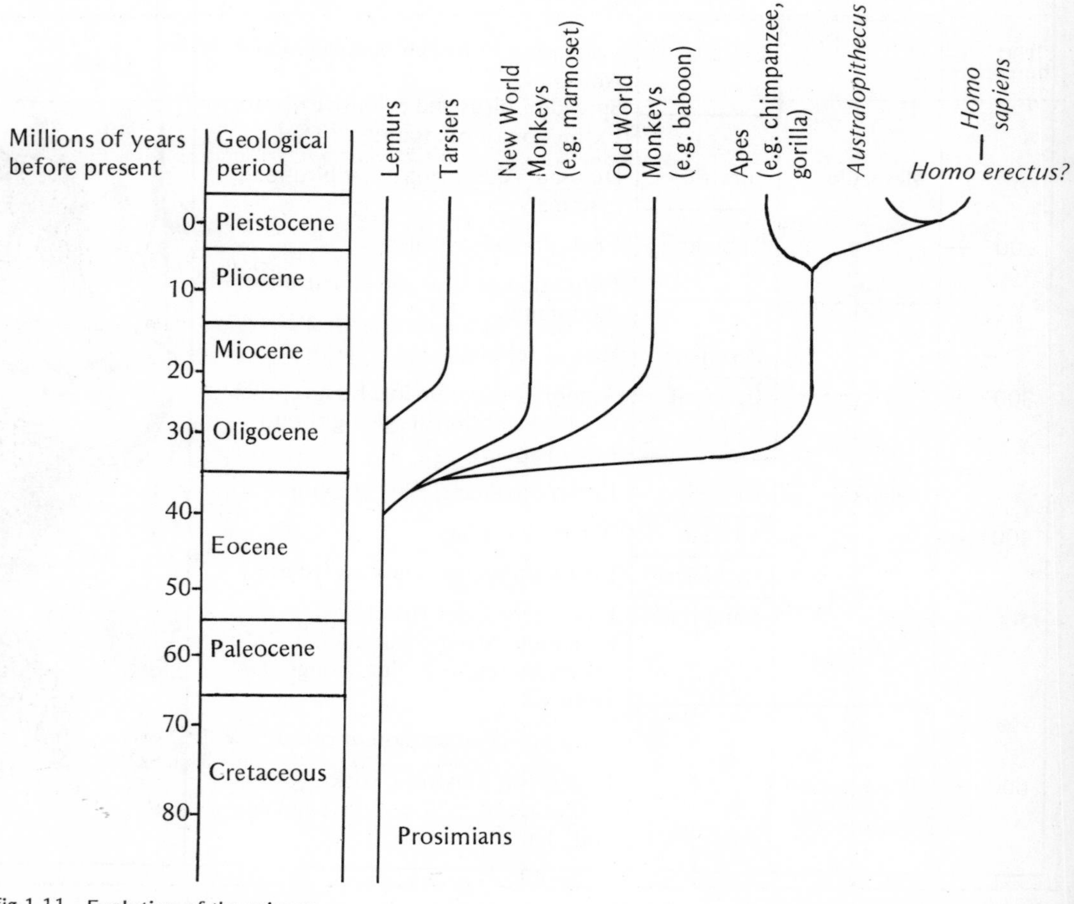

fig 1.11 Evolution of the primates

The date at which the ape-like line (family Pongidae) and the Man-like line (family Hominidae) separated from each other is by no means certain, but it is now thought to have been some time between five and ten million years ago. The possible nature of our ancestors at this stage in evolution has been a matter of great interest over the past hundred years, ever since Charles Darwin challenged conventional mid-nineteenth century views by putting forward his theory of evolution. No fossil evidence has yet been found, but the accepted view is that early Man had a small brain (perhaps only one quarter the size of our own) and long arms. These enabled him to walk on his hind limbs and the knuckles of his forelimbs while holding stones or pieces of wood in his hands (see fig 1.10(b)).

Fossils do exist of a rather more advanced form, *Australopithecus*, in African rocks laid down four million years ago. This hominid had a slightly larger brain and shambled along on his hind limbs. In rocks of the same age were found simple stone tools, made from pebbles with one surface chipped or flaked away.

The first member of our own genus, *Homo*, did not appear until about two and a half million years later. He has been named *Homo erectus* because he walked more or less upright on two legs only. His brain was about three-quarters the size of modern Man's and he produced more advanced tools than *Australopithecus*, made of stones carefully shaped on both sides.

The tools of neanderthal Man (*Homo sapiens neanderthalensis*) where made of flakes of flint and were not very much more elaborate than those of *Homo erectus*, even though they were made by men with brains as large as ours only 70 000 to 100 000 years ago. The skull of neanderthal Man was heavier and more massive than that of modern Man (*Homo sapiens sapiens*) who first appeared in about 40 000 B.C. The earliest known tools of modern Man include bone needles and harpoon heads, as well as some small statuettes and other decorative articles.

Millions of years before present	Era	Period	Major stages in human evolution
0	Quaternary	Recent (post-glacial)	*Homo sapiens sapiens* – modern Man *Homo sapiens neanderthalensis* – neanderthal Man
0.5		Pleistocene (ice ages)	Discovery of fire 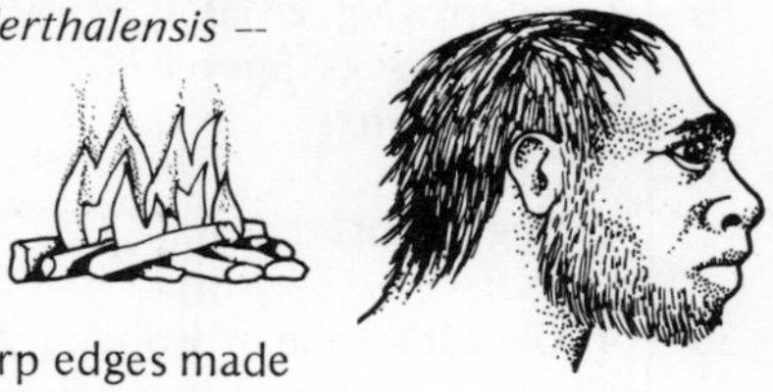
1.0			Stone tools with sharp edges made by flaking on both sides
1.5	Tertiary	Pliocene	*Homo erectus* 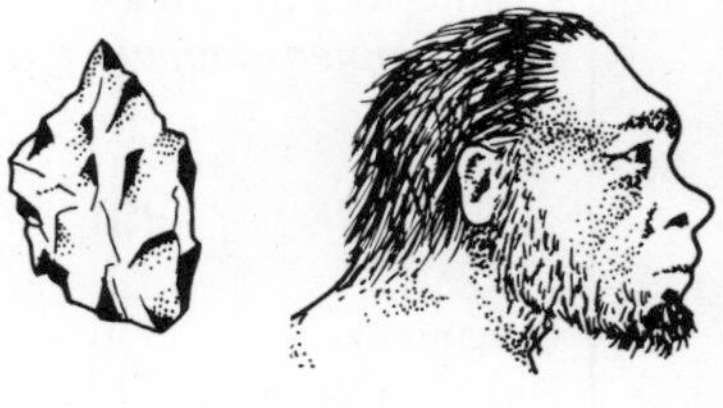
2.0			
2.5			
3.0			
3.5			Simple stone tools *Australopithecus* 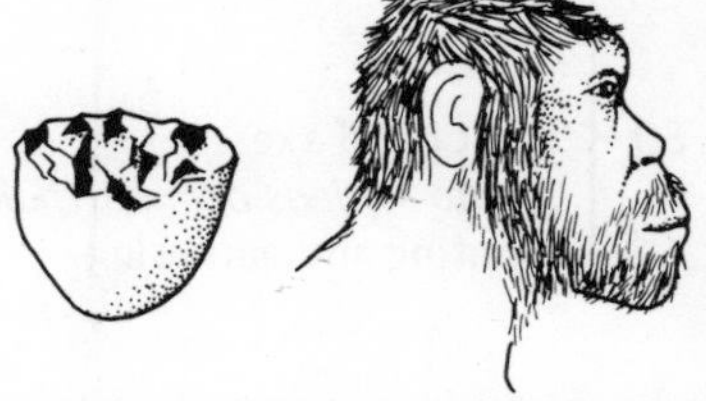
4.0			

fig 1.12 Geological time scale (II)

1.20 Cultural evolution

Examination of figs 1.8, 1.9 and 1.12 will show that Man has been in existence for only 0.001% of the life of the Earth. A fraction so small that it cannot be represented simply on a single diagram in a book of this size. If, for example, the age of Man were to be represented by one centimetre, the page would have to be one kilometre long in order to show the age of the Earth to the same scale. Yet not very long ago it was widely accepted that both were created at the same time.

Physically, by definition, Man has changed little over the past 40 000 years. However, tremendous cultural and technological advances have taken place to transform the primitive hunter into the developer of today's world of space travel, computers and telecommunications. These advances have not taken place at an even rate, but have followed a step-like pattern of major breakthroughs, followed by rapid changes, followed by periods of stabilisation. However, with the passage of time there has been a notable increase in the rate at which these steps have occurred (see figs 1.13 and 1.14).

What are the reasons for these advances, which have made Man the most 'successful' animal in the history of the world?

First of all, he is a relatively unspecialised animal in physical terms. Among other things this means that he has not been restricted to a very limited diet and thus has not been dependent on the success of any one other organism as a source of food. Also, he has been able to cope well with changes in climate and has colonised some of the most hostile environments on

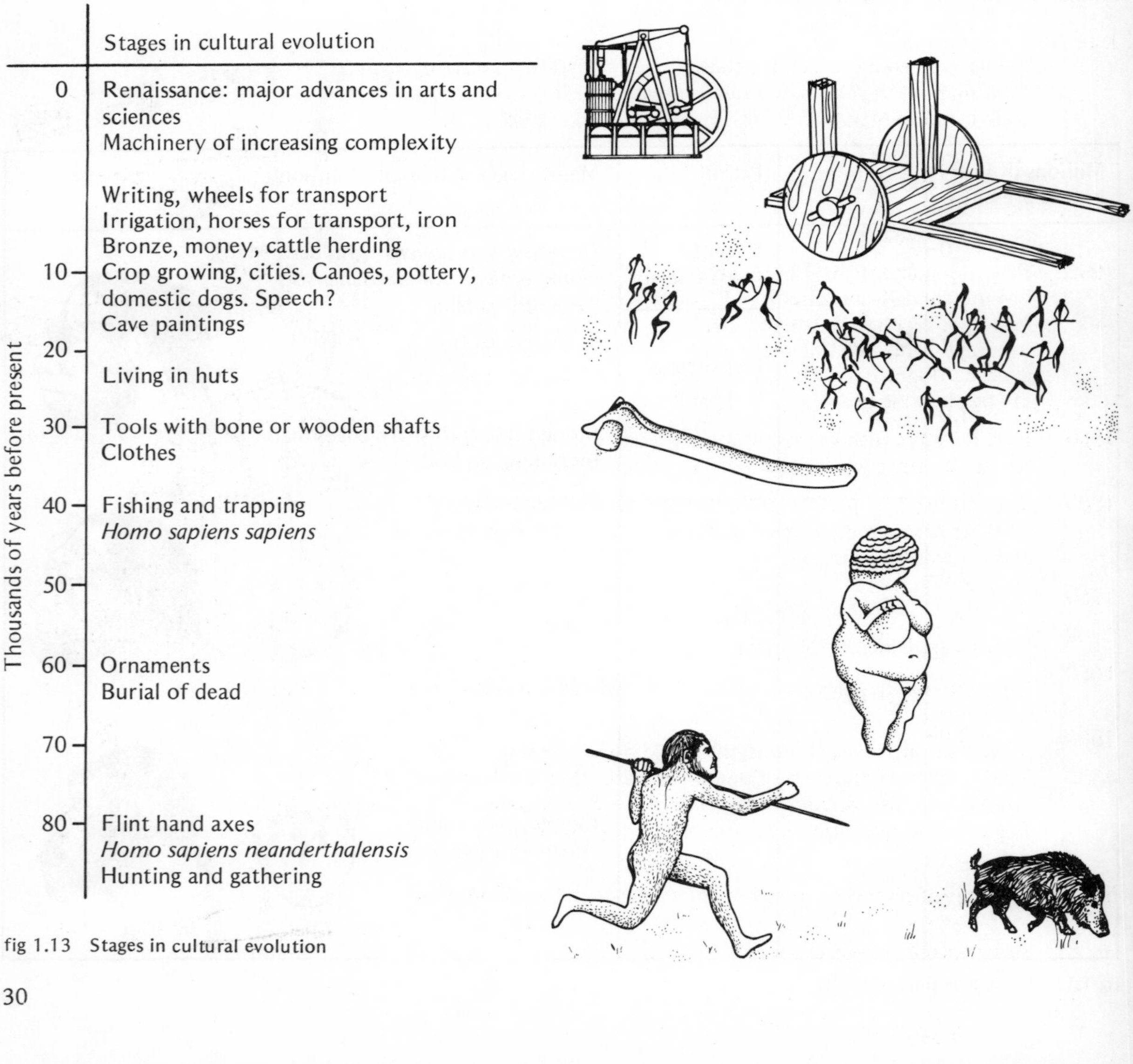

fig 1.13 Stages in cultural evolution

the Earth's surface — from the arctic circle to the Sahara desert to the tropical rain forests of Africa, Asia and South America.

Part of this ability to cope is related to the second major reason for man's success: the relatively great size (15 cm^3) and complexity of his brain (particularly the cerebral cortex), which gives him his high level of intelligence. This has helped man to adapt to conditions for which he is not physically suited by, for example, making warm clothes and developing means of travel in water and in air. It has also enabled him to carry out large projects, such as the building of roads and dams, by sharing them between several individuals, none of whom would be able to complete the tasks unaided.

Both of the above factors, unspecialised structure and a large brain, were present in the earliest modern man some forty thousand years ago. Why then was there so little cultural and technological advance over the following thirty thousand years, compared with the strides that have been made in the last ten thousand? (See fig 1.13.) One answer could lie in the development of *language,* which enabled abstract ideas and experience to be passed from individual to individual.

A breakthrough of similar importance occurred some five thousand years later with the advent of writing, which allowed experience and knowledge to be passed from generation to generation and from place to place without distortion.

Finally, the industrial revolution of the early 19th century (see fig 1.14) has led ultimately, by the harnessing of power sources, to the enormous advances in communications and other technology that we see today.

Date	
	Rocket powered spacecraft achieve speeds of 40 000 km/hour. First man on the Moon. World-wide live television service. Space craft travel to Mars and Venus. First man-made satellite.
1950	Aeroplane flies faster than sound. Discovery of DNA. First atomic bomb and first jet aeroplane. Invention of radar. Invention of television. First transatlantic flight. Transmission of speech by radio.
1900	First aeroplane flight. First motor cars. Wireless telegraphy invented and X-rays discovered. Telephone and electric light invented. Electric motor and generator invented. Transatlantic telegraph cable laid. Mendel's work on heredity. Railway speeds exceed 100km/hour. Darwin writes *Origin of Species.* Invention of steel-making process.
1850	First railways with speeds of about 25km/hour. Steamship crosses Atlantic with mean speed of about 10km/hour.
1800	Vaccination first performed by Jenner. Steam engine invented by Watt. Oxygen discovered by Priestley. Classification of plants and animals by Linnaeus.
1750	
1700	Gravity explained by Newton.
1650	Invention of the telescope. Circulation of the blood discovered by Harvey.
1600	Voyages of discovery around the world by sailing ship.
1550	Theory of the universe by Copernicus states that Earth moves around Sun. Researchers in human anatomy by Vesalius. Fastest communication by horse or sailing ship: about 150km/day.
1500	Voyages of discovery from Europe to America by sailing ship.

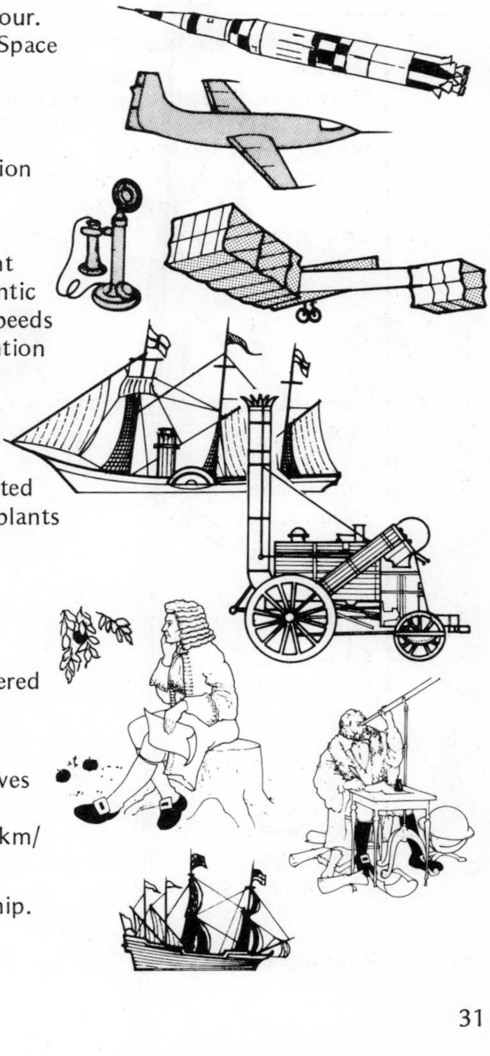

fig 1.14 Major scientific advances since the Renaissance

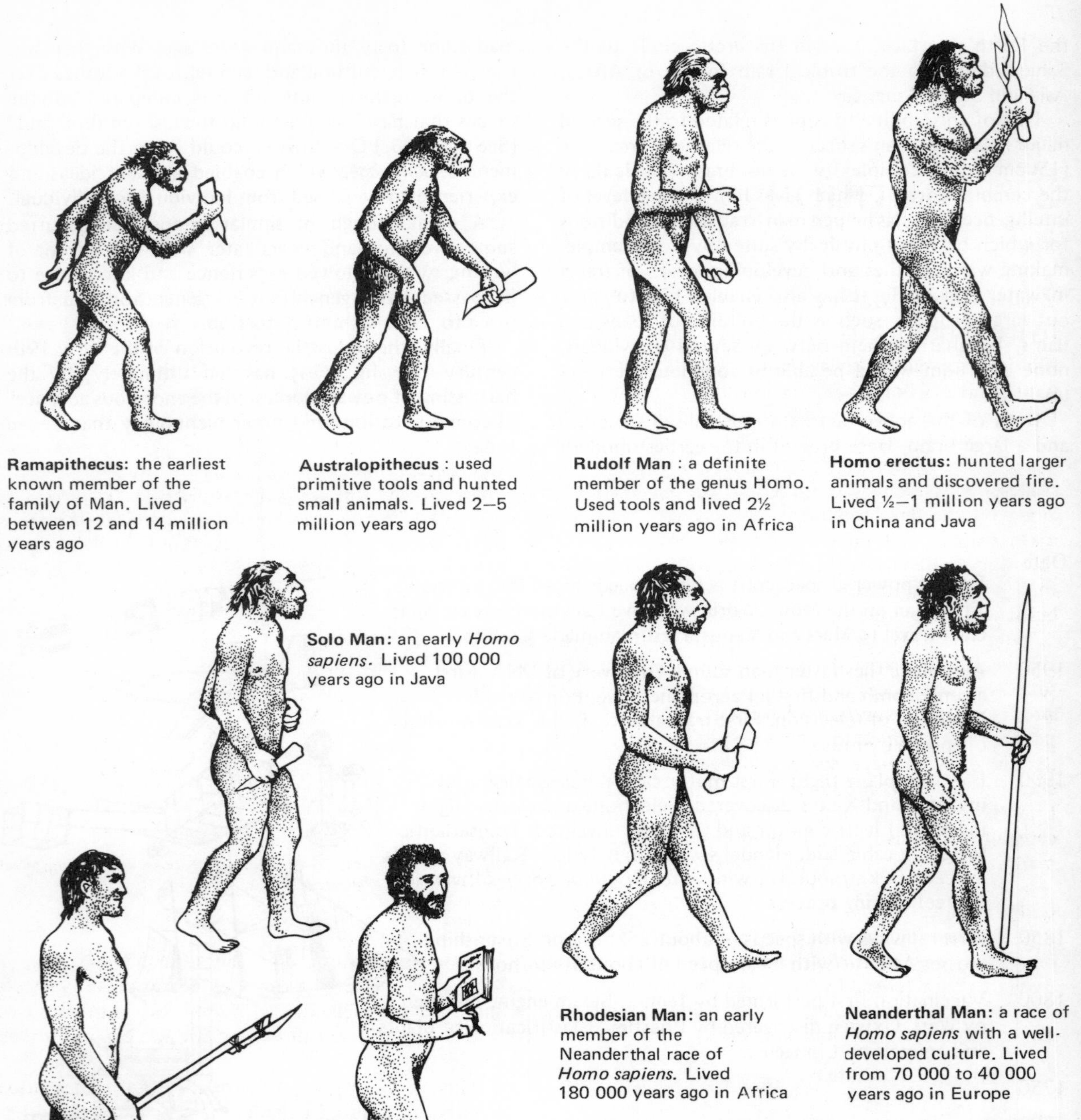

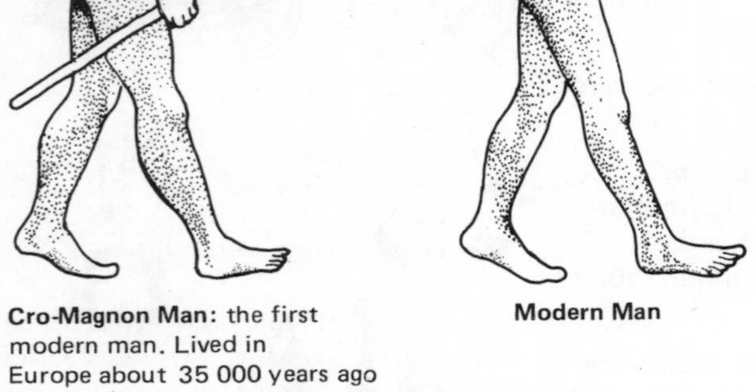

Ramapithecus: the earliest known member of the family of Man. Lived between 12 and 14 million years ago

Australopithecus : used primitive tools and hunted small animals. Lived 2–5 million years ago

Rudolf Man : a definite member of the genus Homo. Used tools and lived 2½ million years ago in Africa

Homo erectus: hunted larger animals and discovered fire. Lived ½–1 million years ago in China and Java

Solo Man: an early *Homo sapiens.* Lived 100 000 years ago in Java

Rhodesian Man: an early member of the Neanderthal race of *Homo sapiens.* Lived 180 000 years ago in Africa

Neanderthal Man: a race of *Homo sapiens* with a well-developed culture. Lived from 70 000 to 40 000 years ago in Europe

Cro-Magnon Man: the first modern man. Lived in Europe about 35 000 years ago

Modern Man

fig 1.15 A pictorial summary of the evolution of Man

Questions requiring an extended essay-type answer

1 (a) State three vertebrate and three mammalian characteristics of Man.

(b) Describe briefly the cultural evolution of Man.

2 Construct a key to classify the following:

i) A six-wheeled, double decker, yellow bus
ii) A six-wheeled, single decker, white bus
iii) A four-wheeled, two door, blue car
iv) A four-wheeled, four door, red car
v) A two-wheeled, motor-cycle
vi) A two-wheeled, pedal cycle.

2 Nutrition

2.00 Chemical compounds made by living organisms

Biology has moved away from being the study of natural history only, that is a purely descriptive science, to an emphasis on experimentation. This change is reflected in biology at all levels, from current research interests in universities to first biology courses in secondary schools. Experimental biology has been aided by the development of physical, chemical and mathematical techniques whereby processes can be followed and analysed in a more quantitative fashion than they have been in the past. One relatively new branch of study is biochemistry which is the chemistry of living things. This involves investigating not only the chemical structure of the compounds of life but also such essential processes as energy exchange in *respiration* and *photosynthesis*, synthesis of protein, the transport of compounds and so on. The first priority therefore, is to have some knowledge of the basic chemistry of the compounds found in living organisms. Life as we know it on this earth has made use of the element *carbon* as the basic component of all complex materials that circulate throughout living things.

Carbohydrates

The simplest of these carbon compounds in living organisms are the *carbohydrates*, so called because they are characterised by having carbon, hydrogen and oxygen in their molecules, with the ratio of hydrogen to oxygen the same as in water. Thus:

	carbo		hydrate
=	C	+	H_2O
	carbon		water

= $C_n(H_2O)_m$ where n and m may be the same or different numbers.

They are extremely important compounds because:

i) they are *sources of energy*, and

ii) they have *structural uses* in plant and animal tissues.

Carbohydrates have a basic molecule (building brick) which can be put together to form much more complex molecules (houses). The building brick molecule of carbohydrates is a *simple sugar* or *saccharide*.

The carbohydrate group of substances can be subdivided according to the number of simple sugar molecules:

1 Monosaccharides – one molecule
2 Disaccharides – two molecules
3 Polysaccharides – many molecules

Monosaccharides

These are the simplest sugar molecules. *Glucose* is an example of a monosaccharide. Its chemical formula is $C_6H_{12}O_6$ and it is commonly found in living cells. *Ribose* is another example of a monosaccharide, with the formula $C_5H_{10}O_5$.

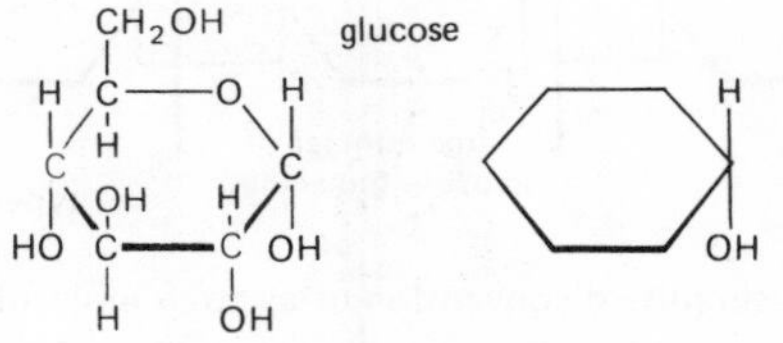

fig 2.1 The structural formula of glucose and a simplified convention of a glucose molecule

Disaccharides

Sucrose (cane sugar or table sugar), $C_{12}H_{22}O_{11}$, is formed by the joining together of a fructose molecule and a glucose molecule, with the loss of a molecule of water. This method of joining two molecules is called *condensation.*

Maltose $C_{12}H_{22}O_{11}$ is formed by the condensation of two molecules of glucose.

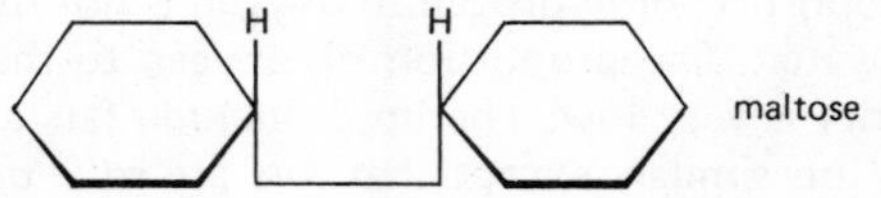

fig 2.2 A simplified convention of a maltose molecule

Lactose (milk sugar) is formed by the condensation of a glucose molecule and another monosaccharide called galactose. Since a molecule of water is lost during condensation the number of atoms in the disaccharide is not exactly twice the number of atoms in a monosaccharide:

$$C_6H_{12}O_6 + C_6H_{12}O_6 = C_{12}H_{22}O_{11} + H_2O$$

glucose fructose sucrose

Maltose and lactose are reducing sugars, but sucrose is a non-reducing sugar.

Polysaccharides

These are the largest carbohydrate molecules and show particularly the two main functions mentioned above:

i) *energy source* – starch and glycogen are used for storage in plants and animals respectively.

ii) *structural molecules*, e.g.: *cellulose* forms the cell walls of plants and *chitin* forms the exoskeletons of arthropods.

Starch has the formula $(C_6H_{10}O_5)_n$ where n is a large number. Starch is formed by the condensation of molecules of simple sugar to give chains several thousand sugars in length.

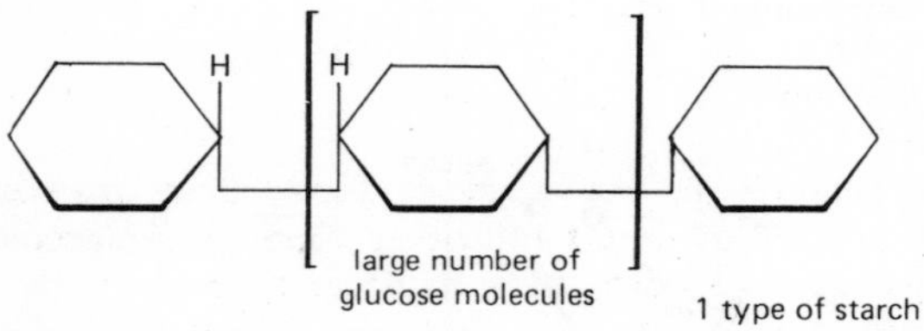

fig 2.3 A simplified convention of a starch molecule

Cellulose, which makes up the cell walls of plants contains many thousands of monosaccharide molecules forming a single long chain molecule.

Chitin is found in the exoskeletons of insects and other animals, as well as in the cell walls of fungi, and again is made up of long chains of monosaccharide units.

Inulin is a storage polysaccharide formed by the condensation of fructose molecules.

Lipids

These are organic compounds which contain carbon, hydrogen and oxygen, as do the carbohydrates, but the proportion of hydrogen to oxygen is not the same as in water. The proportion of oxygen to the other elements is very low. The lipids include fats and oils which are similar, except that *fats* are *solid at room temperature* and *oils are liquid at room temperature.* Plants tend to have liquid oils which contain unsaturated *oleic acid,* $C_{18}H_{32}O_2$, whereas animals store solid fats which contain *palmitic acid,* $C_{16}H_{32}O_2$ or *stearic acid,* $C_{18}H_{36}O_2$.

Fats are made from two types of chemical: *fatty acids* and the alcohol *glycerol*: such a combination is called a *glyceride.* Three fatty acids and glycerol form a *triglyceride* such as *tristearin,* $C_{57}H_{110}O_6$. *Waxes,* such as those found on the surface of certain leaves of plants and in beeswax, are combinations of fatty acids and other alcohols. Fats, like carbohydrates, are important sources of energy and because of their high energy content and chemically inert nature they are the most economical storage materials for living organisms. Some mammals accumulate food reserves of fat; insects have fat bodies and most plant fruits and seeds have food reserves of oil e.g. oil palm fruits, groundnuts, sunflowers, coconut and castor oil plants. Humans eating more food than their bodies can use, also accumulate fat. Animals that live in cold climates have fat reserves under the skin to provide insulation against heat loss e.g. polar bears, penguins and seals. Tropical animals generally lack fat reserves.

Proteins

The living substance, *protoplasm,* is essentially a solution of *proteins in water.* In addition to carbon, hydrogen and oxygen, all proteins contain *nitrogen. Phosphorus* and *sulphur* may also be present. As in polysaccharides, proteins are made up of smaller units forming a long chain molecule. These units are called *amino acids. Twenty three* different amino acids are found, and twelve of these are particularly common.

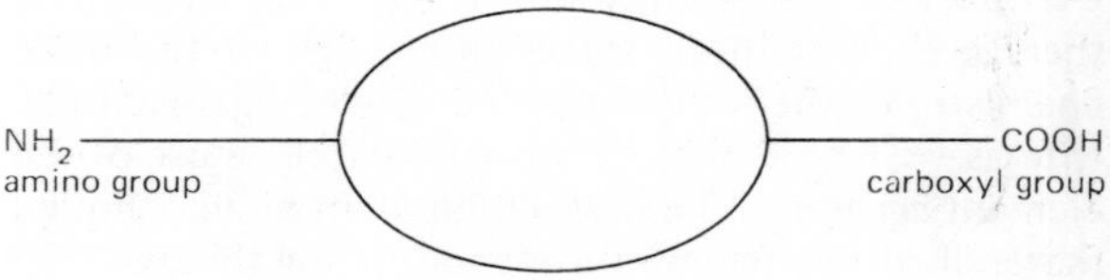

fig 2.4 A simplified convention of an amino acid molecule

Proteins are formed by the condensation of amino acids so that the negatively charged end of one amino acid is attracted to the positively charged end of another amino acid. A molecule of water is eliminated during the process. The bond between two amino acids is called a *peptide bond.* When many amino acids are joined together by this peptide linkage the product is a *polypeptide.* These are joined to form the protein molecule. Below is a list of some amino acids.

Name	Abbreviation
Glycine	Gly
Alanine	Ala
Valine	Val
Leucine	Leu
Aspartic acid	Asp
Glutamine	Glu
Cysteine	Cys
Lysine	Lys

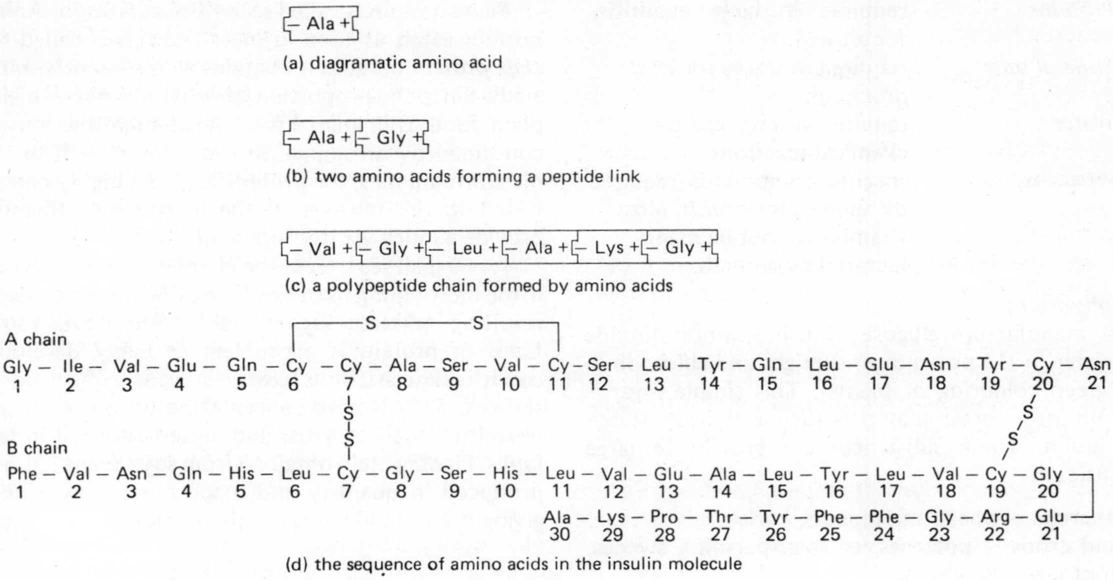

fig 2.5 Polymerism showing the amino acid sequence in an insulin molecule

Since amino acids can be joined in many different combinations an almost infinite number of proteins can exist. It is fairly certain, however, that living organisms have been economical in the development of new proteins and the number in existence may be much lower than that which is possible theoretically. The amino acids in a protein molecule can form long chains, or the chains can be folded in a complicated fashion to form a *globular molecule.* Recent research has demonstrated the structure of a few of these complex protein molecules, such as *insulin* and *haemoglobin.*

Living processes are controlled by enzymes, and each of these is a protein molecule. The importance of this group of substances in forming the structure and controlling the working of living organisms is therefore very great. Amino acids and proteins are very sensitive to certain physical changes such as heat. When heated, protein molecules lose their special properties and become *'denatured'.* The coagulated egg-white in a boiled egg is the protein *albumen,* which has become denatured.

2.10 Food

All living organisms require organic compounds for their living processes. The most obvious use of these compounds is for **growth**. Plants are able to manufacture organic compounds from raw materials (carbon dioxide, water and mineral salts containing nitrogen etc.), whereas animals must be supplied with organic compounds in the form of food. Once growth is completed the adult still needs food for replacement and repair of worn out or damaged tissues.

In addition the living organism requires **energy** for its chemical processes, movement and heat production. This energy is produced (released) by burning (combustion) food in the process of **respiration.** Like a machine, every organism must be maintained in good working order. This condition we refer to as **health.**

We can summarise the reasons why plants and animals require food:

1 Growth – the initial increase of cells in organisms to grow to adult size, and the repair of damaged or worn-out tissues.

2 Energy – to provide energy to drive the chemical processes, for mechanical work of muscles, and for the maintenance of body temperature in warm-blooded animals.

3 Health – to afford protection against disease and to provide raw materials for the manufacture of secretions such as hormones and enzymes.

2.11 Classes of food

Animals and plants must be supplied with food or be capable of making their own food from raw materials. The process of making or obtaining food is called *nutrition.*

There are six classes of food:

1 *Carbohydrates* } required in large quantities for *energy* production.
2 *Fats* }

3	*Proteins*	required in large quantities for *growth.*
4	*Mineral salts*	required in traces for *vital processes.*
5	*Water*	required as a *solvent* for chemical reactions.
6	*Vitamins*	specific compounds required by animals for *health.* Most vitamins cannot be manufactured by animals.

Carbohydrates
Plants manufacture glucose by using carbon dioxide and water in the presence of sunlight and chlorophyll (the green colouring of plants). This simple sugar is built up into storage materials such as sucrose, starch and inulin. These substances are present in large quantities in:

Cereal crops – wheat, maize, oats, barley.
Ground crops – potatoes, carrots, parsnips, swedes, beetroot.
Leguminous crops – beans, peas.

Man and animals can make use of these food stores to provide their own carbohydrate requirements. The carbohydrates built up or taken in by living organisms are broken down to release energy. If large quantities of carbohydrates are consumed by animals over and above their energy needs, the carbohydrates are changed into fats and stored in the body. Man throughout the world generally has enough carbohydrate in his diet, except in special circumstances such as drought or flooding, which result in famine.

Fats and oils
Simple carbohydrates manufactured by plants can be changed into oils for storage purposes in fruits and seeds. The food reserves of oil in these structures are used for growth and energy when the seeds germinate and produce the young plants. Animals can feed on oily fruits and seeds and so obtain energy-rich foods.

Fruits and seeds rich in oils include:
peanuts, coconut, maize, cashew nuts, soya bean, sunflower seed.

Proteins
The living material (protoplasm) of plants is manufactured from simple carbohydrates, with the addition of elements obtained from soil salts through the roots. Thus all plant food consumed by animals contains a certain amount of protein, but some plants have a particularly high concentration of protein in certain organs. *Leguminous crops* have a high protein content in their seeds, e.g. peas, peanuts, haricot beans and soya beans. In addition sunflower seed, maize, cashew nuts and wheat all contain considerable amounts of protein.

Plant protein is called *second class protein.* Animal protein eaten as food in meat and fish is called *first class protein* because it contains all the essential amino acids. Since the proportion of protein is not as high in plant food as in animal food, large quantities must be consumed by an animal in order for growth to take place. In animals, the protein becomes highly concentrated in the muscles of the body which therefore provide a much greater source of protein.

As we shall see later in the chapter, a lack of protein in the diet of humans causes serious deficiency problems resulting in diseases such as *kwashiorkor* and *marasmus.* Lack of protein is a problem in many developing countries and attempts have been made to produce new sources. Efforts have concentrated on quick growing organisms such as yeast and algae cultivated in large tanks. Plant protein obtained from soya beans has been produced in quantity and made more appetising by giving it a taste like meat such as chicken.

Mineral salts
Chemical analysis of the content of animal tissues produces the following percentages of elements by weight.

Oxygen	65%	93% of the total weight
Carbon	18%	
Hydrogen	10%	
Nitrogen	3.0%	6.45% of the total weight
Calcium	1.5%	
Phosphorus	1.0%	
Potassium	0.35%	
Sulphur	0.25%	
Sodium	0.15%	
Chlorine	0.15%	
Magnesium	0.05%	

(All together: 99.45%)

Iron, copper, iodine, manganese, zinc, fluorine, molybdenum and others are present only in minute traces – 0.55%.

The diet of an animal must therefore contain these elements. We have seen how carbohydrates, fats and proteins do contain the three major elements constituting 93% of the total weight. The remainder are also present in the diet, particularly in the amino acids, and although they make up only a small percentage they can represent a considerable quantity in a man.

A man weighing 70 kg will contain:

Calcium	1.5%	1050 g
Phosphorus	1.0%	700 g
Sodium	0.15%	105 g
Magnesium	0.05 %	35 g

An analysis of plant structure will give very similar percentages for the same elements: these are absorbed by the plant from the soil.

Element	Source from normal food	Importance to the mammalian body
Nitrogen N	Protein foods, lean meat, fish, eggs, milk.	For synthesis of protein and other complex chemicals, formation of muscle, hair, skin and nails.
Sulphur S	As for nitrogen.	As for nitrogen.
Phosphorus P	As for nitrogen.	For synthesis of protein and other complex chemicals, formation of bones, and teeth, formation of ATP.
Iron Fe	Liver, green vegetables, yeast, eggs, kidney.	Forms haemoglobin in red blood cells. Absence causes anaemia.
Calcium Ca	Milk, cheese, green vegetables.	Formation of bones and teeth, necessary for muscle contraction and blood clotting. Absence causes rickets.
Iodine I	Sea fish, and other sea foods, cheese, iodised table salt.	Formation of hormone in the thyroid gland, absence causes goitre and reduced growth.
Sodium Na	Table salt, green vegetables.	Maintenance of tissue fluids, blood and lymph, transmission of nerve impulses.
Potassium K	Vegetables.	Transmission of nerve impulses.
Chlorine Cl	Table salt.	Maintenance of tissue fluids, blood and lymph.

Table 2.1 Mineral elements: their sources and characteristics

Mineral elements are not taken in as elements but as salts in their ionic form, e.g. sodium chloride (common salt, table salt) as sodium and chlorine ions.

The formation of proteins and other complex compounds in the plant body incorporates the elements into the structure of the plant. Animals acquire their elements by feeding on plants. The facts about mineral salts are given in table 2.1.

Water

The chief constituent of living matter is water, which is so familiar that its importance in structure and functioning is often overlooked. Below is the approximate composition of a man weighing 70 kg:

Water	70%	49 kg	85% fluid
Fat	15%	10.5 kg	
Protein	12%	8.4 kg	15% solid parts of cells and support structures
Carbohydrate	0.5%	0.35 kg	
Minerals	2.5%	1.75 kg	

Water is most important as a solvent in living organisms and thus plays a fundamental part in cellular reactions. It has a *high heat capacity*, so that in an animal or a plant, chemical reactions producing considerable heat make little alteration in the temperature

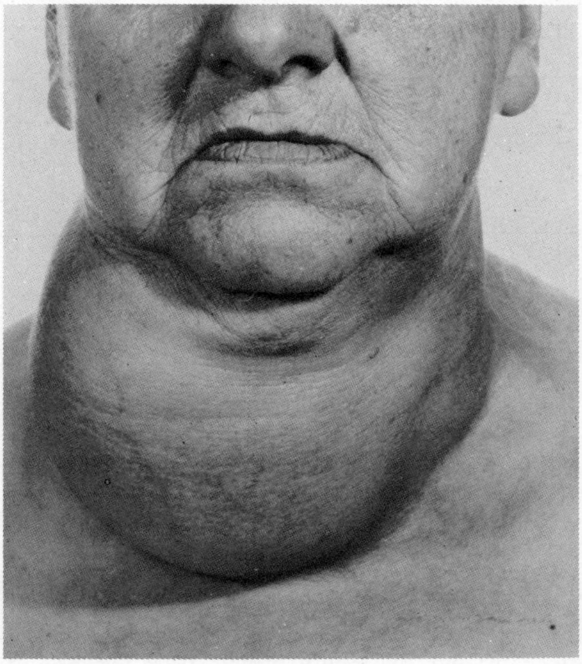

fig 2.6 An enlarged thyroid: goitre

Vitamin	Source from normal food	Special source	Symptom of deficiency	Special notes
A Retinol	Liver, egg-yolk, green vegetables, cocoa, carrots.	Butter, margarine, cod-liver oil.	Sore eyes, reduced night vision, colds and bronchitis, unhealthy skin.	Carotene from plant pigment converted to vit A in intestine walls.
B_1 Thiamine	Unpolished cereals, beans, lean meat, egg yolk.	Bread, milk, kidney.	Retarded growth, lack of appetite in children, nervous inflammation and weakness, paralysis, the disease called beri-beri.	Likely in rice-eating peoples of Asia.
B_2 Riboflavin	As for B_1 plus green vegetables.	As above for B_1 plus Marmite and liver.	Skin disorders, eye and mouth membrane sores, the disease called dermatitis.	
Nicotinic acid	As above for B_1 plus green vegetables.	As above for B_1.	Digestive disorders, mental disorders and skin disease, called pellagra.	Likely in maize-eating people of Africa.
C Ascorbic acid	Fresh fruit, citrus fruit (e.g. oranges, lemons, grapefruit), raw vegetables.	Prepared concentrated juices.	Bleeding from gums and from membranes, teeth disorders, reduced resistance to infection, the disease called scurvy.	Common fatal disease on old sailing ships where sailors had no fresh fruit.
D Calciferol	Liver, fat, fish, egg yolk, formed in the skin by sunlight.	Butter, margarine.	Weak bones, particularly leg bones, poor teeth, the disease called rickets.	Young mammals susceptible to disease.
K	Liver, green vegetables, egg yolk, unpolished cereals		Prolonged bleeding, essential for blood clotting.	Made by bacteria in the gut.

Table 2.2 Vitamins: their sources and characteristics

of the organism. Large amounts of water are lost daily from animals and plants and so a corresponding amount must be taken in to maintain the *water balance.* Water is gained from two main sources:

i) Most water is *absorbed by the roots* of plants, or taken through the mouth of animals in the form of *food or drink.*

ii) Small amounts are formed in the tissues by *oxidation of the hydrogen* in food.

Vitamins

It was not until the beginning of the twentieth century that it became clear that humans and other animals could not be kept healthy on a diet of pure carbohydrates, fats, proteins, mineral salts and water. Small amounts of a number of other substances were found to be necessary. Gowland Hopkins, an English chemist, first gave the clear proof of these facts by his experiments on rats. He showed that the addition of milk to food consisting of pure protein, starch, sucrose, lard, inorganic salts and water brought about greatly increased growth in a group of rats compared with a control group. He deduced that certain 'accessory factors' which are essential for health were present in milk. These factors are today called *vitamins* and the facts about them are given in table 2.2

Their absence from the diet causes certain *deficiency diseases.* This was a strange new idea at the time – that disease could be caused by absence of food, coming so shortly after Pasteur had found that most disease is caused by microbial organisms in the body.

The vitamins are only required as traces in the diet, but once inside the body their importance is considerable. They act as *co-enzymes* in certain reactions. These co-enzymes cannot be synthesised in the body cells as are all true enzymes, hence they must be otained from foods. If they are not present in the diet then the body functions are disrupted.

Each vitamin was first named with a letter before its chemical nature had been established. Once this had occurred then the vitamin was given a chemical name. In some cases the vitamin has been synthesised in the laboratory.

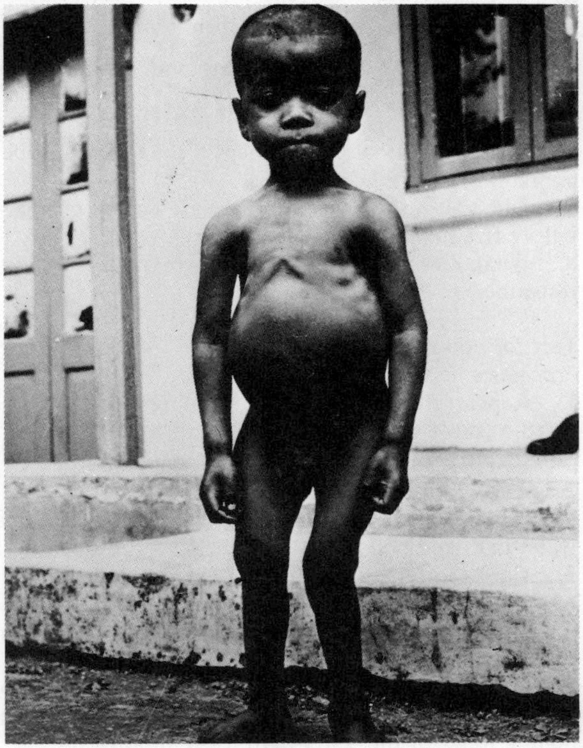

fig 2.7 Rickets

Experiment

Each class of food substance can be detected by a chemical test. The following investigations establish each test on a pure food substance:

Carbohydrates

Test for starch

Procedure

Take 1 cm^3 of starch suspension in water and add two drops of *iodine solution* (aqueous solution of iodine in potassium iodide), in a test-tube. This can also be performed using a small quantity of powdered starch.

Question

1 What colour change do you see?

Test for reducing sugar

Procedure

1 Take 1 cm^3 of glucose solution and place in a test-tube. Add 2 cm^3 of *Benedict's* or *Fehling's* solution.

2 Place the test-tube in a water-bath (beaker or tin) of boiling water for five minutes.

3 Take 2 cm^3 of Benedict's or Fehling's solution and place in a second test-tube with 1 cm^3 of distilled water. Place the tube in the water-bath as in 2 above.

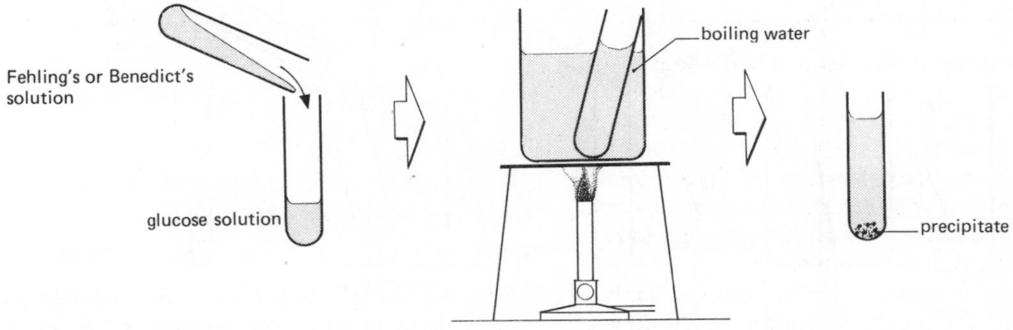

fig 2.8 A flow sequence for the reducing sugar test

This is the control experiment, for comparision with the first tube.

4 Shake each tube at intervals and examine the colour of the liquid.

Questions

2 What is the final colour seen in the first test-tube containing glucose?

3 Did you notice any intermediate colour changes before five minutes elapsed?

4 What colour changes are shown by the second test-tube?

Test for non-reducing sugar

Procedure

1 Repeat the Benedict's or Fehling's test with a sucrose (table sugar) solution. Set up a control tube. Compare the two test-tubes.

2 Take 1 cm^3 of sucrose solution in a test-tube and add three drops of dilute hydrochloric acid. Place in a water-bath and boil for 2–3 minutes.

3 Cool the test-tube under a cold tap, or in beaker of cold water. Add solid sodium bicarbonate slowly until the fizzing stops. This indicates that the solution has been neutralised.

4 Add 2 cm^3 of Benedict's or Fehling's solution to the *neutral* (or *slightly alkaline*) solution. Place in the boiling water of the water-bath for 5 minutes.

Questions

5 What colour changes (if any) take place in procedure no. 1?

6 What colour changes take place after the completion of procedure no.4 ?

7 What action does the hydrochloric acid have on the sucrose during boiling?

Test for fats and oils Test No. 1

Procedure

1 Take 2 cm^3 of *ethanol (ethyl alcohol)* in a test-tube and add to it a small quantity of oil (e.g. *palm oil* or *castor oil*). Shake thoroughly with the thumb over the end of the test-tube.

2 Pour the mixture into a second test-tube containing about 2 cm^3 of water.

Questions

8 What do you observe when shaking the ethanol with oil?

9 What do you observe when adding the mixture to water?

Test No. 2

Procedure

1 Place a spot of oil on a clean sheet of absorbent paper.

2 Alongside the oil spot place a spot of water.

3 Leave the sheet of paper for ten minutes, then hold the paper to the light.

Question

10 What differences do you observe in the appearances of the oil and water spots?

Test for proteins – the Biuret test

Procedure

1 Place 2 cm^3 of egg albumen into a test-tube and add 1 cm^3 of *sodium hydroxide* solution.

2 Add 1% *copper sulphate* solution drop by drop,

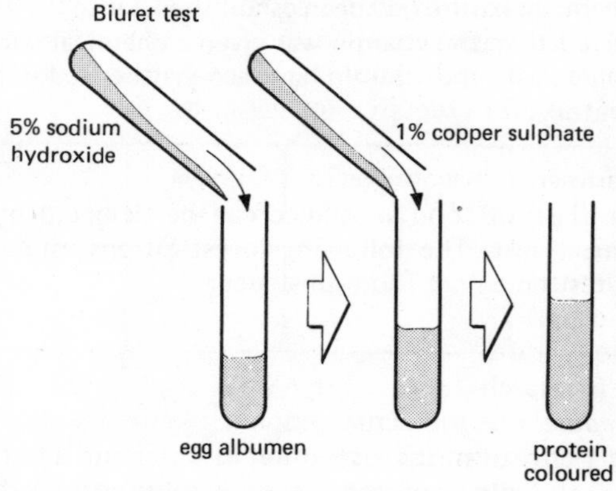

fig 2.10 A flow sequence for a protein test

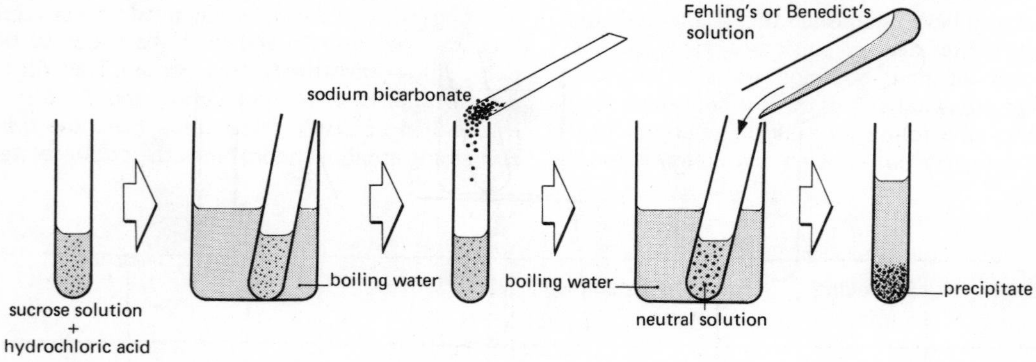

fig 2.9 A flow sequence for a non-reducing sugar test

shaking after each drop (do not add the copper sulphate too rapidly).

Question

11 What colour changes occur when the copper sulphate solution is added?

Test for vitamin C (ascorbic acid)

Procedure

1 *Ascorbic acid* is a powerful reducing agent. It can be detected by using the dye *Dichlorophenolindophenol, DCPIP.* The dye is blue in colour and is decolourised on *reduction* by ascorbic acid.

2 Place 2 cm^3 of 0.1% DCPIP solution into a clean test-tube. This must be an exact amount. Use a syringe or pipette to measure the quantity required.

3 Make up a 0.1% solution of ascorbic acid (1 mg per cm^3). Using a syringe, add it to the DCPIP drop by drop until the blue colour disappears. This happens quite suddenly. Record the number of drops of ascorbic acid required to decolourise the DCPIP.

4 Repeat the test with the same amount of DCPIP solution using the juices of raw fruit and vegetables. For example, squeeze an orange so that the juice is collected in a beaker and add it drop by drop until the DCPIP in a test-tube is decolourised. For each juice count the number of drops.

5 Record your results in the form of a table with the name of each juice tested, together with the number of drops required to change the dye.

Juice	Ascorbic acid 0.1%	Orange juice	Lemon juice
Number of drops			

Questions

12 Work out the actual amount of ascorbic acid in one mg of each of the various juices.

13 Leave some of the juices exposed to air for several days and test again for ascorbic acid. What do you find?

Now that you have established the various tests for the different classes of food substance, carry out the same tests on different common food materials in order to find out what classes of food substance they contain. Draw the following table in your practical book and then carry out tests on any common foodstuffs such as cabbage, carrot, broad bean, fish, milk, peanuts, etc.

Each food should be *cut or ground into small pieces* in order to break up the cell structure so that the organic compounds can be released for the tests. Grinding the food in a pestle and mortar is the best method.

Note that a non-reducing sugar test should only be carried out if the reducing sugar test gives a negative result.

Check your results with table 2.18 which shows the actual quantities of each class of food per 100 g. Carbohydrates are not subdivided into sugars and starch.

Food tested	Reducing sugar	Non-reducing sugar	Starch	Fats or oils	Proteins
Soya bean	nil	nil	present	present	present

2.20 Nutrition in green plants

All animals obtain food either directly or indirectly from green plants. (The green colour of plants is caused by a pigment called *chlorophyll.*) Large meat-eating animals feed on smaller animals but these smaller animals feed on vegetation. For example, lions eat antelopes and antelopes graze on grass. This kind of relationship is called a *food chain.* Every food chain includes an animal which takes in plant matter for food. These animals are called *herbivores* and must be able to obtain all their requirements for energy production, growth, repair and health from plant material. The herbivores are then consumed by *carnivores.*

Let us consider therefore what types of food material are present in plants.

2.21 What foods are present in the green leaf?

Experiment

Testing leaves for starch

Procedure

1 Take a green broad-bladed *dicotyledonous* leaf, e.g. geranium, balsam, lilac, that has been exposed to sunlight. (This experiment can be done on a small scale by using discs cut from the leaf with a cork borer.)

2 Dip the leaf or the discs into boiling water. This kills the living substance of the leaf.

3 The green colouring matter must now be removed. Prepare a water-bath and heat the water to boiling point. **Extinguish the lighted bunsen.** Half fill a test-tube with *ethanol* (ethyl alcohol) and place in it the leaf (rolled up) or the leaf discs. Place the test-tube containing ethanol and leaf into the boiling water.

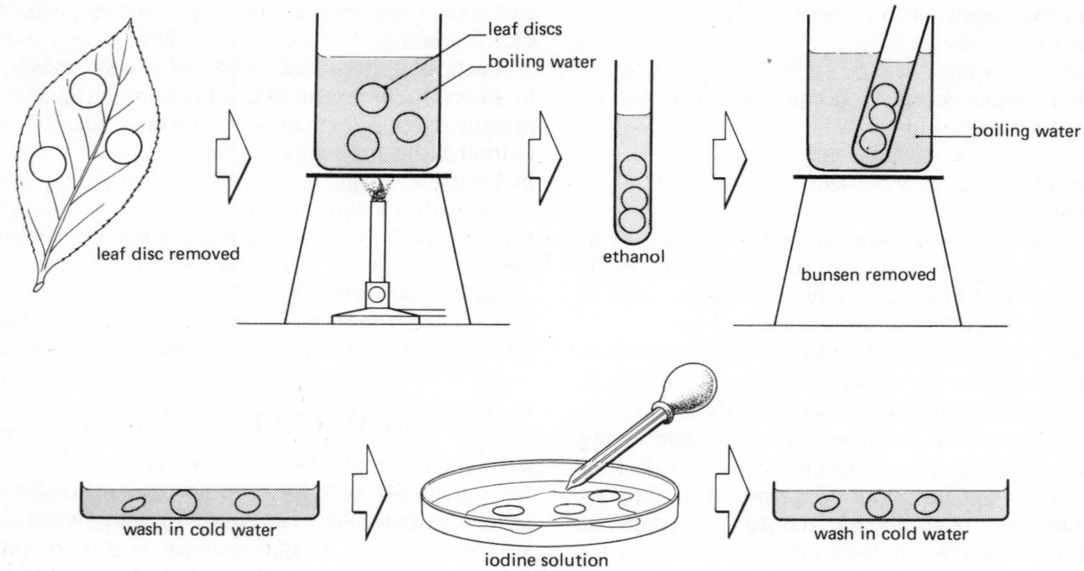

fig 2.11 A flow sequence for decolourising a leaf, followed by the starch test

4 The ethanol will boil and dissolve out the green matter. When the leaf or discs are *colourless,* remove them from the ethanol with forceps and wash them in cold water. This *softens* the leaf structure by returning water removed by the ethanol. A brittle leaf will disintegrate.
5 Place the softened leaf into a petri dish and cover with *iodine solution* for several minutes. Wash away the iodine solution with tap water.
6 Repeat the experiment with a leaf from a shoot of a similar plant which has been kept in the dark for 24 hours.

Questions
1 What do you observe regarding the ethanol that remains in the test-tube? Explain your observation.
2 What do you observe regarding the first leaf after testing with iodine solution? What do you conclude from this result?
3 What do you observe regarding the second leaf after testing with iodine solution? What do you conclude?

Testing leaves for reducing sugars

Procedure
1 Take a *monocotyledonous* leaf such as onion or lily and grind up the leaf with a little sand using a pestle and mortar. Add a small quantity of water.
2 Pour off the water and test for reducing sugar with Benedict's or Fehling's solution.

Question
4 What coloured precipitate was obtained as a result of the reducing sugar test? What do you conclude?

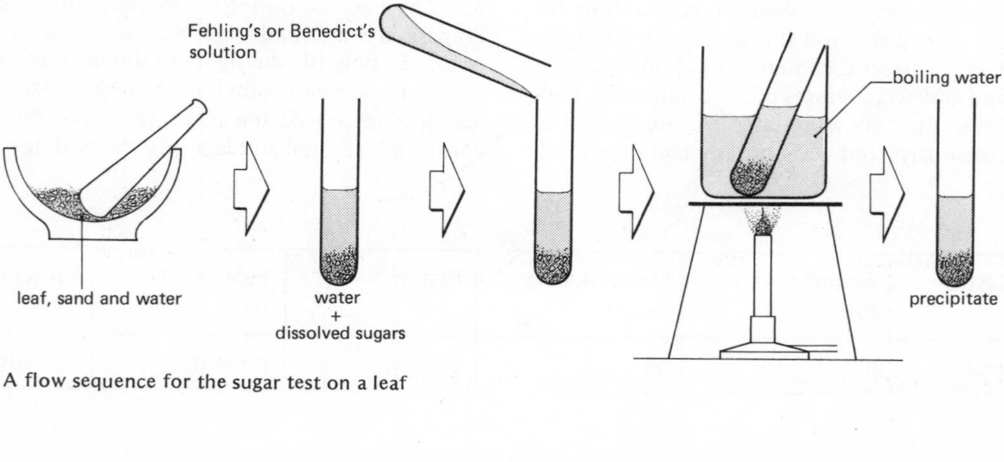

fig 2.12 A flow sequence for the sugar test on a leaf

Testing a variegated leaf for starch

Procedure

1 Take a *variegated* leaf (e.g. *Tradescantia, Coleus,* variegated maple or ivy) which as been exposed to sunlight and make a *sketch* of the distribution of green and white areas of the leaf.

2 Carry out a starch test on the complete leaf.

3 Compare the distribution of starch in the leaf with your original sketch of the green and white areas.

Questions

5 Where, in terms of green and white areas, is starch produced in the leaf?

6 What conclusions can you draw from these observations?

You have shown that a carbohydrate (starch) is present in the leaf, and this has been produced by the plant for its own use. It is of course available to animals when the leaves are eaten as part of the diet. It is also clear from the experiments that light is necessary for starch formation and also that chlorophyll is necessary for starch formation.

2.22 Carbon dioxide

The average *carbon dioxide* content of the atmosphere has been found to be fairly constant at about 0.03% or *300 parts per million* (ppm). It is possible to measure the carbon dioxide content of air accurately, and fig 2.13 shows the measurements for the air in a forest over a period of twenty-four hours. The results of this experiment show that in this particular forest the carbon dioxide concentration was always greater than 0.03%. Perhaps a little later you will be able to explain why the atmosphere of the forest investigated in this experiment contained an average carbon dioxide concentration higher than is normally found in other vegetation.

Examine the graph carefully and answer the following questions.

Questions

1 Did the carbon dioxide concentration fall below 0.03%?

2 Describe briefly how the concentration changed during the twenty-four hour period.

3(a) At what time was the carbon dioxide concentration greatest, and what was its value?

(b) At what time was the concentration of carbon dioxide least, and what was its value?

4 The main environmental change over the twenty-four hour period is the presence and absence of light. Can you produce an hypothesis to suggest an explanation for your observations in terms of the forest vegetation, the carbon dioxide and the light changes?

In order to test this hypothesis we can use an indicator which is extremely sensitive to concentrations of carbon dioxide. It is called a bicarbonate indicator (see Introductory Chapter) and it has a dull red colour established by drawing air through it from outside the laboratory. We can draw through this indicator air which has had the carbon dioxide removed by an absorbent substance called soda lime (see Introductory Chapter).

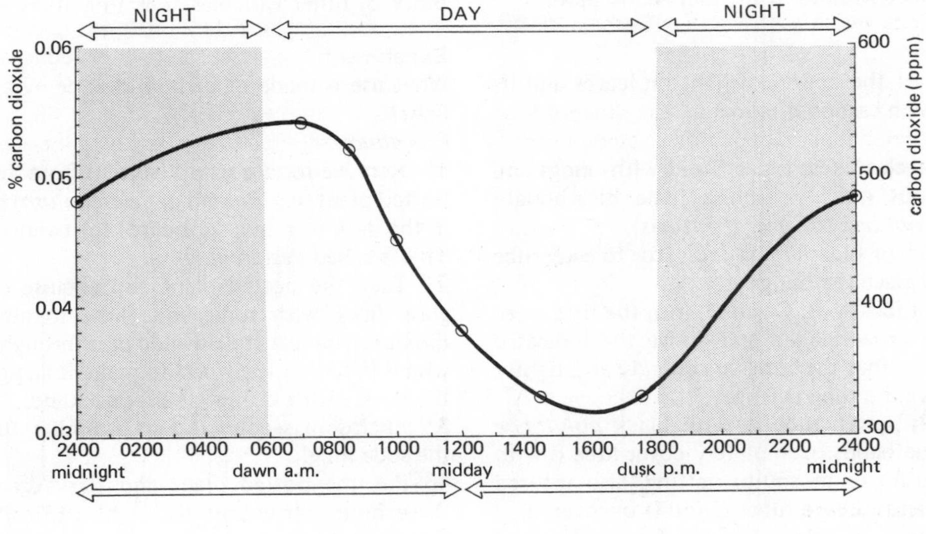

fig 2.13 A graph of CO_2 concentration in the air in a forest over a 24-hour period

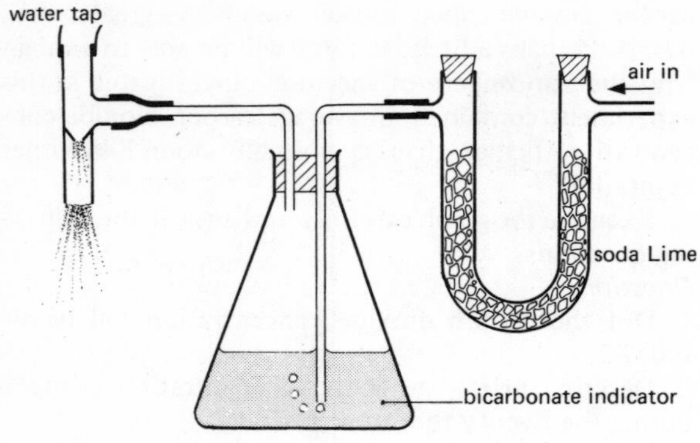

fig 2.14 The absorption of CO_2 from air and its effect on bicarbonate indicator solution

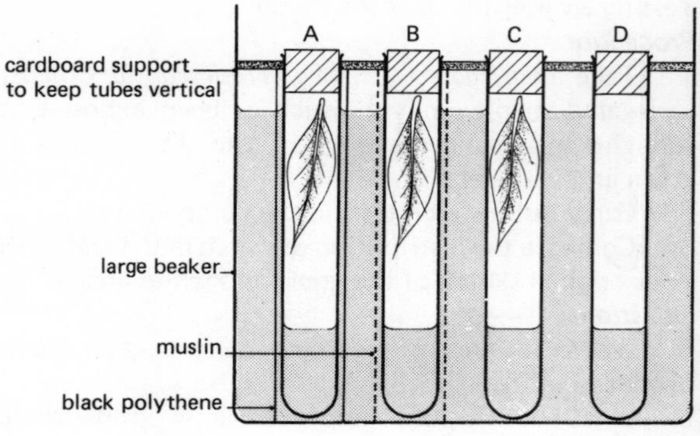

fig 2.15 Leaves in different light conditions enclosed with bicarbonate indicator solution

Procedure

1 Set up the apparatus as shown in fig 2.14.
2 Turn on the tap in order to extract air from the flask. Fresh air will thus bubble into the flask, having been drawn through the soda lime. Keep the tap running until the colour of the indicator changes.

You will see that the indicator becomes *purple* in colour. The air with no carbon dioxide in it must have drawn carbon dioxide from the indicator. Carbon dioxide in solution forms *carbonic acid* and if it comes out of solution there is less acid present. Thus the solution has become more alkaline and less acid, so that the change in colour is really an indication of the pH. This is clearly a very sensitive indicator since it can detect changes in the carbon dioxide content below that of the concentration in air, that is 300 ppm.

Experiment

Investigation of the action of light on leaves and its relationship with carbon dioxide.

Procedure

1 Take four clean test-tubes fitted with bungs and wash them out quickly with a little bicarbonate indicator (*do not breathe near the tubes*).
2 Add 2 cm^3 of bicarbonate indicator to each tube and quickly replace the bungs.
3 Label the tubes A, B, C and D. Into the first three tubes place a broad-bladed leaf above the indicator solution. Ensure that the bungs are quickly and tightly replaced. Leave the tube D empty.
4 Completely cover tube A with black *polythene* fixed by elastic bands. Completely cover tube B with some thin muslin or mosquito netting, also secured with rubber bands. Leave tubes C and D uncovered.
5 Put the tubes in a clean beaker of tap-water, and place a powerful lamp next to the beaker. Alternatively, place them in strong sunlight.
6 Record the time at which the experiment was set up. Shake the tubes at frequent intervals. Observe the tubes for any colour change in the bicarbonate indicator (this experiment will take between one and three hours).

Questions

5 What colour changes do you observe in the indicator solution in tubes A, B, C and D?
6 What explanation can you give in terms of uptake or release of carbon dioxide?
7 What is the function of tube D?
8 Do your results agree with your hypothesis in answer to question 4 of section 2.22?
9 Why were the tubes placed in a beaker of water? (Consider the hypothesis you are testing and the possibility of other variables operating in the experiment.)

Experiment

What use is made of carbon dioxide by a plant in the light?

Procedure

1 Remove food reserves (starch) from the leaves of a potted plant (e.g. *Balsam* or *Pelargonium*) by keeping it in the dark (e.g. in a cupboard) for twenty-four hours. This is called *destarching.*
2 Take the potted plant and enclose one leaf in a glass flask with some soda lime to absorb carbon dioxide. The leaf stalk should pass through a split cork which has been *vaselined* to make it airtight. Support the flask with a clamp on a retort stand.
3 Enclose a second leaf in a similar flask *without* the soda lime.
4 Put the potted plant and flasks in sunlight for three hours (or expose to bright artifical light in the laboratory).
5 Test the two enclosed leaves and another leaf off the plant for the presence of starch (see section 2.21).

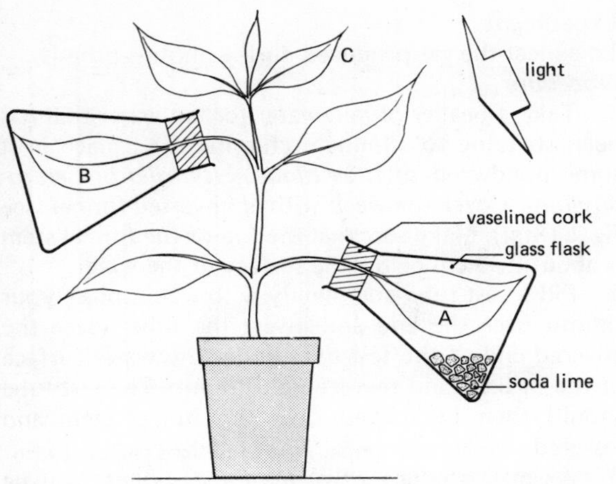

fig 2.16 The experimental apparatus to investigate the use of CO_2 by the leaves of a green plant

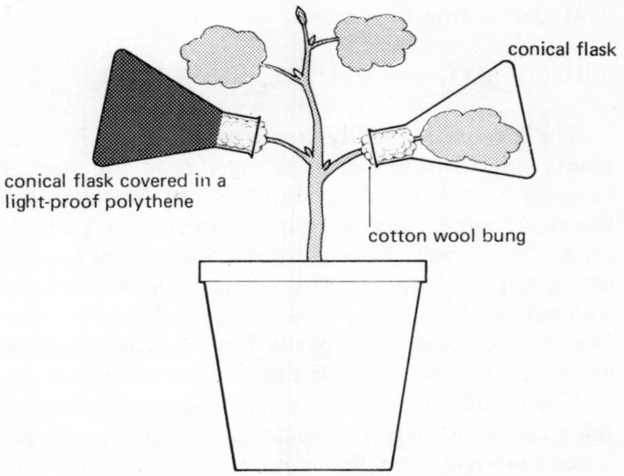

fig 2.17 The experimental apparatus to investigate the use of light by the leaves of a green plant

Questions
10 What were your results with the three leaves after testing with iodine solution for the presence of starch?
11 What do you think happens to the carbon dioxide absorbed by the plant in daylight?
12 Since we are looking at the uptake of a gas from air, how could you criticise the apparatus used, and suggest a method of improving the apparatus to overcome your criticism?

Experiment
Is light necessary for starch formation?
Procedure
1 Remove the food reserves from (destarch) a potted plant (*Balsam* or *Pelargonium*) as in the previous experiment.
2 Take the potted plant and enclose one leaf in a glass flask. Lightly surround the leaf stalk with a small cotton wool bung (this allows gas to circulate). Cover the flask *with light-proof material,* e.g. a black polythene bag.
3 Enclose a second leaf in a similar flask *without the light-proof cover,* and again surround the leaf stalk with a small cotton wool bung.
4 Put the potted plant and flasks in sunlight for three hours (or expose to bright artificial light in the laboratory).
5 Test the two enclosed leaves and another leaf off the plant for the presence of starch (see section 2.21).
Questions
13 What were your results on testing the three leaves with iodine solution for the presence of starch?
14 What do you conclude about the relationship between light and starch formation?
15 In the light of your experimental work in sections 2.21 and 2.22, criticise the apparatus used, and suggest a method of improving the apparatus to overcome your criticism.

2.23 Formation of carbohydrates in plants

Your experiments have shown that, in order to form starch in the leaves of a plant, *light, carbon dioxide* and the green pigment *chlorophyll* are necessary. Carbohydrates consist of the elements C, H and O so that clearly the third element H must be supplied in addition to the C and O provided by carbon dioxide. Scientists have discovered that the *function of light* absorbed by the leaf is to *split* water.

$$2H_2O \longrightarrow 4H + O_2 \ \ldots\ldots \text{(i)}$$

The carbon dioxide molecules are *reduced* by the *hydrogen* produced as a result of the breakdown of water.

$$CO_2 + 4H \longrightarrow (CH_2O) + H_2O \ \ldots\ldots \text{(ii)}$$

where (CH_2O) stands for a simplified carbon compound. (The first carbon compound produced is glucose $C_6H_{12}O_6$.)
Thus equation (ii) can be written:

$$6CO_2 + 24H \longrightarrow C_6H_{12}O_6 + 6H_2O$$

Notice that if we write equation (ii) to include glucose, six molecules of carbon dioxide are reduced to produce a molecule of glucose plus six molecules of water. If the two equations (i) and (ii) are put together to include glucose they are written:

$$12H_2O \longrightarrow 24H + 6O_2$$

$$6CO_2 + 24H \longrightarrow C_6H_{12}O_6 + 6H_2O$$

This can be simplified as:

$$12H_2O + 6CO_2 \longrightarrow C_6H_{12}O_6 + 6H_2O + 6O_2$$

The glucose is rapidly converted to starch in most plants but in others such as *sugar cane* the storage carbohydrate is *sucrose.* Starch is not found in the leaves of other monocotyledonous plants such as onion and cereals. You will note that oxygen is given off during this process. This oxygen *comes from the water molecules,* whereas the oxygen in the carbon dioxide molecules goes partly into the carbohydrate molecule and partly into the new water molecules.

The reduction of the carbon dioxide to form sugars, which are much larger molecules, is called a *synthesis.* Since the sugar formation can only take place in light, the term *photosynthesis* is applied to the process. Another term often used is 'carbon assimilation' referring to the build up of carbon compounds.

The energy required to drive the reduction process comes from the light absorbed in the leaves by the green chlorophyll.

Remember photosynthesis in plants requires:
two raw materials – carbon dioxide and water,
two conditions – light and chlorophyll,
and has **two products** – sugar and oxygen.

Experiment
To collect the gas produced during photosynthesis.
Procedure
1 Take a beaker of rain water (or tap-water that has been standing to eliminate chlorine) and place in it some pondweed such as *Elodea, Hydrilla* or *Ceratophyllum.* Cover the weed with an inverted funnel (see fig 2.18(a)). Make sure that the top of the funnel stem is about 2–3 cm below the surface of the water.
2 Fill a test-tube completely with water, place your thumb over the end and invert the tube. Place the covered end of the test-tube under the water surface in the beaker, and remove your thumb. The test-tube should then be moved over the funnel stem and lowered.
3 A small amount of *sodium bicarbonate* can be placed in the water of the beaker to ensure that sufficient carbon dioxide is released to enable the water weed to photosynthesise.
4 The apparatus can now be placed on a bench near the window of the laboratory so that it receives strong sunlight. At night it should be kept illuminated with a powerful electric light bulb.
5 Gas will start to collect in the top of the test-tube. Leave the apparatus until the test-tube is about half full of gas.

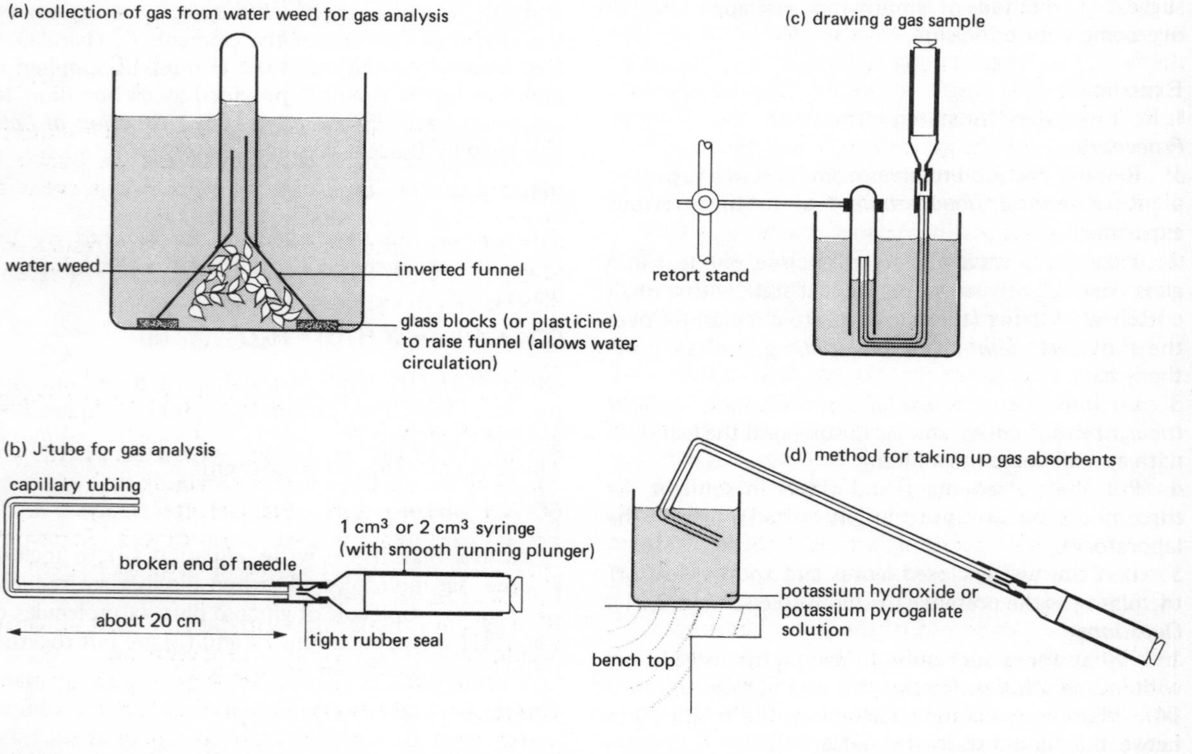

fig 2.18 The experimental apparatus to investigate the gas evolved by water weed, followed by the use of a J-tube

The gas which collects in the test-tube must be *analysed* in order to determine its constituents, particularly with regard to carbon dioxide and oxygen. In order to do this you will need a J-tube (made of capillary tubing with a fine bore) attached to a syringe of 1 or 2 cm^3 capacity. The syringe should have a needle attached which has been cut off near its base. This gives a finer adjustment to the syringe (see fig 2.18(b)).

Experiment

To confirm that oxygen is given off by pondweed in the light.

Procedure

1 Lift the test-tube from the funnel and place a thumb over the open end. Transfer the tube to another large beaker of water and hold the tube in the clamp of a retort stand (see fig 2.18(c)).

2 Push the plunger of the syringe in fully, and place the open end of the capillary tube into the water of the beaker. Gently pull out the plunger and draw water into the capillary tube. The length of the water enclosed in the J-tube should be about 5 cm. Throughout the experiment, try to handle the J-tube as little as possible, since the heat of the body will cause expansion changes in the gas sample.

3 Without removing the J-tube from the water, place the open end up into the test-tube until it is in the gas at the top of the tube (see fig 2.18(c)). Gently pull out the plunger in order to draw a sample of the gas into the tube. The length of this gas bubble should be about 10 cm, twice the length of the first sample of water. Now lower the J-tube and draw in another 5 cm of water to seal off the gas bubble. These measurements of the water and gas bubble need not be accurate.

4 Remove the J-tube from the beaker and place it in a trough of water that has been standing for a while at laboratory temperature. This will bring the J-tube and its contents to a constant temperature. Leave for 5 minutes and then measure the length of the gas bubble to the nearest millimetre.

5 Gently push down the plunger to eliminate most of the water from the seal at the end of the capillary tube. **Ensure that no gas escapes.** Now dip the end of the capillary tube into a 10% solution of *potassium hydroxide* in a beaker. The beaker must be at the edge of the bench since the J-tube must be inverted to perform this operation (see fig 2.18(d)). Draw in about 2–3 cm of potassium hydroxide solution by pulling back gently on the plunger.

6 Take the J-tube out of the potassium hydroxide and by pushing and pulling the plunger move the contents of the tube backwards and forwards several times in order to mix the gas bubble with the chemical. Leave the apparatus in the water bath for 5 minutes to stabilise the temperature.

7 Measure the length of the gas bubble again. Any shortening of its length will be due to the carbon dioxide absorbed.

8 Repeat steps 5–7 using *potassium pyrogallate* solution. This is kept in a beaker with a layer of liquid paraffin over its surface to prevent the absorption of oxygen from the air. Make sure that the open end of the J-tube is **pushed through** the paraffin into the pyrogallate solution. Push out nearly all of the potassium hydroxide solution and pull in about 2–3 cm of pyrogallate solution. Move the contents backwards and forwards as in step 6 above. Place the J-tube in the water trough and leave for 5 minutes.

9 Measure the length of the gas bubble in the J-tube. This time any shortening of its length will be due to the oxygen absorbed.

10 Write down your results as follows:

Initial length of gas bubble	= A cm
Length of bubble after absorption by potassium hydroxide	= B cm
Length of bubble after absorption by potassium pyrogallate	= C cm
Therefore length of carbon dioxide absorbed	= A–B
Therefore percentage of carbon dioxide present	$= \frac{A-B}{A} \times 100\%$
Therefore length of oxygen absorbed	= B–C
Therefore percentage of oxygen present	$= \frac{B-C}{A} \times 100\%$

Note Since the bore of the capillary tube is uniform we can assume that the length of the gas bubble is proportional to its volume. The percentage needs to be expressed to the nearest whole number.

11 In a class situation *tabulate* the class results and obtain the *mean value* for the percentages of carbon dioxide and oxygen respectively.

Questions

1 Why was the J-tube placed in the water trough after each operation?

2 Why was the gas bubble moved backwards and forwards several times with each absorbent?

3 What results were obtained after analysis of the gas bubble? What is their significance? (Remember that air in the laboratory has about 21% oxygen and 0.03% carbon dioxide).

4 Do you think that variation in light intensity would have any effect on this experiment?

Further evidence of the exchange of gases by plants can be obtained by examination of data obtained from water weed growing naturally in a habitat such as a small river.

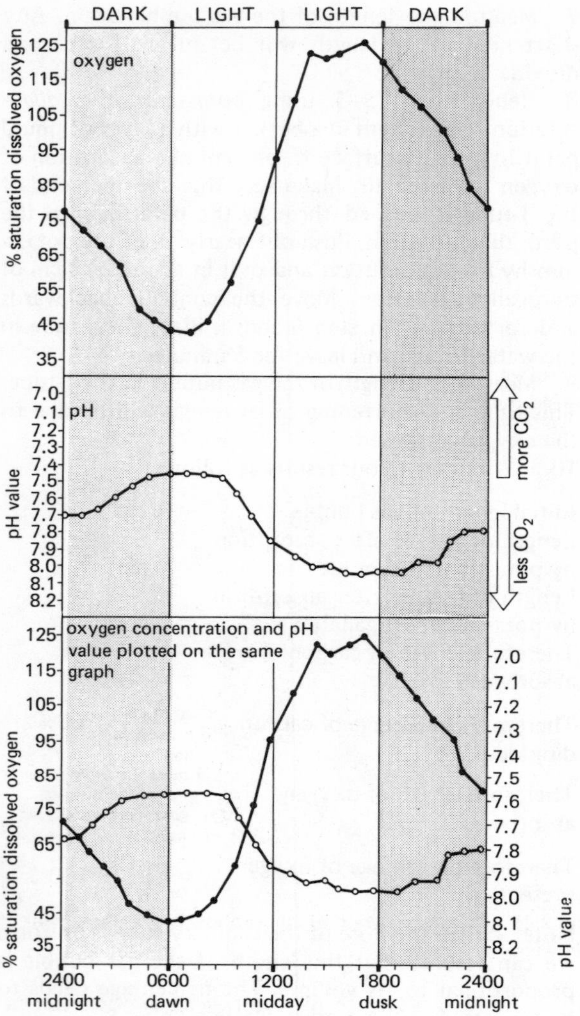

fig 2.19 Graphs of changes in river water over a 24-hour period

Examine fig 2.19 and answer the following questions.
5 Look at the curve for *oxygen saturation* against time. At what times was the oxygen present at (a) its greatest concentration and (b) its lowest concentration?
6 During what period of the twenty-four hours was the oxygen concentration increasing?
7 The *hydrogen ion concentration* (pH) is related to carbon dioxide concentration because dissolved carbon dioxide forms carbonic acid. In the middle graph pH value and carbon dioxide concentration are shown. What is the pH value at the highest concentration of oxygen? Can you suggest why this value coincides with the lower CO_2 level as shown on the graph?
8 What is the pH value at the lowest concentration of oxygen? Why does this value coincide with the greater CO_2 level as shown on the graph?
9 Which of the figures so far shown in this chapter corresponds with the middle graph of fig 2.19?

The bottom of fig 2.19 shows the other two graphs put together, indicating a *clear relationship* between oxygen and carbon dioxide concentrations. At night oxygen is absorbed and carbon dioxide released, while during the hours of daylight, when photosynthesis occurs, carbon dioxide is absorbed and oxygen released. The significance of the absorption of oxygen and release of carbon dioxide at night will be investigated later, but it should be noted now that gas *exchanges* during night and day *compensate* partially for each other.

2.24 Is all light important?

If you look at a detached leaf against the light, you will notice some light is passing through it. This light is a pale green colour. The surface of a leaf on which the sun is shining has a bright green colour, and this light must have been reflected. Thus the leaves can only be absorbing part of the available sunlight for use as energy during photosynthesis (see fig 2.20(a)). A leaf extract can be produced by grinding green leaves with ethanol, or with a mixture of 80% acetone and 20% water. If white light from a projector is passed through a $60°$ *glass prism,* the white light is split into its *constituent colours.* When these are projected onto a screen they are seen as the colours of a rainbow. Their order can be distinguished as red, orange, yellow, green, blue, indigo and violet. If the green leaf extract is poured into a flat-sided container and placed between the prism and the screen, the change in the colours on the screen is quite striking, for some of the colours are *absorbed by the green extract.* The greatest absorption occurs in the blue and red parts of the spectrum (see fig 2.20(b)). The colours have been extracted by the green pigments (chlorophyll) in the leaf extract. It is probable that the plant uses the *light energy* from these particular colours and converts it to *chemical energy.* Thus it is enabled to split the water molecules at the beginning of the photosynthetic process. It takes a great deal of energy to split a molecule of water so that our first equation in section 2.23 can now be written to take account of this fact.

$$2H_2O + \text{energy} \longrightarrow 4H + O_2$$

Further energy is incorporated into the glucose molecules during their formation and when they are built up into starch molecules. It is this energy which is released in animals when the food is utilised in the body (see section 2.11). The chlorophyll which absorbs the light is located inside the leaf cells in the *green disc-shaped chloroplasts* which line the internal cell walls. They are very numerous in the *palisade cells* of the upper parts of the leaf and for this reason the upper surfaces of most leaves are darker than the lower

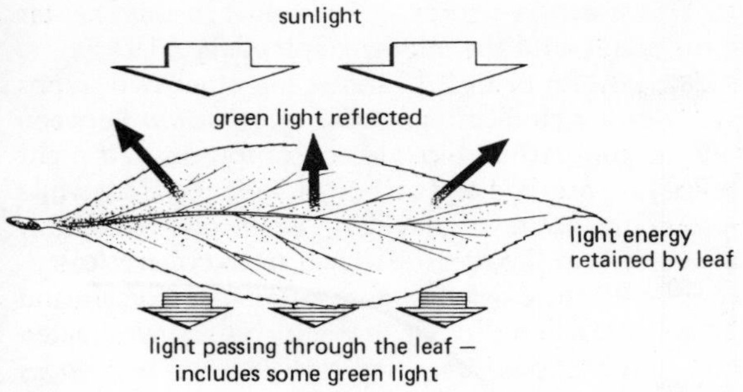

(a)

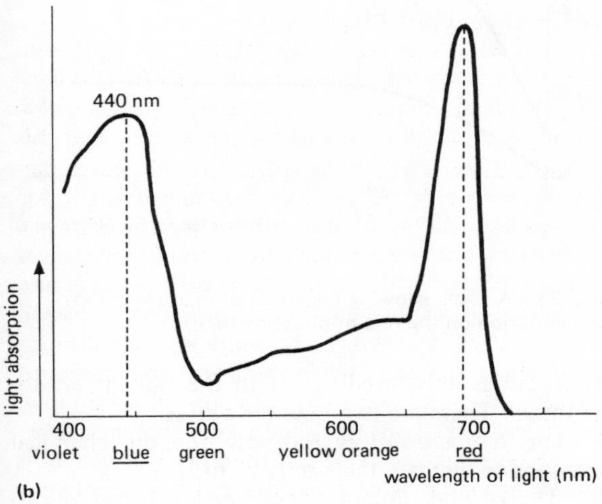

(b)

fig 2.20
(a) The relationship between a leaf and light
(b) The absorption spectrum of chlorophyll

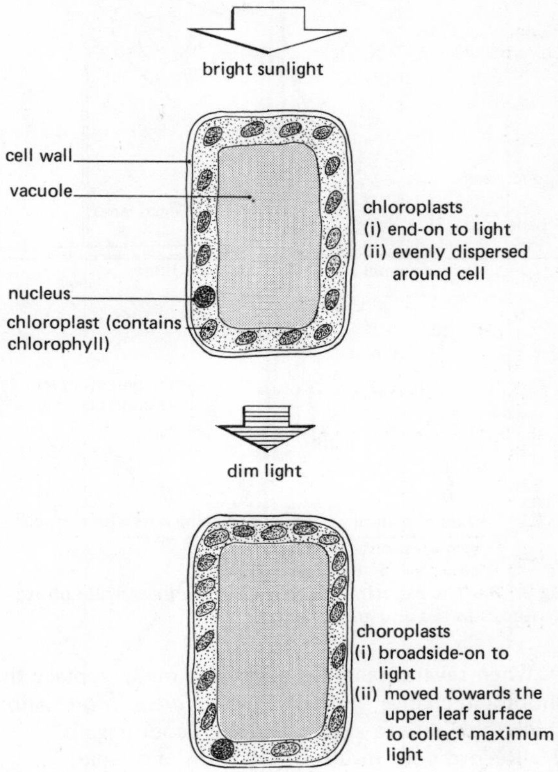

fig 2.21 The distribution of chloroplasts in the upper cells of a leaf

surfaces. In this position the chloroplasts receive most light, and so the major part of photosynthesis takes place in the cells of the upper surface. An interesting confirmation of the relation between chloroplasts and light absorption is the fact that the chloroplasts can adjust their position to absorb maximum light. In *bright* light they are *end-on* to the incident light, but in *dim* light they move *side-on* at right angles to the light rays. Furthermore they move closer to the surface of the leaf and towards the light during dull conditions (see fig 2.21), enabling more light to be absorbed.

Experiment
Do differences in the wave length of light have an effect on photosynthesis?
Procedure
1 Take a beaker or glass jar filled with water and add a small amount of sodium bicarbonate. Stir to dissolve.
2 Place a bunch of water weed (*Elodea* or *Ceratophyllum*) in another jar near to an electric light, so that photosynthesis and gas production are stimulated.
3 Take a small piece of water weed about 5 cm in length and weight the uncut end (near the growing point) with a small paper clip. Place the pond weed in the beaker prepared in no. 1 above with the cut end upwards, but well beneath the water level. The paper clip will cause the weed to float vertically.
4 Place a *red light filter* between a bench lamp (100 watt) and the beaker containing the pond weed (see fig 2.22). Switch on the lamp so that the light shines through the red filter. The apparatus should be away from bright sunlight.
5 Bubbles will soon appear through the cut end of the stem. If they do not, then cut off a small portion of the stem, and if this fails then try again with a fresh piece of water weed.
6 When the weed is bubbling regularly, count the number of bubbles over a period of time, say 5 minutes.

The counting is made easier if each time a bubble appears, you tap with a pencil on a piece of paper, thus obtaining a series of dots which can be counted later (see fig 2.22).

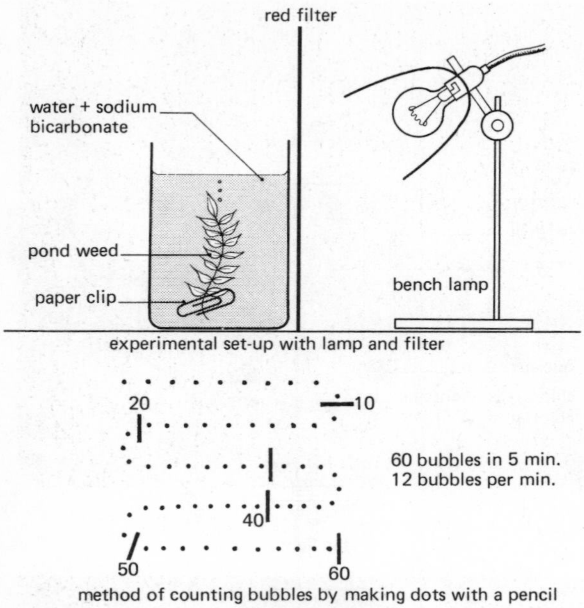

fig 2.22 The experimental apparatus to investigate photosynthesis in red and green light

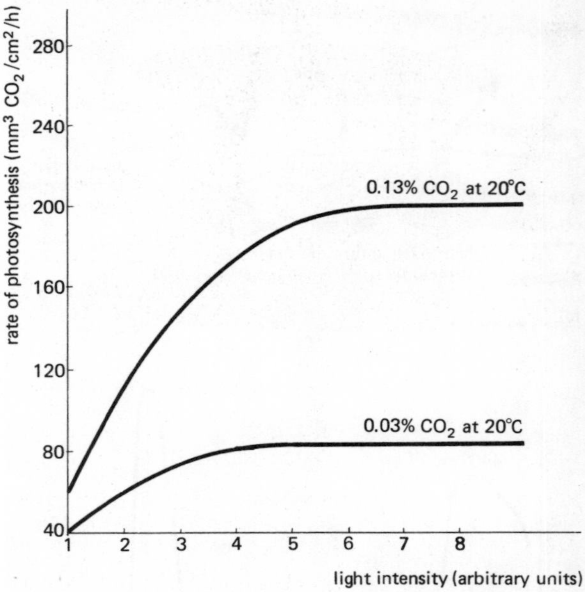

fig 2.23 A graph showing the effect of increase in CO_2 concentration on the rate of photosynthesis

When several readings have been made, replace the shoot and change the red filter to a *green filter.* Allow the bubble rate to adjust, and then count again.
7 Record your results in the form of a table.

Number of dots (therefore number of bubbles) in 5 min.		Rate of bubbling per min.	
red light	green light	red light	green light

mean rate of bubbling in red light =
mean rate of bubbling in green light =

Questions
1 In what part of the shoot is the gas produced? Why is it bubbling from the cut end?
2 Which light filter produced gas at the faster rate?
3 What explanation can you give for this difference in the rate of bubbling?

2.25 Limiting factors

When a plant is exposed to *increasing light intensity,* the *rate* of photosynthesis steadily *increases.* If the temperature is kept constant at 20°C and the carbon dioxide concentration at 0.03% (average for the atmosphere), the rate of photosynthesis can be plotted against increasing light intensity (see fig 2.23, graph A). It can be seen from the graph that, at a certain light intensity, the rate of photosynthesis levels off.

What causes this levelling off in the rate of photosynthesis? There are two possible causes.
i) The *temperature* is *too low* for the chemical reactions to increase their activity further.
ii) The *carbon dioxide concentration* is *too low* to allow any further increase in the rate of photosynthesis.

If the temperature is raised to 30°C it makes little difference to the rate of photosynthesis. This rules out cause (i). However, if the experiment is repeated at 20°C with an additional 0.1% of carbon dioxide, there is a startling increase in the rate of photosynthesis, as shown in graph B in fig 2.23. Carbon dioxide is therefore a *limiting factor* in this process. Any chemical reaction dependent on a number of factors will always proceed at a rate controlled by the factor that is operating at full capacity. In the graphs in fig 2.23 as long as the *rate of photosynthesis is rising,* the *light intensity* is a limiting factor, but directly the graph levels out *some other factor* must be limiting the rate of photosynthesis.

In the natural environment, *light and temperature* can also be *limiting factors* for photosynthesis, especially early in the morning and in the evening. Plants living in shady places in a forest have only a short period of the day when light intensity is not a limiting factor. In such places, there is intense competition between plants for the available light. Since it is advantageous for a plant to be taller than its neighbour, most shoot systems tend to grow upwards, towards the light. In tropical rain forests the trees with their great

fig 2.24 A methane burner for CO_2 production in glass-houses

fig 2.25 Lettuce grown in air (CO_2 at 0.03%)

fig 2.26 Lettuce grown in enriched air (CO_2 at 1.00%)

height and foliage dominate the vegetation below. Epiphytes are able to grow high up in the trees, thus capturing some of the light essential for photosynthesis.

Carbon dioxide remains constant in the atmosphere at an average of 0.03%, and for this reason it is a considerable *limiting factor under natural conditions.* As seen from the graph, fig 2.23, the rise of 0.1% in the concentration of carbon dioxide increases the rate of photosynthesis and therefore the productivity of the plant. In the search for increased food production, scientists have devised a method of increasing carbon dioxide concentration in glass houses in temperate climates. By the use of *methane burners* as in fig 2.24 additional carbon dioxide is produced and this, combined with a slight increase in temperature, increases considerably the productivity of the plants in the glass houses (see fig 2.25 and fig 2.26). In temperate climates, light and temperature are more likely to be limiting factors than they are in tropical countries, and therefore glass houses are in common use. In such enclosed conditions, increasing carbon dioxide concentration to 1.0% can mean *greater production.* It is of course an expensive procedure and furthermore great care has to be taken to ensure that no dangerous fumes are produced by the burners

2.26 Protein

In many plants the sugars formed by photosynthesis are immediately changed into starch grains in the leaf cells. At night this starch is changed back into sugar and transported around the plant to where it is needed for growth and energy production. The movement of the products of photosynthesis is called **translocation.** The sugar may be used immediately to produce energy or it may be stored for long periods in swollen plant organs (e.g. sweet potato) where it is changed back into starch. In other organs such as fruits and seeds, it may also be changed to oils or carbohydrates for storage purposes. At growing points in roots and shoots new protein is required and the sugars contribute towards its production.

Prior to the formation of *6-carbon sugars* (e.g. glucose $C_6H_{12}O_6$) in photosynthesis, a *3-carbon* compound *phosphoglyceraldehyde* is formed in the leaf. It is from this basic compound that a whole series of rapid reactions occurs as part of photosynthesis.

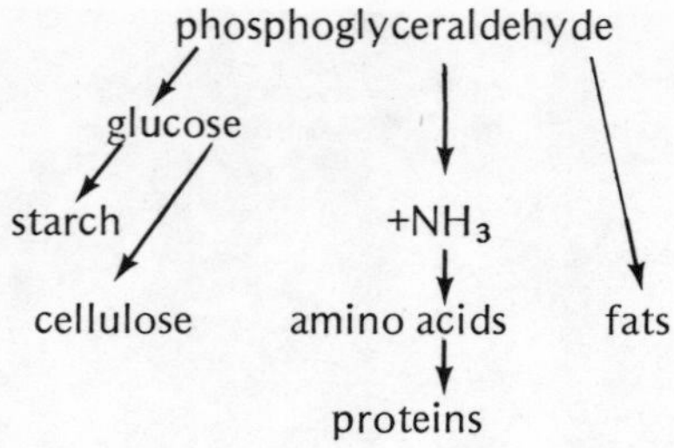

Similar syntheses of complex compounds occur in other parts of the plant to which sugar has been translocated. The sugar is changed to phosphoglyceraldehyde and from this the other substances can be built up. Phosphoglyceraldehyde already contains the C, H and O required for carbohydrate and fat synthesis, but proteins also contain *nitrogen.* This element enters the root system most commonly as the *nitrate ion* NO_3^- and goes through a series of reductions until it becomes *ammonia.*

NO_3^- (nitrate) $\longrightarrow$ NO_2^- (nitrite) $\longrightarrow$ NH_2OH
NH_2OH (hydroxylamine) $\longrightarrow$ NH_3 (ammonia)

Ammonia is used in the formation of *amino acids,* the building blocks of proteins. The first amino acid formed is *glutamic acid* and from this the other essential amino acids can be produced. The nitrate ion is the form in which most plants absorb their nitrogen. The extensive root system in the soil takes up the nitrogen salts through the root hairs. Complex plant protein is built up, but ultimately the plant dies and decays, or is eaten by animals and turned into animal protein. Animals likewise will die and decay, but before this they produce nitrogenous excretory matter (urine) and faeces. *Putrefying bacteria* break down the protein of dead organisms, and also their excretory matter into *ammonium compounds. Nitrifying bacteria* in the soil change the *ammonium compounds* to *nitrites* and then to *nitrates,* so that once again these ions are available to the plant (see fig 2.27). *Nitrosomonas* sp. act on ammonium compounds, and *Nitrobacter* sp. on nitrites.

Some of the nitrates are broken down by *denitrifying bacteria* and the nitrogen is released into the atmosphere, but *nitrogen-fixing bacteria* in the soil can carry out the reverse process and restore nitrates to the soil by using nitrogen taken from the air. One particular group of plants, the *Leguminosae* (including peas, beans and clover), can utilise *atmospheric nitrogen* by means of nitrogen-fixing bacteria which live in swollen *nodules* on their roots. The great majority of plants, however, cannot make direct use of the vast store of nitrogen in the air. Scientists are trying to introduce nitrogen-fixing bacteria into plants such as cereals.

During thunderstorms some nitrogen is added to the soil by the action of lightning which forms *nitric acid* and *nitrous acid* in the rain. These acids enter the soil and combine with the metallic parts of salts to

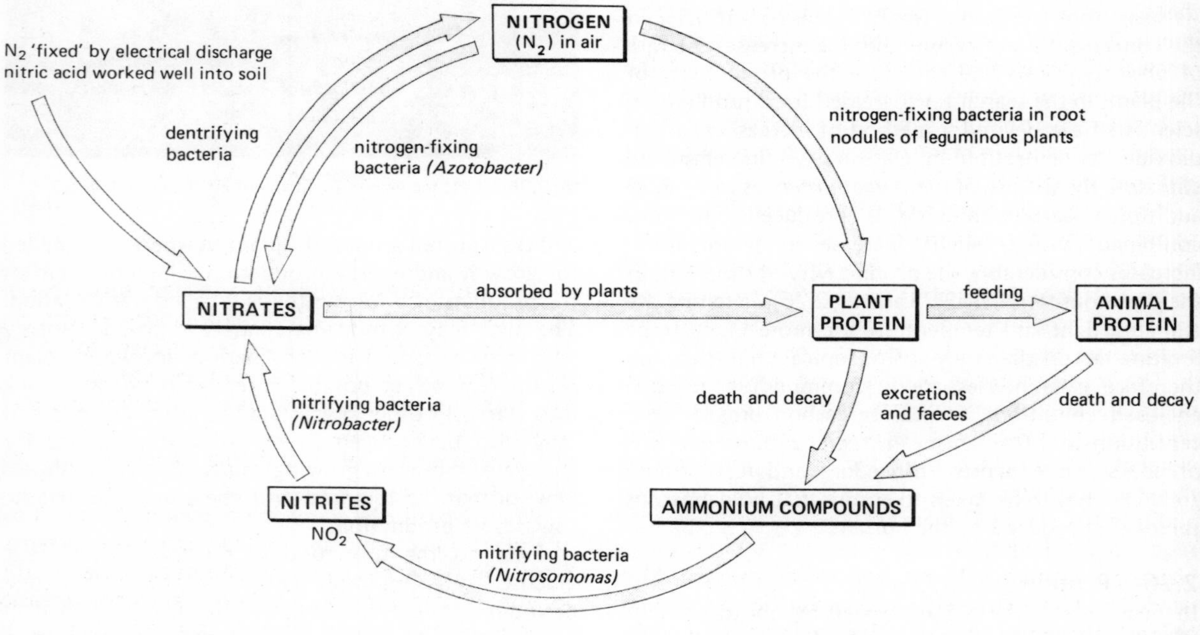

fig 2.27 The nitrogen cycle

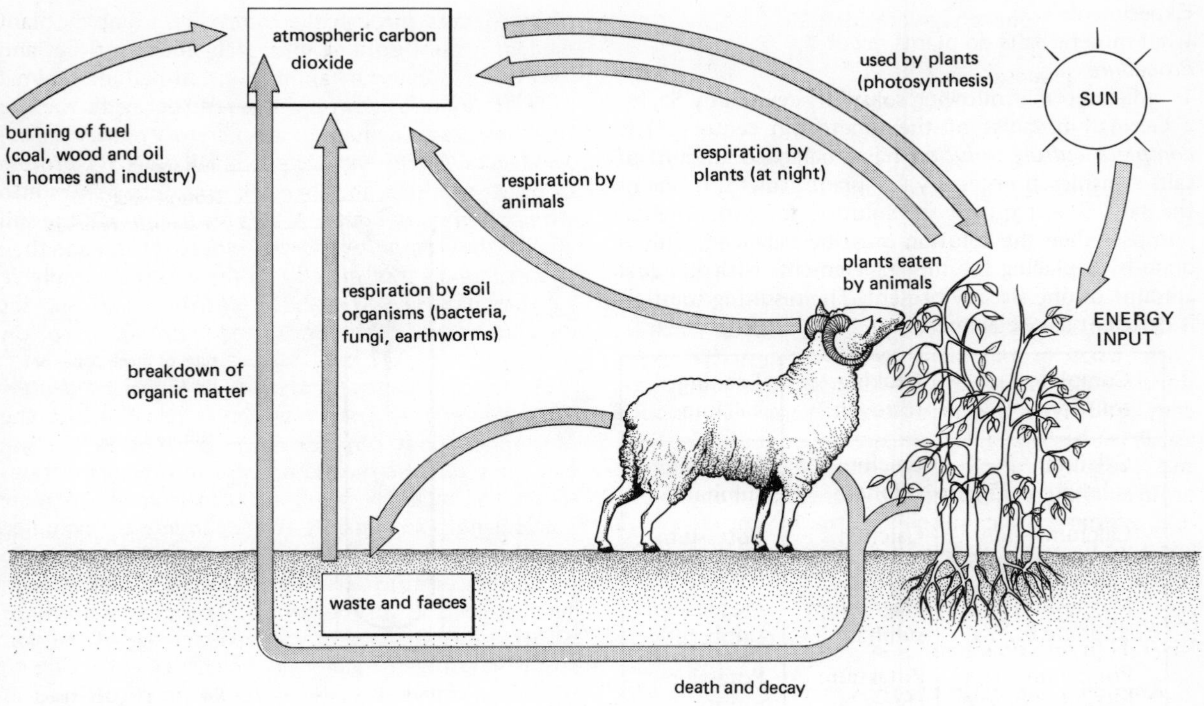

fig 2.28 The carbon cycle

form nitrates. This is an important source of nitrogen for tropical soils. This build-up of compounds of nitrogen and their eventual breakdown and return to the soil or air can be seen as a continuous process which is called a *cycle.* Nature provides continual renewal of this element, which is so important for the manufacture of protein. Unfortunately, this cycle can be broken where natural habitats are replaced by Man's agriculture, which in many cases does not return plant and animal material to the soil.

There are a number of elements in nature in addition to nitrogen that travel through a cycle. Fig 2.28 shows the *carbon cycle* which involves photosynthesis and respiration (see Chapter 4). Both of these processes exchange carbon dioxide with the air. Combustion of plant tissue releases carbon dioxide into the air.

The manufacture of plant protein always requires nitrogen, but other elements, such as *sulphur* and *phosphorus,* may also be necessary. These, and other elements are obtained from the soil in the form of mineral salts which are absorbed by the roots. In addition to C, H and O there are seven 'major' elements required in plant nutrition: nitrogen, sulphur, phosphorus, potassium, calcium, iron and magnesium. Also needed but in extremely small quantities are certain 'minor' elements, such as: zinc, manganese, cobalt, boron, copper, chlorine and silica. Analysis of a plant such as maize shows that the following elements are present:

Elements as a percentage of dry weight

Carbon	43.5	Nitrogen	1.5	Silica	1.2
Oxygen	44.4	Sulphur	0.2	Chlorine	0.1
Hydrogen	6.2	Phosphorus	0.2	Aluminium	0.1
	94.1%	Calcium	0.2		1.4%
		Iron	0.1		
		Magnesium	0.2		
		Potassium	0.9		
			3.3%		

The three basic elements and the seven major elements constitute 97.4% of the total dry weight. The major elements are needed by the plant for:

i) synthesis of proteins (including enzymes),

ii) formation of the middle lamella (*calcium pectate*) between the cell walls,

iii) maintenance of a suitable medium for the functioning of enzymes.

Where a particular element is missing in the nutrition of a plant, the plant may develop a mineral deficiency disease recognisable by its symptoms as being specific for the missing element. Similarly, *deficiency diseases* in animals may occur through lack of vitamins or minerals (see tables 2.1 and 2.2).

Experiment

What mineral salts do plants need?

Procedure

1 Make up the following solutions devised by Sachs, a German botanist of the nineteenth century. The *complete culture solution* has a balanced amount of salts considered necessary for plant growth. If one of the salts is left out of the solution for experimental purposes then the solution must be balanced. This is done by replacing the missing elements with an equal amount of one already present. The resulting solution is deficient in one element.

2

Complete solution	Lacking nitrogen	Lacking calcium
Calcium sulphate 0.25 g	Calcium sulphate	Magnesium sulphate
Calcium phosphate 0.25 g	Calcium phosphate	Potassium phosphate
Magnesium sulphate 0.25 g	Magnesium sulphate	Magnesium sulphate
Potassium nitrate 0.75 g	Potassium chloride	Potassium nitrate
Iron(III) chloride 0.005 g	Iron(III) chloride	Iron(III) chloride

Each culture solution to be dissolved in one litre (dm^3) of distilled water.

3 Take four test-tubes, place in a test-tube rack and fill each tube with one of the following culture solutions.

Tube 1 Complete solution – all mineral elements present for healthy growth.

Tube 2 All mineral elements present except nitrogen (nitrate).

Tube 3 All mineral elements present except calcium.

Tube 4 Distilled water (no dissolved salts).

4 Select four maize seedlings at about the same stage of development (these should have been germinated about seven days previously). Root systems should be of equal length. By means of cotton wool, wedge the seedling in the mouth of each tube so that the roots are well covered. Do not wet the cotton wool. Cut off the endosperm from each grain (see fig 2.29). This removes any alternate food supply.

5 Note the date and time, and set aside the tubes where the shoots will receive adequate light. The roots should be shielded from light by a black paper or black polythene tube.

6 Inspect the seedlings every day. Top up falling levels in each tube with *distilled water,* taking care each time not to wet the cotton wool.

7 At the end of two or three weeks examine the seedling carefully and draw up a table of comparison showing:

A leaf colour – greenness, yellowness, dead-patches, etc.

B leaf length – cut off each leaf at the base, measure its length and record the total lengths of all the leaves measured.

C root length – cut off the root system below the grain, separate each major root and cut off at the base. Measure the lengths of these roots and record the total root length. Construct the table at the foot of the next page for each culture tube:

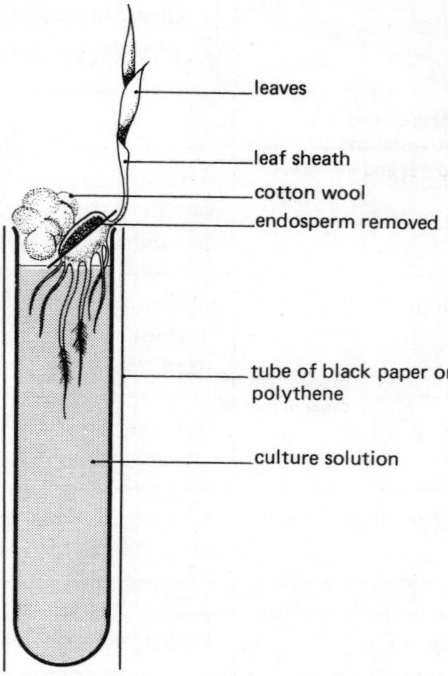

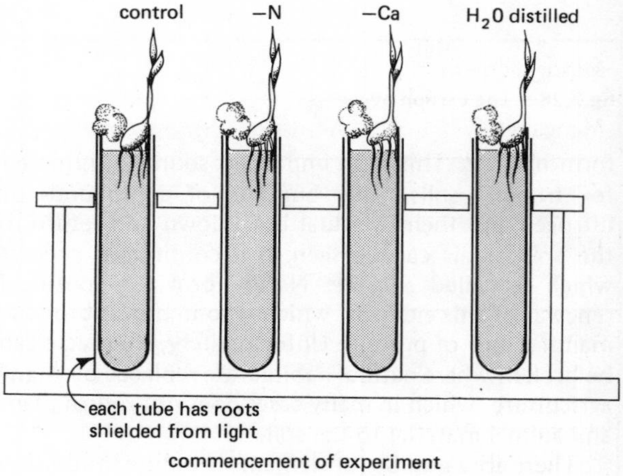

fig 2.29 The experimental apparatus for the culture of seedlings

Major element	Function in the plant	Effect of deficiency	Notes
Nitrogen	Amino acid, protein, nucleotide synthesis.	Small size plants – yellow underdeveloped leaves.	Frequently deficient in heavily used soils – nitrogenous fertiliser required.
Potassium	Amino acids and protein synthesis and particularly cell membrane formation.	Leaves have yellow edges – premature death.	Soils need potassium after heavy nitrogen manuring.
Calcium	Cell wall (middle lamella) development at root and stem apex.	Very poor root growth.	Little present in acid soils – helps aeration of clay soils.
Phosphorus	Formation of high energy phosphate compounds.	Small plants – leaves dull and dark green.	Frequently deficient – little present in soils over pH 7.
Magnesium	Part of the chlorophyll molecule – activator of enzymes.	Leaves turn yellow – veins remain green.	Deficient in acid soils.
Sulphur	Protein formation.	Leaves turn yellow.	
Iron	Chlorophyll synthesis.	Leaves turn yellow – veins remain green.	Not available in clay soils.
Minor element			
Manganese	Activator of some enzymes.	Shoots die back.	
Zinc	Activator of some enzymes.	Leaves not formed properly.	Often deficient in acid soils.
Boron	Influences uptake of calcium ions.	Brown colouration appears in the shoot.	Easily washed out of soil by heavy rain.
Silicon	Cell wall formation in grasses, not essential in most plants.	Decrease in weight of cereal straw.	
Aluminium	Not essential, but absence can cause cell division to be upset.		

Table 2.3 The major elements and their importance

	Tube 1 Control All minerals present	**Tube 2** Less nitrogen	**Tube 3** Less calcium	**Tube 4** Distilled water
A Leaf colour				
B Leaf length				
C Root length				

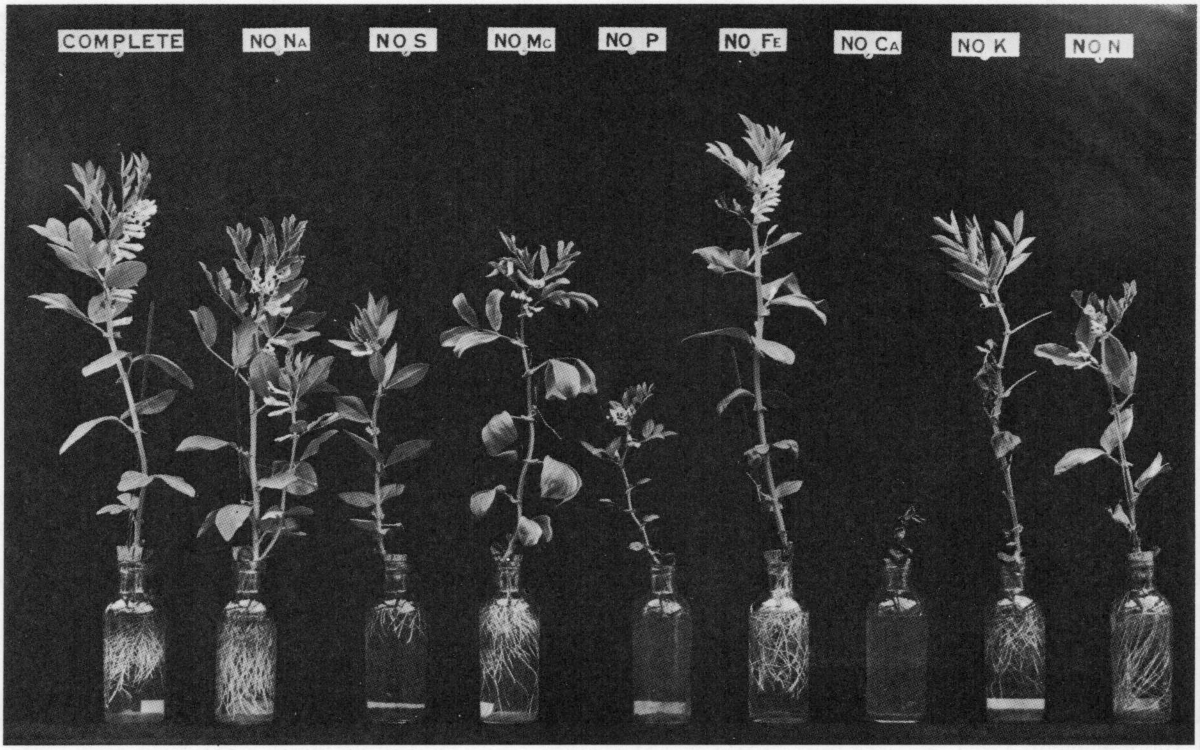

fig 2.30 Bean plants in culture solutions

8 Present the results shown in the table as a series of histograms for leaf length and root length in tubes 1, 2, 3 and 4.

Questions

1 Why should the cotton wool be kept dry?

2 What disadvantage have the roots of the seedlings in the culture solutions compared with roots in soil?

3 Which solution provided:

a) the most growth judged on leaf length;

b) the least growth judged on leaf length?

4 Which seedling had the least growth of root system? Check table 2.3 and explain this result.

5 Why is it difficult to determine whether the absence of a particular element did more harm to root growth or to shoot growth?

6 Why are the solutions topped up with distilled water rather than with the mineral salt solutions?

2.30 Soil

Plants grow in the outermost layers of the Earth's crust. These are called *soil,* and were formed, originally, from the breakdown of rock. The exposed rock was broken into smaller and smaller pieces because of the action of heat and cold. The expansion of water as it changes to ice, the expansion of rock under a hot sun and the friction of stones and water are factors which all contribute to this *weathering*. Gradually, small plants (lichens, mosses and ferns) are able to grow amongst the pieces of rock and, after a long period of time, the growth and death of living organisms provide another important part of soil, the *humus*.

Investigation

To obtain samples of topsoil and subsoil and to expose a soil profile.

Procedure

1 In an isolated spot, well covered with vegetation, dig a hole in the ground.

2 Clean off the side of the hole so that it is vertical and shows a section through the soil.

3 Measure and record the thickness of any clearly defined layers that you can see. Make a simple diagram to show the thickness of the layers drawn to the correct scale.

4 Note and record in your diagram the details of the appearance of each layer, e.g. colour, presence of roots, presence or absence of other living organisms, presence or absence of stones, compactness, etc.

5 Remove soil carefully from a grass root system and observe the spread of the roots. Repeat this for a larger herb. Make sketches of these root systems.

6 Take a sample of the soil from every layer and enclose each in a plastic bag. Seal the bags, by tying up the opening. Collect also a sample of the dead material (litter) from the surface of the soil, and seal in a plastic bag. Label each bag. The soil layers or horizons progress upwards from the parent rock (*C horizon*), through the subsoil (*B horizon*) to the organically enriched surface layer (*A horizon*). The total is referred to as the *soil profile*. This profile may be altered by water movement; in the tropics this is mainly due to excessive rainfall washing out (*leaching*) soluble substances from the A horizon and depositing them lower down by drainage of the water through the soil. Leaching is most pronounced when the rainfall exceeds surface evaporation, especially on soil exposed by heavy grazing which has removed the transpiring plants. Thus leaching and consequent acidity are particularly common in sandy soils (see fig 2.31).

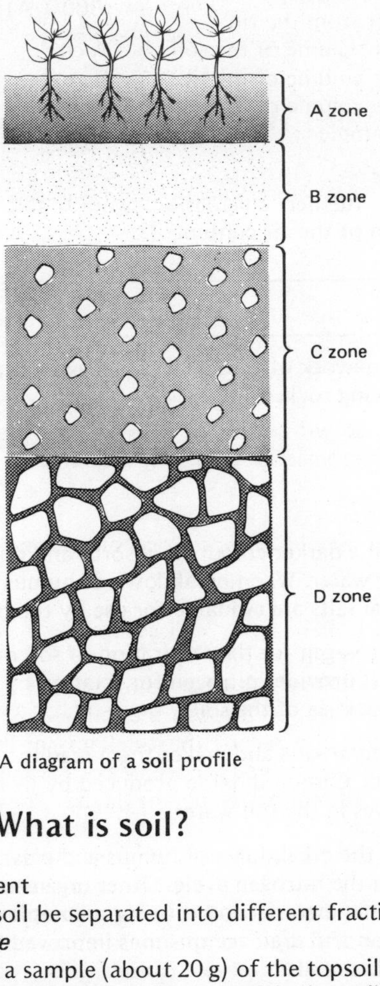

fig 2.31 A diagram of a soil profile

2.31 What is soil?

Experiment

Can the soil be separated into different fractions?

Procedure

1 Take a sample (about 20 g) of the topsoil from the plastic bag and press it into a weighed crucible (metal or porcelain). Weigh the crucible plus the soil.

2 Place the crucible and soil into a steam oven or an oven kept at 105°C in order to dry the soil. Heating should be continued for several days, weighing each day until there is no further loss in weight (allow the crucible and soil to cool before weighing).

3 Place the crucible and soil on a pipeclay triangle and a tripod and heat strongly with a hot flame. At first the soil will turn black, and there will be a smell of burning organic material, but gradually this will disappear. Allow to cool and then weigh. Continue heating until there is no further weight loss. By this time the soil will appear red in colour.

4 Work out your results as follows:

Weight of crucible	= A g
Weight of crucible + soil	= B g
Weight of crucible + soil less water	= C g
Weight of crucible + soil less water and less humus	= D g
Therefore original weight of soil	= (B–A) g
Weight of water in the soil	= (B–C) g
Weight of humus in the soil	= (C–D) g

Weight of mineral matter in the soil
= (B–A)–(B–C)–(C–D) g

Calculate the percentage of each fraction (water, humus and mineral matter) present in the soil sample, e.g.:

$$\% \text{ of water in the topsoil} = \frac{(B - C)}{(B - A)} \times 100\%$$

5 Repeat this experiment with topsoil from different sites, e.g. forest soil, garden soil, hard-packed soil from a footpath, etc.

Questions

1 The amount of water in the soil will vary considerably from sample to sample. What factors can cause this variation?

2 What general differences regarding humus content do you find in soils?

3 What constituent is missing from hard-packed soil such as is found in a footpath?

4 What general differences do you find between topsoil and subsoil?

Experiment

To estimate, quantitatively, the amount of air in a topsoil.

Procedure

1 Take a small tin (volume about 200 cm³) and make one or two holes in its base (with a hammer and

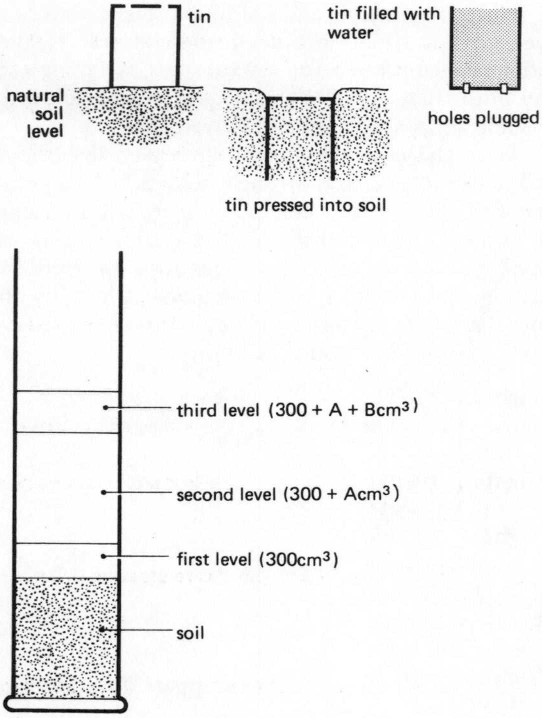

fig 2.32 The experimental apparatus to determine the quantity of air in soil

nail). Press its open end into some topsoil having first removed the surface litter.

2 Scrape the soil away from the sides of the tin and then remove the tin carefully so that it is completely filled with soil.

3 Pour or scrape all of the soil into a large measuring cylinder (e.g. 1 000 cm^3) containing 300 cm^3 of water. Shake or stir the soil until no more air bubbles appear.

4 Measure the new volume of the water.

5 Plug the holes in the tin with clay or gum. Fill the empty tin with water, then pour this into the measuring cylinder and measure the new volume of water (see fig 13.2).

6 Work out your results as follows:

Initial volume of water in the measuring cylinder	= 300 cm^3
Volume of water plus soil	= (300 + A) cm^3
Volume of water plus soil plus water from the tin	= (300 + A) + B cm^3
Therefore volume of the soil	= A cm^3
Therefore volume of the tin	= B cm^3
Therefore volume of air in the soil sample	= (B – A) cm^3

Questions

5 What further calculation would give a better indication of the air content?

Fraction	Constituents	Function
Rock particles	Insoluble – e.g. gravel, sand, silt, clay and chalk. Soluble – e.g. mineral salts, compounds of nitrogen, phosphorus, potassium, sulphur, magnesium, etc.	Provides the framework of the soil and is mainly derived from the underlying rock.
Humus	Decaying plant and animal matter.	This gives the soil a darker colour. It absorbs and retains large amounts of water. When breakdown of humus is complete, mineral salts are available for use by the plant.
Air	Principally oxygen and nitrogen.	The air provides oxygen for the respiration of soil organisms and plant roots. It provides nitrogen for fixation by the nitrogen-fixing bacteria of the soil.
Water	Rain water seeping downwards or capillary water from underground sources.	Dissolves the mineral salts and makes them readily available to plants. Carbon dioxide produced by living organisms dissolves in the soil water.
Living organisms	Bacteria and many other small organisms (see table 2.5).	Bacteria assist in the breakdown of humus and play an important part in the nitrogen cycle. Other organisms feed on plant and animal material and thus form complex food webs. The aeration and drainage of soil is improved by the burrowing of soil animals.

Table 2.4 Soil fractions and their functions

6 Where are the likely sources of error in this experiment?
7 What do you consider to be the importance of air in the soil?

Experiment
Does soil contain a population of living organisms?

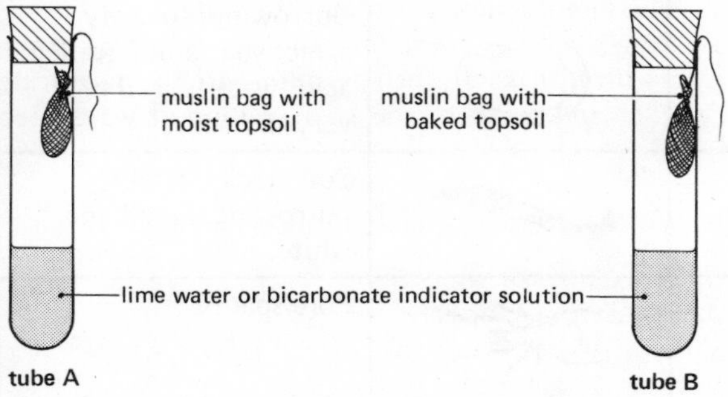

fig 2.33 The experimental apparatus to investigate the presence of living organisms in the soil

Procedure
1 Set up the test-tube as shown in fig 2.33.
2 Tube A contains moist topsoil and tube B contains a similar amount of sterilised soil. Soil can be sterilised by placing it in a steam oven, or any oven at 105°C, for about thirty minutes.
3 Leave the tubes for several days and observe the appearance of the lime water or bicarbonate indicator. You may shake the tubes gently.
Questions
8 What change takes place in the indicator (or lime water)?
9 What do you conclude from your observations?

Investigation
Can the living organisms in soil be extracted and identified? Do topsoil and subsoil differ in their populations of living organisms?
Procedure
1 Assemble two sets of apparatus as shown in fig 2.34. This is known as a *Tullgren funnel* and can be constructed fairly simply.
2 In one funnel, place a sample of topsoil and in the other, a sample of subsoil.
3 Switch on the lamp of each funnel and leave for 24 to 48 hours.
4 Soil organisms will move from the soil, through the grid and into the preservative.
5 Identify the organisms as broadly as possible from the key shown in table 2.5.
6 Make a comparative table of the organisms found in the topsoil and subsoil, indicating type and numbers of each organism (see table 2.5).

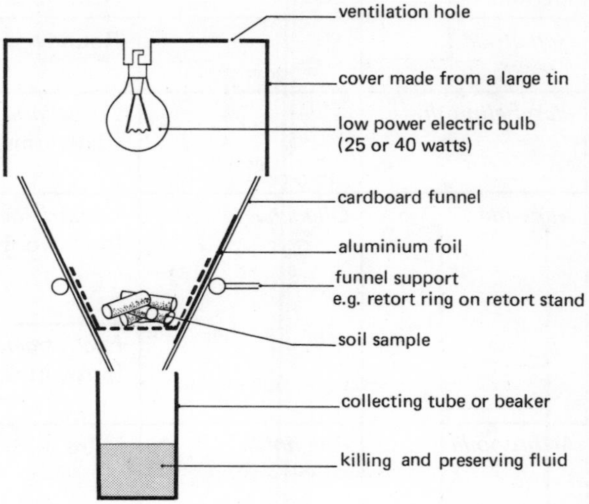

fig 2.34 The experimental apparatus to separate living organisms from the soil

7 Use the same apparatus to compare different types of soil.
Questions
10 What environmental factors cause the soil organisms to move out of the soil?
11 Why is the aluminium foil placed inside the funnel?
12 What general differences do you observe between the topsoil and subsoil in respect of the soil organisms collected? Account for these differences.

2.32 Characteristics of different types of soil

Experiment
To investigate the rise of water (capillarity) through different types of soil
Procedure
1 Set up the apparatus as shown in fig 2.35. The cotton wool helps to retain the soil in the tubes.
2 Note the time and date.
3 At hourly and then daily intervals, note the rise of water up through the soil, and measure the height above the free water surface. Note the time and date.
4 Record your observations in the form of a table and draw a graph for each tube.
Questions
1 Which soil shows the greatest rise of water in the early stages of the experiment?
2 Explain this result.
3 Which soil shows the highest rise of water at the end of the experiment?
4 Explain this result.

Phylum	Class	Name of animal	Structure	Habitat
Nematoda		Roundworms		Water film. Occasionally trapped by fungi.
Platyhelminthes		*Turbellaria* (flatworms)		Under logs, decaying vegetation, damp moss. Water film.
Annelida	*Oligochaeta*	*Lumbricidae* (earthworms)	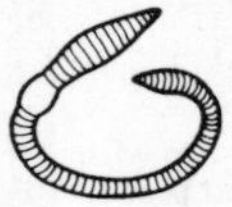	Burrowing; strongly affect soil fauna. Rare in acid, waterlogged or very dry soils.
		Enchytraeidae (potworms)		Soil cracks, rarely burrowing. Small and white.
Arthropoda	*Arachnida*	Mites		Pore spaces.
		Pseudoscorpions	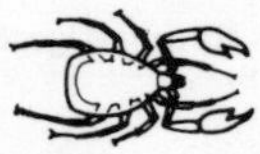	Pore spaces.
		Spiders		Leaf litter, soil crevices, under stones.
	Crustacea	*Isopoda* (woodlice)		Leaf litter, soil crevices, under stones.
	Insecta	*Collembola* (springtails)		Pore spaces.
		Orthoptera (grasshoppers)		Active burrower.
		Coleoptera (beetles)		Active burrowers and channellers.
		Diptera (flies)		Pore spaces.
		Hymenoptera (Ants)	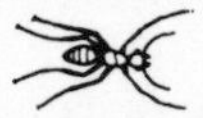	Active burrowers.
		Lepidoptera (Butterflies and moths)	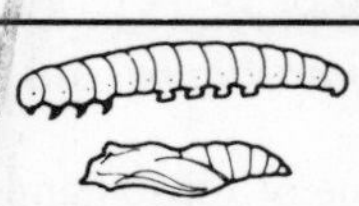	Crevices and pore spaces.
	Myriapoda	*Diplopoda* (millipedes)	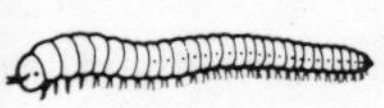	Leaf litter, pore spaces, crevices.
		Chilopoda (centipedes)		As above.
Mollusca	*Gastropoda*	*Pulmonata* (slugs and snails)	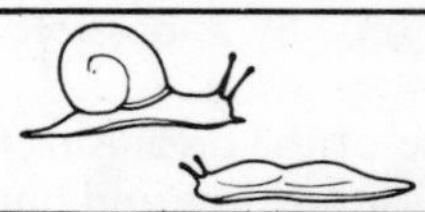	Leaf litter, larger crevices, of upper soil, under stones.

Table 2.5 Key to soil organisms

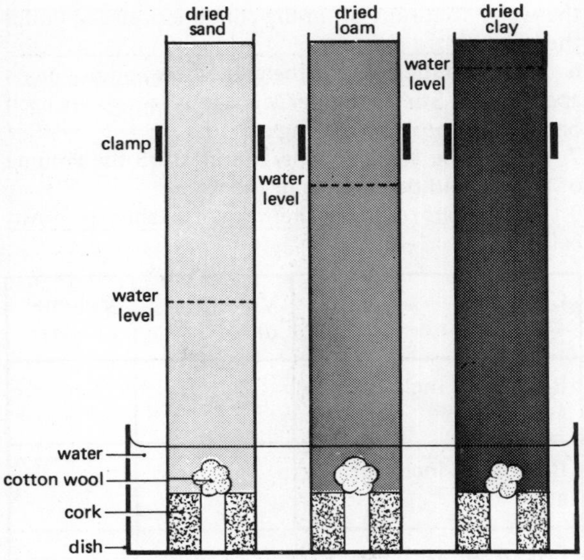

fig 2.35 The experimental apparatus to compare the rates at which water rises in sand and clay soils

Capillarity is an important factor in soil fertility. In times of drought, water can rise from the water table to a high level in some soils.

Experiment
To investigate the drainage of water in different types of soil
Procedure
1 Set up the apparatus as shown in fig 2.36, with the filter papers containing dried samples of sandy and clay soils.
2 Pour 50 cm^3 of water onto each soil sample and leave to drain through. When the water has stopped

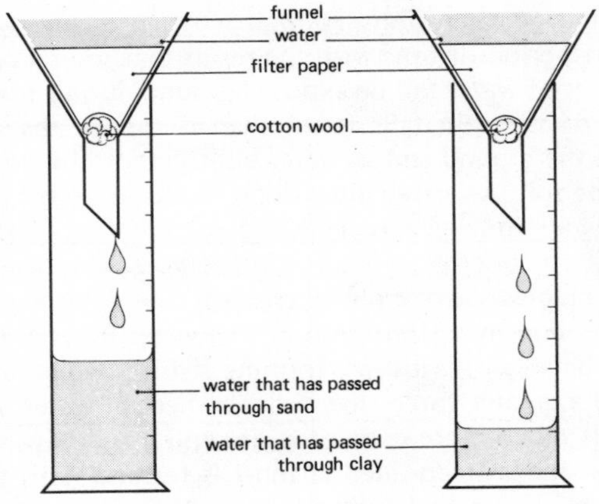

fig 2.36 The experimental apparatus to compare the drainage of sand and clay soils

dripping from both funnels, record the quantity of water in each measuring cylinder.
3 Calculate the quantity of water retained by each soil.
Questions
5 Which soil has the best drainage?
6 Explain this result.
7 Which soil retains most water?
8 Explain this result.

The characteristics of *sandy* and *clay* soils are shown in table 2.6. Loam soils are the best all-round soils. They contain a mixture of humus and both coarse and fine rock particles, thus they have fairly high powers of water retention while allowing the free movement of air and water. Their proportions of mineral matter are: sand 40–70%, silt 20–30% and clay 10–30%.

Clay soils	Sandy soils
1 Very small air spaces between the particles.	Larger air spaces between the particles.
2 Water does not drain well and is easily absorbed.	Water drains well and is not so easily absorbed.
3 The soil remains wet – retains water.	The soil dries out easily – cannot retain water.
4 Water can rise to a high level by capillarity.	Water cannot rise to a high level by capillarity.
5 Rich in dissolved salts.	Poor in dissolved salts.
6 The soil requires a long time to warm up after a rainy season and is heavy to cultivate.	The soil warms up quickly after a rainy season and is easy to cultivate.
7 More than 30% clay and less than 40% sand.	More than 70% sand and less than 20% clay.

Table 2.6 Properties of clay soils and sandy soils

Experiment
To investigate the effects of lime on clay
Procedure
1 Place a teaspoonful of finely-powdered clay in each of two 100 cm^3 beakers, labelled A and B.
2 Add water to both beakers until they are about four-fifths full and stir vigorously.
3 Add to beaker A, 10 cm^3 of 5% calcium hydroxide (slaked lime) suspension in water and stir both beakers again.
4 Leave for 20 minutes, then examine both beakers and note any difference in their appearances.

Questions

9 How does beaker A differ in appearance from beaker B after 20 minutes?

10 In areas where the soil has a high clay content, farmers and gardeners treat it by adding lime. What can you deduce from your results about the principle underlying this practice?

2.33 Soil acidity and alkalinity

In the Introductory Chapter section I 5.40, the measurement of pH value was discussed. Soil acidity and alkalinity can be measured on the pH scale. The amount by which the value varies from neutral (pH7) gives a measure of the strength; thus, pH3 is strongly acid, pH6 mildly acid, pH8 is mildly alkaline and pH10 is strongly alkaline.

Investigation

What is the pH value of a soil sample?

Procedure

1 Place about 2 g of soil in a Petri dish and soak it with BDH Universal Indicator. Leave for at least 2 to 3 minutes.

2 Tilt the Petri dish so that the indicator drains out of the soil.

3 Compare the colour of the indicator with the chart supplied with the indicator solution.

4 (Alternative to steps 1, 2 and 3 above) Soak the soil sample with distilled water, drain off and test with wide range Universal Indicator Test Papers. Obtain a more accurate result using a second test paper of more limited range. Compare with the colour chart supplied with the papers.

2.40 Loss of fertility

Experiment

What effects do slope and plant cover have on the soil?

Procedure

1 Set up three seed boxes (about 40 cm x 20 cm x 5 cm). Put equal amounts of soil in two of the boxes and for the third cut a turf from a well grassed area and place it in the box so that it fits exactly.

2 Each box should have a V-shaped opening in one end.

3 Set up the two boxes of soil inclined at 20° and 40° to the horizontal and the box of turf at 20°.

4 The V-shaped opening should be at the lower end of the box, and directly below a container should be placed to catch water.

5 Obtain a tin can, and pierce some holes in the bottom. Hold the can over one of the boxes and pour in a measured amount of water. Move the can around sprinkling water evenly over the soil. Repeat for the other two boxes, using equal amounts of water. Collect the water that runs off in the containers placed under the boxes.

6 Examine the water in the containers and notice its appearance. Stir the water vigorously and pour each one in turn into a measuring cylinder.

7 Allow the soil to settle and measure the volume of the soil and the volume of the water.

8 Enter your results in the form of a table, as shown below.

Condition	Volume of soil	Volume of water
Box of soil inclined at 20°		
Box of soil inclined at 40°		
Box of soil with grass turf		

Questions

1 Which box lost most soil and water? Explain your findings.

2 Which box lost least soil and water? Explain your findings.

3 What name is given to this loss of soil from the surface of the earth?

When natural vegetation covers the soil, heavy rain is deflected by leaves and thus drops lightly onto the soil. The soil is also *bound together* by the root systems and humus, so that it is able to absorb vast quantities of water. When natural vegetation is removed, by the felling of trees and shrubs, by over-cultivating or by overgrazing with sheep or cattle, then the sun's rays fall directly on the soil and dry out the humus. When the rain comes, erosion results.

In tropical regions, rainfall is often heavy over a short period of time with the result that several centimetres of water fall on exposed ground during a single storm. The rain falls faster than it can be absorbed into the ground and so water builds up on the surface of the soil. A very shallow slope results in a *large mass of water* moving downhill and *carrying soil with it.* This type of erosion is called *sheet erosion.*

The practice of growing crops in round heaps or on ridges does not stop erosion. The water flows around and between the ridges, forming rivulets which often have a greater force than a single sheet of water. As a result much more damage is done; the soil is washed away and a steep-sided channel is formed. This type of erosion is called *gully erosion* (see fig 2.37(a)). The storm water has an increasingly powerful action as the gully deepens, through the washing away of subsoil. In the dry season the gully hardens and, since there is no topsoil, new vegetation cannot become established. This process, if continued, can cause large areas of land to become devastated (see fig 2.37). This is not usually a great problem in the United Kingdom, since rainfall is moderate and distributed throughout the year, and the rate of evaporation is low.

Another effect of heavy rainfall is to dissolve out the soluble salts from the soil and wash them away. This process, known as *leaching*, poses particularly difficult problems for hill farmers.

Depletion of salt content also results, of course, from over-cultivation. If crop plants are grown in a field year after year, the salts are used to build plant tissues and are not replaced naturally.

fig 2.37
(a) Gully erosion
(b) Sheet erosion

Any *trampling* of soil which *destroys the plant cover*, provides a ready made situation for gully erosion to occur. Footpaths, animal tracks and dirt roads become the starting point of erosion, especially where they wind up a hill or mountain. This is becoming a particularly severe problem in some of our areas of greatest natural beauty, such as the Pennine Way and Snowdonia, where thousands of tourists tread the same paths every year.

Any vegetation removed from slopes must be replaced as quickly as possible. Logging operations removing timber should be quickly followed by a reafforestation programme. The growing of crops in hilly country is often carried out on terraces, or with planting along strips that follow the contour lines and that alternate with strips of grass or 'cover crops'.

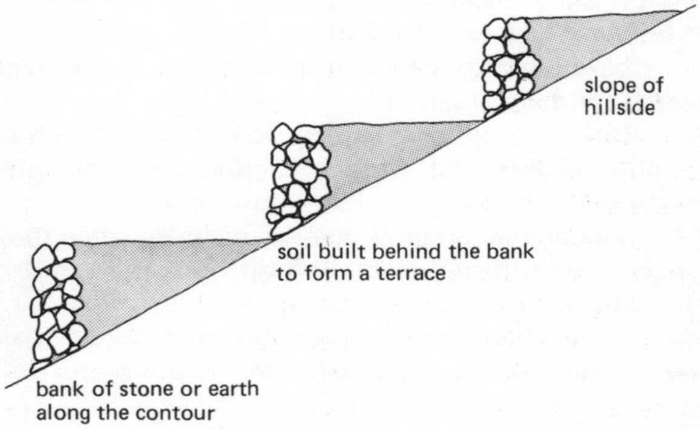

fig 2.38 A diagram of a section through a terraced hillside

Another type of erosion occurs in flat areas of country, where the topsoil has been impoverished by overcropping or overgrazing. The humus no longer binds the soil, and in arid conditions, the wind blows the topsoil away as dust. The Harmattan wind blowing from the Sahara in West Africa carries large quantities of soil and dust. The United States of America has also experienced this *wind erosion* which has resulted in the formation of dust bowls. Again the problem is relatively mild in Britain, but some wind erosion of light, sandy soils and of East Anglian 'black fen' does occur.

2.41 Conserving and renewing soil fertility

Rotation of crops

The method of crop rotation allows the maximum possible use of land because fields are kept under constant cultivation. The method needs careful planning, bearing in mind the following points.

1 The length of time through which the crops may be rotated is variable (1–5 years).

fig 2.39 Strip cropping

fig 2.40 A shelter belt of trees

2 Crops included in the rotation:
i) make different demands on the soil, and
ii) have varying root depths to draw on different levels for mineral salts.
3 Rotation does not increase fertility and therefore regular addition of animal manure, green manure, compost or artificial fertilisers must be made.
4 Leguminous crops should be included since they make little demand on the soil nitrogen and can also be ploughed in as a green manure.
5 An adequate cover of vegetation maintained at all times helps to prevent excessive leaching and soil erosion.
6 Crop rotation prevents the establishment of a crop disease or insect pest, which might flourish on a single crop repeated year after year.

Selection of crops should include one or two *main crops* such as wheat or barley, alternating with *legumes* and *root* crops such as beet. *Deep-rooted* crops should alternate with *shallow-rooted* crops. Deep-rooted crops help to increase soil depth by breaking up the subsoil.

Continuous crop production in a crop rotational cycle is only possible with adequate manuring replacing lost minerals, salts and humus. If this is not possible, then, for some period of the cycle, the ground must be allowed to recover, that is, lie *fallow*. This can be achieved in various ways, by putting down to grass and feeding cattle or by growing a short-term legume which can be ploughed in as a green manure. In the former example the dung and urine of the animals replace the nitrogenous matter.

Reafforestation or shelter belts

The accelerating effect of wind over flat areas of country can be stopped by planting shelter belts of trees, between farms or between large areas of cultivation. These trees reduce the wind speed and thus prevent erosion of topsoil, and in addition, they provide shade for cattle.

Renewal of humus

The function of humus in binding soil is most important in preventing erosion, and furthermore, its breakdown in the soil provides valuable mineral salts. It is essential, therefore, to maintain the humus fraction of the soil. This is usually achieved by one of the following two methods.

a) *Manure* Animal manure is sometimes available, and another source is human manure. Sludge from sewage farms can be used as manure, but with this source great care must be taken that careless disposal does not spread intestinal infections.

b) *Compost* All waste plant and animal materials can be collected into a heap and allowed to rot down as compost. There is a large variety of these materials which can be used, e.g. weeds, kitchen waste, feathers, animals' intestines, urine and plant stalks. The heap should be prepared in layers with some shallow layers of soil added at intervals. Moisture and air are essential in the first stages of decomposition, and as a result, high temperatures are produced within the heap. This first stage of decay, induced by bacteria is followed by fungal decay as the temperature falls. The heap can be turned over at this stage to let air in for a further bacterial stage which also needs to be moist. The final product should be dug into the soil and not left exposed to the sun so that it dries out.

Artificial fertilisers

Adding salts directly to the soil leads to increased growth and a bigger yield of crops. Salts added in this way are called artificial fertilisers and are produced in factories or formed as a waste product of certain industrial processes. The most common elements lacking in highly cultivated soil are nitrogen, phosphorus and potassium, known by their symbols N,P,K. They are supplied in the form of their salts: potassium sulphate; ammonium sulphate and calcium phosphate. Great care should be taken to *avoid continual application* of artificial fertilisers, for they do not replace humus, and this alone can bind the soil together and prevent erosion. Over-use of fertilisers may also result in excessive nitrate concentrations in the ponds and lakes that receive run-off water from the fields. In some instances this may have an adverse effect on drinking-water supplies.

Irrigation
Irrigation is the controlled supply of water to soils in order to increase crop yield. It has been used for over 6 000 years by many different civilisations throughout the world, particularly along river systems and around river deltas, which have thus been able to produce sufficient food to support populations of high density.

In the last twenty-five years many governments in Africa have financed and developed methods of irrigation. Such countries as Kenya, Egypt, Nigeria, Ghana, Zambia and Senegal have all taken advantage of such methods to improve their food supply. The system needs careful planning in order to ensure that the water is used economically and to avoid over-saturation of the soil and possible erosion. The earliest types of irrigation, and that practised to-day on the flat lands of the world, utilised seasonal field waters of the land alongside rivers. Alternatively, water can be drawn from a river by mechanical means (pumps) or by gravity, for distribution through land channels. Modern developments include man-made lakes, land reclamation and dual purpose hydro-electric schemes to direct waters for controlled irrigation. Irrigation by sprinklers is used in the production of fruit, tobacco, vegetables and sugar in countries such as Zambia, Tunisia, Morocco, Mauritius and the Caribbean. This method demands the use of modern pumps and reservoirs. Unfortunately, irrigation, whilst improving food production, increases the spread of water-borne diseases such as bilharzia.

fig 2.41 Irrigation projects in Africa showing plant growth

2.50 Storage of food reserves in the plant

In this chapter we have considered the manufacture of food and its uses in organisms. Plants make their own food from raw materials. This food is not all used immediately after formation. In most plants it is transported to some organ in which the organic compounds can be stored. The stored material may remain for a short or long period of time, after which it is used for a variety of purposes. Very commonly it is used during a period of growth when food manufacture is not always possible. In many areas of the world there is a season of the year when plants find environmental conditions so difficult that they cannot go on living actively. In temperate climates this season is the winter, when low temperatures are the limiting factor, but in the tropics it is the dry season when lack of rain is the limiting factor. Plants store up food material before this cold or dry season, and then use it for rapid growth at the end, when conditions are becoming more favourable.

Storage of food material is most important in perennial plants which live on year after year.

Perennials
These are plants which do not die after producing flowers, fruits and seeds, but persist from year to year. The growth is reduced during the dry or cold season, and in some cases, the parts above ground die off and disappear. These are *herbaceous perennials* which are able to survive throughout the difficult season by underground storage organs. The organs are modifications of stem, root or leaf which because of their function are called *perennating organs.*

Woody plants (shrubs and trees) do not die off during the cold or dry season. The trunk and branches persist above ground and continue to grow year after year. During the difficult season, however, the woody plant often loses its leaves to prevent excessive water loss by transpiration; for in dry seasons water would not be available to the root system. Evergreen trees retain their leaves, but are usually specially adapted to restrict water loss. These plants are known as *woody perennials.*

Herbaceous perennials: daisy, dandelion, crocus, mint, *Gladiolus,* iris, daffodil.
Woody perennials: hazel, ash, oak, holly, heather.

A second group of plants which make use of storage of food over the dry or cold season are the biennials.

Biennials
These are plants which grow during their first year, producing root and shoot, and at the end of their growth period store food material. The food is used during the rapid growth at the beginning of their second year and to produce new shoots, flowers, fruits and seeds. The life cycle is completed in two years, after which the plant dies. Man often grows biennials as a food crop and harvests them at the end of the first year of growth, when the plant is swollen with food reserves. He must allow some to grow through into the second year in order to obtain seeds for planting.
Biennials: carrot, radish.

Thirdly, there are the *annuals*, a group of plants which do not store food for long-term use, but complete their life cycle within a short space of time.

Annuals
These are plants which grow to maturity, produce flowers, fruits and seeds within the space of one year, and then die. In many annuals, the whole life cycle from germination to seeding lasts only a few weeks. The seeds of these plants germinate and the cycle is repeated again and again within a single growing season. Annuals can only survive the cold or dry season as seeds. Instead of food materials stored in the vegetative structures of the parent plant, the food is present in the seed structure.
Annuals: willowherb, groundsel, poppy, chickweed, shepherd's purse.

2.51 Storage in vegetative organs

In the plant kingdom we can see varying degrees of storage of manufactured food.
i) *Initial storage* of soluble sugars and amino-acids occurs in the cell vacuole during photosynthesis. Temporary storage of starch in the cells of the leaf occurs during the day-time as the products of photosynthesis build up. (*Translocation* of soluble foods takes place particularly at night, to areas where the food material is required.)
ii) *Long term storage* occurs in perennating organs or fruits and seeds, to be used at a much later date for rapid growth. This food is manufactured in excess of normal daily requirements for growth and other living processes.

2.52 Storage in modified stems

The long term storage of food may result in the modification of a plant structure. The stem is often used in perennials, and since the storage structure must be below ground in order to escape the extremes of the cold or dry season, this results in the unusual phenomenon of an *underground stem.* We must look at these storage organs in detail in order to understand why they are stem rather than root structures.

Rhizome
The rhizome is a *horizontal* underground stem. The rhizome of iris, for example, at first sight appears to be a root. Close examination reveals the following typical stem characteristics.

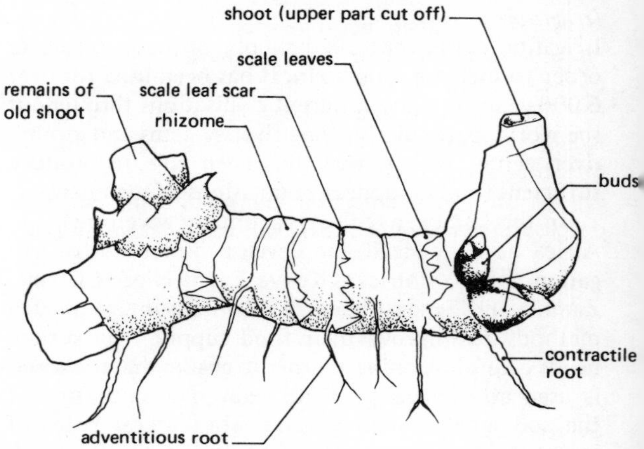

fig 2.42 A rhizome of iris

i) *Scale leaves*, or scars where leaves were attached.
ii) *Buds* growing in the axils of these leaves. The buds develop into green shoots which will grow above the ground.
iii) *Adventitious* roots grow from the rhizome.
iv) *Contractile* roots grow from the rhizome; these by their contractions pull the rhizome down into the soil.
v) A cross-section of the rhizome reveals a distribution of tissues which is *stem-like* rather than root-like.

The rhizome of iris is thick because it is swollen with food reserves. During the summer these are passed down from the leaves which are photosynthesising. Some of the translocated food is used for the growth of lateral buds into new branches of the rhizome. At the end of each branch, the terminal bud grows upwards, and produces the aerial shoot with its leaves and flowers. During the winter the aerial parts die down and the plant exists only as the underground rhizome.

Rhizomes are a common method of perennation, but they are not always swollen to the size of that of the iris. The perennial grasses, e.g. couch grass, have long slender rhizomes, but during the winter they still contain food reserves which enable new green shoots to appear early in spring.

Stem tuber

Stem tubers are the *swollen ends* of underground stems. The plant produces a number of slender rhizomes, and it is the terminal part of these structures which becomes swollen with food reserves. The 'European' or 'Irish' potato (*Solanum*) came originally to Europe from South America. The plant is noted for the large, edible stem tubers that it produces. Examine fig 2.43 and note the stem features shown by each tuber.

i) The presence of *scale leaves* or leaf scars with a bud in the axil of the scale leaf. These are called the 'eyes' of the tuber, because each has a curved mark resembling an eyebrow, and below this a tiny bud in the position of the eye.

ii) The *buds* are arranged spirally around the tuber facing towards one end where a terminal bud is present.

iii) If the tuber remains in the ground the bud will develop *green shoots* which will grow above ground.

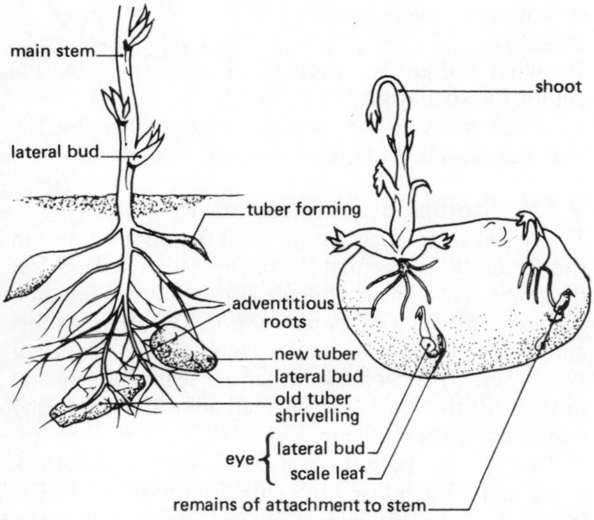

fig 2.43 A stem tuber of potato (*Solanum*)

iv) *Adventitious* roots will grow from the base of the shoot.

v) A cross-section of the tuber reveals a *stem-like* distribution of tissues, while on the outside are *lenticels* which allow gaseous exchange. These are normally found on the stems of woody plants. If the tubers are placed in the ground the 'eyes' begin to develop new shoots, using up the starch reserves in the tuber. A new plant is formed and eventually the original tuber withers and decays. New stems from the axils of lower leaves grow down into the soil and begin to form new tubers. In the dry season, the above ground portion dies and the tubers remain in the soil ready to grow when more favourable conditions return.

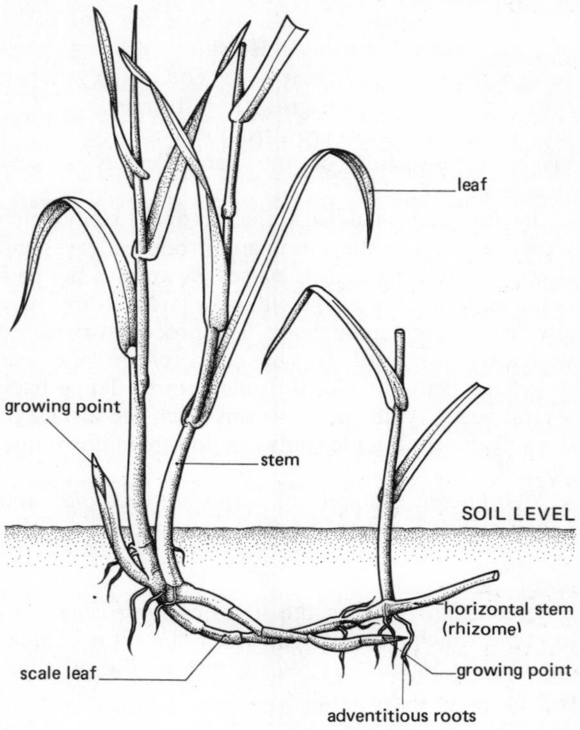

fig 2.44 A rhizome of couch grass (*Agropyron*)

Corms

The corm is a *short, swollen stem base* positioned *vertically* in the soil. It is encircled by leaf bases which form a protective scaly covering, and looks like a condensed vertical rhizome. It can be identified as a stem because it shows the following characteristics.

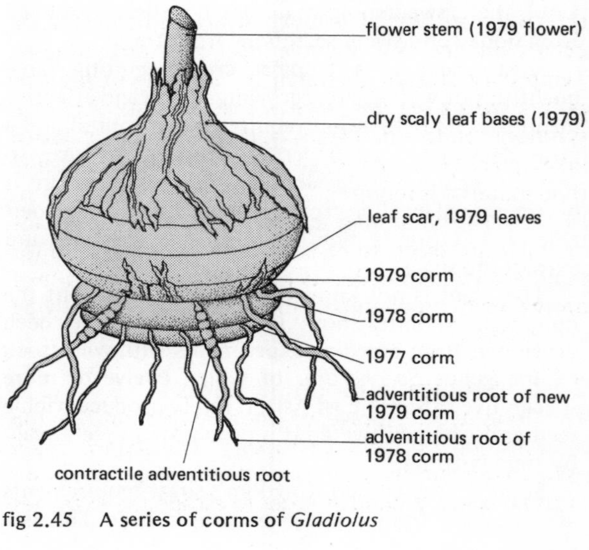

fig 2.45 A series of corms of *Gladiolus*

i) A series of *leaf bases* arises from it.
ii) *Buds* are present in the axils of the leaf bases, and the buds can develop green shoots which give rise to new corms at their base at the end of the summer.
iii) *Adventitious* roots grow from its base.
iv) *Contractile* roots grow from its base.
v) A cross-section of the corm reveals a *stem* structure.

There is a terminal bud at the top of the stem which grows upwards in the spring and produces leaves and flowers. Food material is passed down and the new corm forms on top of the old. New lateral corms may also arise from lateral buds. The process is repeated year after year, and the new corm for any one year may have below it old, shrivelled corms dating back several years. As the newer corms reach the surface of the soil the contractile roots can pull them down into the soil.

Well known examples of corms are *Gladiolus*, and *Crocus* which represent horticultural species.

Sugar cane
The storage of sugar in the stems of sugar cane is an example which must be considered, since it is of great commercial importance, being one of the principal export crops of the tropics.

It is obtained from the stem of a member of the grass family, *Saccharum officinarum.* It does not fall into the same class as the other examples in this section of the book, since the stem is not a perennating organ, nor is it underground. Nevertheless, the stem has become enlarged, and it does have much greater reserves of sugar than are normally found in plants.

There are very few plants that store their carbohydrates as sugar, and therefore it was not until the discovery of sugar cane (and sugar beet) that sugar came to be the important commodity that it is today. Until then, sweetening materials had been obtained from honey and sweet secretions from trees.

Sugar cane is a tropical crop requiring large quantities of water, a long growing season and effective manuring. Its cultivation requires much capital and a great deal of labour at certain times of the year, so that its economic development has resulted in it becoming a plantation crop. After cropping, a considerable organisation is required to process the stems and extract the sugar.

The plant is not known in the wild state, but the numerous varieties now in cultivation have been developed from breeding experiments with wild stock of the genus *Saccharum*, of which twelve or more species live in South East Asia. It rarely produces viable seed and when cultivated it is propagated vegetatively by stem cuttings.

Some of the United Kingdom's sugar requirements are obtained from home-grown sugar beet. This is a swollen tap-root in which food reserves are substantially in the form of disaccharide sugar. Modern varieties of sugar beet produce about 16 tonnes per acre, with a yield of 17% to 18% sugar.

Experiment
What foodstuffs are present in perennating organs and how are they distributed?
Procedure
1 Cut thin (2 mm) slices across a rhizome (e.g. iris), a corm (e.g. *Gladiolus*) and a stem tuber (e.g. potato *Solanum*).
2 Flood the surface of the slice with aniline hydrochloride. Leave for five minutes and then wash off with water. Examine the colour distribution in the section. Use a hand lens.
3 Flood the surface of the same slice of each plant with iodine solution. Leave for five minutes and then wash off with water. Examine the colour distribution in the section. Use a hand lens.
4 Make a drawing of the distribution of tissues shown up by the reagents.
Questions
1 What colour is produced in each slice by the aniline hydrochloride?
2 What parts of the section were stained by the iodine solution? What food reserves are present?

2.53 Storage in modified leaves
The leaves of all plants store food temporarily, but in some plants a perennating organ called a *bulb* has developed. This is an underground structure resembling a *large bud* or a *condensed shoot.* The stem is short and triangular in cross-section, while the leaves are of two types: (i) outer leaves which are scaly and dry and protect (ii) the inner leaves which are thick and fleshy containing stored food material.

In the onion bulb the storage leaves are cylindrical, completely encircling the central terminal bud. The latter grows up to produce the new plant at the end of the difficult season. This growth is aided by the stored food in the fleshy leaves. In the lily family, the bulb is formed by the bases of the leaves which completely encircle the stem. New bulbs can be formed by buds in the axils of the scale leaves.

The food reserves in bulbs such as the onion are sugars. These taste very sweet when cooked. In the lily bulbs the reserves are starch.

2.54 Storage in modified roots
Tap root
The tap root is a swollen main root with a very short stem at the top. A vertical section through a tap root such as a carrot shows the conducting strands, the outer cortex and the central pith areas. The food reserves are sugars which are stored in the cortex of

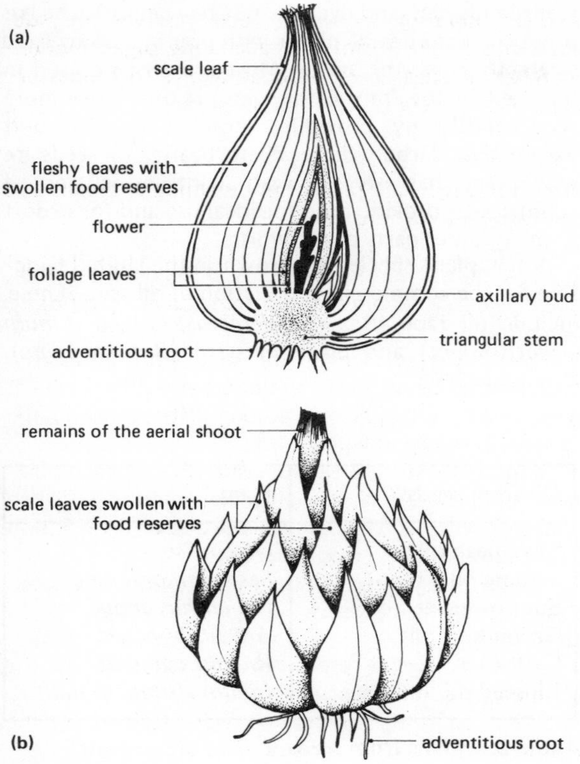

fig 2.46
(a) A vertical section of an onion bulb (*Allium*)
(b) A scaly bulb of a lily

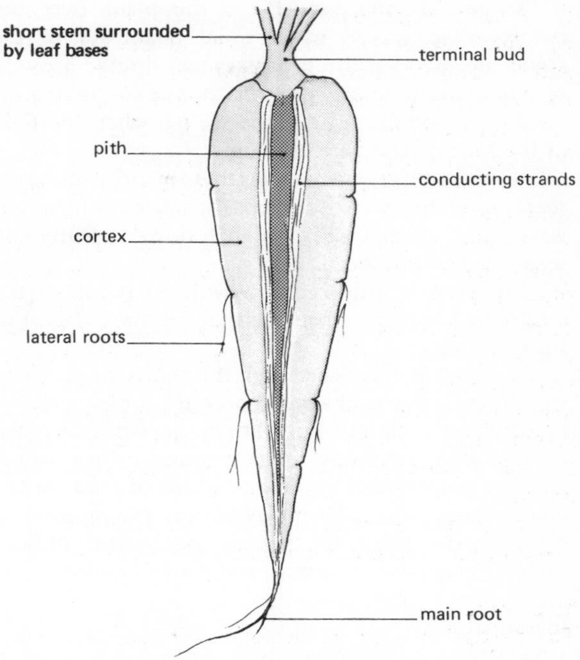

fig 2.47 A vertical section through the tap root of a carrot

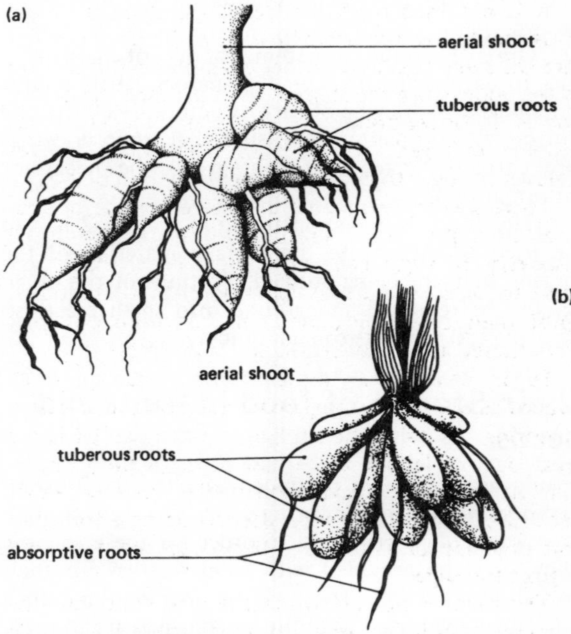

fig 2.48 Root tubers of (a) dahlia (b) lesser celandine (*Ranunculus ficaria*)

the carrot and in the phloem of the beetroot.

The tap root is a common means of storage for biennials. The main root swells at the end of the growing season of the first year, and these stores are utilised to complete the life cycle in the second year. Tap roots can also act as perennating organs in herbaceous perennials.

The radish (*Raphanus*) is grown as an annual plant. It produces a short stem which elongates rapidly and flowers develop amongst the leaves. The tap root grows to some depth, but only the upper portion swells, to form the edible portion. Storage is mainly in the form of starch.

Root tuber

Plants can also store food in roots other than tap roots, particularly where they have a *fibrous root system.* The tubers develop in the swollen ends of the roots, and can be distinguished from stem tubers by their lack of buds, scale leaves and scale leaf scars. Their internal anatomy is also different. Food produced by the leaves in the growing season is passed downwards to be stored in the tubers. Since roots do not produce buds they do not usually act as reproductive structures. They are simply food storage organs, e.g. dahlia.

2.55 Advantages of food storage in perennials and biennials

The *advantages* of food storage to the plants are as follows:

i) It permits the *survival* of the plant over the *unfavourable* season when small quantities of the stores are used for life processes and limited growth.
ii) It permits *rapid growth* at the beginning of favourable conditions, e.g. the spring, when the bulk of the food reserves are used up.
iii) This rapid growth enables these plants to *outgrow their competitors* in the intense struggle for light, air, water and mineral salts, so that flowers, fruits and seeds may be produced.
iv) It permits this rapid growth to occur in the established habitat where the plant is already part of the ecosystem.

As far as Man is concerned, the ability of plants to store food is tremendously important, for by selective breeding and careful cultivation, he has developed strains with relatively huge storage organs which produce correspondingly bigger yields of food for the human population. Examples of this exploitation of plant food storage for human use include potato, carrot, onion, swede, turnip, parsnip and radish.

Investigation
Vegetative food storage organs.
Procedure
1 All types of vegetative food storage organs in biennials and perennials should be examined and drawn.
2 Observe carefully the stem, leaf or root characteristics in each structure. Use explanatory labels to give an annotated drawing of each structure. (Refer to the Introductory Chapter to ensure that your drawings are correctly made and contain all the necessary information.)
3 The food tests demonstrated in section 2.10 can be used to investigate the storage materials of the organs that are being studied. Portions of the organ should be removed and ground into small pieces so that the test reagents can act quickly.

2.60 Storage of food in fruits and seeds

Fruits and seeds are produced as a result of *sexual reproduction* in the flowering plant. The seeds contain the young plant embryos. In order for these to grow they must be provided with food reserves for their initial development. They will be unable to manufacture their own food until the green leaves are formed. All seeds contain greater or lesser amounts of stored carbohydrates, fat and protein, and indeed many fruits carry large quantities of food stored in the fruit walls. In the case of fruits, the reserves are an attraction to animals, and are thus an aid to dispersal rather than growth.

From the earliest times, Man has used fruits and seeds in his diet, and over the last few centuries he has produced varieties of plants with ever-increasing food stores to satisfy his needs. Most plants store food in the seed in the form of oil, since it uses space more economically by providing more energy per unit weight than carbohydrate. Plants bearing oil seeds are grown and harvested in vast quantities in tropical countries to provide food for local use and for export to many other parts of the world.

A few plant species are grown in the United Kingdom for the commercial production of oil seed. These include oil rape (*Brassica oleracea*), linseed (*Linum visitatissimum*) and black mustard (*Brassica nigra*). (See table 2.7.)

Oil from seeds	Plant
Soya bean oil	*Glycine soja*
Sesame oil	*Sesamum orientale*
Sunflower seed oil	*Helianthus annua*
Groundnut oil	*Arachis hypogea*
Castor oil	*Ricinus communis*
Linseed oil	*Linum visitatissimum*

Table 2.7 Oils from seeds

Seed	Plant
Haricot bean	*Phaseolus vulgaris*
Field or broad bean	*Vicia faba*
Soya bean	*Glycine soja*
Runner bean	*Phaseolus coccineus*
French or kidney bean	*Phaseolus vulgaris*
Butter bean	*Phaseolus lunatus*
Lentil	*Lens esculenta*

Table 2.8 Legumes rich in starch and protein

Carbohydrates, particularly starch, are stored in two families of flowering plants, the *Leguminosae* and the *Gramineae*. The leguminous crops of the tropics are many and varied. Their importance as human food is far greater than in temperate regions. They provide a great deal of the protein for the human diet in tropical communities where meat and fish are in short supply.

Not only are these crops used by Man but they also supply the proteinaceous fodder for tropical livestock. In addition to the seeds other parts of the plant are rich in nitrogenous material.

Fruit	Plant
Maize	*Zea mays*
Rye	*Secale cereale*
Barley	*Hordeum sativum*
Oats	*Avena sativa*
Wheat	*Triticum* spp.

Table 2.9 Cereals rich in starch and protein

The term cereal is given to members of the *Gramineae* which are cultivated for their fruits. The cereal crops of the world form the most important source of the world's food supply and are extensively cultivated. It is surprising therefore that the number of species of cereal crops is small. *Oats* and *rye* are the dominant cereals of the colder parts of the world. *Wheat* and *barley* are most important in warm temperate regions, while in the tropics *rice, maize* and *millet* form the bulk of the diet. In temperate climates, such as that of the United Kingdom and of Europe, they are important in the diet as energy-rich foods (see table 2.8).

Many non-cereal plants in the tropics produce edible fruits containing food reserves (mostly sugars), but relatively few are cultivated. Only five or six of those cultivated have been commercially exploited or have received attention from the plant breeder. Attention has been focused mainly on bananas, citrus fruits, dates, mangoes, pineapples and avocadoes, to the exclusion of most others. In addition to these, guava, durian, pawpaw and pomegranate are common in the diet of tropical countries, but there is considerable local variation in quality and quantity. Many of these fruits are exported from tropical areas into European markets. Cold storage and gas storage can ensure that they are available throughout the year. Areas with temperate climates also produce their own varieties of fruit during the summer season and modern storage methods ensure that they are readily available in greengrocer's shops throughout the winter (see table 2.10).

Seeds are classified broadly into two groups, according to the structure storing the food reserves.

i) *Endospermic seeds*

These develop an additional tissue in the seed apart from the embryonic tissue. This is the *endosperm.* Its food reserve is used up on germination of the seed, e.g. wheat, oats and maize fruits and coconut.

ii) *Non-endospermic seeds*

No endosperm develops, and the food is stored in the cotyledons (seed leaves) which are part of the plant embryo (see fig 2.50), e.g. peas and beans.

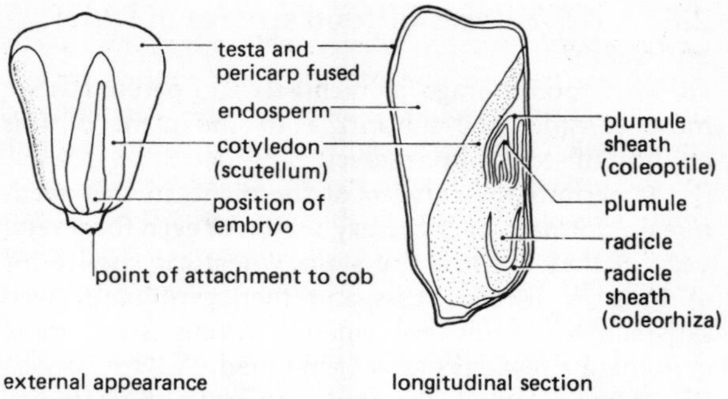

fig 2.49 A fruit of maize (*Zea*) —starch and protein food reserves mainly external to the cotyledon in the endosperm

Fruit	Plant
Banana	*Musa* spp.
Citrus fruits	*Citrus* spp. (includes oranges, lemons, grapefruits)
Pineapple	*Ananas comosa*
Date	*Phoenix dactylifera*
Tomato	*Solanum melongera*
Red or green peppers	*Capsicum annuum*
Plum	*Prunus domestica*
Cherry	*P. avium*
Peach	*P. persica*
Almond	*P. amygdalus*
Pear	*Pyrus communis*
Apple	*Malus sylvestris*

Table 2.10 Tropical and temperate fruits

In fruits the food reserves can also be found in two different types of tissue.

i) The *fruit wall* or *pericarp* becomes swollen. Different parts of this wall can be utilised e.g. the mesocarp in fruits of the drupe type, or the mesocarp and endocarp in fruits of the berry type (see fig 2.51), e.g. lemon, orange, tomato, cucumber, marrow.

ii) The *receptacle,* on which the flower develops, becomes swollen and surrounds the fruit proper. The fruit is then known as a 'false fruit' since its food store is not part of the fruit wall (see fig 2.51), e.g. apple.

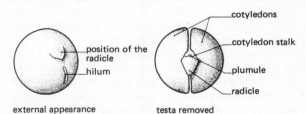

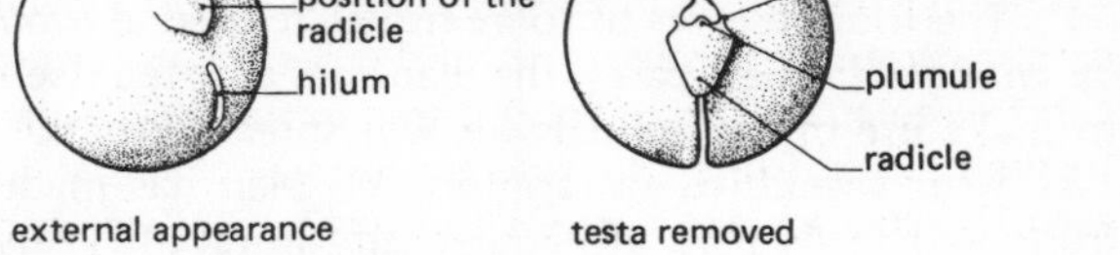

fig 2.50 A seed of pea (*Pisum*) — food mainly in the cotyledons as no endosperm is present

2.61 Advantages of food storage in fruits and seeds

As with food storage in biennials and perennials we must consider the advantage to the plant of this method of storing up reserves.

i) It permits the *survival* of the plant, in that seeds remain dormant over the dry season or even for several years if they are in a dry state. Wheat can survive for as long as fifteen years and then germinate, given satisfactory conditions, whereas rubber seeds must germinate a few days after being shed.

ii) After dispersal, the seed may reach an unfavourable habitat, but even so growth can proceed for a while until all the reserves have been used, or until favourable conditions return.

(a)

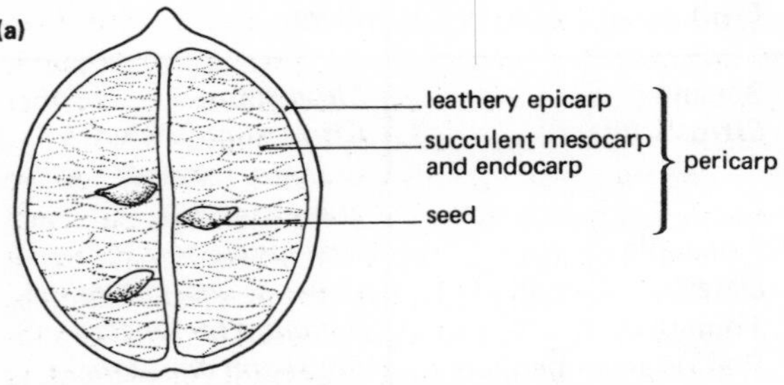

Radial longitudinal section through a lemon (fruit – a berry)
Food reserves in the mesocarp and endocarp of the fruit wall

(b)

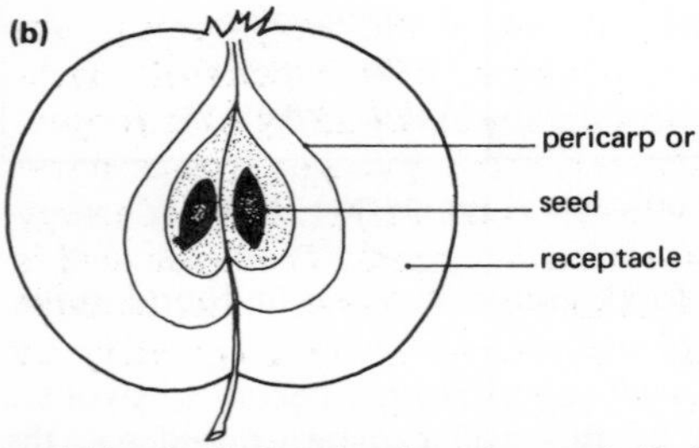

Radial longitudinal section through an apple (fruit – a pome)
Food reserves in the receptacles, not the pericarp, hence a 'false-fruit'

fig 2.51
(a) A radial L.S. through a lemon (fruit – a berry)
(b) A radial L.S. through an apple (fruit – a pome)

iii) The *reserves* of the seed are used in the *rapid growth* of the plant embryo when environmental conditions are at their best. When leaves and roots have become established, then the growing plant is able to make its own food.

iv) The food reserves of some fruits are used as *food by animals.* In these cases, the plant has adapted itself to make use of the animals for widespread dispersal.

As with vegetative storage organs, Man has made fruits and seeds part of his diet, and in recent years has greatly improved the productivity of many plants that he now cultivates.

2.62 Storage of food reserves in Man

We have seen in section 2.00 that the *lipids* (fats) are important to animals as a source of energy, and furthermore that lipids can be stored as a food reserve in many groups of animals. After digestion of fat, the fatty acids and glycerol formed are carried around the body. These building blocks may be used to form new fats which can be stored under the skin and around the internal organs. Each animal stores a form of fat characteristic of its species, so that fat from sheep, for example, is different from beef fat. Not all of the fat in mammals and other vertebrates comes from dietary lipids, for fat can be synthesised in the body from carbohydrates. Excess carbohydrate, not used immediately, is converted to fats for long term storage. In Man, the fattening properties of a diet over-rich in sugars and starch, are well known. Herbivorous animals synthesise their fat from cellulose which is broken down into its constituent sugar molecules by bacteria in the gut. These molecules are then either utilised by the body for metabolism, or, if unused, are changed into fat.

Glycogen, a complex carbohydrate, is another important storage substance. It has great significance in the immediate supply of energy when required by an animal. It is stored in the liver and the muscles. Certain vitamins can also be stored in the liver. Thus the liver has a most important role to perform as controller of the level of a number of substances in the body.

Animals that live in cold climates have fat reserves under the skin which provide insulation against heat loss. Birds and mammals in particular with their high body temperatures, lose heat continuously to their surroundings. In Arctic or Antarctic regions, their insulating coverings of hair or feathers are supplemented by the fat layers below the skin. Large mammals, such as seals and whales, which live in icy water have an extremely thick layer of fat called blubber below their skin. It is this fat which has resulted in their being hunted and consequently the considerable depletion of their numbers.

Tropical animals do not generally store large quantities of fat.

2.63 The liver

The liver, in spite of its relatively simple structure, has a multiplicity of functions. Above all it is an organ which controls the levels of many of the organic substances circulating in the blood, the tissue fluids and the lymph. In this function it is aided by the kidneys.

It is the largest gland in the body of Man, having an average weight of 1.25 kg, and is situated on the right side of the upper abdomen. The upper surface is convex, fitting closely under the diaphragm, while the lower surface is concave. It has a dark red colour and is divided into lobes. There is a large amount of blood circulating through the organ from the hepatic artery and the hepatic portal vein. The latter collects all the blood from the intestines and passes it through the liver.

We are concerned here with the *storage functions* of the liver. The stored material lies in the liver cells and exchanges take place between these cells and the blood circulating.

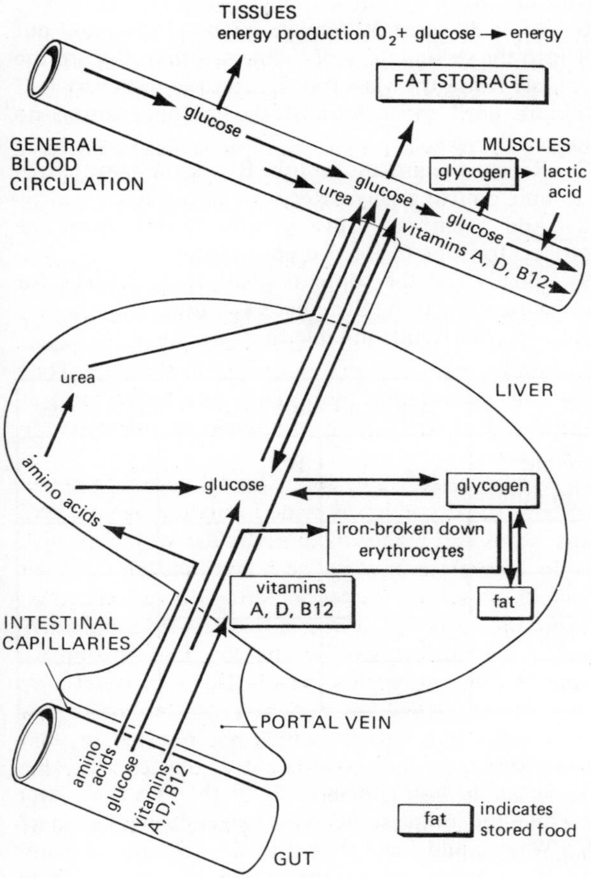

fig 2.52 The relationship between the liver as a storage organ and the blood circulation

i) The liver has a special role in carbohydrate metabolism. It contains a substantial reserve of glycogen which is converted to free glucose in order to maintain the level of glucose circulating in the blood (known as blood sugar). The glycogen is thus a carbohydrate reserve. After a meal the digested carbohydrate passes into the blood as glucose, the blood sugar level rises, and the liver takes up glucose to convert it to glycogen. Glucose is continually removed from the blood, during the normal activities of the body and is used by the muscles, the nervous system and other organs. In this way blood sugar level falls, and so liver glycogen is changed into glucose to raise the level back to normal. Thus blood sugar levels are kept within fairly narrow limits.

ii) The liver has a special role to play in fat metabolism. It contains about 4% lipid, and if starvation occurs then the fat content of the liver falls. This will only happen after the exhaustion of all other body fats. Excessive eating of fats will result in the liver considerably increasing its fat content.

iii) *Proteins* are *not* stored in the body and so excess amino-acids must be eliminated. The amino-acids are broken down (*deaminated*) by the liver to *carbohydrate* and *urea*. The urea is excreted through the kidneys, but the carbohydrate can contribute to further formation of glycogen or fat in the body.

iv) *Vitamins A and D* are stored in the liver. They are available to carnivores in the livers of animals on which they feed. The livers of fish are relatively much richer in vitamins (particularly D) than the livers of mammals.

v) *Vitamin B_{12}* is found in liver. This is the anti-anaemic factor which is most important for preventing the symptoms known as *pernicious* anaemia.

vi) The liver also stores *iron* from broken down erythrocytes.

Skeletal muscle, as well as the liver, stores a substantial amount of *glycogen*. The muscles of a well-nourished man can contain up to 250 g of glycogen. During vigorous excerise the glycogen breaks down into lactic acid with the consequent release of energy for the immediate muscular needs. The lactic acid is released into the blood and then can either be built up into glycogen or broken down completely to carbon dioxide and water.

2.70 Nutrition in animals

We have seen in section 2.20 that plants can manufacture food to provide for their energy requirements, growth and health. In order to take carbohydrates, fats and proteins into their bodies, animals must break down these complex molecules into their original building blocks. This is partly because food substances must be *transported* by the blood to where they are needed. In order to be *absorbed* by the blood, they need to be *soluble* and *small* enough to pass through a blood vessel wall. This process of food breakdown is called *digestion* and occurs in the alimentary canal (gut) of an animal. Any undigested food is eliminated from the end of the alimentary canal as *faeces*.

Experiment

A model gut and the movement of molecules

Procedure

1 Take a beaker (or tin) of tap water and warm to about 37°C (body temperature of a mammal).

2 Take about 15 cm of *visking (cellulose) tubing*, wet it under a tap and tie a knot very tightly in one end. Open up the other end and pour in a mixture of 5% starch and 10% glucose solution. Close the open end with a paper clip.

3 Wash the outside of the tubing with tap water to remove any external traces of starch and glucose.

4 Place the tubing inside a boiling tube containing a little tap water. Place the boiling tube in the water bath.

5 Test some of the water from the boiling tube. Extract a small sample with a pipette, place in a small test-tube and add iodine solution. Test another sample with Benedict's solution. Boil this in a test-tube.

6 Repeat the two tests after twenty minutes. Record your results in the form of a table.

Time of test	Result with iodine solution	Deduction

Time of test	Result with Benedict's solution	Deduction

Questions

1 Was any starch or glucose present in the water in the first tests?

2 Was any starch or glucose present in the water after the second tests, after twenty minutes?

3 If one of the substances has passed through the tubing, what explanation can you give for this result?

4 What control should be set up in this experiment?

The visking tubing was considered to be a model for the gut and we can suppose that the gut might behave in the same way, permitting small molecules to pass, but not large ones. Starch, however, could also be broken down into smaller maltose or glucose molecules. We must look for evidence of this breakdown.

Experiment

What action does saliva have upon starch?

Procedure

1 Heat a beaker or tin of tap water to 37°C.

2 Wash out your *buccal cavity* (mouth) with distilled water and then *suck* a small smooth stone or a rubber band. This results in the collection of *saliva* which should be discharged into a small beaker. Pour the saliva into a test-tube and place the test-tube in the water-bath.

3 Take three test-tubes and label them A, B and C. Add 1 cm^3 of 2% starch solution to tubes A and B. Add 1 cm^3 of distilled water to tubes A, B and C. Measure the quantities carefully.

4 Prepare a spotting tray or white tile with three rows of iodine solution drops as in fig 2.53.

5 Record the time, then add 1 cm^3 of warmed saliva to tube A. Stir the tube contents with a glass rod. Immediately remove one drop of the mixture, by means of the glass rod, and add it to the first drop of iodine solution. Record the colour of the iodine solution. Wash the end of the glass rod.

6 After 30 seconds, remove a second drop and put it into the second drop of iodine solution. Record the colour. Rinse the glass rod. Repeat the test every half minute until the colour of the iodine solution no longer changes.

7 Repeat the process for tube B over the same length of time, omitting saliva. Record your results.

8 Add 1 cm^3 of saliva to tube C and repeat the process for tube C. Record your results.

9 Finally test the contents of tubes A, B and C for reducing sugar using Benedict's solution.

Record your results in table form.

	Colour of iodine solution in test-tubes		
Time (minutes)	A	B	C
0			
0.5			
1.0			
etc.			

Questions

5 What do you conclude about the fact that, after a while, the iodine solution no longer changed colour?

6 What could have caused this? Give your reasons.

7 What happened to the starch in tube A? State which test supports your answer.

8a) How does this experiment help to explain digestion?

b) What control would confirm your conclusion?

From the last two experiments you can see that:

i) the glucose molecules which are smaller than the starch molecules can pass through minute pores in the visking tubing,

ii) saliva can cause the breakdown of starch. Thus there must be 'something' in saliva which speeds up chemical change. Starch can be hydrolysed by boiling

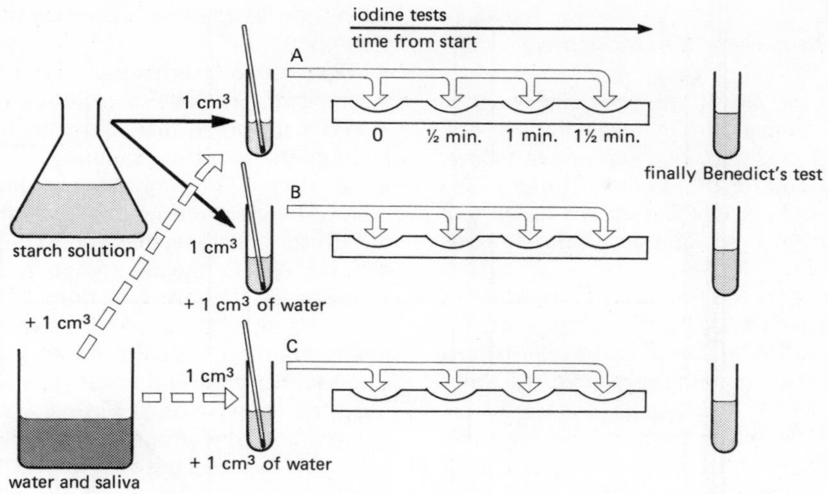

fig 2.53 A flow sequence for the saliva experiment

with acid in the same way that sucrose was hydrolysed to test it for reducing sugar, but this hydrolysis by acid is slow compared with the action of saliva. The 'something' which speeds up the reaction is called a *catalyst*, and organic catalysts are called *enzymes*.

Enzymes occur widely in plants and animals, particularly within the protoplasm, where they bring about many chemical changes. Such changes, resulting from the action of enzymes are most easily studied when they occur in the gut. The enzyme in saliva is an example. It is called *salivary amylase* or *ptyalin* for it is responsible for the breaking down of starch to maltose. The latter is both a disaccharide and a reducing sugar. Let us examine another digestive enzyme called *pepsin* which is present in the stomach.

Experiment
The action of pepsin on egg albumen
Procedure
1 Take a beaker (or tin) of tap water and warm to about 37°C.
2 Take four test-tubes and into each place about 5 cm³ of *egg albumen suspension* (1% dried albumen in water heated to 90°C.) Label the tubes, A B, C and D.
3 To each of tubes B, C and D add three drops of *hydrochloric acid* (2M or 10%).
4 Place 1 cm³ of 1% pepsin in a test-tube and heat until it boils, then add it to tube D. Place 1 cm³ of 1% pepsin (unboiled) in tubes A and C only.
5 Place all four tubes in the water bath.
6 Examine the tubes at 2 minute intervals, and after 6–8 minutes remove the four tubes from the water bath and place them in a test-tube rack. Examine the contents of each tube and record your observations in the form of a table.

Tube	Contents	Appearance at beginning of experiment	Appearance at end of experiment
A	albumen + pepsin		
B	albumen + HCl		
C	albumen + pepsin + HCl		
D	albumen + boiled pepsin + HCl		

Questions
9 In which tube have the contents a different appearance after six minutes or more? What has happened to the egg albumen suspension?
10 How is the enzyme affected by boiling? What evidence leads you to this conclusion?
11 What other hypothesis could you advance to account for the results in tube C?
12 What are the best pH conditions in which the enzyme pepsin can act?

Pepsin is a *protein-digesting enzyme* which occurs in the *stomach*. The experiment shows that it can digest egg albumen which is mainly protein. It is not possible to generalise about all enzymes from this one experiment. Nevertheless we have now examined two enzymes occurring in two different parts of the alimentary canal, the buccal cavity (mouth) and the

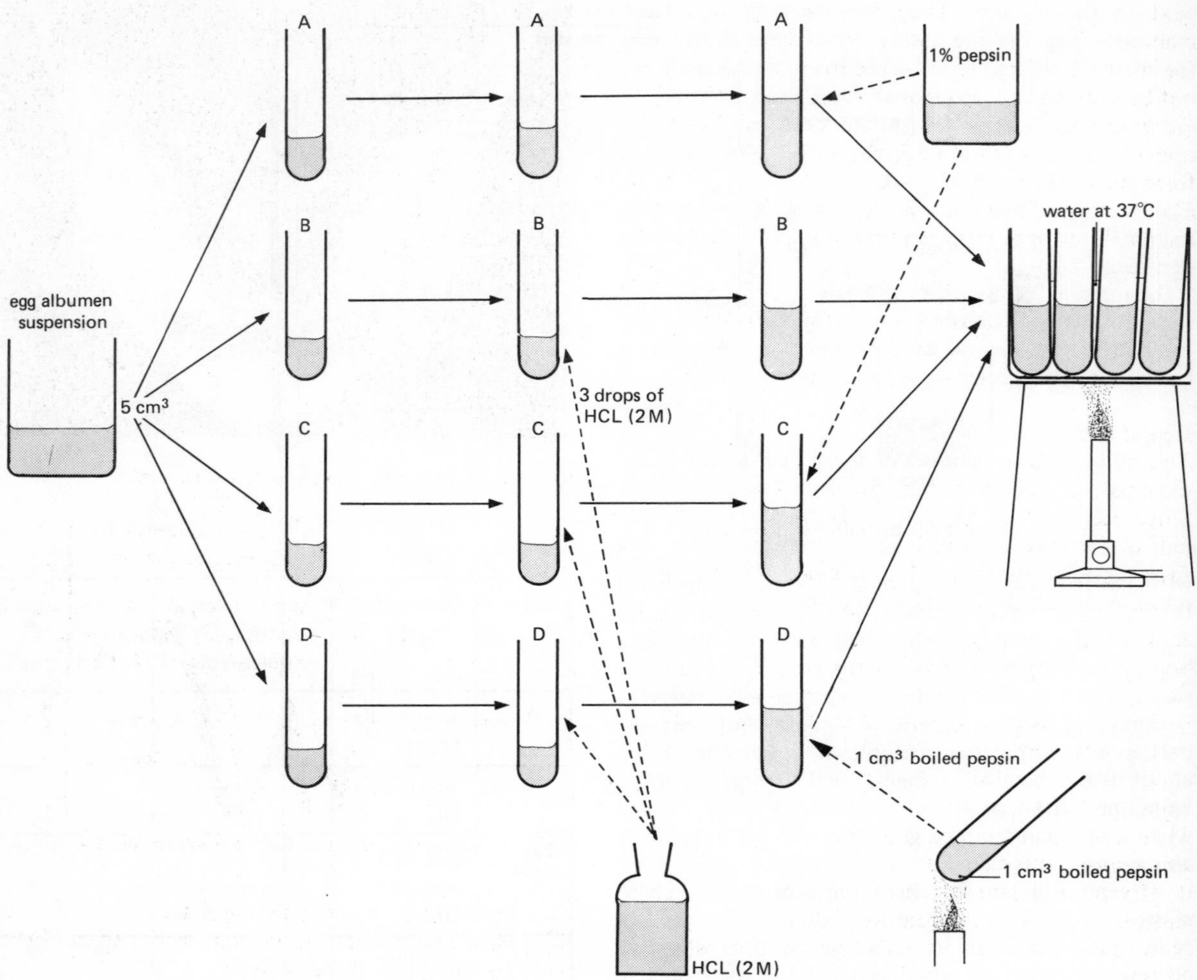

fig 2.54 A flow sequence for the pepsin experiment

stomach. One enzyme digests starch (carbohydrate) and the other digests egg albumen (protein). Digestion is essentially the *breaking down of food* into smaller and smaller parts. This may begin, even before we eat food, by cutting it with a knife into a size we can get into the mouth. Once in the buccal cavity the teeth and the tongue help to cut the food into even smaller pieces, and finally the process is completed by enzymes. We can therefore describe the two types of digestion as (i) mechanical (ii) chemical.

2.71 Mechanical digestion

Larger animals like amphibia and reptiles capture their food and immediately swallow it down into the gut where strong digestive juices can bring about the chemical breakdown of food. Nevertheless they often have to rest for several days to allow their food to be digested, and during this time they are very vulnerable. Examples of this are:

a large python swallowing a pig,
a grass snake swallowing a frog,
a toad swallowing a worm.

The teeth of these organisms are all alike and are simply sharp backward-pointing cones. Mammals are unique in that they cut and grind their food into small pieces before it is swallowed. They are able to do this because of their complex *dentition* (i.e. several types of teeth) and the arrangement of their *jaw bones, jaw muscles* and *cheek muscles.* All of these, together with the *lips and tongue,* help in their different ways. However, the most important are the *teeth.* Each type is different in structure and each performs a special function.

There are four main types of mammalian tooth.

Incisors (i) These are in the front of the buccal cavity.
Canines (c) These are single teeth in each half jaw

next to the incisors. They may be large in certain mammals (e.g. the dog family, which gave its name to the teeth), but large teeth at the front of the jaw need not be canines (e.g. an elephant tusk is an incisor).

Premolars (p) They lie behind the canines on the upper and lower jaws. Together with the molars they form the cheek teeth.

Molars (m) These are not present in the young mammal, but appear for the first time in the permanent dentition.

The mammal has two sets of teeth in its lifetime. The first (milk) set consists of incisors, canines and premolars; the second or permanent set consists of incisors, canines, premolars and molars.

Dental formula

The number and position of teeth in the jaw of the mammal can be described by means of a *formula.* Since each half of the jaw is *symmetrical,* only one half of each jaw need be included in the formula. The four types of tooth can be indicated by their initial letter as shown above. Thus for Man:

$$\begin{matrix}\text{(upper jaw)}\\ \text{(lower jaw)}\end{matrix}\ i\frac{2}{2}c\frac{1}{1}p\frac{2}{2}m\frac{3}{3} \text{ or } \frac{2}{2}\frac{1}{1}\frac{2}{2}\frac{3}{3}$$

$$\text{The total number of teeth} = 2\left(\frac{2}{2}\frac{1}{1}\frac{2}{2}\frac{3}{3}\right) = 32$$

Investigation 1

How are human teeth distributed?

Procedure

1 Examine a fellow pupil's teeth or those of your brother or sister. How many teeth can you see? Record the following: Age of subject and total number of teeth.
Complete the table below.

2 Ask your subject to eat a fruit, such as an apple. Observe how the lips, tongue and teeth are used.

3 Look at fig 2.55 and compare your information in 1 above with the number of teeth shown in the figure.

4 Try to find the answer to the following questions by examining your young brothers and sisters and by discussion with your fellow students.

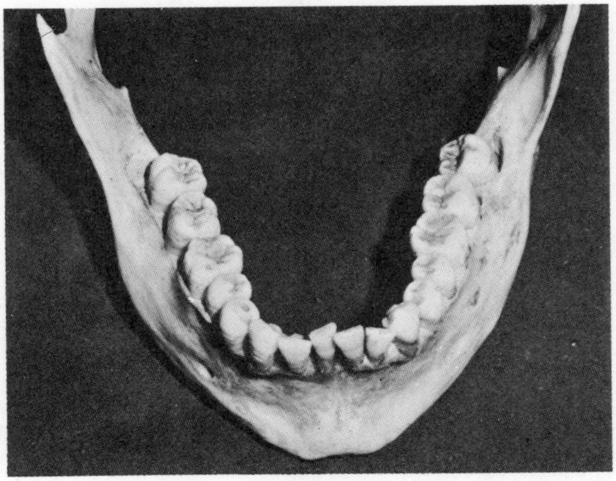

fig 2.55 An adult human lower jaw with teeth

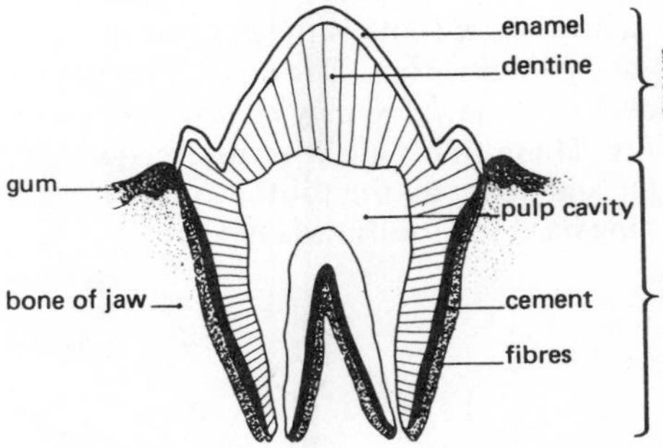

fig 2.56 A vertical section of a premolar tooth of a carnivore

i) At what age do teeth first appear?
ii) How many teeth are present in upper and lower jaws by the age of three years?
iii) Which teeth appear first?
iv) Which are the last teeth to appear?

Internal structure of the tooth

If a tooth is sawn in half the internal structure can be seen as shown in fig 2.56. The outer white layer of the *crown* is the *enamel,* which is the hardest organic substance known. Within this lies the *dentine* which is also hard but unlike the enamel contains living tissue.

	Number in upper jaw	Number in lower jaw	Teeth lost or extracted	General shape of tooth	Probable function
Incisor					
Canine					
Premolar					
Molar					

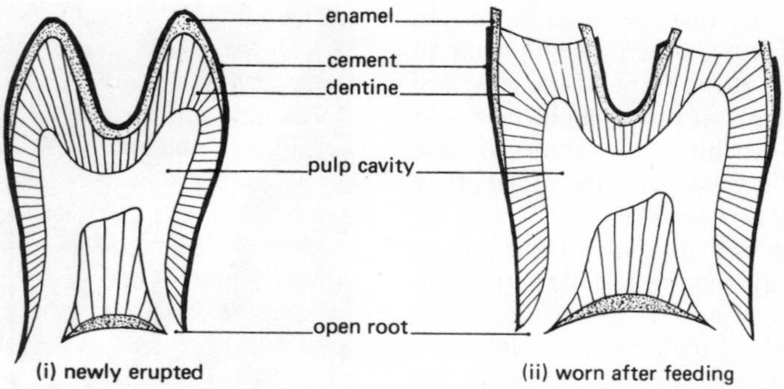

fig 2.57 A vertical section of a premolar tooth of a herbivore

However, it is soft enough to be carved, for the ivory of the elephant's tusk is dentine. Dentine occurs in the crown and in the *root* below the surface of the gum. In the centre of the tooth is the *pulp cavity* containing nerves and blood vessels. The root is buried firmly in the *socket* of the jaw, fixed by *cement* and surrounded by *fibres.* These fibres allow some freedom of movement up and down by the tooth, so that if we bite on something hard the tooth does not break.

Animals can be divided into three groups, according to the type of food they eat:

i) *Herbivores* – feed on plant material, e.g. deer on bark and leaves; cattle on grass.

ii) *Carnivores* – feed on the flesh and bone of animals, e.g. lion on zebra; stoat on rabbits; fox on chickens.

iii) *Omnivores* – feed on both animal and plant material; e.g. Man on meat and vegetables; wild boar on grass, shoots, leaves and small animals.

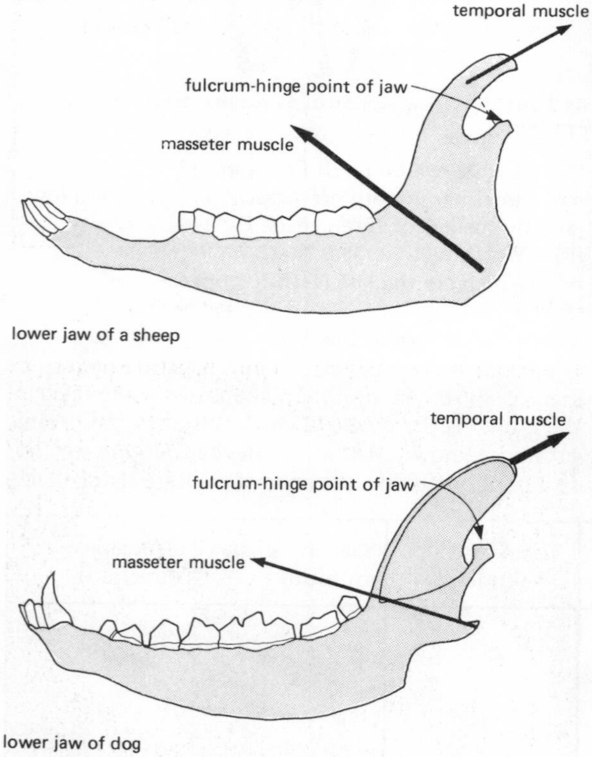

fig 2.58 The lower jaws of a sheep and a dog showing the relative sizes of the temporal and masseter muscles

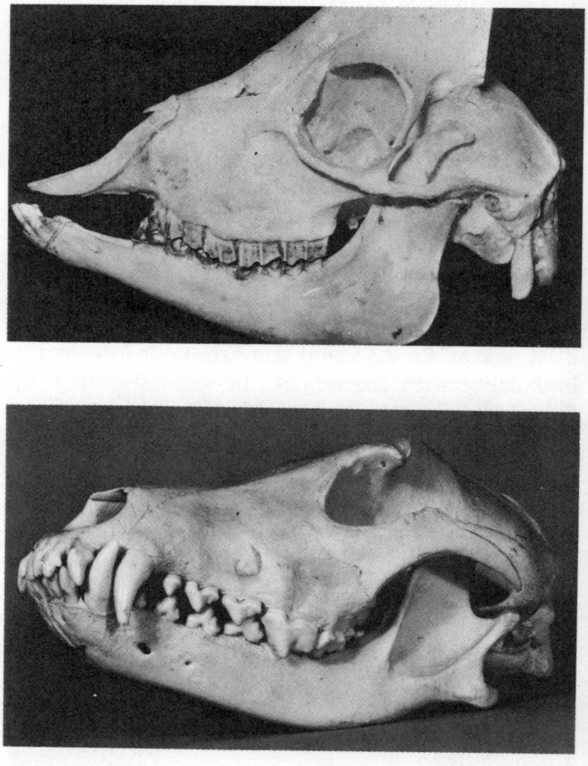

fig 2.59
(a) The skull of a sheep
(b) The skull of a dog

Investigation 2

The sheep as a herbivore

Procedure

1 Examine the upper and lower jaws of the prepared skull of a sheep noting:

i) the types of tooth,

ii) the gaps in the teeth on each jaw,

iii) the hinging of the jaws.

Questions

1 List the types and numbers of the teeth in the upper jaw.

2 List the types and numbers of the teeth in the lower jaw.

3 How does the sheep eat its food?

4 What is the function of the large gap between the teeth of the lower jaw?

5 What is the dental formula for the sheep?

6 How does the sheep chew?

7 Describe the shape and probable functions of each type of tooth.

8 Extract a molar and look at the end of the root. What do you observe?

Investigation 3

The dog as a carnivore

Procedure

1 Examine the upper and lower jaws of the prepared skull of a dog noting:

i) the types of tooth,

ii) the hinging of the jaws.

Questions

9 List the types and numbers of the teeth in the upper jaw.

10 List the types and numbers of the teeth in the lower jaw.

11 How do such teeth help to obtain food?

12 Describe the action of the hinge and the movement of the lower jaw.

13 What is the dental formula for the dog?

14 Describe the shape and the probable functions of each type of tooth.

The jaw movements of these two animals can be seen if you watch them feeding. The sheep moves the lower jaw from side to central so that the premolar and molar teeth grind the grass. The dog jaw moves only in an up and down plane so that the animal cannot grind, but cuts or crushes flesh and bone. These different actions are possible because of the different ways in which the jaws are hinged. In the sheep, the hinge is loose so that the jaw can move in a circular fashion, whereas in the dog, it is tightly hinged to permit only up and down movement. The jaws are moved by two muscles, the *temporal* and *masseter*, and their differences in the two animals are shown below.

	Temporal muscle	Masseter muscle
Sheep	A small muscle.	A large muscle, the high jaw joint gives this muscle greater leverage.
Dog	A large muscle, applies greater leverage for cutting and crushing, and also for the stabbing movement of the large canines.	A small muscle.

Table 2.11 Jaw muscles of sheep and dog

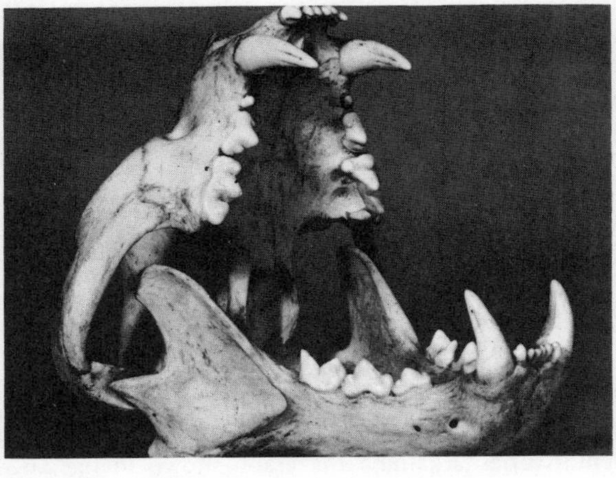

fig 2.60 The skull of a leopard

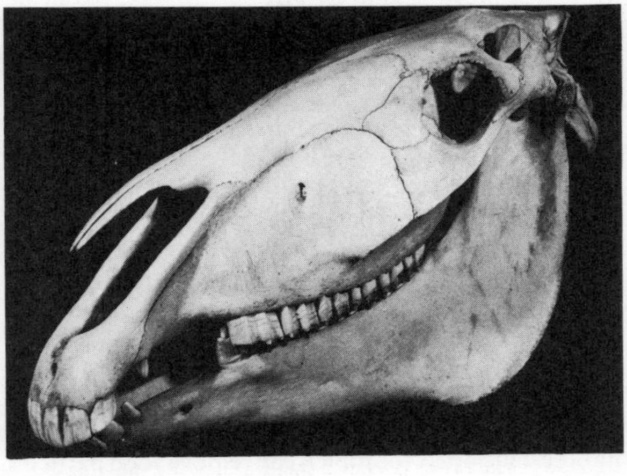

fig 2.61 The skull of a horse

Herbivores can be divided into

i) those that *cut* roots, bark, etc., using very sharp incisors, e.g. rabbits, rats, guinea pigs,

ii) those that *pull up* grass using the lower front teeth and a pad on the front of the upper jaw (or upper and lower incisors), e.g. cows and sheep,

iii) those that use their lips to pull up grass or leaves off trees, e.g. hippopotamus,

iv) those that use a specialised organ to pull leaves, bark and branches off trees, e.g. the trunk of the elephant.

Carnivores do not have the variety of feeding methods of the herbivores, but use their prominent canines and powerful jaws to kill their prey, and then tear off flesh.

Experiment

Tooth decay and its cause

Procedure

1 Take a tooth extracted from a mammal (e.g. sheep) and poke it in the crown region with a sharp needle or forceps.

2 Place the tooth in *dilute hydrochloric acid* for two or three days.

3 Wash the tooth and examine it. Using a needle or forceps again, poke the tooth in the crown region.

Questions

15 What differences can you see in the tooth? Compare it with a similar tooth still in the jaw of the sheep.

16 What happens when you poke the tooth with the sharp instrument, after it has been in the acid?

17 What part of the tooth has become exposed?

Dental care in Man

In order to keep teeth healthy, there are three points to consider.

i) They must be *used* in the *correct* way.

ii) They must be kept *clean.*

iii) They must receive *regular* expert attention.

i) Teeth should never be used for improper purposes such as cracking nuts, opening bottle tops or even breaking cotton. They must be given the right kind of work by chewing hard food, for modern diets are often too refined and prepared to need chewing. Small babies should be provided with a bone ring or some hard object on which to chew. Fibrous tough foods, e.g. apples, raw carrots, nuts, oranges and other fruits, should be included in the diets of children and adults.

ii) Teeth must be kept clean to prevent food residues causing decay. *Bacteria* growing on these pieces of food *produce acids* which can eat away the enamel and the dentine. A toothpick can remove food particles from between the teeth, but they should also be brushed after each meal with a toothbrush or a tooth stick. If no toothbrush is available, a small piece of rag and some salt can be equally effective. The salt acts as an antiseptic and an abrasive.

iii) In spite of care, decay and other diseases may attack the teeth. Treatment will be most effective if the disease can be discovered and controlled in the early stages. If possible, the teeth should be examined at least twice a year by a dentist. Small cavities in teeth can be filled, while badly decayed teeth can be extracted.

Dental diseases

The two most important diseases are *dental caries* and *periodontal disease.*

i) Dental caries is caused by:

1 lack of hard food;

2 too much sweet food;

3 lack of calcium in the diet;

4 lack of vitamin D;

5 lack of cleaning;

6 general ill-health.

In the earliest stages only *enamel* is involved and since the process is painless, it escapes notice. If this early stage is neglected it spreads *to the dentine* and this is noticeable when one is eating sweet foods or drinking hot or cold drinks. This stage can be easily halted by a 'filling' performed by a dentist. The third stage involves invasion of the pulp cavity by bacteria and can be extremely painful when the nerves are attacked. The final result could be an abscess, and then the tooth may have to be extracted.

ii) Periodontal disease is caused by:

1 lack of vitamins A and C;

2 lack of massage of the gums;

3 imperfect cleaning.

This disease is more common among adults than among children. It causes the *gums* to become *soft and flabby* so that they do not properly support the teeth. The reddening of the gums, bleeding and the presence of *pus* are symptoms of the disease described as *pyorrhoea.* It does not occur if the diet is correct, if the gums are brushed regularly to encourage blood circulation and if particles of food are not allowed to accumulate. When food accumulates, it leaves a film on the teeth which builds up to form a hard layer called 'plaque'. Plaque is the main cause of periodontal disease.

Fluoridation

Recent research has shown that *fluoride salts* in the drinking water may have some effect on reducing dental caries. Examine the graph shown in fig 2.62 and answer the following questions.

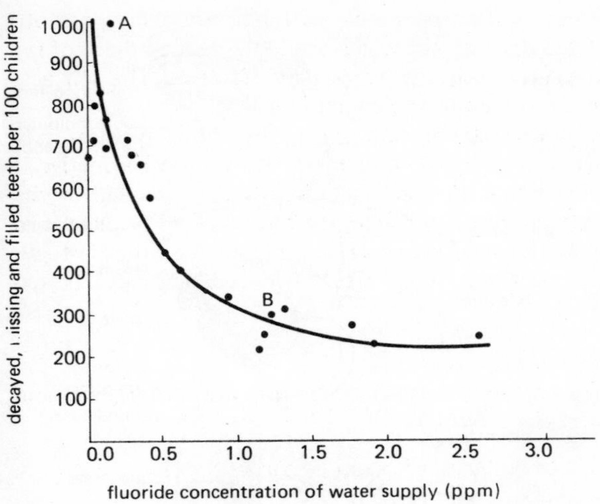

fig 2.62 A graph showing the effect of fluoride concentration on tooth decay in children

Questions

18 Do you think there is a relationship between the incidence of dental caries and the fluoride content of water? If so, in what way?

19 What is the number of cavities per child where the water contains about 0.2 ppm of fluoride (point A on the graph)?

20 What is the number of cavities per child where water contains about 1.3 ppm of fluoride (point B on the graph)?

21 If your drinking water contains 0.1 ppm of fluoride, how much more fluoride would you need to add to the water in order to reduce the incidence of caries by 75%?

Scientific opinion is divided on the question of whether or not this chemical should be added to the public water supplies. Those against say that it is too imprecise a way of ensuring that safe limits will not be exceeded. The amount of water consumed by people varies considerably and it is now known that in polluted industrial areas the amount of fluoride contained in food alone may already be too high. Certainly it does involve a risk to add 1 ppm of fluoride to a water supply system, and many countries have now had second thoughts about the safety of such a scheme. In Sweden fluoridation is banned for medical reasons, in France scientific and medical opinion has always been against it; Holland, which has had fluoride in its water supply for a long time, has now decided against it.

Caution would seem to be necessary, particularly in highly industrialised countries, in order to avoid over-dosage.

Muscular action on food

The stomach has a muscular coat of four layers arranged in different directions. The *churning action* brought about by the contraction of these muscles, and the secretion of *gastric juices* by the walls of the stomach, results in the food becoming a liquid called *chyme.*

The movement of food through the *oesophagus* towards the *stomach* and onwards through the small intestine after leaving the stomach, is also brought about by the muscular action of the gut wall. When a portion of food is swallowed it stretches the wall of the oesophagus. This contains circular and longitudinal muscles which, by their contraction, force the food on down the oesophagus. Similar waves of contraction and relaxation flow along the small intestine, pushing food in front of the contracting muscles. This process of moving food is called *peristalsis.*

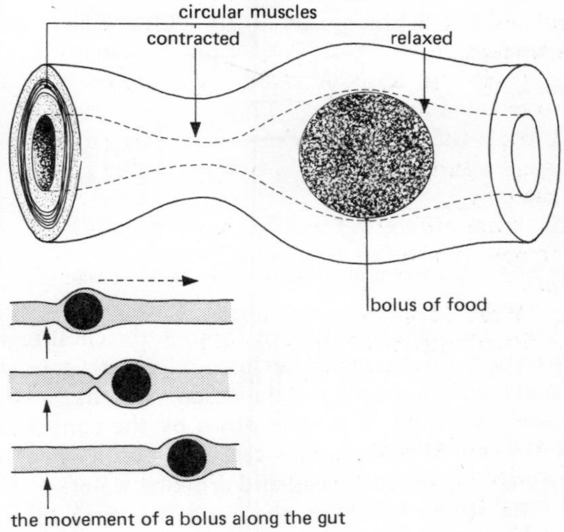

fig 2.63 A diagram showing peristalsis and how food is moved along the oesophagus and intestines

2.72 Structure of the alimentary canal (gut)

Examine fig 2.64 and fig 2.65 which shows the comparable structures of the gut in Man and the rabbit. In all mammals there are corresponding sections of the gut, but in different species there may be different emphasis on certain sections. This is particularly noticeable in the case of the *caecum* and *appendix* of the rabbit, compared with those of Man.

Food is taken in through the mouth and its digestion begins in the buccal cavity where it is chewed and moved around. As the food is pushed to the back of the buccal cavity an automatic swallowing action (a reflex) causes it to be pushed into the oesophagus.

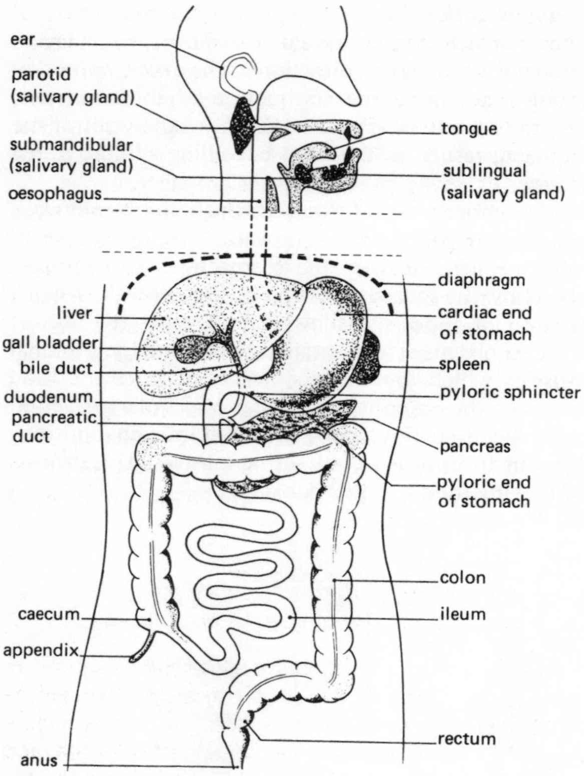

fig 2.64 A diagram of the alimentary canal of Man

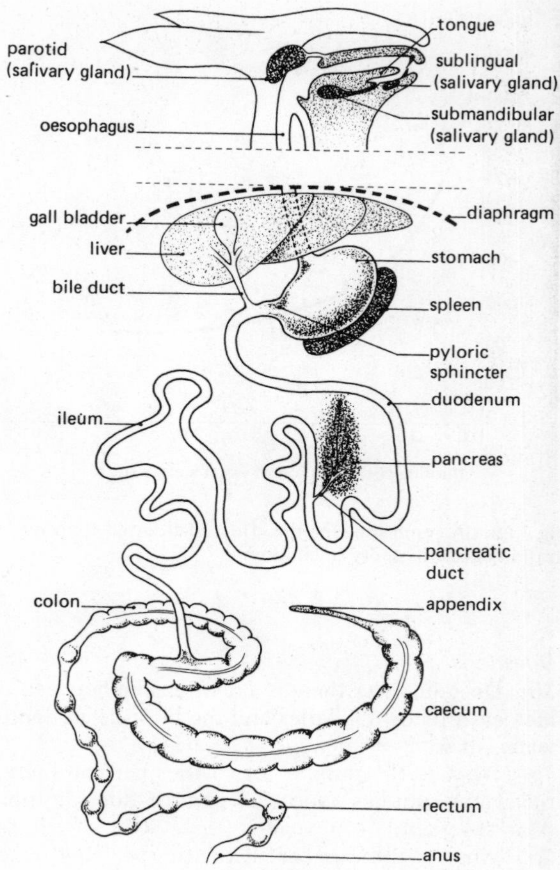

fig 2.65 A diagram of the alimentary canal of a rabbit

This is a straight tube passing through the chest region into the stomach, which lies just under the *diaphragm.* Food does not drop down the oesophagus, like a stone down a well, but is pushed down by the contraction of the muscular wall. This can even occur when you are standing on your head and drinking water!

The stomach retains the food for three to four hours, churning it into *chyme.* Digestion, begun in the buccal cavity, is continued by mechanical and chemical means. The exit from the stomach is surrounded by a circular muscle (sphincter) called the *pylorus.* This opens at intervals to allow chyme to pass into the small intestine. Some mammals have more complicated stomachs than Man or the rabbit. The sheep, for example, has a stomach with four compartments.

The first part of the small intestine is the *duodenum.* Two ducts enter this section, the *bile duct* from the liver and the *pancreatic duct* from the *pancreas.* In Man these two ducts join to form a common duct, whereas in the rabbit they open quite separately in different parts of the duodenum. The rest of the small intestine is divided into the *jejunum* and the *ileum.* In both Man and the rabbit these two sections form the longest section of the gut. In Man it is about 7 m in length and in the rabbit about 1 m. In order to compress these lengths into the abdominal cavity, the small intestine forms a series of loops and folds held in position by a tough membrane, the *mesentery.* The next section of the gut is the *large intestine.* At its junction with the small intestine is a side branch, the *caecum,* which narrows into the blind ending *appendix.* In Man, the large intestine is divided into the ascending *colon,* the transverse colon, descending colon and *rectum.* These form almost four sides of a square from the bottom right quadrant of the abdomen, up the right side, across to the left and down the left side. The rectum opens to the exterior at the *anus,* which is surrounded by a circular muscle, the *anal sphincter.*

In the rabbit the caecum is very large and also ends in a blind appendix. The colon is small and corrugated leading into the rectum. The latter shows lumps which are the characteristic faecal pellets of the rabbit. The rectum in both organisms contains the food residue for some twenty-four hours before it is egested as *faeces* by muscular action.

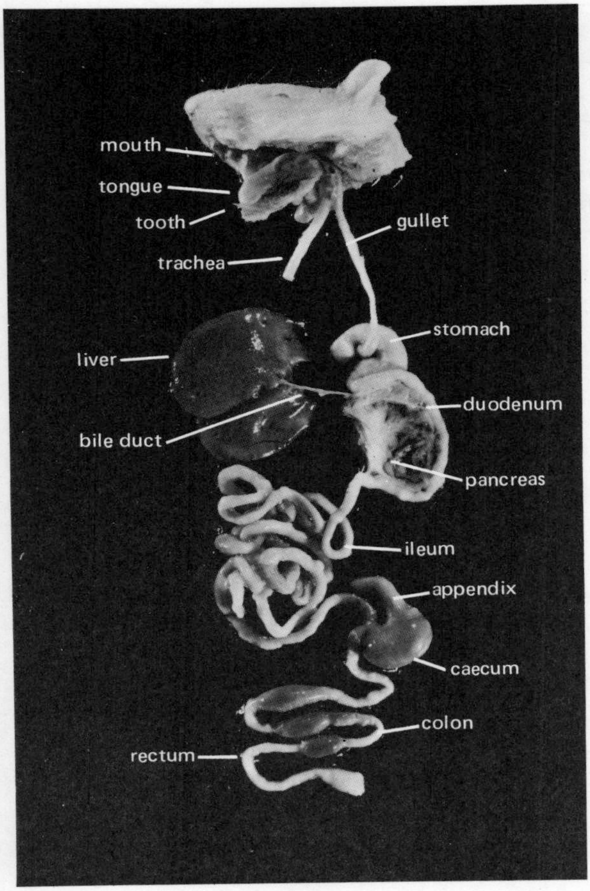

fig 2.66 The alimentary canal of a rat

2.73 Chemical digestion

Functions of the gut

Digestion in the buccal cavity. *Saliva* is discharged from the salivary glands (three pairs in Man, four pairs in the rabbit). The saliva of Man contains *amylase,* which begins the *digestion of starch.* The rabbit does not possess this enzyme. The saliva also contains *mucus* to soften the food and to lubricate its passage. In addition, salts provide a neutral to alkaline medium suitable for the action of the enzyme *ptyalin* (amylase). The food is formed into a ball or *bolus* by the tongue and then swallowed.

Digestion in the stomach. The stomach wall secretes *gastric juice* containing *rennin* which clots the soluble protein of *milk,* (particularly important in juveniles), and *pepsin,* which begins the breakdown of protein molecules into smaller units called *polypeptides.* The acidity of the juice is provided by *hydrochloric acid,* which kills some bacteria and also provides an acid medium (pH 1.5 to 2.0) suitable for the action of pepsin. The acid stops the activity of ptyalin. The food becomes very liquid in the stomach, owing to mucus, the fourth constituent of gastric juice. The chyme begins to leave the stomach through the pylorus after two or three hours. Chyme is only allowed through the pylorus in small, easily digestible amounts.

Digestion in the small intestine. *Bile* is secreted from the *liver* through the bile duct. This juice contains *salts* which break up the liquid fats into an *emulsion,* (emulsification) such that the tiny globules provide a larger surface area on which the fat-digesting enzymes can act. The alkaline bile also *neutralises* the acid chyme.

Pancreatic juice is alkaline and contains *three enzymes* acting on peptides, fats and starch. The intestinal wall secretes a group of enzymes called the *succus entericus,* which help the pancreatic juice to finally complete digestion of:

proteins to form amino-acids,
carbohydrates to form glucose,
and fats to form fatty acids and glycerol.

These are the final breakdown products which are soluble and small enough to be absorbed through the intestinal wall into the bloodstream. The vegetable diet of the rabbit contains a great deal of *cellulose,* and although Man cannot digest this carbohydrate, the rabbit must do so or its food intake would be very limited. The digestion of cellulose occurs in the caecum with the aid of large numbers of bacteria which produce *cellulase,* the required enzyme.

Enzymes have the following important properties.

i) They act as *organic catalysts, speeding up* the rate of food breakdown.

ii) Each enzyme causes the breakdown of the *one substance* (that is they are *specific*) to form a particular end-product.

iii) They are affected by *temperature.* They work best at 37°C to 40°C which is about body temperature of birds and mammals.

iv) They are affected by the *pH of their surroundings.* Each enzyme works most effectively at its own specific pH.

Enzyme	**Source**	**Optimum pH**
amylase	saliva	6.8
pepsin	gastric juice	2.0
rennin	gastric juice	2.0
trypsin	pancreatic juice	7.0 to 8.0

v) They are required only in *small quantities.*

vi) The *rate* of an enzyme reaction will *increase* with increasing *substrate concentration,* up to a certain limit.

Gland	Secretion	Enzymes	Action substrate	Products
Salivary glands	Saliva	Amylase	Starch	Maltose
Stomach wall – gastric glands	Gastric juice	Pepsin + hydrochloric acid + renin	Protein Soluble milk protein	Polypeptides Insoluble milk protein
Pancreas	Pancreatic juice	Trypsin Lipase Amylase	Protein Fats Starch	Peptides and *amino-acids* *Fatty acids* and *glycerol* Maltose
Wall of small intestine	Succus entericus	Sucrase Maltase Peptidase Lipase	Sucrose Maltose Polypeptides Fats	*Glucose* and *fructose* *Glucose* *Amino acids* *Fatty acids* and *glycerol*

The substances in italics are the final soluble breakdown products of digestion.

Table 2.12 The major enzymes concerned with digestion

2.74 Absorption of digested foods

It can be seen from table 2.12 that food is broken down by digestion into the end-products of amino-acids, glucose, fatty acids and glycerol. These together with mineral salts, vitamins and water must now be absorbed through the walls of the alimentary canal. This occurs in the region of the small intestine, where a large surface area for absorption is available because of:

i) the great length of the small intestine,
ii) the folded inner surface,
iii) the small projections, called *villi*, which cover the inner folds. Each square millimetre of surface has about twenty to forty villi.

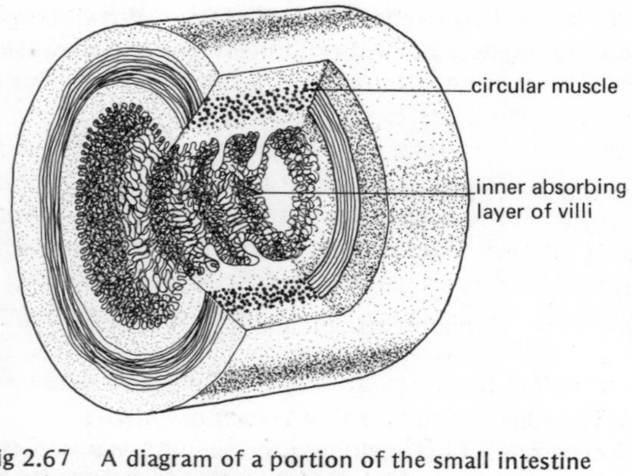

fig 2.67 A diagram of a portion of the small intestine showing the internal folds and the villi

It has been shown that even the surface of each villus has tiny projections (*microvilli*) of the cell surfaces which must increase greatly the total internal surface area of the small intestine.

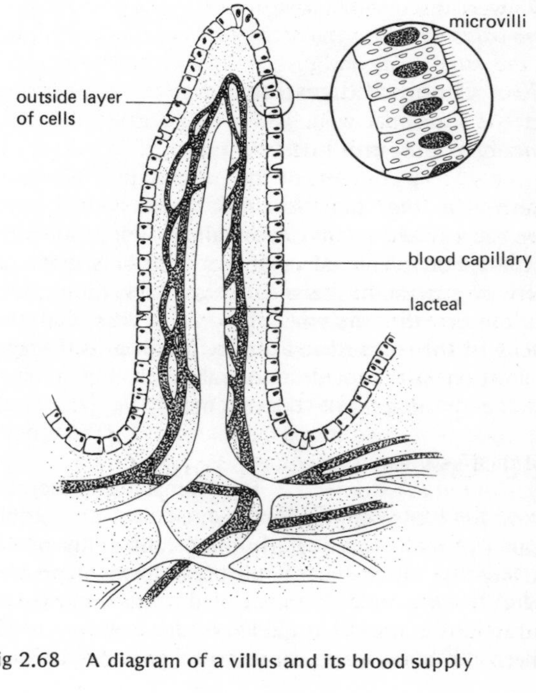

fig 2.68 A diagram of a villus and its blood supply

fig 2.69 The internal wall of the small intestine

What happens to the digested food materials when they have passed through the wall of the gut? This important question will be dealt with in Chapter 3 where we consider the problem of transporting digested food around the body. Once in the blood-stream, however, the food is available to perform the functions which were discussed in section 2.10. The *glucose* and *amino-acids* pass into the *blood capillaries* of the villus, and the *fatty acids* and *glycerol* pass into the *lacteal*. Both of these vessels carry the digested food away from the intestinal wall. Water is absorbed through the walls of the small intestine and large intestine, so that the watery contents of the gut gradually become solidified to form the faeces. Many substances pass along the gut and remain undigested. *Cellulose* is the principle example in Man, and this carbohydrate is largely present in the fruits and vegetables of the diet. Cellulose provides the *roughage* or bulk on which the muscles of the rectum can act. The faeces passed from the anus contain, in addition to cellulose, a good deal of mucus, broken down cells and bacteria.

2.80 Energy value of food

One of the functions of food is to produce energy for all our conscious activities and, in addition, for all of the activities of the body over which we have no control, for example, the beat of the heart, muscular contractions of the gut etc. Some of this energy is used to keep the body warm, some for chemical reactions and some for the muscular contractions that produce body movement. We can classify the *energy* we need as follows:

i) To maintain the functioning of the ordinary living processes; e.g. heart beat, maintenance of body temperature, breathing, circulation of blood, etc. All of these functions can be called the *Basal Metabolism*. The amount of energy required for this is known as the *Basal Metabolic Rate* (BMR).

ii) All other activities including: (a) our everyday activities such as moving, standing, washing, combing hair, etc., and (b) the work which we perform. This can vary from sitting at a desk and writing all day, (bank clerk), to cutting down trees for eight hours a day (forester). The amount of energy needed for these two very different activities will vary considerably. Further factors will affect the two divisions above.

The BMR will depend upon:

Age In actively growing children the basal metabolic rate is high and decreases rapidly in the first twelve years of life. Thereafter it falls more slowly, and at the age of twenty years decreases very gradually. With increasing age, also, people become less physically active.

Size The body size is important because of the relation of surface area to volume. The heat loss from a tall thin person is much greater than from a short fat one. Babies and children require more energy-rich food because they have (i) a greater surface area in relation to their volume and thus they lose heat more quickly, and (ii) they are growing rapidly and use up a great deal of additional energy when playing games and sports. Women use less energy since (i) they have less body weight to carry around than men, (ii) their more rounded shape means that their surface area is relatively smaller and (iii) they often do lighter work.

Climate In a warm climate the individual body demands less energy to maintain body temperature. Taking a reference temperature of 10°C, it is estimated that energy requirements are decreased by 5% for every 10°C rise in temperature. Because Man uses up energy to protect himself against cold, it is estimated that there is an increase of 3% for every 10°C fall below the reference temperature. This will help him to maintain his body temperature.

All energy is transferable so that, for whatever purpose it is required, the energy value of foods and energy requirements of Man can be calculated in terms of heat energy. Heat energy for many years has been measured in calories which are defined as follows:

1 *calorie* is the heat required to raise the temperature of 1 g of water through 1 degree C.

For practical purposes in nutrition this unit was too small so that the *kilocalorie* was used.

1 *kilocalorie* (= 1 000 calories), is the heat required to raise the temperature of 1 000 g of water through 1

degree C (or 100 g of water through 10 degrees C, and so on).

With the introduction of SI units, the unit of heat used is now the *joule* which is defined in terms of individual work done. James Joule showed that *mechanical work* can be converted into an *equivalent amount of heat.* Therefore both can be measured in the same units. For our purposes all we need to consider is that:

1 calorie (cal) = 4.2 joules (J)
1 kilocalorie (kcal) = 4 200 joules
= 4.2 kilojoules (kJ)

In many books on nutrition the energy value of foods is still described in terms of kilocalories, but kilojoules and megajoules are the current units.

joule	J	1 000 J = 1kJ
kilojoule	kJ	1 000 kJ = 1 MJ
megajoule	MJ	1 MJ = 1 000 000 J

1 MJ = 238 kcal
1 kcal = 4.2 kJ

Remember that 1 g of water has a volume of 1 cm^3, thus 10 g of water has a volume of 10 cm^3.

Questions

1 How many kilojoules are required to heat 50 cm^3 of water through 1 degree C?

2 How many kilojoules are required to heat 50 cm^3 of water from 20°C to 25°C?

You can use this knowledge to find out how much heat energy is released from food when it is burned. Carbohydrates and fats in the food are the principal sources of energy for the body, but protein can also be used for this purpose. If these substances are burned under controlled conditions in the laboratory the energy produced can be calculated. Typical results:

1 g of carbohydrate (as glucose) produces in the body about 16 kJ.
1 g of fat produces in the body about 38 kJ.
1 g of protein produces in the body about 17 kJ.

Notice that *fats* have about 2½ times the energy per unit mass of *carbohydrates.*

Using these values the energy content of any food can be calculated from the proportions of nutrients that it contains. For example, 100 g of maize (whole grain) contains 10 g of protein, 4.5 g of fat and 70 g of carbohydrate. Thus, the total energy value can be worked out as shown:

70 x 16 kJ = 1 120 kJ from carbohydrate
4.5 x 38 kJ = 171 kJ from fat
10 x 17 kJ = 170 kJ from protein
1 461 kJ

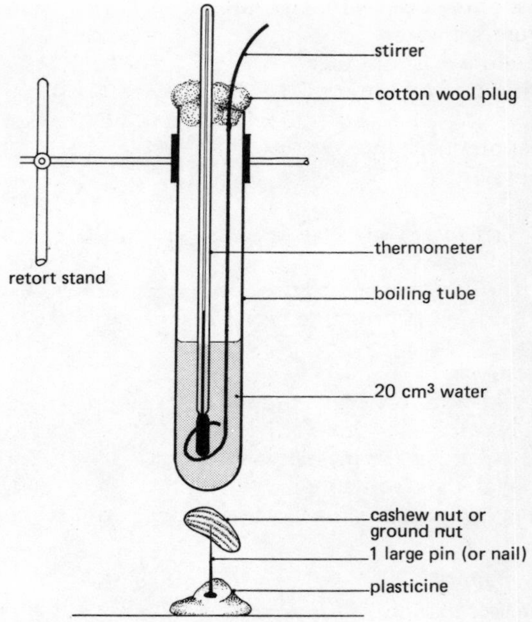

fig 2.70 The experimental apparatus for measuring heat energy produced by the burning of food

Experiment

To measure the heat energy produced by burning food

Procedure

1 Take a small test-tube, place in it 20 cm^3 of water and fix the test-tube in the clamp of a retort stand.

2 Weigh a piece of cashew nut or groundnut (about 0.5 g) and attach the nut to a pin held upright by a piece of plasticine (see fig 2.70).

3 Position the nut immediately beneath the test-tube. Place a thermometer in the test-tube together with a small piece of bent wire to act as a stirrer. Plug the mouth of the test-tube with cotton wool.

4 Record the temperature of the water.

5 Mark the position of the plasticine on the bench with a piece of chalk, then remove the plasticine, pin and nut. Hold the nut in a bunsen flame until it starts to burn and then place it beneath the test-tube in its original position.

6 Move the stirrer up and down as the nut is burning. Adjust the height of the tube on the retort stand so that the flame is beneath it, and no flame flows around the side of the tube.

7 When the nut has burned out continue stirring the water until the temperature reaches its highest point, then *record this temperature.*

8 Work out your results as follows:

temperature of the water at the start = A°C
highest temperature reached by the water = B°C

mass of the cashew (or ground) nut = C g
volume of water = 20 cm^3
therefore mass of water = 20 g
heat gained by water = 20 g x (B–A) deg. C x 4.2
= 20 x (B–A) x 4.2 J

Heat produced by 1 gm nut $= \frac{20 \times (B-A) \times 4.2}{1\,000 \times C}$ kJ/g

9 *Tabulate* your class results and work out the *mean value.*

The principle behind this experiment is the transfer of heat energy from the food to the water in the test-tube.

Questions

3 This is a very crude experiment. Can you give any criticisms of the method? How could you improve the experiment to make it more accurate?

4 Will the food produce more; the same, or less energy when it is eaten and used in the body?

Occupation

The occupation followed by an individual man or woman has the greatest effect on the rate at which energy from food is utilised. *Occupations* can be grouped broadly into the following classes.

i) *Sedentary* Men involved in these occupations need about 3 400 kJ for an 8-hour day. This class includes office workers, journalists, teachers, doctors, lawyers, clergy and secretaries. (= 3.4 MJ)

ii) *Moderately active* Men involved in these occupations need about 4 600 kJ for an 8-hour day. Often their work is not very strenous, but involves a lot of walking or weight lifting. This class includes railway workers, shop assistants and fishermen. (= 4.6 MJ)

iii) *Very active* Men involved in these occupations need about 6 800 kJ for an 8-hour day. This class includes steel workers, forestry workers, army recruits, farm labourers and builders labourers. (= 6.8 MJ)

Most women working at home or doing light work would require 3 400 kJ per day, while those involved in great physical activity would require about 4 600 to 5 000 kJ per day. (= 4.6 to 5.0 MJ)

The energy expenditure, for these three groups (including BMR and everyday activities) is analysed below. Total energy requirement for one day will be equal to total energy expenditure. Thus a sedentary man will require 10 300 kJ (10.3 MJ) in his daily diet.

	Sedentary	Moderately active	Very active
BMR	1 900	1 900	1 900
Everyday activity	5 000	5 000	5 000
Work (occupation)	3 400	4 600	6 800
	10 300 kJ	11 500 kJ	13 700 kJ
	= 10.3 MJ	11.5 MJ	13.7 MJ

Women need to expend *comparatively less energy* because of their *smaller size.* If more food is eaten than an individual needs for his activities then some of the *excess food* will be converted into *fat.* All three classes of nutrient can be converted in the body into fat. The most fattening foods are those containing most energy.

2.81 Balanced diet

The individual cells of the body tend to have a short life compared with the length of life of the organism as a whole. In Man, the total body protein is replaced about every six months. *A balanced diet* must also take this into account together with energy requirements and health requirements. Daily food intake should be a mixed diet of:

1 meat, eggs, fish or milk
2 fresh vegetable foods such as fruit, citrus fruit juice, tomato
3 grain cereals, potatoes, bread.

Each of these foods provides certain essentials, but may be deficient in others. For example vitamin C, lacking in 1 and 3 above, is present in 2. Vitamin D is lacking in 2 and 3, but present in 1.

A diet consisting mainly of carbohydrate-rich food such as in 3 would be considered *unbalanced*, because it does not contain enough protein, mineral salts and vitamins present in 1 and 2.

In a balanced diet the essential nutrients are eaten in the correct amounts for the age and activity of the individual. When calculating the correct proportions of food in a balanced diet, the following should be taken into consideration.

1 The amount of energy required.
2 The minimum protein requirements, taking into account that one-fifth of the protein should be animal protein.
3 The fat content, which should supply one-sixth of the total energy requirements.
4 The carbohydrate content which should supply five-sixths of the total energy requirement.
5 The other essential nutrients, vitamins and mineral salts although only needed in very small quantities, are absolutely vital.
6 Include sufficient fluids to complement that contained in the 'solid' foods.
7 Include an element of roughage to stimulate bowel action.

Questions

Estimate the weights of protein, fat and carbohydrates to be eaten by a 16-year-old boy (weight 40 kg) whose energy requirement per day is 12 000 kJ.
This can be calculated as follows on page 89:

Age range and occupational category	Body wt kg	Energy MJ	Protein[1] g	Thiamin mg	Riboflavin mg	Nicotinic acid mg equivalents	Ascorbic acid mg	Vitamins A µg retinol equivalents	Vitamins D µg cholecal-ciferol	Calcium mg	Iron mg
BOYS and GIRLS											
0 up to 1 year	7.3	3.3	20	0.3	0.4	5	15	450	10	600	6
1 up to 2 years	11.4	5.0	30	0.5	0.6	7	20	300	10	500	7
2 up to 3 years	13.5	5.9	35	0.6	0.7	8	20	300	10	500	7
3 up to 5 years	16.5	6.7	40	0.6	0.8	9	20	300	10	500	8
5 up to 7 years	20.5	7.5	45	0.7	0.9	10	20	300	2.5	500	8
7 up to 9 years	25.1	8.8	53	0.8	1.0	11	20	400	2.5	500	10
BOYS											
9 up to 12 years	31.9	10.5	63	1.0	1.2	14	25	575	2.5	700	13
12 up to 15 years	45.5	11.7	70	1.1	1.4	16	25	725	2.5	700	14
15 up to 18 years	61.0	12.6	75	1.1	1.7	19	30	750	2.5	600	15
GIRLS											
9 up to 12 years	33.0	9.6	58	0.9	1.2	13	25	575	2.5	700	13
12 up to 15 years	48.6	9.6	58	0.9	1.4	16	25	725	2.5	700	14
15 up to 18 years	56.1	9.6	58	0.9	1.4	16	30	750	2.5	600	15
MEN											
18 up to 35 years											
Sedentary	65	11.3	68	1.1	1.7	18	30	750	2.5	500	10
Moderately active		12.6	75	1.2	1.7	18	30	750	2.5	500	10
Very active		15.1	90	1.4	1.7	18	30	750	2.5	500	10
35 up to 65 years											
Sedentary	65	10.9	65	1.0	1.7	18	30	750	2.5	500	10
Moderately active		12.1	73	1.2	1.7	18	30	750	2.5	500	10
Very active		15.1	90	1.4	1.7	18	30	750	2.5	500	10
65 up to 75 years } assuming a sedentary life	63	9.8	59	0.9	1.7	18	30	750	2.5	500	10
75 and over } assuming a sedentary life	63	8.8	53	0.8	1.7	18	30	750	2.5	500	10
WOMEN											
18 up to 55 years											
Most occupations	55	9.2	55	0.9	1.3	15	30	750	2.5	500	12
Very active		10.5	63	1.0	1.3	15	30	750	2.5	500	12
55 up to 75 years } assuming a sedentary life	53	8.6	51	0.8	1.3	15	30	750	2.5	500	10
75 and over } assuming a sedentary life	53	8.0	48	0.7	1.3	15	30	750	2.5	500	10
Pregnancy, 2nd and 3rd trimester		10.0	60	1.0	1.6	18	60	750	10	1200	15
Lactation		11.3	68	1.1	1.8	21	60	1200	10	1200	15

[1] Recommended intakes calculated as providing 10% of energy.

Table 2.13 Recommended daily intakes of energy and nutrients for the UK (Department of Health and Social Security (1969))

Protein requirement

Total weight of protein required (see Table 2.14) = 1.5 x 40 = 60 g

Weight of animal protein = $\frac{1}{5}$ of 60 g = 12 g

Weight of plant protein = $\frac{4}{5}$ of 60 g = 48 g   Total 60 g

Fat and carbohydrate requirement

Energy to be provided by fat = $\frac{1}{6}$ of 12 000 kJ = 2 000 kJ

Weight of fat to provide this energy (see page 86 section 2.80 1 g of fat produces 38 kJ) $= \frac{2\,000}{38}\text{ g} = 53\text{ g}$

Energy to be provided by carbohydrate = 12 000 – 2 000 kJ = 10 000 kJ

Weight of carbohydrate to provide this energy (1 g of carbohydrate produces 16 kJ) $= \frac{10\,000}{16} = 625\text{ g}$

Now work out the following for yourself.

1 What is the weight of each major food class (protein, fat and carbohydrate) to be eaten by a 30 year old man (weight 60 kg) whose energy requirement, because of very active employment, is 13 500 kJ?

Group	Babies	Children	Adults	Pregnant or lactating women
Weight of protein per kg of body weight	2 g	1.5 g	1 g	2 g

Table 2.14 Minimum protein requirements

Experiment

To calculate the energy value of your daily diet.

Procedure

1 Use Table 2.18. Note that the last column gives the amount of energy per 100 g of food. Thus haricot beans have 1 073 kJ per 100 g or 10.73 kJ per g.

2 Weigh as accurately as possible each portion of food consumed throughout the day. Use kitchen scales or a spring balance to weigh the food in grams. If the balance is calibrated in Imperial units revalue as follows:

1 lb = 0.453 592 kg
= 453.592 or approx. 454 g

1 oz = 32.399 g or approx. 32 g

3 From your measurements complete the following table.

Examples of balanced diets

(i) Daily requirement for a child of 4 to 6 years to provide 6 300 kJ

Breakfast:

cornflakes, tea or coffee with milk, toast   1 470 kJ

Lunch:

potatoes, vegetables and fish   2 310 kJ

Supper or tea:

tea with milk and sugar, bread or other carbohydrate source   420 kJ

Evening meal:

meat, potatoes or rice, vegetables, fruit   2 100 kJ

Total 6 300 kJ

Food item	breakfast	morning snack	lunch	tea	dinner	other food	total
	g kJ	g kJ	g kJ	g kJ	g kJ	g kJ	

(ii) Daily requirement for an active man to provide 11 760 kJ

Breakfast:	
bread, eggs, tea or coffee with milk and sugar	2 520 kJ
Lunch:	
meat, potatoes, vegetables and fruit	4 200 kJ
Supper or tea:	
bread or other carbohydrate source, tea with milk and sugar	1 680 kJ
Evening meal:	
meat, eggs, vegetables and fruit	3 360 kJ
Total	11 760 kJ

The diet for this man can be set out in another fashion as follows:

500 g of potatoes	7 330 kJ
100 g of lean meat	672 kJ
100 g of beans	139 kJ
25 g of fat	926 kJ
600 g of other vegetables	504 kJ
300 g of fruit	480 kJ
3 eggs	500 kJ
100 g of milk (sweetened)	1 300 kJ
Total	11 851 kJ

	Weight g	Energy kJ	Protein g	Fat g	Calcium mg	Iron mg
Meal 1						
Roll	98	1 058	8.8	1.1	109	1.8
Butter	8	273	0.0	7.3	0	0.0
Cheese	56	983	14.2	19.6	460	0.4
Tea (with milk and sugar)	280	84	1.0	1.0	30	0.0
Total	442	2 398	24.0	29.0	59.9	2.2
Meal 2						
Mutton	67	945	8.9	21.1	7	1.4
Cabbage	84	88	1.2	0.0	54	0.9
Potato	112	353	2.4	0.0	8	0.8
Apples	84	151	0.3	0.0	3	0.3
Custard	56	260	1.8	2.2	70	0.0
Total	403	1 797	14.6	23.3	142	3.4

Table 2.15 Analysis of two meals

Meal	Energy kJ	Protein g	Fat g	Calcium mg	Iron mg	Vitamin A i.u.	Thiamine mg	Riboflavin mg	Ascorbic acid mg
Breakfast	2 982	21	26	240	4	630	0.51	0.30	5
Snack	882	9	8	140	1	230	0.12	0.12	0
Lunch	3 486	20	27	350	4	1 100	0.87	1.90	30
Snack	1 008	5	9	50	1	50	0.06	0.05	0
Evening meal	2 394	19	21	100	4	560	0.51	1.12	31
Total	10 752	74	91	880	14	2 570	2.07	3.49	66

Table 2.16 Daily diet of woman doing light factory work

To obtain a balanced diet it is necessary to eat a *wide variety of foods*, since a shortage of any of the main classes will result in deficiency diseases.

In the developed countries today, these diseases are rare, as food of sufficient quality and quantity is available. Most people are educated to understand which foods are necessary to keep themselves and their children healthy. In certain developing countries deficiency diseases are common for *three reasons* which are listed below.

i) Shortage of the right kind of food. In many areas foods rich in protein and vitamins are not available. The normal diet consists mainly of cereals, starch foods and sugar. Animal protein especially is very scarce.

ii) *Low incomes.* Families are often too poor to buy the right kinds of food even if they were available. Foods rich in proteins and vitamins, such as milk, meat, fish, eggs, and many of the fruits are usually the most expensive.

Examine fig 2.71 that shows a child suffering from kwashiorkor.

Questions

2 What symptoms can you see?

3 Why has the child developed this disease?

iii) *Superstition and lack of education.* Some people may not eat the right foods because they have not been educated in the concepts of a balanced diet. Local custom can dictate what is eaten and, even if suitable foods are available, people may be unwilling to try them.

In some countries it is thought that certain foods possess magical properties and that to eat them would bring bad luck.

There are a number of diseases caused by protein deficient diets.

a) *Kwashiorkor* This is common in large parts of South East Asia and throughout Africa, particularly in West Africa from where it gets its name. It affects young children particularly when they are put on a starchy adult diet after weaning. The *stomach distends* because of fluid retention, the muscles waste away and the skin becomes discoloured. *Diarrhoea* and *anaemia* are also symptoms. If prolonged, the disease retards physical and mental growth.

Early treatment, with a special *high-protein* fluid containing skim-milk powder and vegetable protein is essential.

b) *Marasmus* The symptoms are similar to those of kwashiorkor and are due to *general starvation* rather than just protein deficiencies. There is no swelling of the abdomen or skin rash. All body tissues exhibit *general wasting* and a severe *loss of fluid* often results in death. The treatment is the same as that for kwashiorkor, though more carbohydrate may be needed in some cases.

c) *Vitamin deficiency* is still widespread in many countries of the Far East, Latin America and Africa, and is mainly due to insufficient intake of green vegetables, milk, butter and eggs.

In the developing world, the death rate for young

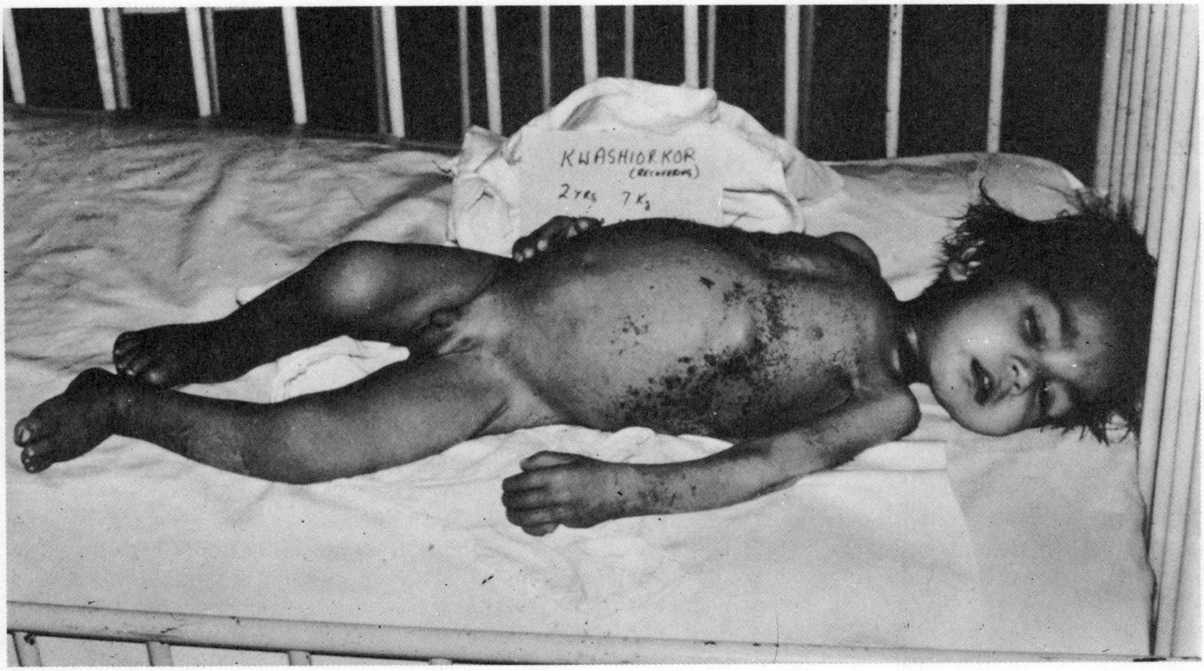

fig 2.71 Kwashiorkor

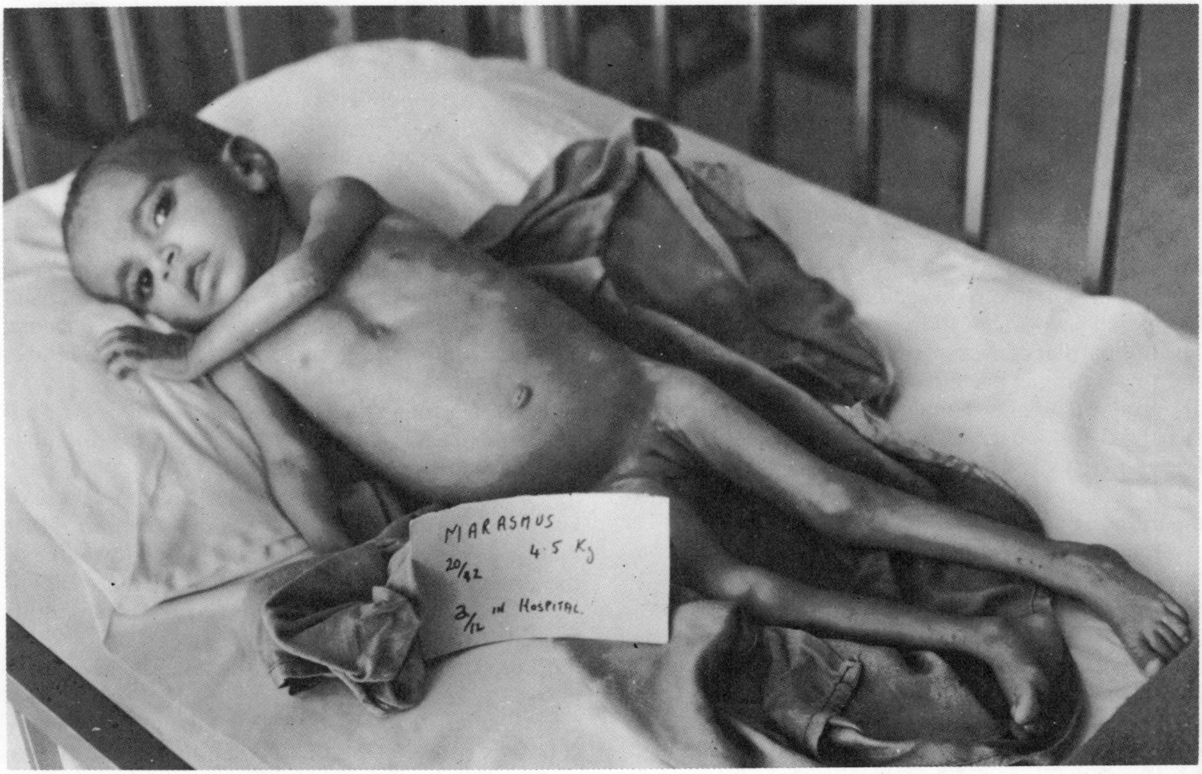

fig 2.72 Marasmus

children is ten times the rate for the same age group in industrialised nations. Eleven million children die every year—30 000 every day. Many of these die not from the initial malnutrition but from bacterial and virus diseases which they would survive if properly fed. Pneumonia and measles are great killers under conditions of malnutrition.

2.82 Over and under-eating

In some children, over and under-eating can be viewed as a bid to establish an individual identity. Over-eating may result from rivalry between brothers and sisters, feelings of insecurity and simply overfeeding by the parents. A diet too rich in fats or carbohydrates and the taking of snacks between meals can lead to obesity and, more rarely, this condition may be due to glandular or genetic problems.

Under-eating by children in an affluent society may have similar common roots to over-eating, but is less frequent. *Anorexia nervosa* is a pathological form of food refusal which can lead to loss of weight and sometimes death by starvation. It is a pschychiatric disorder often affecting adolescent girls and is only effectively treated in a specialised hospital unit. The adolescent girl often thinks she is too fat and that she must diet, and so becomes obsessed with avoiding carbohydrate intake. Eating becomes repulsive and she will often refuse to eat with the family, will indulge in excessive exercise, induce vomiting, etc.

In considering over and under-eating in children it is important to:

i) regard ideal body weight as a target (this to be determined by standard tables for age, height and thus appropriate weight),

ii) encourage a standard and balanced diet,

iii) encourage co-operation where there are likes and dislikes in the diet so that adjustments can be made,

iv) establish a weight check at least once per week.

Accumulation of fat in adults results from eating more food than the body needs for its energy expenditure. The body of a 20 year old man is about 10% fat, while that of a woman of the same age is about 25% fat. Thus the softer body of a woman produces her characteristic shape after puberty as a result of a greater proportion of fat (see fig 2.73(a)). As the body ages, however, the proportion of fat rises, in a man at 50 years to 25%, and in a woman at the same age to 45%. In women particularly these fat centres are concentrated in certain areas, notably breasts, buttocks and legs (see fig 2.73(b)) and the table opposite.

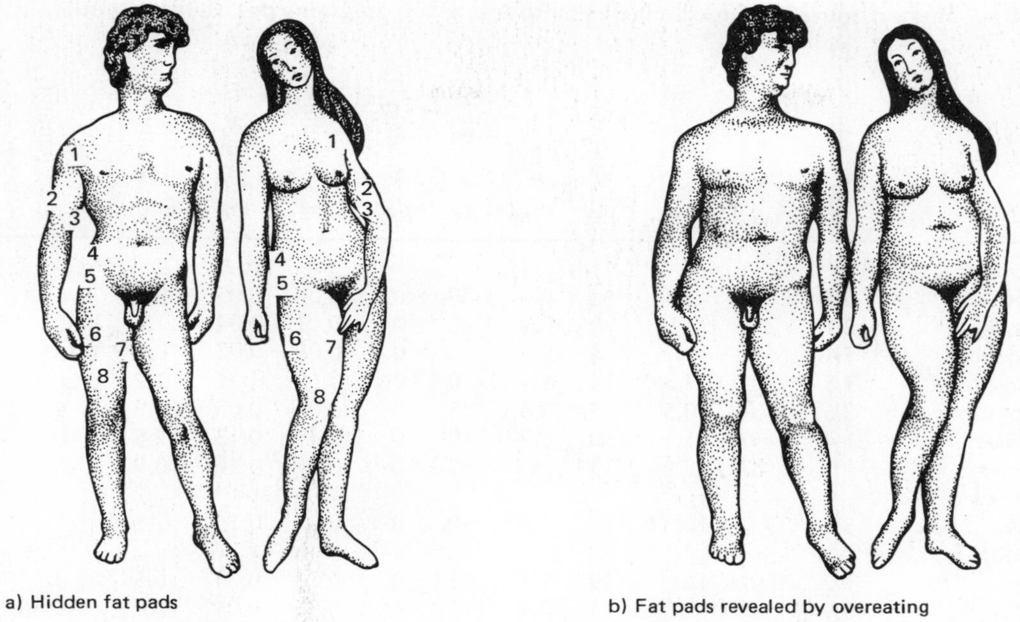

fig 2.73
(a) A twenty-year-old man and woman and hidden fat pads numbers 1 to 8
(b) Excess of fat reveals position of fat pads

Fat areas numbered in fig 2.73	Desirable average thickness of each fat pad (mm)	
	Men	Women
1 shoulder	18.0	17.8
2 outside arm	4.4	6.2
3 inside arm	3.5	6.6
4 hip	11.2	19.0
5 top of thigh	15.6	28.1
6 outside leg	4.8	7.4
7 inside leg	6.0	10.9
8 front of leg	7.0	13.0

Table 2.17 Fat pads

2.90 Food additives

The use of chemical additives in food is an ancient practice. Wine is fruit juice preserved by alcohol. Sugar, in the form of honey, salt and sulphur-rich smoke have extended the life of food for centuries. The rapid development of chemistry in the nineteenth century enabled a much more sophisticated approach, but unfortunately, at the same time, the adulteration of food provided profit for the unscrupulous. Bones were ground to fine powder and added to flour, copper was added to beer, iron filings to tea and alum to bread. Food was often coloured by the addition of toxic salts of heavy metals such as antimony, lead, copper, and arsenic. By the second half of the nineteenth century however, food legislation in many countries had stopped such practices. Inspectors and public analysts were appointed to ensure that legislation was strictly observed.

Foods have many different chemicals added to them today for a variety of reasons. These are broadly as follows.

i) To improve the nutritional value of food. For example margarine must have minimum levels of vitamins A and D which are similar to that of butter. Bread flours must have calcium and iron added and conform to certain minimum requirements for the B-group vitamins.

ii) To improve colour and flavour. This is a doubtful practice according to pure food specialists but it is true that food must be attractive and tasty if it is to be widely accepted.

iii) To improve texture. This can include a number of substances used, for example, to form emulsions in salad creams and fruit juices, to act as raising agents in baking and to thicken sauces and gravies.

iv) To preserve food. Certain substances are added to increase its shelf life which aid the wholesaler and retailer rather than the purchaser. It does mean of course that foods are available at times of the year when otherwise they might be scarce. Such additives include bleaches, anti-oxidants, acidifiers, bases and buffers.

Chemical contamination of food can occur in other ways. Agricultural chemicals used on farms may be

Table 2.18 Common sources of food classes (amounts given as grams per 100g of food)

Name of food	Protein g	Fat g	Carbohydrate g	Calcium mg	Iron mg	Vitamin A μg	Vitamin D μg	Thiamine mg	Riboflavin mg	Nicotinic acid mg	Vitamin C mg	Energy value kJ
Vegetables												
Carrots, old	0.7	0	5.4	48	0.6	2000	0	0.06	0.05	0.6	6	96
Potatoes, raw	2.1	0	18.0	8	0.7	0	0	0.11	0.04	1.2	8-30*	318
Potatoes, boiled	1.4	0	19.7	4	0.5	0	0	0.08	0.03	0.8	4-15*	331
Potato chips, fried	3.8	9.0	37.3	14	1.4	0	0	0.10	0.04	1.2	6-20*	989
Sweet corn, canned	2.6	0.8	20.5	5	0.5	35	0	0.03	0.05	0.9	4	398
Beans, haricot	21.4	0	45.5	180	6.7	0	0	0.45	0.13	2.5	0	1073
Beans, runner	1.1	0	2.9	33	0.7	50	0	0.05	0.10	0.9	20	63
Peas, fresh raw or quick frozen	5.8	0	10.6	15	1.9	50	0	0.32	0.15	2.5	25	264
Peas, canned, processed	7.2	0	18.0	29	1.1	67	0	0.06	0.04	0.5	2	402
Tomatoes, fresh	0.9	0	2.8	13	0.4	117	0	0.06	0.04	0.6	20	59
Brussels sprouts, boiled	2.4	0	1.7	27	0.6	67	0	0.06	0.10	0.4	35	67
Cabbage, boiled	0.8	0	1.3	58	0.5	50	0	0.03	0.03	0.2	20	34
Lettuce	1.1	0	1.8	26	0.7	167	0	0.07	0.08	0.3	15	46
Onions	0.9	0	5.2	31	0.3	0	0	0.03	0.05	0.2	10	96
Parsnips	1.7	0	11.3	55	0.6	0	0	0.10	0.09	1.0	15	205
Fruit												
Apple	0.3	0	12.0	4	0.3	5	0	0.04	0.02	0.1	5	193
Bananas	1.1	0	19.2	7	0.4	3	0	0.04	0.07	0.6	10	318
Black currants	0.9	0	6.6	60	1.3	33	0	0.03	0.06	0.3	200	117
Grapefruit	0.6	0	5.3	17	0.3	0	0	0.05	0.02	0.2	40	92
Oranges	0.8	0	8.5	41	0.3	8	0	0.10	0.03	0.2	50	147
Orange juice, canned unconcentrated	0.8	0	11.7	10	0.4	8	0	0.07	0.02	0.2	40	197
Peaches, canned	0.4	0	22.9	3.5	1.9	41	0	0.01	0.02	0.6	4	369
Pineapple, canned	0.3	0	20.0	13	1.7	7	0	0.05	0.02	0.2	8	318
Raspberries	0.9	0	5.6	41	1.2	13	0	0.02	0.03	0.4	25	105
Rhubarb	0.6	0	1.0	103	0.4	10	0	0.01	0.07	0.07	10	25
Strawberries	0.6	0	6.2	22	0.7	5	0	0.02	0.03	0.4	60	109
Nuts and cereals												
Peanuts, roasted	28.1	49.0	8.6	61	2.0	0	0	0.23	0.10	16.0	0	2455
Biscuits, chocolate	7.1	24.9	65.3	131	1.5	0	0	0.11	0.04	1.1	0	2082
Biscuits, plain, semi-sweet	7.4	13.2	75.3	126	1.8	0	0	0.17	0.06	1.3	0	1806
Bread, brown	9.2	1.8	49.0	92	2.5	0	0	0.28	0.07	3.2	0	993
Bread, white	8.3	1.7	54.6	100	1.8	0	0	0.18	0.02	1.4	0	1060
Crispbread, Ryvita	10.0	2.1	69	86	3.3	0	0	0.37	0.24	1.4	0	1332
Rice	6.2	1.0	86.8	4	0.4	0	0	0.08	0.03	1.5	0	1504
Spaghetti	9.9	1.0	84.0	23	1.2	0	0	0.09	0.06	1.7	0	1525
Meat												
Bacon, average												
Bacon, average	11.0	48.0	0	10	1.0	0	0	0.40	0.15	1.5	0	1994

* vitamin C falls during storage

Name of food	Protein g	Fat g	Carbohydrate g	Calcium mg	Iron mg	Vitamin A µg	Vitamin D µg	Thiamine mg	Riboflavine mg	Nicotinic acid mg	Vitamin C mg	Energy value kJ
Beef, average	14.8	28.2	0	10	4.0	0	0	0.07	0.20	5.0	0	1311
Beef, corned	22.3	15.0	0	13	9.8	0	0	0	0.20	3.5	0	939
Chicken, roast	29.6	7.3	0	15	2.6	0	0	0.04	0.14	4.9	0	771
Ham, cooked	16.3	39.6	0	13	2.5	0	0	0.50	0.20	3.5	0	1768
Lamb, roast	25.0	20.4	0	4	4.3	0	0	0.10	0.25	4.5	0	1190
Liver, fried	29.5	15.9	4.0	9	20.7	6000	0.75	0.30	3.50	15.0	20	1156
Pork, average	12.0	40.0	0	10	1.0	0	0	1.00	0.20	5.0	0	1710
Sausage, pork	10.4	30.9	13.3	15	2.5	0	0	0.17	0.07	1.6	0	1546
Steak and kidney pie, cooked	13.3	21.1	16.2	37	5.1	126	0.55	0.11	0.47	4.1	0	1274
Fish												
Cod, haddock, white fish	16.0	0.5	0	25	1.0	0	0	0.06	0.10	3.0	0	289
Fish fingers	13.4	6.8	20.7	50	1.4	0	0	0.12	0.16	1.8	0	804
Herring	16.0	14.1	0	100	1.5	45	22.25	0.03	0.30	3.5	0	796
Kipper	19.0	16.0	0	120	2.0	45	22.25	0	0.30	3.5	0	922
Sardines, canned in oil	20.4	22.6	0	409	4.0	30	7.50	0	0.20	5.0	0	1194
Milk, eggs, fat												
Cream, single	2.8	18.0	4.2	100	0.1	155	0.10	0.03	0.13	0.1	1	792
Milk, liquid, whole	3.3	3.8	4.8	120	0.1	44(a) 37(b)	0.05(a) 0.01(b)	0.04	0.15	0.1	1	272
Milk, whole, evaporated	8.5	9.2	12.8	290	0.2	112	0.12	0.06	0.37	0.2	2	696
Cheese, Cheddar	25.4	34.5	0	810	0.6	420	0.35	0.04	0.50	0.1	0	1726
Eggs, fresh	11.9	12.3	0	56	2.5	200	1.50	0.10	0.35	0.1	0	662
Butter	0.5	82.5	0	15	0.2	995	1.25	0	0	0	0	3122
Margarine	0.2	85.3	0	4	0.3	900(c)	8.00	0	0	0	0	3222
Honey	0.4	0	76.4	5	0.4	0	0	0	0.05	0.2	0	1207
Jam	0.5	0	69.2	18	1.2	2	0	0	0	0	10	1098

(a) Summer value (b) Winter value (c) some margarines contain carotene

incorporated into foods through the roots and leaves of plants. As a result of the widespread use of pesticides on crops every reader of this book has traces of these substances in his body fat. DDT, Dieldrin and Aldrin have all been used on farms and have been traced through food chains to the human body and to mammal or bird predators (Chapter 11 section 11.02).

Storage can result in contamination by pests such as insects and rodents. Factory processing can also result in contamination from metals, plastic and even machine lubricants.

Questions requiring an extended essay-type answer

1 a) How would you test maize meal for
   i) reducing sugar
   ii) starch
   iii) fat
   iv) protein
   b) Describe a comparative test for the amounts of vitamin C in lemon juice and tomatoes.

2 a) Why does Man require food?
   b) List the major classes of foodstuff. Give examples of food in which each of these main classes is present in relatively large amounts.

3 a) What do you understand by the term a balanced diet? (In your answer indicate the types of food required in such a diet.)

b) How would a balanced diet for a child of five differ from that of an adult leading an active life?

4 Describe what happens to a meal of rice and beef (cooked in oil) from the time that it is placed in the mouth until the products of digestion enter the blood stream.

5 a) For each of the following constituents of a fertile soil, describe its importance to the growth of the plant.

i) air ii) water iii) micro-organisms

b) How would you determine the degree of acidity or alkalinity of a soil? Why is such information important to a farmer?

6 a) What are the advantages to the gardener of vegetative reproduction in green plants?

b) A vegetative storage organ of a plant is thought to contain non-reducing sugars, protein and oil. How would you confirm that these substances are present?

7 a) State the exact position of the liver in relation to other organs in the body of Man.

b) Describe the functions of the liver in Man.

3 Circulation

3.00 Why is transport necessary?

All animals need to take in oxygen and foodstuffs from the environment and to eliminate carbon dioxide and other waste products from their bodies. Small animals such as protozoa, jellyfish and flatworms carry out these exchanges by simple *diffusion* through their body surfaces, but with *large organisms* some cells become *too far away* from the external surface of the body or the internal surface of the gut for simple diffusion to be effective. Thus oxygen would never reach the innermost cells while digested food moving through the gut wall would never supply the outermost cells of the body. There is a *limiting factor* present; this is the rate at which gases and dissolved substances can diffuse through living material.

As animals evolved and increased in size a *transport system* became essential, in order that dissolved materials could be moved rapidly around the organism. In many cases this function is fulfilled by a blood system.

3.10 Surface area to volume ratios

We can examine more closely the need for a transport system if we look at the vital factor of increasing size and the relationship between *surface area* and *volume* in the living organism. This factor affects the rates of gaseous exchange, heat loss, and loss of excretory matter in every animal. In order to understand this problem it is easier to think of the body shape, not as an irregular structure, but as a symmetrical box shape. Consider a simple animal as though it were a cube of size 1 cm: its surface area (S.A.) is 6 x 1 cm^2 and this will be the total area through which oxygen and carbon dioxide passes in and out. The volume (V) of such a cube is 1 cm^3.

The surface area: volume ratio is thus

$$\frac{S.A.}{V.} = \frac{6}{1} = 6$$

If the length of the cube is now doubled to 2 cm, then:

$$\frac{S.A.}{V.} = \frac{6 \times 2^2}{2^3} = \frac{24}{8} = \frac{3}{1} = 3$$

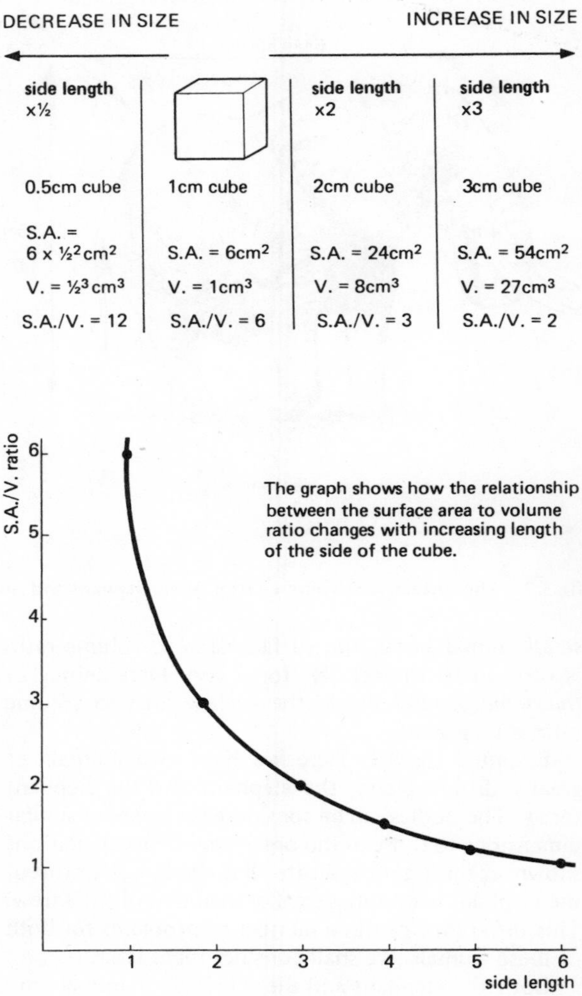

fig 3.1 A graph showing the relationship between the surface area/volume ratio and length

Now examine fig 3.1 and you will see this illustrated in two ways. Taking the 1 cm cube as the reference size it is clear that, as the cube size increases, the surface area:volume ratio decreases, whereas a decrease in cube size brings about an increase in this ratio. This principle is shown in a different fashion in the graph. You will now appreciate that, for a *very*

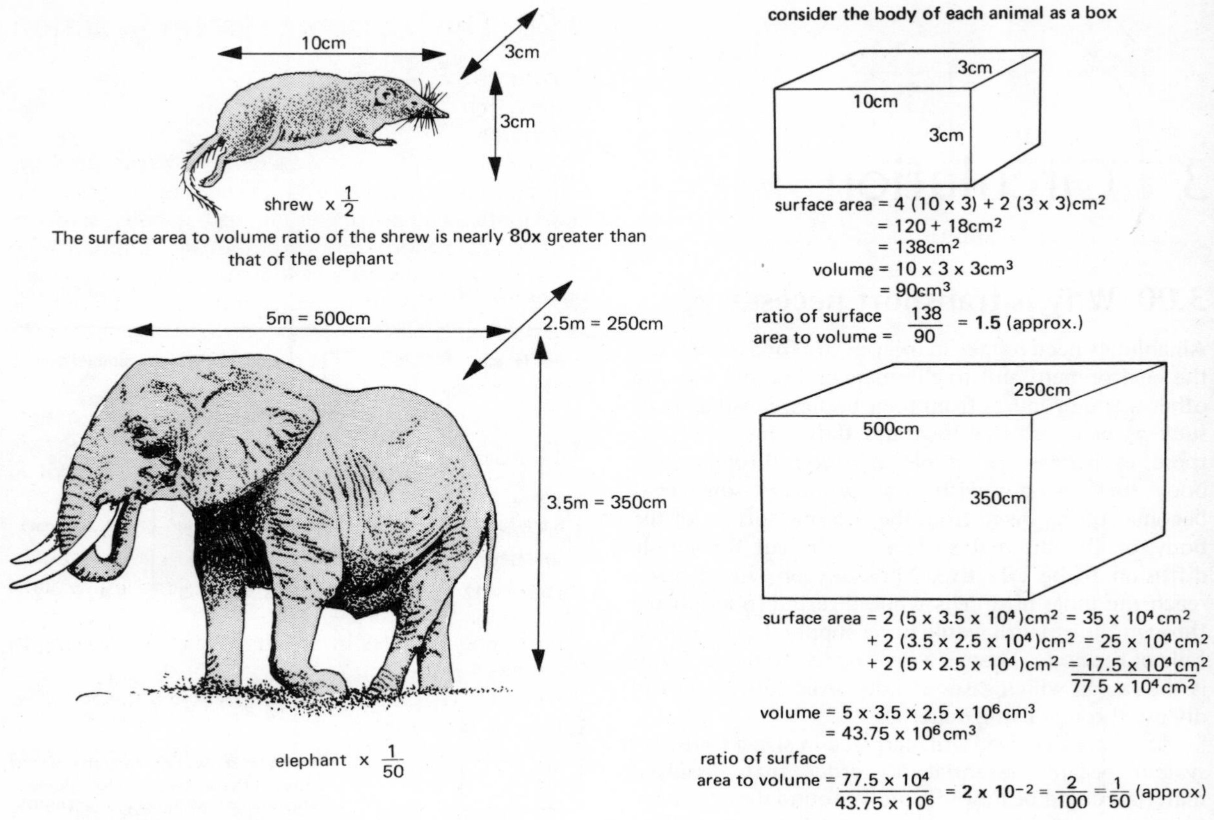

consider the body of each animal as a box
10cm
3cm
3cm
shrew x 1/2
The surface area to volume ratio of the shrew is nearly **80x** greater than that of the elephant
10cm
3cm
3cm
surface area = 4 (10 x 3) + 2 (3 x 3)cm²
= 120 + 18cm²
= 138cm²
volume = 10 x 3 x 3cm³
= 90cm³
ratio of surface area to volume = 138/90 = **1.5** (approx.)
5m = 500cm
2.5m = 250cm
3.5m = 350cm
elephant x 1/50
250cm
500cm
350cm
surface area = 2 (5 x 3.5 x 10⁴)cm² = 35 x 10⁴cm²
+ 2 (3.5 x 2.5 x 10⁴)cm² = 25 x 10⁴cm²
+ 2 (5 x 2.5 x 10⁴)cm² = 17.5 x 10⁴cm²
77.5 x 10⁴cm²
volume = 5 x 3.5 x 2.5 x 10⁶cm³
= 43.75 x 10⁶cm³
ratio of surface area to volume = 77.5 x 10⁴ / 43.75 x 10⁶ = **2 x 10⁻²** = 2/100 = 1/50 (approx)

fig 3.2 The surface area/volume ratios of an elephant and an elephant shrew

small animal body, the surface area to volume ratio is *very great.* Conversely, for a *very large* animal of *the same general shape,* the surface area to volume ratio is *very small.*

Examine fig 3.2. Here we have two animals of greatly differing size, the elephant and the elephant shrew. The bodies can be considered as boxes of similar dimensions to those of the organisms. The calculations shown in the figure indicate that the S.A./V. ratio of the elephant is 80 times smaller than that of the shrew. This difference causes a number of problems for both of these animals. We shall consider these later.

Both the elephant and the shrew are mammals and possess transport systems. However, if we compare the elephant shrew with a single-celled amoeba in the same way, then the latter will have an enormous surface area to volume ratio compared with that of the elephant shrew.

Experiment

Investigation of S.A./V. ratio in agar blocks

Procedure

1 Make up an agar solution in hot water in a beaker, and then pour it into the base of a petri dish to a depth of 1 cm. Allow it to solidify.

2 Place a ruler over the agar and cut parallel lines 1 cm apart. Repeat at right angles, so that the agar is now in 1 cm cubes (see fig 3.3).

3 Take some 1 cm blocks and cut them into 0.5 cm blocks by making three cuts at right angles to each other across the 1 cm blocks.

4 Put three blocks of each size in a beaker and cover with a potassium permanganate solution (0.5%).

5 Remove a small and a large block every three minutes.

6 Cut vertically through each block and observe the penetration of colour.

7 Repeat the experiment using different strengths of solution and shorter time intervals.

Questions

1 What do you observe, concerning the penetration of colour into the blocks?

2 What difference will it make if the experiment is repeated (i) using the same solution but with a shorter time interval, or (ii) using a stronger solution with the same time interval?

From the experiment it is clear that, with increasing

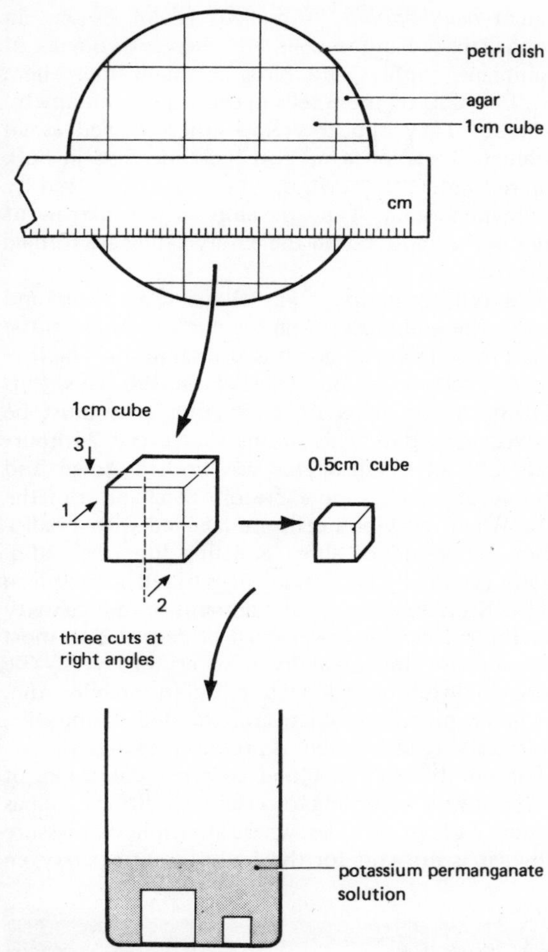

fig 3.3 Agar blocks in potassium permanganate solution

size of organisms, the ability to absorb through the *body surface* (related to surface area) will decrease in relation to the *needs of the organism* (related to volume). The process of diffusion will need to move dissolved substances over greater and greater distances. Remember the potassium permanganate in the experiment. Thus the S.A./V. ratio and diffusion will become *limiting factors* for activities which depend upon supplies of oxygen and food, and upon the removal of carbon dioxide and waste products.

Large organisms need three changes in their structure:

1 i) Additional surface area externally for the absorption of oxygen, e.g. lungs and gills.

ii) Additional surface area internally for the absorption of food, e.g. a long gut, folds and villi.

2 A system whereby a *circulating fluid* carries the absorbed substances at a faster rate than can be effected by diffusion. This is a *mass-flow system.*

3.20 The transport system in action

Experiment

Observation of a transport system

Procedure

1 Place a living tadpole on a cavity microscope slide, or a small fish in a petri dish. (These organisms can be anaesthetised by putting briefly into a 0.03% solution of MS-222 Sandoz (tricaine methane sulphonate)). They will generally remain still enough without anaesthetic, if (i) the water surrounding the tadpole is absorbed with blotting paper so that it is almost dry; (ii) the fish is wrapped in wet cotton wool so that it cannot move.

2 Place the slide (or dish) on the stage of the microscope and focus on the tail. Observe the thinnest part of the tail, towards the edge, where the pigment is lacking.

3 Observe the channels containing moving particles.

Questions

1 Are the particles all going in the same direction in any one channel?

2 Are the particles in different channels flowing in the same direction?

3 Would you describe the particle movement as steady or pulsating?

4 How is the speed of particle related to channel size?

Let us now take a closer look at the particles that are moving within the channels of the fin. We can assume that there is a liquid carrying the objects along just as a stream carries mud particles after heavy rain. The channels are small *blood vessels* enclosing the blood and forming a *mass-flow system transporting material* around the body of the tadpole or the fish. All vertebrates have blood systems and the most readily observable blood system is our own.

Experiment

What does our blood contain?

Procedure

1 Sterilise the end of the little finger of the left hand by using cotton wool moistened with alcohol.

2 Use a sterile lancet or a needle sterilised in a bunsen flame and, having shaken the left hand downwards several times (to force blood into the finger), jab the point into the skin at the side of the sterile little finger. This is the least sensitive part.

3 Now squeeze along the little finger with the thumb of the left hand, and, as a drop of blood comes from the puncture in the skin, collect it near the end of a clean dry slide.

4 Take a second slide, place its edge on the margin of the drop and draw the blood along the first slide. The blood will be drawn out into a thin film or smear.

5 Place the slide with the blood smear on the bench to dry. Sterilise the puncture on the little finger again with the alcohol-soaked cotton wool.

Precaution: Do not use the lancet or needle if it has been placed on the bench top. You must sterilise it in a flame before using it again.

6 When the blood smear is dry, place it on a petri dish and add one or two drops of Leishmann's stain; then add one or two drops of distilled water. Rock the slide gently to mix the two liquids. Leave for five minutes.

7 Wash off the stain with distilled water. After shaking off the remaining distilled water, leave for five minutes to dry.

8 Examine under a microscope.

Question

5 Draw the different types of cell visible. One type will be very numerous but the others will be hard to find. Look carefully. What colour are the majority of cells? What colour are the cells of the less numerous type?

3.30 Structure of blood

Blood consists of a fluid matrix, the plasma, in which float two types of cell. In mammals it constitutes about 10% of the body weight. Man has about 6 litres of blood.

1 Red blood cells (erythrocytes)

These are biconcave discs which contain the red pigment *haemoglobin.* A nucleus is *not* present in mammalian red blood cells but the erythrocytes of amphibians, reptiles and birds are nucleated. There are 5 000 000 of these cells in every cubic millimetre of blood. They appear yellow when viewed as an unstained smear on a microscope slide, but in bulk their red colour is obvious. They are stained red by Leishmann's stain. They develop in the marrow of bones in the adult, but in the embryo they are formed in the liver and spleen.

Red cell formation takes about seven days and towards the end of this time the nucleus visible in the immature cell breaks down and disappears. The life-span of each red blood cell is about 120 days, thus 1/120th of the mass of circulating cells must be removed each day. This means that every 24 hours about 25g of erythrocytes are broken down and disposed of through the excretory mechanism of the body. When red blood cells are labelled with a radioactive compound of short emitting life, the radioactivity eventually becomes concentrated in the spleen and the liver. Since the greater amount of radioactivity is in the spleen this organ must be regarded as most important for the breakdown of erythrocytes. The haemoglobin is broken down to bilirubin and biliverdin, two green pigments which are excreted in the bile. About 250 mg of bilirubin is produced each day.

The number of red blood cells per cubic mm of blood can vary according to certain conditions such as altitude. At high altitudes, where atmospheric pressure is low, it is difficult for the body to extract oxygen

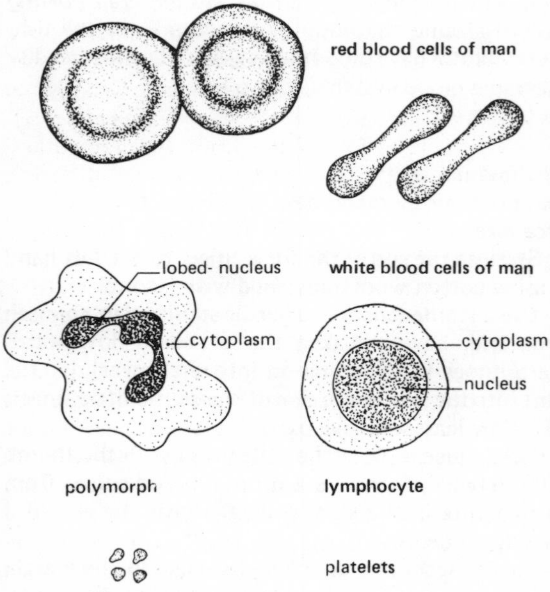

fig 3.4 The different types of mammalian blood cell

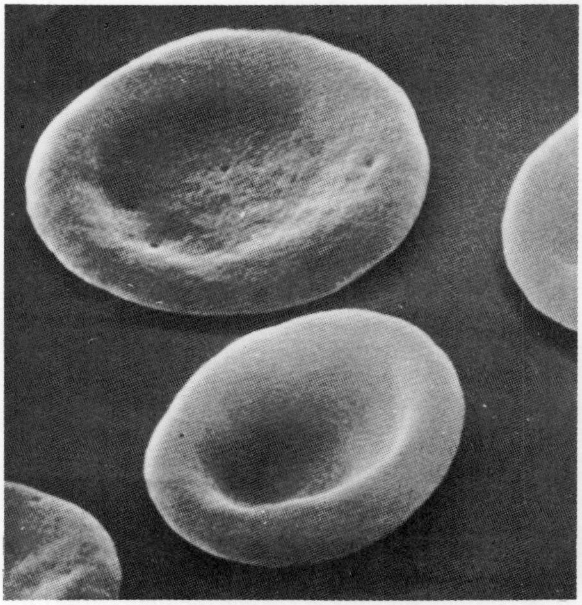

fig 3.5 A photomicrograph of mammalian red blood cells, taken using a scanning electron microscope

from the air, so the red bone marrow compensates by making more red blood cells.

Investigation

Examine fig 3.6 and answer the following questions.

1 How many days were spent continuously above 15 000 ft (4.6 km)?

2 How many days (approximately) had been spent above 10 000 ft (2.8 km) before the red cells and haemoglobin reached their maximum level?

3 What was the maximum red blood cell count?

4 The descent to sea-level in about three weeks restored the red blood cell count to near the original sea-level values. What was the change in red blood cell count from maximum to sea-level value?

5 Could this decrease in number, shown in question 4 above, be accounted for by normal red blood cell destruction? Give reasons for your answer.

6 Suggest probable blood cell counts for people in the Andes and Himalayas who live and work in permanent settlements at 15 000 ft (4.6 km).

Inadequate or improper functioning of the bone marrow can result in insufficient production of red blood cells. The patient suffers from fatigue, shortness of breath with exercise, headache and pallor. This condition is known as *anaemia* and it can be of a number of different types. Deficiencies in production of erythrocytes can be due to lack of vitamin B_{12}, iron

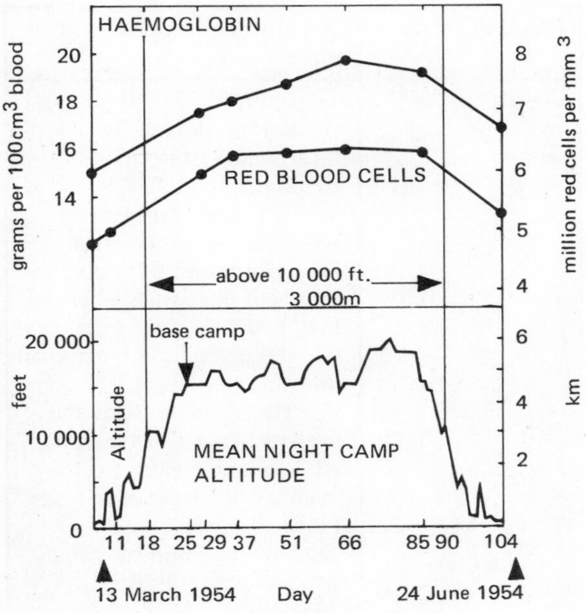

fig 3.6 Mean red blood cell count and blood haemoglobin level and mean night camp altitude of ten climbers during an attempt in 1954 to reach the summit of Mount Makalu in Nepal

or in some cases certain vital proteins. *Iron deficient anaemia* can be caused by blood loss, for example through excessive menstrual bleeding or during child delivery in women. *Haemolytic anaemias* are due to the destruction of fully formed erythrocytes. *Pernicious anaemia* is characterised by the presence of larger than normal erythrocytes and is caused by the absence of vitamin B_{12}. Here again the symptoms are fatigue, weakness and breathlessness, also a waxy complexion, small appetite and a smooth tongue. The lack of vitamin B_{12} in the body can be due to diet deficiency or the inability of the stomach to absorb the vitamin.

Sickle cell anaemia is due to an abnormal form of haemoglobin which causes distortion of the erythrocytes. The haemoglobin-S is unable to take up sufficient oxygen so chronic anaemia is present punctuated by crises brought about by periodic breakdowns of large numbers of red blood cells (see Chapter 9).

2 White blood cells (leucocytes)

You will have noticed amongst the red blood cells on your blood smear, certain blood cells which are stained blue. These are the white blood cells or leucocytes. They do not have a fixed shape and are capable of a type of amoeboid movement brought about by changing shape. The nucleus is present, often with a lobed appearance. It is the nucleus which takes up the blue stain. They are larger than erythrocytes and about 6 000 are present in one cubic millimetre of blood, giving a ratio of approximately 800 red to 1 white blood cells. The number of leucocytes will be greatly in excess of 6 000 in one cubic millimetre during times of infection. As many as 200 000 in one cubic millimetre may be produced to combat micro-organisms. Development takes place in the *bone* and the *lymphatic tissue*. There are five different types of leucocytes each with its characteristic appearance and functions, but two forms (*polymorphs* and *lymphocytes*) constitute 95% of the white cell population.

Leukemia is a name given to several different diseases of white blood cells which are all considered to be malignant. The disease is brought about by a continuous division of the daughter cells producing the leucocytes, but the resulting cells never become mature. As a result they are never able to assume their full functions in the body.

Acute leukemia occurs in children and adults and it may appear very suddenly. *Anaemia* is often a first indication accompanied by a blood white cell count that is less than normal. Abnormal bleeding may occur from the gums or rectum and minor operations such as the removal of teeth or tonsils may also produce severe haemorrhage. Blood counts are made using bone marrow samples obtained by punctures into the bone structure, and these generally confirm the disorganisation of white cell production.

Chronic leukemia has a much slower appearance over a period of years, and although the same basic process is operating, it is slower and furthermore many more cells reach abnormal maturity than in acute leukemia. Anaemia may develop and the patient is unable to cope with body infections. There may be enlargement of the spleen and liver. As a result of the increase in the number of white blood cells they are often as numerous as red blood cells hence the term leukemia which literally means 'white blood'. There is no real cure for the disease but there are ways of lessening the symptoms by the use of drugs.

3 Plasma

This is the fluid matrix of blood which is almost colourless with a slight yellow tinge. 9.7% of the plasma is water, the remaining 3% consisting of dissolved substances. Plasma is essentially a means of transport for these dissolved substances, supplying them to the tissues that it bathes when it passes out of the minute blood vessels. It then collects waste products.

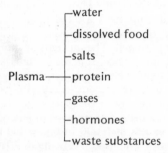

Plasma
- water
- dissolved food
- salts
- protein
- gases
- hormones
- waste substances

4 Blood platelets (thrombocytes)

These are very small particles which are discoid in shape, but they assume a star-shaped appearance in extracted blood. They are especially concerned with blood clotting and the size of the clot or *thrombus* depends upon the numbers of platelets present. Platelets have no nucleus, and they are believed to originate from detached portions of the cells lining the blood vessels.

Functions of blood

1 Red blood cells

Oxygen diffuses into the blood in the lungs and *combines with* haemoglobin to form *oxyhaemoglobin.* The blood cells are transported around the body. When tissues that require oxygen are reached, the oxyhaemoglobin releases its load and the resulting haemoglobin can now pick up small amounts of carbon dioxide to return to the lungs. The vast numbers of red blood cells provide a sufficiently large surface area to pick up all the oxygen that the body needs. The structure

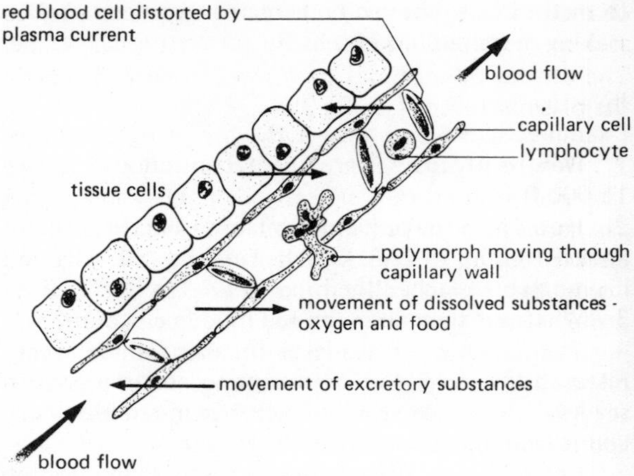

fig 3.7 The movement of blood cells through a capillary

Blood component	Functions
Red blood cells	Transport oxygen as oxyhaemoglobin. Transport carbon dioxide (very small amount).
White blood cells	Attack and engulf bacteria (polymorphs). Produce antibodies (lymphocytes).
Blood platelets	Aid clotting (together with plasma components).
Plasma	Transports: i) Carbon dioxide as bicarbonate from tissues to lungs. ii) Waste matter from tissues to excretory organs. iii) Hormones from ductless glands to the tissues where they act. iv) Digested food from small intestine to tissues. v) Heat from muscles and liver to all parts of the body. vi) Ions and water to maintain the balance of body fluids. vii) White blood cells and antibodies to sites of infection. viii) Platelets and serum proteins.

Table 3.1 Summary of functions of the blood

of the red cells enables them to increase their surface area in the blood capillaries by forming a bell shape. The central thinner part of the cell is pushed out by the plasma current (see fig 3.5).

2 White blood cells

The polymorphs can pass through the walls of the capillaries into the surrounding tissues and there *engulf* disease-causing bacteria. The bacteria are digested inside the blood cells. In response to the foreign proteins (*antigens*) of bacteria, the lymphocytes produce other proteins, called *antibodies*, which immobilise disease organisms by causing them to clump together. These *agglutinated bacteria* cannot reproduce and eventually die.

3 Plasma

This fluid carries the proteins *fibrinogen* and *prothrombin* which, together with the thrombocytes, help in blood clotting. An interlacing network of fibres is formed which entangles blood cells and forms a clot. The excretory materials such as *urea* and *uric acid* formed by the liver are carried to the kidneys by the plasma. *Hormones* are transported in minute quantities from the ductless *endocrine glands* to their sites of action. Gases such as *oxygen* and *carbon dioxide* are carried in chemical combination or in solution. This is in addition to gases carried in the red blood corpuscles.

3.31 Blood groups

In the past many attempts were made to save lives after severe injury by transfusing blood from one human to another. Some of these efforts were successful while others resulted in death. Any attempts with non-human blood, such as that of sheep, were always fatal to the recipient. It was not until 1900 that an Austrian, Karl Landsteiner, discovered the reason for these apparant inconsistencies.

He discovered that death was caused by the transfused red blood cells clumping together (*agglutinating*) and blocking the small blood vessels. His investigations showed that two separate factors were operating, one involving the red blood cells and the other involving the plasma. By mixing together red blood cells from certain people with the plasma from other people, he developed the idea that there are four types of blood based on the cell factor (antigen).

Type O – no cell factor
Type A – A type cells
Type B – B type cells
Type AB – both A and B type cells

These cells are contained in plasma, which contains anti-cell factors (antibodies) according to the following pattern:

Type O blood has anti-A and anti-B factors,
Type A blood has anti-B factors,
Type B blood has anti-A factors, and
Type AB blood has no plasma factors.

The facts can be summarised thus:

Blood group	**Antigen on the red blood cells**	**Antibody in the serum of the same individual**
O	no antigen	antibody a and b
A	antigen A	antibody b
B	antigen B	antibody a
AB	antigen AB	no antibodies

Thus, for transfusion purposes we need consider only the effect that the **recipient's plasma** will have on the **donor's red blood cells.** The plasma of the donor will be so diluted that it will not affect the red blood cells of the recipient.

The proportion of these blood groups varies in different parts of the world, but groups O and A are always the two largest groups. Group AB is the smallest

Patient's (recipient's) blood group	Antibody present in serum	Donor's blood group O	A	B	AB
O	ab	√	x	x	x
A	b	√	√	x	x
B	a	√	x	√	x
AB	o	√	√	√	√
Group AB can receive all other groups. Group AB called **universal recipient**		Group O can be given to all other groups. Group O called **universal donor**			

√ = compatible with recipient
x = incompatible with recipient i.e. agglutinated

Table 3.2 Blood group compatibilities

(about 10% of most populations), so that comparatively few people belong to this group, which can receive blood from any other person.

Other factors present in blood have been discovered since the time of Landsteiner which affect only small numbers of people. A particularly interesting example is the *rhesus factor,* so named because it was first isolated in rhesus monkeys. About 85% of the population possess this antigen and are described as *rhesus positive* (Rh+), while the remaining 15% lack the antigen and are referred to as *rhesus negative* (Rh–). No antibodies to the rhesus factor are normally present in Rh– blood; but if Rh+ blood is introduced by transfusion or by 'leakage' from foetal blood during birth, then they *are* formed. This latter fact aroused considerable interest because it provided an explanation of why certain newborn babies suffered a potentially fatal jaundice known as *haemolytic disease of the newborn.* The problem arises only when the baby's mother is Rh– and his or her father is Rh+. The children's blood will usually be Rh+ and some of it leaks into the maternal circulation. This causes the production of *rhesus antibodies* in the blood of the mother, and because of their small size, these antibodies pass into the blood of the foetus during the next pregnancy. The red blood cells of the foetus are progressively destroyed because of the antigen–antibody reaction on their surface. With successive pregnancies, the antibody level builds up in the bloodstream of the mother and, consequently, it becomes more likely that the child will be born suffering from haemolytic disease. Providing that the mother and father have been diagnosed for their rhesus factors then the child can be saved by a complete blood transfusion shortly after birth. In recent years a vaccine has been developed to prevent destruction of foetal red blood cells by the mother's antibodies. This vaccine contains a small amount of the rhesus+ factor. The mother produces a small number of antibodies, and these destroy any foetal cells which cross the placenta. This stops the production of further antibodies by the mother.

Question

1 A woman of child-bearing age requires a blood transfusion before she has had any children. Her blood is group A and Rh–. What blood group or groups can she safely be given, bearing in mind that she is married to a man who is Rh+?

3.40 Essentials of a transport system

From the time of the ancient Greeks to the seventeenth century the blood system was thought to contain the basic principle of life – a spirit, drawn in by the act of breathing. It was thought to enter the body through the wind pipe and then pass to the heart by way of the lungs, encountering other spirits during its passage from the heart to the rest of the blood system. It was not until the seventeenth century, that William Harvey (1578–1657), an Englishman, finally disproved this theory. By 1615, Harvey had developed the concept

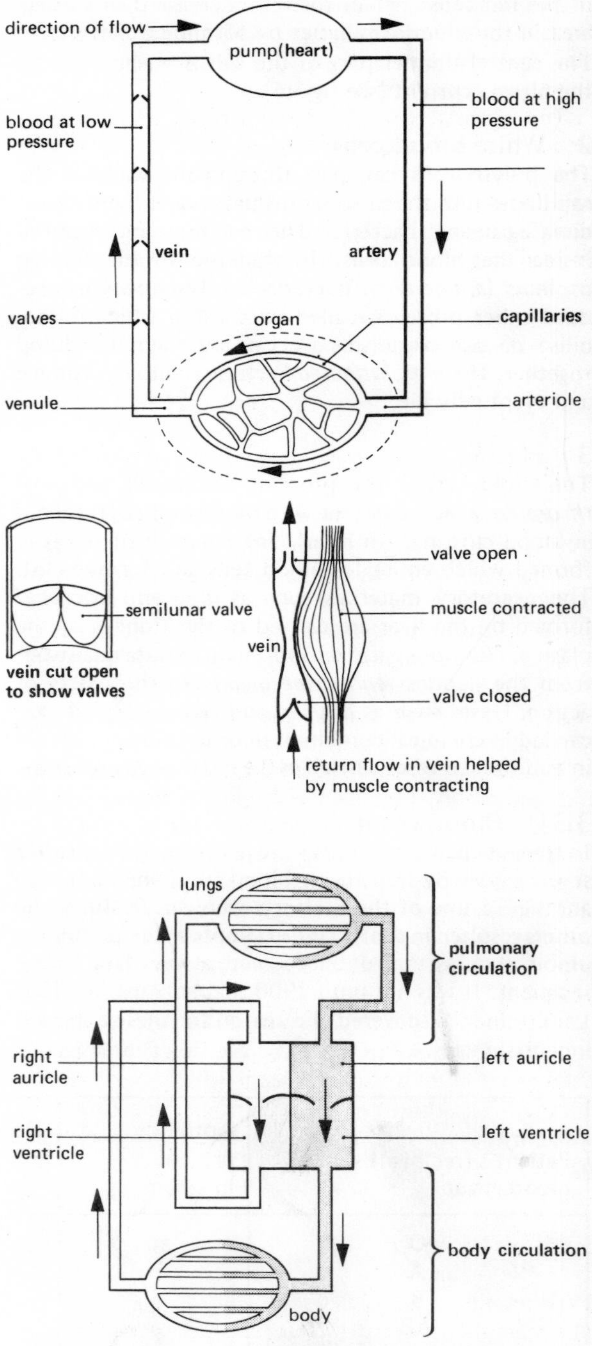

a simple plan of the double circulation of a mammal

fig 3.8 Simple diagrams of circulatory systems

of the **circulation of blood.** The story of that discovery illustrates two important aspects of scientific investigation, namely **observation** and **measurement.** From these, two deduction can be made.

The original theory of the Greeks implied that blood was being *continually formed* in the body. Harvey destroyed this idea with brilliant simplicity by pointing out that the *weight of blood required* to be manufactured by the body *was far too great.* He showed that the blood can only leave the ventricle of the heart in one direction, and then he measured the capacity of the heart. He found that the heart contained about 56 g of blood. It beats 72 times in one minute, so that **in one hour** it throws out:

$(72 \times 60) \times 56$ g
$= 241\,920$ g
or 242 kg

This is about **three times** the average weight of a man!!

Where does all this blood come from? Where does it go to? The answer can only be that it is the **same blood** returning to the heart, time and time again, to be recirculated around the body. This knowledge that blood circulates formed the foundation on which has been built a great deal of the interpretation of the physical activities of living organisms.

You will note that the application of **observation** and **deduction**, together with **simple measurement,** helped to disprove the misguided theories which had been handed down from the Greeks. Harvey confirmed these observations by many dissections of the heart and blood vessels of men and animals. He was not however, able to see the finer blood vessels (capillaries), and it was left to Malpighi to discover these forty years later.

Thus in all vertebrate animals, the transport system consists essentially of a pump (the heart) and a connecting series of vessels through which the blood is pumped. The fluid keeps circulating in one direction, aided by valves in the pump and in the vessels (see fig 3.8).

3.41 The heart

Experiment
The mammalian heart — external examination
Procedure
1 Take the heart of a sheep with the main blood vessels intact (the hearts of rabbit or rat can be used, but these are on a much smaller scale). Note the general shape and use fig 3.9 to help you identify the various features.
2 Distinguish the two auricles from the ventricles. How would you describe their external appearance?

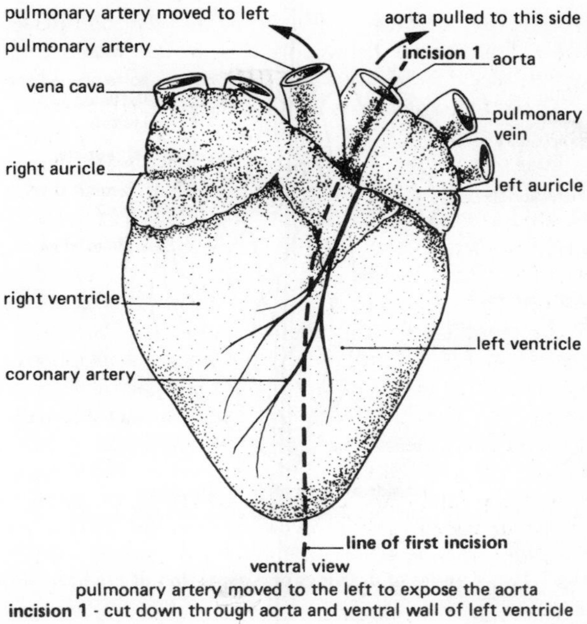

fig 3.9 The external structure of the heart shown in ventral view

3 Look for the remains of the arteries and veins, and examine the thickness of their walls in each case. The arteries have the thicker walls.
4 Look to see if there are any valves inside the vessels. Which have valves at the base, arteries or veins?
5 Look for the coronary artery on the wall of the ventricle. What do you think is the function of this vessel?

Experiment
Internal structure of the heart
Procedure
1 *Incision 1* (see fig 3.9). Cut down through the aorta with a pair of scissors, and carry on cutting through the ventral wall of the left ventricle.
2 Pull the edges of the aorta apart and also the walls of the left ventricle (see fig 3.10).
3 Look for (i) the semilunar valves at the base of the aorta, (ii) the exit of the aorta from the left ventricle, (iii) the thick muscular walls of the ventricle, (iv) the bicuspid valve, which prevents back flow of blood into the auricle on contraction of the ventricle, (v) the strings (tendons) and small muscles attached to the bicuspid valve to prevent it from being turned inside out.
4 *Incision 2* (see fig 3.10). Cut through the bicuspid valve with scissors, and continue cutting along the ventral wall of the left auricle.
5 Look for (i) the thin walls of the auricle (compare them with the walls of the ventricle), and (ii) the veins

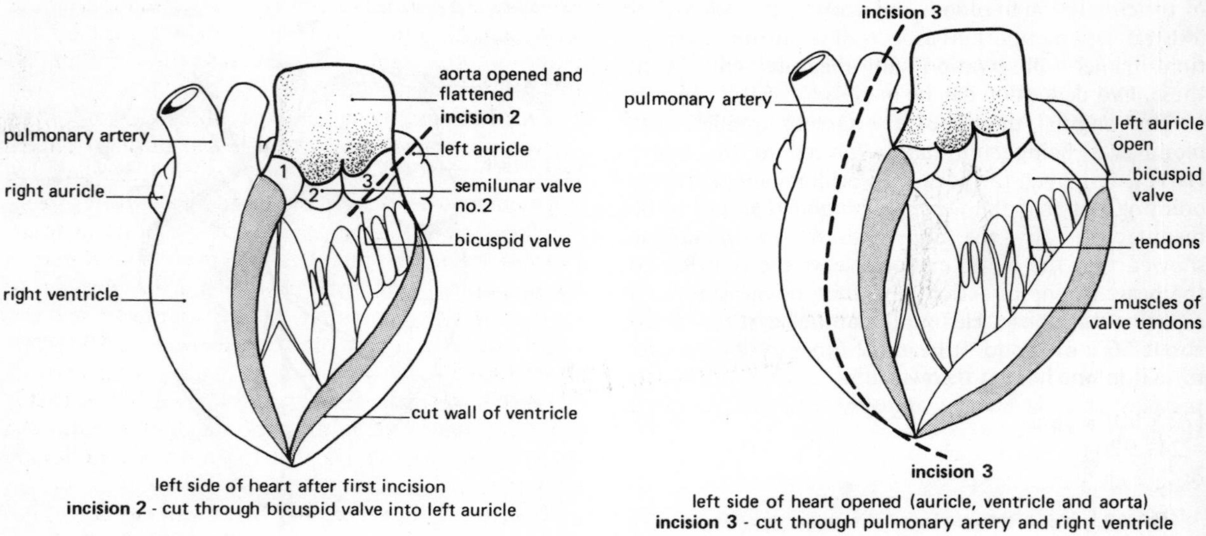

fig 3.10 A series of drawings of a dissection of the heart displaying its internal structure

which open wide into the left auricle (compare the thickness of their walls with those of the aorta).

6 *Incision 3* (see fig 3.10). Cut through the pulmonary artery and along the ventral wall of the right ventricle (*incision 3*). Cut through the tricuspid valve and along the ventral wall of the right auricle (*incision 4*).

Questions

1 Is there any opening or passage between right and left sides of the heart?

2 Which has the larger cavity, the right ventricle or the left ventricle? Which has the thicker muscular walls?

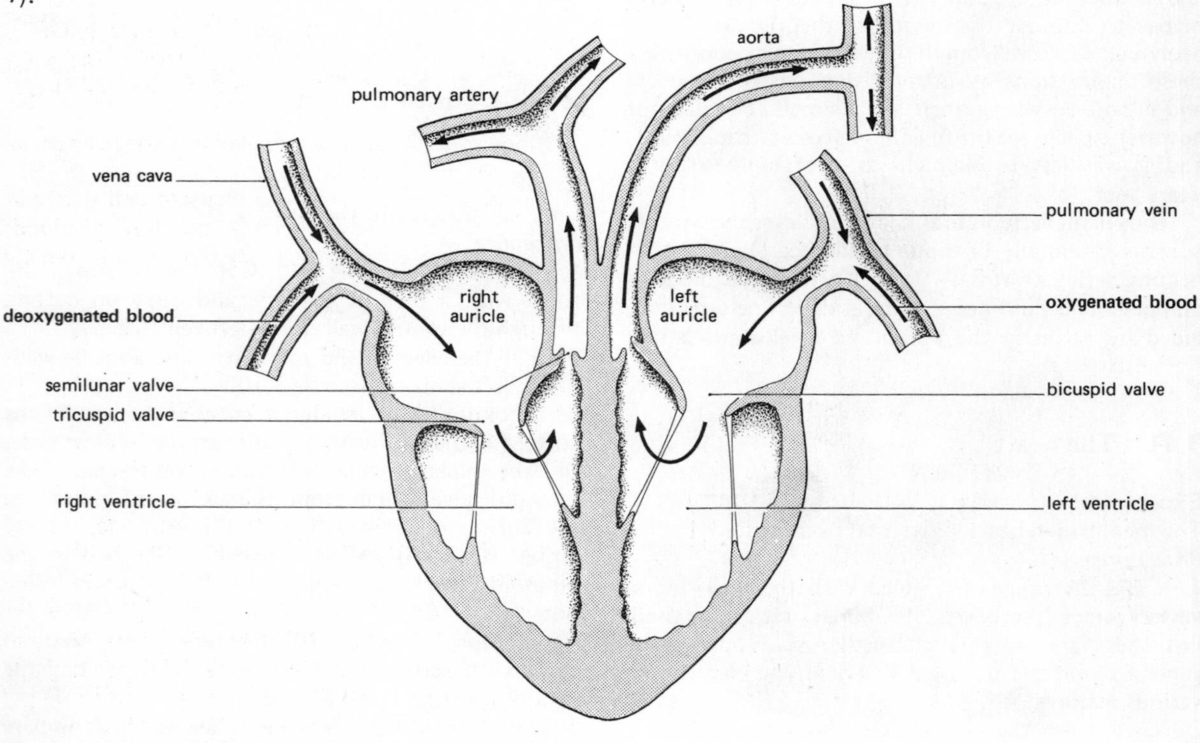

fig 3.11 The circulation of blood through the mammalian heart

3 Which has the larger cavity, the right auricle or the left?

4 Blood enters the heart through the pulmonary veins and through the venae cavae. Which of the following paths can the blood follow after entry?

A right auricle to right ventricle
B left auricle to right ventricle
C right auricle to left ventricle
D right auricle to left auricle
E left auricle to left ventricle

5 What is the function of the semilunar valves at the exit of the ventricles?

6 What is the function of the small tendons and muscles which attach the tricuspid and bicuspid valves to the ventricle walls?

Experiment

Listening to heart sounds

Procedure

1 Use a stethoscope, or place an ear close to the chest of a fellow pupil and listen to the heart sounds.

You will hear two sounds at each beat, often described as 'lub-dupp'. The first sound is a dull thud caused by the closure of the bicuspid and tricuspid valves as the ventricles contract. The second sound is more abrupt and of a higher pitch, caused by the closing of the semilunar valves and the vibrations of the walls of the arteries.

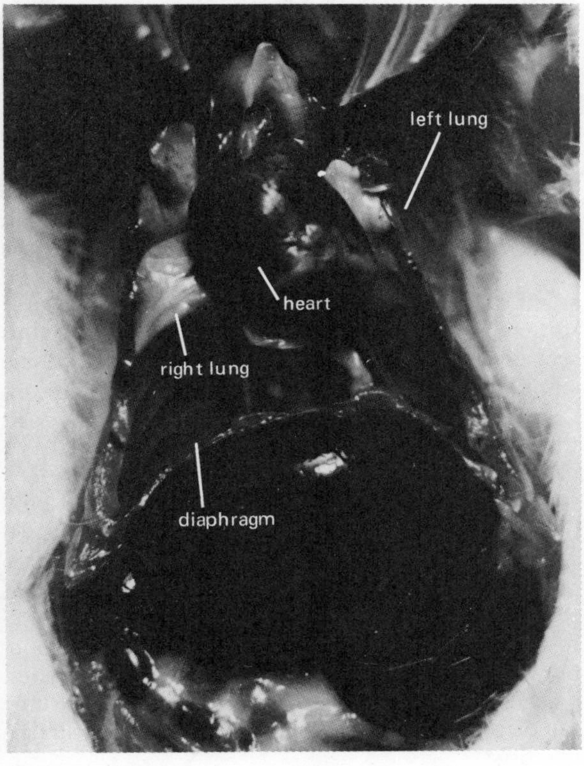

fig 3.12 A dissection of the rat thorax

The heart beats an average of 72–75 times per minute. At higher temperatures and during exercise this can increase up to 200 beats per minute. The rate is generally faster during fevers and slower during sleep. The heart of a child beats faster than that of an adult. Emotion can cause an increase in heart rate. A person reading an exciting book, watching a thrilling film or even looking at pictures of the opposite sex can experience a faster rate of heart beat.

The use of the term 'normal' as applied to rate of heart beat can mean one of two things, either (i) that it has the value found in the majority of people, i.e. the mean or average, or (ii) that the heart is functioning properly, without fault.

Confusion between these two meanings can sometimes cause alarm, especially when people find out that their pulse rate is different from what is said to be 'normal'. Normal in the case (i) above can cover a wide range. The American Heart Association *accepts as normal,* values between 50 and 110 beats per minute.

The heart is supplied with nerves from the brain to control its beating, but because of its own inherent muscular activity the heart will continue to beat after the nerve supply is cut.

The sequence of the beat is as follows:

auricles contract simultaneously (*auricular systole*);
blood in the auricles is expelled into the ventricles;
ventricles contract simultaneously (*ventricular systole*);
blood in the ventricles expelled into the pulmonary arteries and the aorta;
blood is forced against the bicuspid and tricuspid valves which close preventing back flow of blood; they are prevented from being blown inside out by the tendons;
pause (*diastole*) during which all parts of the heart relax;
cycle recommences.

The muscles of the heart must be supplied with oxygen and food materials. This function is performed by the coronary artery branching from the aorta and spreading through the heart muscle.

3.42 Heart disease, arteriosclerosis and hypertension

Diseases of the arteries and the heart are among the major causes of death in the industrialised countries of North America and Europe. Together with cancer and diseases of the respiratory system, they account for two-thirds of all deaths in England and Wales (see Table 3.3).

Ten major causes of death (% value of all causes) 1965 England and Wales

Causes of death	Percentage value of all causes
1 Arteriosclerotic heart disease (including coronary disease)	21
2 Cancer	19.5
3 Vascular disease affecting the central nervous system	14
4 Diseases of the respiratory system	12
5 Accidents (including suicide)	4.5
6 Diseases of the digestive system	2.5
7 Diseases of the uro genital system	
8 Congenital malformations	1
9 Diabetes mellitus	
10 Tuberculosis	0.5

Death from various diseases of the heart and blood vessels in 1954 and 1964, England and Wales

Causes of death	1954	1964
Rheumatic fever	299	61
Chronic rheumatic heart disease (the aftermath of an attack of rheumatic fever)	8 596	6 171
Arteriosclerotic and degenerative disease of the heart		
1 arteriosclerosis of coronary arteries	67 884	106 290
2 other causes of degenerative heart diseases	60 162	34 419
Hypertensive disease (raised blood pressure)	20 573	13 195
Disease of arteries	12 043	14 989
Generalized arteriosclerosis	10 035	10 911

Table 3.3 Causes of death

Arteriosclerosis is a general term covering severe degenerative diseases of the blood vessels. The commonest form is called atheroma or atherosclerosis and is due to the degeneration of the lining of the walls of an artery which can cause a blood clot to form. The obstruction thus caused means less oxygen gets to the tissues supplied by that artery. These tissues are thus destroyed or damaged. If the vessel supplies the heart or brain then death may follow fairly quickly. The disease is present in most adults of both industrialised and developing countries. Post-mortem examinations in Jamaica, for example, show that atherosclerosis is present extensively, but that this condition does not give rise to blood clots. In industrial communities, however, there appears to exist conditions which can increase the production of blood clots. Thus heart attacks and 'strokes' (coronary and cerebral thromboses) are much more common. It would seem therefore, that we must look for two factors, the first causing the atheroma to develop and the second resulting in blood clots in the damaged vessels. There is a similar dual problem in Chapter 4, where there is a second factor in the causation of lung cancer (see Chapter 4 section 4.20).

Atheroma shows its presence as yellow lines of fat lying under the thin layer of cells that lines the artery. A second feature is the development of groups of fatty cells surrounded by fibrous tissue and these appear as irregular, white patches on the inner wall of the vessel. As the disease develops, ulcers may form over the atheroma, causing calcium salts from the blood to be deposited and to form hard, calcified patches. Such irregularities in the artery wall mean that the blood is much more likely to clot there, since the smooth flow of blood without clotting is partly due to the smoothness of the inner wall of normal arteries.

The reasons for the formation of deposits on the artery wall is a matter of debate among doctors and scientists. The plasma of human blood contains a fatty substance called cholesterol. The amount of this substance in the blood is particular high in people who are overweight and who have atherosclerosis. Overweight people are more prone to atherosclerosis than thin people. Thus cholesterol seems to be involved, but exactly how is far from clear. The problem is that cholesterol in the blood is not only derived from the diet, since the body can make its own cholesterol. Therefore, even if the diet is entirely free of cholesterol, there can still be a generous amount of the substance in the blood-stream. What matters therefore, is not the presence of cholesterol in the blood, but the individual tendency to deposit it in the arteries.

There are other factors that must be examined if the full picture is to be obtained as to the cause of atherosclerosis.

i) There is no real evidence that changes in the diet

aimed at reducing cholesterol levels in the blood can alter the outcome of possible coronary disease. There are, however, certain drugs which can lower blood cholesterol levels. Current trials are directed towards assessing whether these can reduce significantly the incidence of thrombosis in patients with a history of heart disease.

ii) Heredity seems to be an important factor, since having one parent who died of coronary disease increases the chances of heart attacks by 100%, as compared with a family free of such disease. If both parents suffered from premature coronary disease the risks are increased by a factor of five.

iii) Heavy smokers of cigarettes are much more likely to develop general atherosclerosis than are non-smokers, pipe smokers and those who have stopped smoking.

iv) Psychological and emotional stress is often an important factor in triggering an attack of heart pain and eventual thrombosis.

v) Age is, of course, a factor, since the chance of suffering from coronary atherosclerosis increases the older we become. The disturbing factor in present statistics is that males are having coronary attacks in younger and younger age-groups. The disease is no longer confined to the elderly and middle-aged.

It has been established that continual exercise in young men which is carried on to middle age has an important effect in preventing atheroma, particularly in the coronary arteries. Therefore, in recent years, various methods of exercise have been developed which continue to keep the heart active right up to old age. One of these is 'jogging', a form of running which does not strain the heart or other muscles, yet keeps the heart strong and the arteries clear of fat deposits. It has been estimated that if only half of all adolescents, young adults and middle-aged men became non-smokers and kept physically active, the mortality in men of any one age-group could be reduced by 10%.

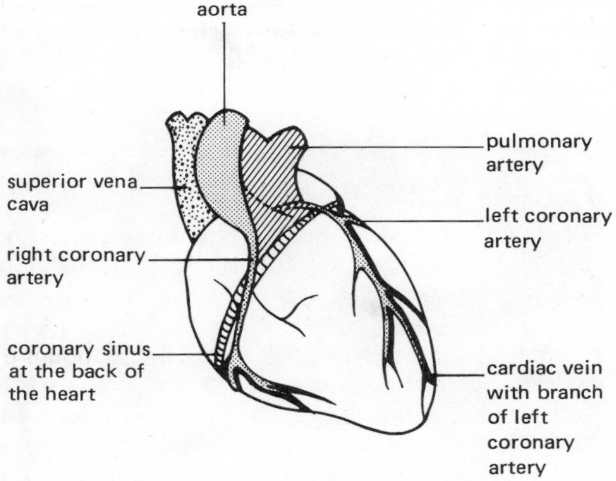

fig 3.13 Blood vessels of the heart

Hypertension is the term given to conditions of high blood pressure, a silent threat to the health of the peoples of the world. Though it presents no symptoms in the early stages and may well pass undetected, the complications are among the most important causes of death in many parts of the world. The risk of strokes, heart failure and kidney failure increases as blood pressure increases. Since this problem is world-wide the World Health Organisation chose hypertension as their theme for World Health Day on April 7th 1978.

Blood pressure is measured by the doctor taking two readings, one high and one low. The high reading is the systolic pressure, which is the maximum pressure of the heart pumping blood into the arteries. The low pressure is the diastolic pressure and is the recovery stage of the heart before the next heart beat. The readings are traditionally expressed in mm of mercury, i.e. the height of a column of mercury that can be supported by the blood pressure. It is written as 120/80, indicating the high and low pressures. The average, normal pressure in children is lower than in adults. For example, in a six-year-old child it is 90/60. In a healthy young adult it can range from 90 to 140 systolic pressure and 60 to 90 diastolic pressure. The average blood pressure can increase with certain diseases and with age. Positive readings above 140/90 are an indication of hypertension or 'high blood pressure'.

The specific cause of hypertension has not been identified, but there are known to be certain contributory factors:

1 Oral contraceptives cause a marked increase in blood pressure in some women. Sometimes merely stopping The Pill will bring down blood pressure.

2 The stress of life in large cities can be another factor. This has been identified in the black population of the U.S.A. and has been described as 'the most serious problem of black people in America today'. A second factor involved with this is that ethnic and racial groups that migrate to other countries are particularly susceptible. Black Afro-Americans in cities have blood pressures 40 to 60% above normal.

3 Obesity associated with fatty diets is also correlated with hypertension.

Hypertension was once thought to be a disease of developed countries with a high standard of living but this is not so. Hypertension is the most common disease amongst African communities and they have blood pressure patterns similar to those of the developed countries, particularly in the over-forty age-group.

GATEWAY SCHOOL
LEICESTER

Altitude seems to correlate with a lack of hypertension. In the Andes of South America the disease is seldom found, but if the people who live at high altitudes come down to sea level to live they suffer at the same rates as those who have always lived there.

The key to avoiding hypertension seems to be (i) stop smoking (ii) eat less salt (ii) keep weight down (iv) exercise regularly.

3.43 The blood vessels

The channels through which the mass flow of blood occurs are of two types; *arteries* and *veins.* Arteries carry blood away from the heart and veins carry blood towards the heart. The force of contraction of the ventricles squeezes the blood from the heart into the arteries. The blood is therefore under considerable pressure in the arteries. The mass of blood pushed into the arteries presses outwards against the artery walls, the elastic tissue stretches and then contracts inwards to force the blood forwards through the arteries. These two processes produce the characteristic pulsating of blood in the arteries. The blood cannot return to the heart, because of the presence of *semilunar valves* at the bases of the main arteries.

The blood returns from the tissues to the heart in a smooth stream at low pressure through the veins. The pressure in the arteries has been lost as the blood flows through the tissues in narrower and narrower channels. Since the pressure has been lost, how does the blood get back to the heart? The veins have semilunar or pocket valves along their length which prevent the back flow, but onward movement is largely maintained by the incidental action of muscles against the large veins. The continual movement of the limbs, the breathing movements of the thorax and the contraction of the abdominal muscles all maintain this flow towards the heart. Regular exercise is therefore essential. Gravity can also play a small part in those parts of the body above the heart.

Man's upright stance places considerable strain on the veins of the legs with the result that he may develop varicose veins. These are most common below the knees, although they may appear from the pelvis downwards. They show up on the surface of the skin as enlarged channels and bumps, usually on the inner side of the thigh and lower leg. Tight garters and bands can be a cause but they can develop as an occupational hazard where the job requires long hours of standing.

The presence of varicose veins can produce a poor circulation of blood and ulcers may develop due to poor tissue nutrition. The ulcers appear just above the ankles. Internal varicose veins may appear in the rectum and these are commonly called haemorrhoids. The slow movement of blood in varicose veins may result in a clot forming in a vein of the thigh or lower leg. This condition is characterised by pain, tenderness and swelling accompanied by a slight rise in body temperature. Massage of the spot where the inflammation is present may cause the clot to pass through the blood system to another part of the body. If it then lodges in the lungs or brain it could cause death. The proper treatment is rest in bed with some light exercise until the clot disperses.

Capillaries are the smallest of the blood vessels. They connect the end of the arterial system with the beginning of the venous system. Their non-muscular walls consist of a *single layer of flattened cells* which allows the plasma to leak out very easily. Thus water,

Arteries	Veins
1 Carry blood away from the heart.	Carry blood towards the heart.
2 Carry blood with a high oxygen content (except the pulmonary artery).	Carry blood with a low oxygen content (except) the pulmonary vein).
3 Walls are thick, muscular and elastic.	Walls are thin (little muscle).
4 Valves are absent (except at the base of larger arteries leaving the heart).	Valves are present throughout their length.
5 Blood flows rapidly under high pressure.	Blood flows slowly under low pressure.
6 Blood flows in pulses.	Blood flows smoothly.
7 Tend to lie deeper in the body.	Tend to lie near the body surface.

Table 3.4 Differences between arteries and veins

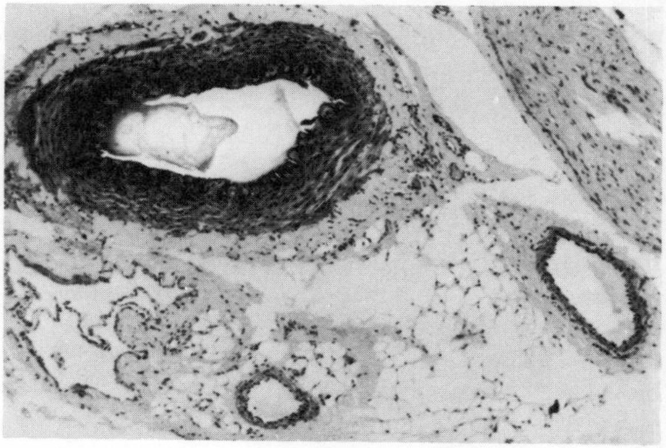

fig 3.14 A transverse section of an artery and a vein

oxygen, carbon dioxide, dissolved foods and other substances move freely between the cells of the body tissues. Some white blood cells (but not red cells) also find it possible to squeeze out between the cells of the capillary wall.

The capillary network is extremely dense throughout the body. Therefore no cell of the body is far from a supply of food materials and oxygen. The capillaries branch from the *arterioles* (small arteries) which have muscular walls, enabling them to contract and shut off blood from certain sets of capillaries. The capillaries join together to form the first *small veins or venules.* Most of the small vessels seen in the tail of the fish or tadpole (section 3.20) are capillaries. It can be seen that they are so narrow as to allow red blood cells to pass in *single file only.*

3.50 Circulation

Fish have a *single* circulation of blood similar to that shown in fig 3.8. Incorporated into the system is a set of capillaries in the gills where the blood receives oxygen and gives up carbon dioxide. The heart in the fish is a much simpler structure than in the mammal, consisting of one auricle and one ventricle. As we saw in section 3.40 it was Harvey who first demonstrated the relationship between the heart and its blood vessels in forming a continuously circulating flow of blood. Unlike the fish heart, the heart of the mammal has four chambers and the circulation is *double* instead of single.

i) *The pulmonary circulation,* in which the blood travels from the heart to the lungs and back to the heart again.

ii) *The body (systemic) circulation,* in which the blood travels from the heart to all the organs of the body (except the lungs) and back to the heart again.

A simple diagram of this circulation is shown in fig 3.15. If we consider any one blood cell commencing its circulation at the left ventricle, then before that same blood cell returns to the left ventricle it must have passed through at least two sets of capillaries. One set is in the lungs and the second set could be in any part of the body. From fig 3.15 it can be seen that the mammalian blood circulation includes a *portal system.* There is only one of these in the mammal, where the blood from the gut capillaries is carried by the *hepatic portal vein* to the liver capillaries, and finally from these back to the heart by way of the *hepatic vein.* The gut capillaries give up oxygen and take up digested food so that the blood flowing to the liver delivers this food to the liver capillaries. Thus the blood flowing through this portal system goes through *three sets of capillaries* before reaching its starting point at the left ventricle again. The blood in the hepatic portal vein is deoxygenated, and so the liver is supplied with oxygen by way of the *hepatic artery.*

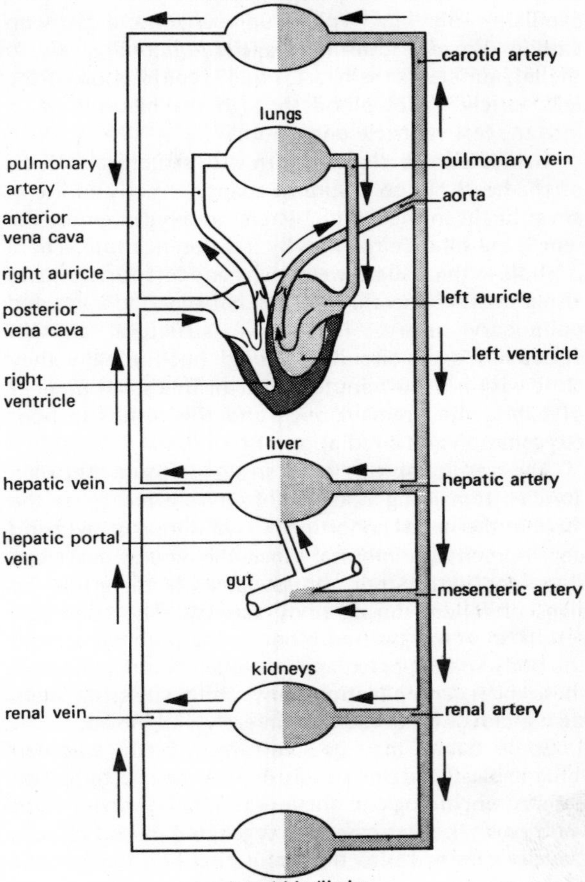

fig 3.15 A diagram of the circulatory system in a mammal

Let us now follow the course of the blood through the double circulation. The oxygenated blood is pumped from the left ventricle into the largest artery, the *aorta.* This distributes the blood to the whole body, from head to toe. Branch arteries carry blood to the gut, kidneys, liver, sex organs, limbs, skin, etc. The blood passes through capillaries in all of these organs where exchange of materials (including oxygen and carbon dioxide) takes place. From the capillaries the deoxygenated blood is collected into venules and then into veins which pour the blood back into the right auricle of the heart. The main veins connecting with this side of the heart are called *venae cavae.* The blood in the veins has been deoxygenated, and is a dull red colour easily distinguished from the bright red arterial blood.

The right auricle then contracts, forcing the deoxygenated blood down through the tricuspid valve into the right ventricle, whence it is pumped to the lungs by way of the *pulmonary artery*. In the lung capillaries the blood picks up oxygen and gives up carbon dioxide, returning by the *pulmonary vein* to the left auricle of the heart. Finally contraction of the left auricle forces blood through the bicuspid valve into the left ventricle once more.

Babies are occasionally born with structural defects of the heart due to failure to adapt correctly to life in air at birth. In the womb, where no oxygen enters the lungs, the blood circulates by a different route. There is a hole in the wall between right and left auricles, and also a small connecting channel between the aorta and pulmonary artery. These two structural changes effectively short-circuit the lungs but normally they close within a few minutes of birth. In a small number of babies they remain open and this results in poor oxygenation and circulation of blood.

The result of this hole in the septum between auricles remaining open is often referred to as the 'hole-in-the-heart' condition. In childhood growth and development are normal since the young heart can stand the extra strain, but in later life there may be signs of failure during body activity. The condition results in deoxygenated blood being pumped around the body so the person appears blue. In order to avoid these later complications any child suffering from such a defect is operated on in early childhood.

Some babies may be 'blue' from birth (so-called 'blue babies') and this condition may be due to several defects, including an aorta which arises from *both* ventricles. In this case deoxygenated blood is continually pumped into the circulation and the amount of oxygen available is considerably reduced.

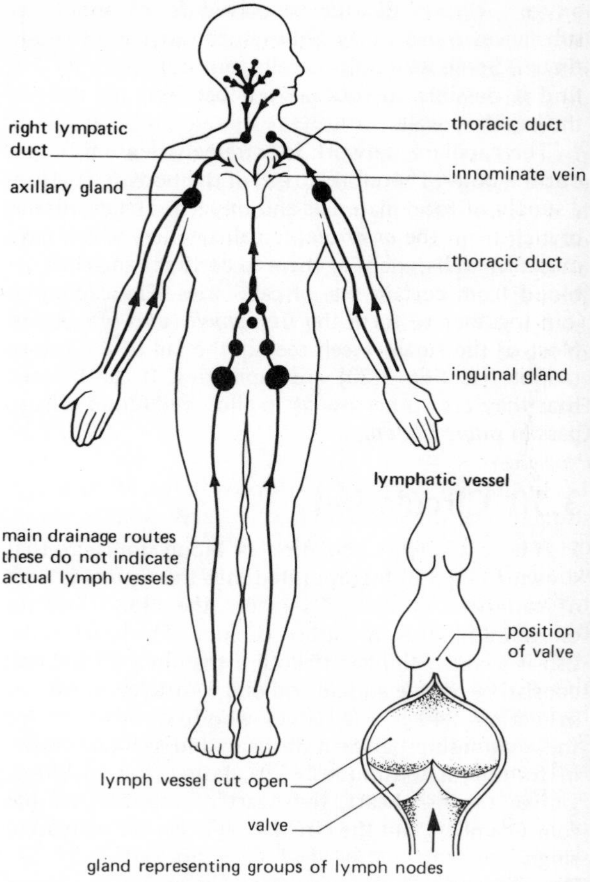

fig 3.16 The position of the main lymph glands and vessels

3.51 Lymph

The functions of the plasma can only be performed if it can be brought into contact with cells in order to exchange dissolved substances. To this end the plasma must pass through the walls of the capillaries and bathe the cells of the tissues. In this phase, plasma becomes known as *tissue fluid*. The pressure of the liquid in the blood vessels prevents the return of the tissue fluid and so it must return by another route. The tissue fluid around the cells drains into the series of vessels making up the *lymphatic system* (see fig 3.16). The tissue fluid, now called *lymph*, is caused to flow along the system by:

i) muscular movement around the lymph vessels which compresses them;
ii) valves preventing back flow of lymph whilst the vessels are compressed;
iii) greater pressure in the lymph capillaries than in the lymphatics (main lymph vessels);
iv) inspiratory breathing movements of the thorax which suck lymph into the thoracic duct, and the expiratory movements which force it into the innominate veins.

Thus the lymph is eventually discharged back into the blood circulation by way of the *innominate veins* near the heart. The lymph is now plasma and contains much of the waste materials to be dealt with by the excretory organs.

Lymph vessels have swellings called *lymph nodes* at certain points. These are glandular structures, which produce white blood cells. In humans they are perhaps most noticeable in the armpit (*axillary glands*) the angle of the jaw, and the groin (*inguinal glands*). Note that the term 'gland' here refers to a group of lymph nodes, and not to an endocrine gland. Any infected wound can cause these glands to swell, as the infection spreads along the lymph vessels. This condition known as *septicaemia.*

3.52 The pulse

The pulse wave in the arteries is dependent on three factors:

i) the *inflow* of blood from the heart,
ii) the *resistance* of the arterioles to outflow of blood into the capillaries, and
iii) the *elasticity* of the arterial walls.

At each beat of the heart (systole), a mass of blood passes down the arteries and at the same time, the elastic tissue is distended. Between the beats of the heart (diastole), the arteries contract because of their elasticity, and thus a steady stream of blood flows along the arteries. The pulse wave is non-existent in the capillaries unless the arterioles are widely dilated. In the veins there is no detectable pulse and the rate of blood flow is uniform.

Experiment
Can the pulse be detected?

Procedure

1 In certain parts of the body, large arteries are situated near the surface. Place the three middle fingers of your right hand over the radial artery of the left wrist. This can be your own wrist or that of your fellow pupil. Move your fingers until you can feel a regular pulse.
2 Count the number of beats in 30 seconds. Repeat this several times and obtain the average rate over a half minute period. The subject must be seated.
3 Calculate the pulse rate per minute for yourself and your partner.
4 Record the pulse rate for a group such as a class of pupils or a group of adults. Draw a histogram to demonstrate the variation of the pulse rate. Find the mean (see Introductory Chapter section 4.10).

Questions

1 What is the spread of your histogram?
2 What is the mean?

Experiment
Does exercise have any effect on the pulse rate (how fit are you)?

Procedure

1 Use a scoring technique which can give an indication of physical fitness. Start with 12 points and deduct points as indicated in the following tests.
2 Sitting pulse. The subject under investigation should be sitting quietly for at least three minutes. Take the sitting pulse at the wrist.
Scoring:
if pulse is over 78/min, deduct 1 mark
if pulse is over 84/min, deduct 2 marks
if pulse is over 96/min, deduct 3 marks.
3 Standing pulse. The person should stand up slowly. Count the number of beats from the fifth to the tenth second after standing, and for every five second period until the rate is constant for three consecutive periods. Convert to beats per minute. This is the standing pulse rate.
Scoring: Increase of pulse rate after standing for fifth to tenth second by:
24 or more – deduct 1 mark
36 or more – deduct 2 marks.
4 Response to exercise. The subject should stand in front of a chair, and then step up on the chair and down again five times in fifteen seconds. Then he should stand still on the floor, and count the number of pulse beats in the first five seconds after exercise (convert to beats per minute). This should be continued every five seconds. Note the time for return of the pulse rate to standing rate.
Scoring: i) if the pulse rate in the first 5 seconds after exercise is:
over 108 beats/min, deduct 1 mark
over 120 beats/min, deduct 2 marks
over 132 beats/min, deduct 3 marks
over 144 beats/min, deduct 4 marks.
ii) If time for return of pulse to constant for standing rate is:
over 30 seconds, deduct 1 mark
over 45 seconds, deduct 2 marks
over 60 seconds, deduct 3 marks.
5 List your class results on the board and indicate whether each person plays team games or takes regular exercise.

Points score	Details of team games or other exercise

Questions

1 Why does the pulse rate increase when a person changes from a sitting to a standing position?
2 Why is there a marked increase in a person's pulse rate after he has been stepping on and off a chair?
3 Is there any correlation between the test score and the standard of fitness of the person performing the test?

Questions requiring an extended essay-type answer

1 a) Explain the importance of a large surface area to volume ratio in
i) the lungs,
ii) the red blood cells.
b) What are the advantages and disadvantages of a small surface area in the body of an elephant?
2 a) Give an account of the structure and functions of blood.
b) Compare and contrast the composition of the blood supplied to the liver with that which leaves the liver.

3 a) With the aid of diagrams describe the structure of:

i) an artery, and ii) a vein.

b) Describe the route taken in the blood system by a carbon dioxide molecule from the time it enters the blood in the muscle of the leg until it reaches the lungs.

4 a) Make a large labelled diagram to show the structure of the heart and its major blood vessels.

b) Describe the sequence of events in systole and diastole and explain how a regular heart beat is maintained.

5 a) Describe the probable causes of coronary thrombosis.

b) State the type of life style that contributes towards this disease and the advice that you would give to a young adult in order to prevent heart attacks.

6 a) Describe what happens to blood plasma between leaving the blood capillaries and returning to the blood circulation.

b) What are the functions of the red and white blood corpuscles?

4 Breathing and respiration

4.00 Breathing

All living things depend upon the presence of air, and we know that any organism will die when deprived of air. What is it about air that is so important? Animals familiar to us, can be seen to move air in and out of their bodies by movements of the chest region. Is the air breathed out (*exhaled air*) different from the air breathed in (*inhaled* air or *atmospheric* air)? Any difference which can be detected between the two must be due to addition or subtraction by the person or animal breathing the air.

In order to examine the two different types of air we must first obtain samples on which experiments can be performed. Gases are easily collected by the method of displacement (see fig 4.1).

Experiment
To obtain samples of inhaled and exhaled air.
Procedure
1 Use small gas jars with ground glass edges, together with glass plates to cover the jars. Fill a large trough with water and place in it a bee-hive shelf.
2 Fill several small gas jars with water, invert them and leave them to stand in the trough.
3 Obtain a length of rubber tubing and place one end through the side opening of the bee-hive shelf. Move one of the gas jars on to the top of the bee-hive shelf above the central hole.
4 Exhale most of the air from your lungs and, with the last remaining air, breathe out through the rubber tubing. The exhaled breath should displace all the water from the gas jar.
5 Place a glass plate over the open end of the gas jar, invert the jar, remove it from the trough and stand it on the bench. Repeat this process to fill three gas jars.
6 Atmospheric (inhaled) air can be obtained by simply opening a gas jar on the bench and then covering with a glass plate.

This technique should be used for any experiment in this chapter which requires a sample of exhaled air. It is advisable to use the last air from an exhaled breath since this ensures that it comes from deep inside the lungs.

Now that we have samples of both types of air, we can begin our investigations.

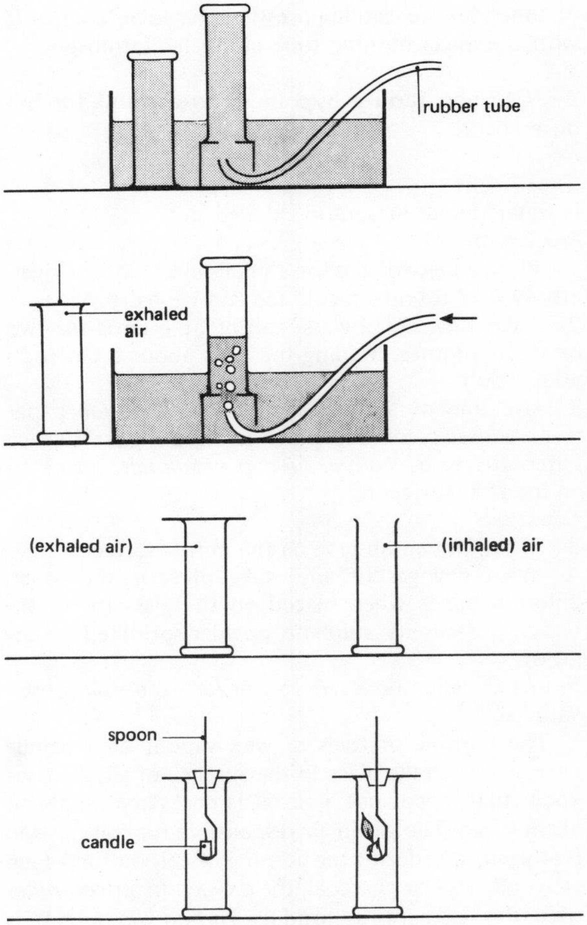

fig 4.1 The method for collecting exhaled gas over water

Experiment
Is the air breathed out different from air breathed in?
Procedure
1 Take two gas jars, one containing a sample of atmospheric air and the other containing exhaled air.
2 Fix a small candle stub on the cup of a *deflagrating spoon*. Light the candle. Obtain a stop watch or a watch with a second hand.
3 Place the burning candle into the jar of exhaled air; time its entry and then observe it carefully. When the

flame goes out record the time again (this is a matter of a few seconds, so very careful observation and timing is important).

4 Repeat this exercise, using the jar of atmospheric air.

5 Carry out both procedures at least three times. Record your results and obtain a mean value for each burning time.

Questions

1 What is the difference between the mean length of time that the candle burns in exhaled air compared with the mean burning time in inhaled (atmospheric) air?

2 Can you form a hypothesis to account for this difference?

Experiment

Is water vapour present in exhaled air?

Procedure

1 Place a piece of glass or a small mirror in a refrigerator or cool it under a cold tap and wipe dry.

2 Take the cooled glass, and breathe on it for two or three minutes, holding the glass about 3 cm from your mouth.

3 Use forceps to take some *cobalt chloride* paper from a *desiccator* and place it on the glass surface (alternatively, use *anhydrous copper sulphate* sprinkled on the glass surface).

Questions

3 What do you observe on the mirror surface?

4 What changes (if any) take place in the cobalt chloride paper when placed on the glass (or in the anhydrous copper sulphate powder sprinkled on the glass)?

5 What deductions can you make from your observations?

The burning of fuels or wax vapour on a candle wick will not take place in the absence of air, also, we know that to put out a fire it is necessary simply to exclude air. The gas in air responsible for combustion is oxygen, and, during the burning of carbon fuels such as wood, wax or charcoal, the *oxygen* in air *combines* with this element to produce *carbon dioxide*. It is possible that lack of oxygen or excess of carbon dioxide could be responsible for the candle flame going out in the air samples. As carbon dioxide is easily detectable, let us examine this component of the air samples.

Experiment

Is there any detectable difference between the carbon dioxide content of inhaled and exhaled air?

Procedure

1 Fit up two boiling tubes or specimen tubes with rubber bungs and glass tubing as shown in fig 4.2.

2 Place equal volumes of *bicarbonate indicator* (or lime water) in each tube.

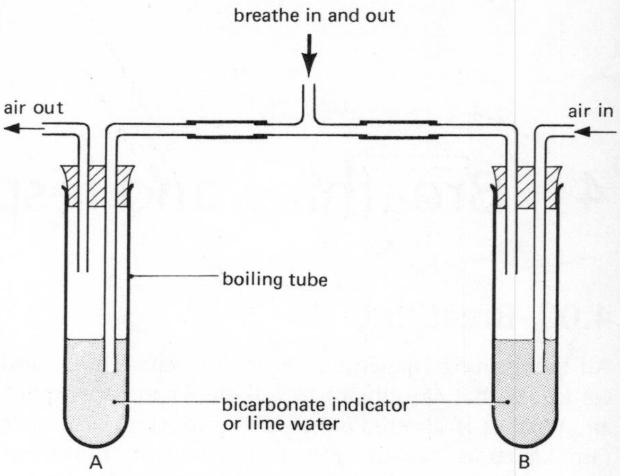

fig 4.2 The experimental apparatus for bubbling inhaled and exhaled air through bicarbonate indicator solution

3 Place the mouthpiece in the mouth and breathe gently in and out through the apparatus.

Questions

6 What happens in the boiling tubes immediately breathing begins, before any colour changes take place?

7 How do you explain these observations?

8 What colour changes take place in the indicator solutions in each tube?

9 What deductions about the inhaled and exhaled air can be made from these colour changes?

10 What function is performed by the boiling tube B?

The experiments have shown that carbon dioxide is present in *greater quantities* in *exhaled* air than in inhaled air. We can now move forward and consider in more detail the gas content of the two types of air. We know that living things need oxygen to survive, and apparently Man produces carbon dioxide when breathing out. These two gases are present in atmospheric air, together with the commonest of all gases, *nitrogen*. In order to estimate the quantities of each gas, we can analyse a number of samples of inhaled and exhaled air.

We can use the same apparatus that was used in Chapter 2 in order to analyse the gases given off by photosynthesising plants. The same reagents, *potassium hydroxide* and *potassium pyrogallate* solutions, are used to absorb carbon dioxide and oxygen respectively. The sample of exhaled air should be collected as shown in fig 4.3.

Experiment

How does the gas content of exhaled air differ from that of inhaled air?

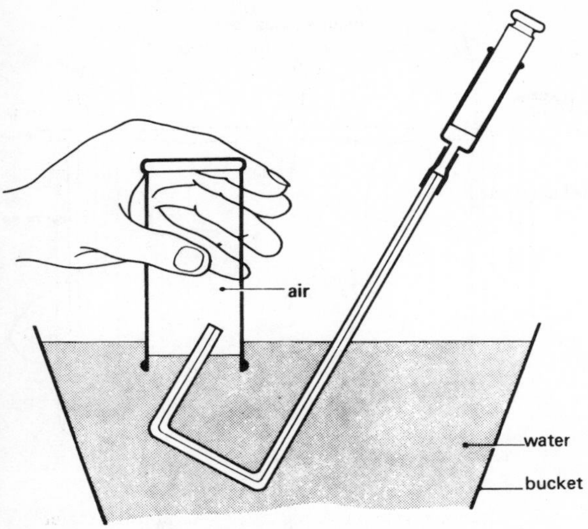

fig 4.3 The method for obtaining a sample of air for analysis

Procedure

Remember that both absorbent reagents are caustic, so take care to avoid drops on the skin, clothing, bench or floor. Be careful when you carry the tube from the reagent to your water trough. Any reagent dropped must be washed away immediately with plenty of water.

2 Collect air samples from the atmosphere and from your exhaled breath as shown in fig 4.1. Smaller samples can be collected in boiling tubes rather than gas jars.

3 Place the container with the gas sample, open and downwards in a bucket of water or a large beaker full of water. The test-tube or gas jars should be held by a clamp and retort stand, but if necessary a person could stand and hold it.

4 Take the J-tube with syringe and place the open end in the water. Draw a thread of water about 5 cm in length into the capillary tube. Then place the open end of the tube up into the gas sample and draw in about 10 cm of gas.

5 Lower the J-tube and draw in another seal of water to enclose the gas bubble. Put the apparatus into a water trough for five minutes. Remove and, without too much handling, measure the length of the bubble (A cm). Record this in your notebook.

6 Push in the plunger of the syringe until the gas bubble is near the open end of the capillary tube. Place the open end of the tube downwards into the potassium hydroxide and draw in about 5 cm of reagent. (See Chapter 2, section 2.23 for further details of precautions to be taken in this experiment.)

7 By pulling out the plunger, move the bubble and reagent round into the long arm of the J-tube, and then move the bubble backwards and forwards in order for the reagent to absorb the carbon dioxide.

8 Place the apparatus into the water trough for 5 minutes. Then remove it and quickly measure the length of the gas bubble (B cm). Record this in your notebook. If the bubble has broken into two or more parts, measure each part and add the lengths together.

9 Push out the potassium hydroxide seal into the sink, so that only a small amount remains next to the bubble. Place the open end of the tube into potassium pyrogallate solution, making sure that it is below the level of the paraffin on the surface.

10 Pull out the syringe plunger to admit about 5 cm of the potassium pyrogallate. The reagent will turn brown as it absorbs oxygen. Pull the bubble into the long arm of the J-tube, and then move it backwards and forwards to absorb the gas.

11 Place the apparatus in the water trough for five minutes. Then remove it and quickly measure the length of the bubble (C cm). Record this in your notebook. Treat a broken bubble as in step 8 above.

12 Work out your results as indicated in Chapter 2 (section 10 of experiment 2.23).

In a class situation, tabulate the class results and obtain the mean values for the percentages of carbon dioxide and oxygen respectively (see table below).

Questions

11 Why was the J-tube placed in the water trough after each part of the experiment?

12 What results were obtained after analysis of (a) inhaled air, and (b) exhaled air? What is their significance?

13 Will exercise have any effect on the results that you have obtained? Suggest how you could investigate this problem.

Name	Inhaled air		Exhaled air	
	% of carbon dioxide	% of oxygen	% of carbon dioxide	% of oxygen
Mean value				

4.01 The content of air

The experimental results in table 4.1, show how the percentage composition of air samples exhaled by humans differ from that of inhaled (atmospheric) air. These figures are constant throughout the world.

Gas	Inhaled air %	Exhaled air %
Carbon dioxide	0.03	3.50
Oxygen	20.93	16.89
Nitrogen	79.04	79.61
	100.00	100.00

Table 4.1 The gaseous content of inhaled and exhaled air

The figures refer to a man in a *resting* condition. After *exercise* the *oxygen consumption rises* slightly and so does the *output of carbon dioxide,* but in spite of this it can be seen that the lungs are not very efficient in extracting oxygen from the atmosphere, since exhaled air still has about 17% oxygen.

For this reason, the most effective modern method of artificial respiration is to blow this exhaled air into the lungs of another person. Anyone who has suffered an electric shock, been rescued from gas or smoke, or saved from drowning, may stop breathing. If such a person is to be saved it is most important that the brain is not deprived of oxygen. Therefore, breathing must be restored as quickly as possible, and often the simplest and best method is by mouth to mouth resuscitation. The rescuer blows air from his own lungs into the patient's mouth. In order that this transfer should be effective, the patient's nose must be held closed and the head pulled well back. The rescuer covers the patient's mouth with his own and blows, watching the chest rise as the air enters. He then allows the chest to fall (or presses it down) and then blows in more air. In this way many lives have been saved by the use of the oxygen remaining in exhaled air (see fig 4.4).

The densest part of the atmosphere is nearest to the Earth's surface, but as the distance away from the Earth increases, the atmosphere becomes thinner and thinner. Although the proportion of oxygen in the atmosphere remains constant at 21%, there are fewer molecules of all gases within a given volume. Thus, a normal inhalation at 6 000 metres will bring only half the amount of oxygen to the lungs brought by a normal inhalation at sea level. Mountain climbing at high altitudes needs a much faster rate of breathing to supply oxygen, and in addition, the breaths become deeper. The conquest of Mount Everest, the highest mountain in the world, has only become possible by the use of a type of oxygen cylinder.

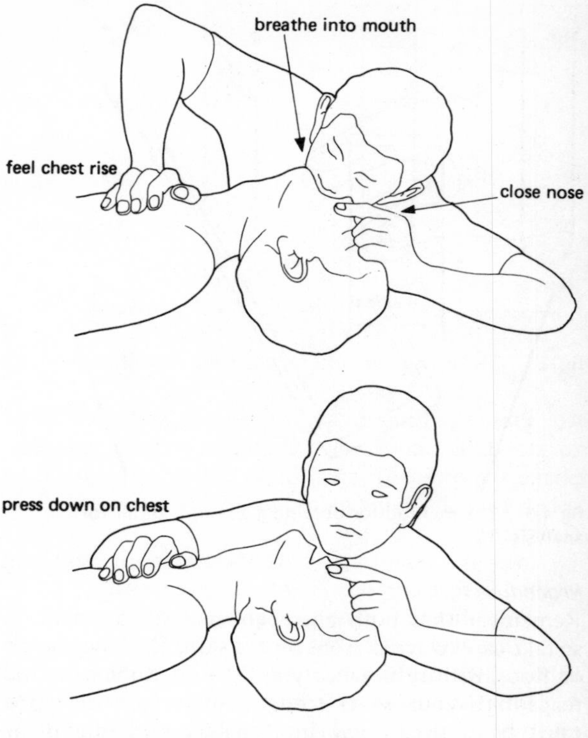

fig 4.4 Mouth to mouth resuscitation

4.02 Gaseous exchange in other living organisms

Experiment

Do all living organisms produce carbon dioxide?

Procedure

1 Take a number of test-tubes and fit them with solid rubber bungs. Each tube should be fitted with a wire or nylon gauze (or cotton wool) platform (see fig 4.5). All pieces of apparatus should be well washed with *distilled water* and then with bicarbonate indicator solution before use.

2 Place about 2 to 3 cm³ of bicarbonate indicator in each tube. Take care not to breathe over the tubes. Put in the rubber bung immediately.

3 Into each tube place some living material so that it rests on the gauze (or cotton wool). Examples of these could be:

i) caterpillars;
ii) beetles;
iii) millipedes;
iv) slugs or snails;
v) grasshoppers;
vi) cockroaches;
vii) woodlice;

a) piece of potato;
b) piece of apple or paw paw;
c) piece of rhizome;
d) piece of corm. (No green plant material should be used.)

Replace the rubber bung immediately.

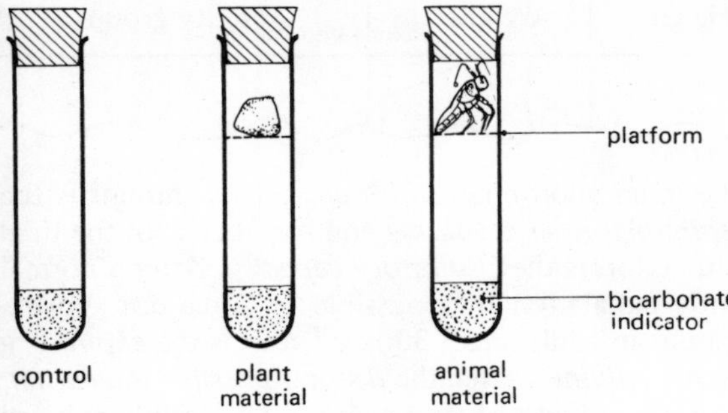

fig 4.5 The production of CO_2 by living organisms

4 Place the tubes in a test-tube rack and allow them to stand. Examine the tubes every thirty minutes. Shake the indicator each time, but do not allow it to splash over the living material.

Questions

1 Why is it essential to have the apparatus clean and washed before commencing the experiment?
2 Why should you not use green plant material?
3 What happens to the bicarbonate indicator in the tubes (a) with animal material, (b) with plant material?
4 What conclusions can you draw from the time taken to change the indicator by the two types of living organism?

Experiment
Do animals remove oxygen from the air?

Procedure

1 Fit up the apparatus as shown in fig 4.6. Insert a glass T-piece. Support the U-tube in the clamp or a retort stand, and use a pipette to put coloured water into it to a depth of about 5 cm.
2 Three-quarters fill a small ignition tube with potassium hydroxide pellets. Attach a thread to the upper end of the tube.

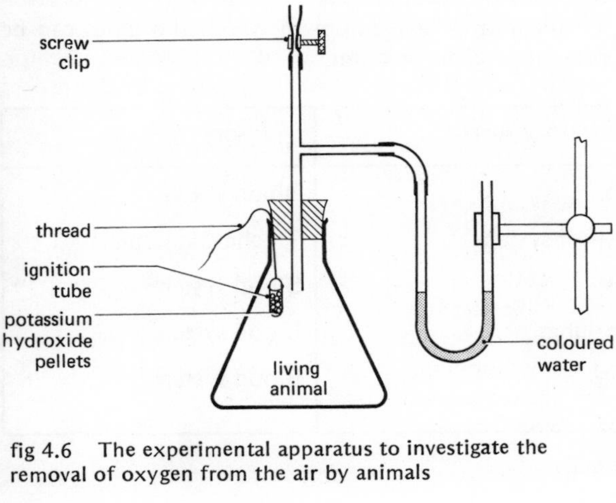

fig 4.6 The experimental apparatus to investigate the removal of oxygen from the air by animals

3 Place the living organism (e.g. mouse, insects or snails) inside the flask, and then suspend the tube of pellets in the flask by inserting the rubber bung securely over the thread.
4 Connect the T-piece with the U-tube leaving the screw clip open. Now close the screw clip. Mark the position of the liquid levels in the U-tube.
5 Set up a control experiment in exactly the same way but without the living organism. Place both sets of apparatus in such a position that they are not subject to temperature changes. Observe the liquid levels in each piece of apparatus.
N.B. when using a mouse or other small mammal ensure that the animal does not become distressed as the oxygen is used up.

Questions

5 What is the function of the potassium hydroxide pellets?
6 What do you observe in the U-tube?
7 Is this experiment a direct measurement of oxygen removal? What is the relationship between oxygen use and carbon dioxide emission?

4.03 What volumes of air are moved during breathing?

Experiment
To find the volume of air breathed out during a forced exhalation and to determine if there is any correlation with the level of activity.

Procedure

1 Obtain a large bell jar, turn it upside down and place in it 1 litre of water. Mark the level of the water on the side of the bell jar with a marking pencil. Pour in a second litre of water, and mark the second level. Continue in this way until seven litres have been marked on the jar (or 5 litres if it is a smaller bell jar).

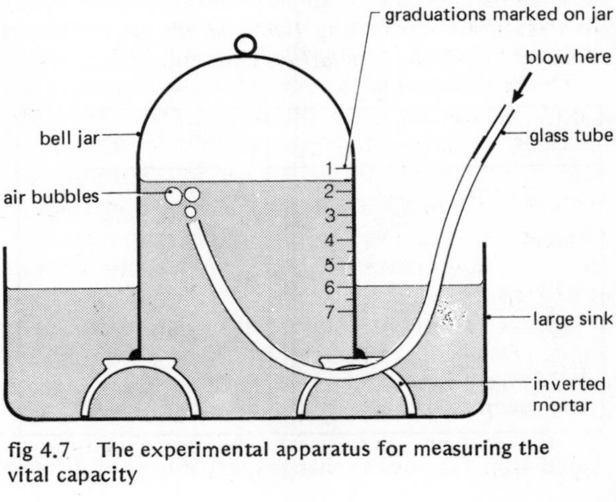

fig 4.7 The experimental apparatus for measuring the vital capacity

Name	Vital capacity	Age	Height	Weight	Activity group

2 Fill a large sink with water, fill the bell jar with water (by lying it on its side) and then stand the bell jar on three supports (metal blocks, evaporating dishes or mortars). Push a rubber tube into the bell jar, between the blocks. Insert a glass mouthpiece into the outer end of the rubber tubing (see fig 4.7).
3 Each pupil should take a deep breath and then exhale fully into the bell jar. Measure the volume of air breathed out. An estimate can be made to the nearest 100 cm^3, that is a tenth division of the distance between each litre mark.
4 Refill the bell jar, and sterilise the mouthpiece with alcohol or a disinfectant.
5 Tabulate the results, indicating volume of air breathed out in a forced exhalation (known as the *vital capacity*) together with the other information for correlation purposes.
Classify individuals as follows according to their ability group:
Very active – always in training for athletics (particularly long distance running)
Active – plays games (football, hockey, swimming, netball)
Average – does some walking and cycling
Inactive – has no interest in games, dislikes walking or cycling

Questions

1 Analyse all class results and any other data that can be obtained. Classify individuals into age, weight, and height groups and work out the average vital capacity for each group. Draw histograms of the results.
2 Do you consider that there is any correlation (relationship) between vital capacity and these factors?

The body under normal quiet conditions moves only a small volume of air into and out of the lungs. This is called the resting *tidal volume* and is about 500 cm^3. A *forced inhalation,* to its fullest extent, can take in an additional 2 000 cm^3. This amount is the *inspiratory reserve volume* and together with the tidal volume forms the *inspiratory capacity.* After a normal gentle exhalation it is possible to force out approximately an additional 1 300 cm^3: this is the *expiratory reserve volume.* After the *deepest possible inhalation,* the total amount of air which can be forcibly exhaled is the *vital capacity.* Even after this forced exhalation there is still a volume of air in the lungs which cannot be driven out. This is called the *residual volume* and amounts to about 1 500 cm^3.

These amounts of air moving into and out of the lungs can be shown as follows:

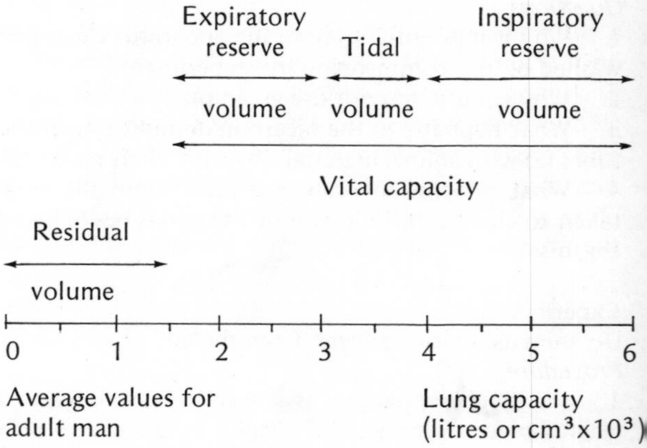

4.10 The structures of the mammal concerned in gaseous exchange

The exchange of carbon dioxide and oxygen in animals demands three conditions.
1 **Medium** – a medium in which the gases can be present, e.g. air or water.

Animal	Medium	Respiratory surface	Transport system
Earthworm	Air	Skin	Blood system
Insect	Air	Tracheal system	Tracheal system
Fish	Water	Gills	Blood system
Toad	Water/air	Skin/lungs	Blood system
Reptile, bird and mammal	Air	Lungs	Blood system

Table 4.2 Gaseous exchange mechanisms of different animals

2 **Respiratory surface** – a large, moist surface area where the gases are dissolved before diffusing into or diffusing out of the cells.

3 **Transport system** – a system to move the dissolved gases around the body to all the cells, and to collect gases which need to be eliminated.

In simple organisms the exchange of gases takes place through the skin and if the surface area/volume ratio is high, no transport system is required. Gases simply diffuse through the outer surface of the body. When the surface area/volume ratio decreases a transport system becomes necessary because some cells will be a relatively long way from the source of oxygen. Also, the outer surface is no longer large enough to absorb sufficient oxygen for the respiratory needs of the proportionately larger bodies. Thus, special structures with a very large area must be developed to absorb oxygen and to eliminate carbon dioxide. These are the gills and lungs of vertebrate animals.

The structures concerned with gaseous exchange in a mammal are two elastic organs called *lungs*, which occupy most of the space inside the *thoracic cavity*. Each lung consists of a series of finely branched tubes (*bronchioles*) terminating in tiny air sacs called *alveoli*. The bronchioles lead into a single wide tube or *bronchus*, and the bronchi from right and left lungs unite with the windpipe or *trachea*. The bronchi and the trachea are strengthened by incomplete hoops of *cartilage* which prevent them from collapsing when bent (see fig 4.8).

The trachea opens into the *larynx* (voice box) which contains the *vocal cords*. The vibration of these structures provides sound which is modified by the buccal cavity, lips and tongue. In the evolution of Man, these sounds have become organised into the complex patterns of language. The larynx opens into the *pharynx* by an opening called the *glottis*. This is closed by a flap, the *epiglottis*, as food passes over it (and down the oesophagus) during swallowing. The pharynx is connected to the nasal and buccal cavities, so that air can pass in and out of the lungs during breathing. With the mouth closed, the air passes into the nostrils, through the *nasal cavity*, and into the pharynx. Thus when the buccal cavity is full of food the mammal can still breathe. The bronchi and trachea have a lining of cells covered with minute protoplasmic projections called *cilia*. *Mucus* is also secreted by this lining so that any dust, bacteria and other foreign particles are collected and swept upwards by the cilia into the pharynx, being finally swallowed into the oesophagus.

The thin, elastic walls of the alveoli consist of a single layer of cells interspersed with elastic cells covered by a dense network of capillaries. In one human lung there are approximately 300 million alveoli which provide a *very large surface area*, about the size of a tennis court (see fig 4.8).

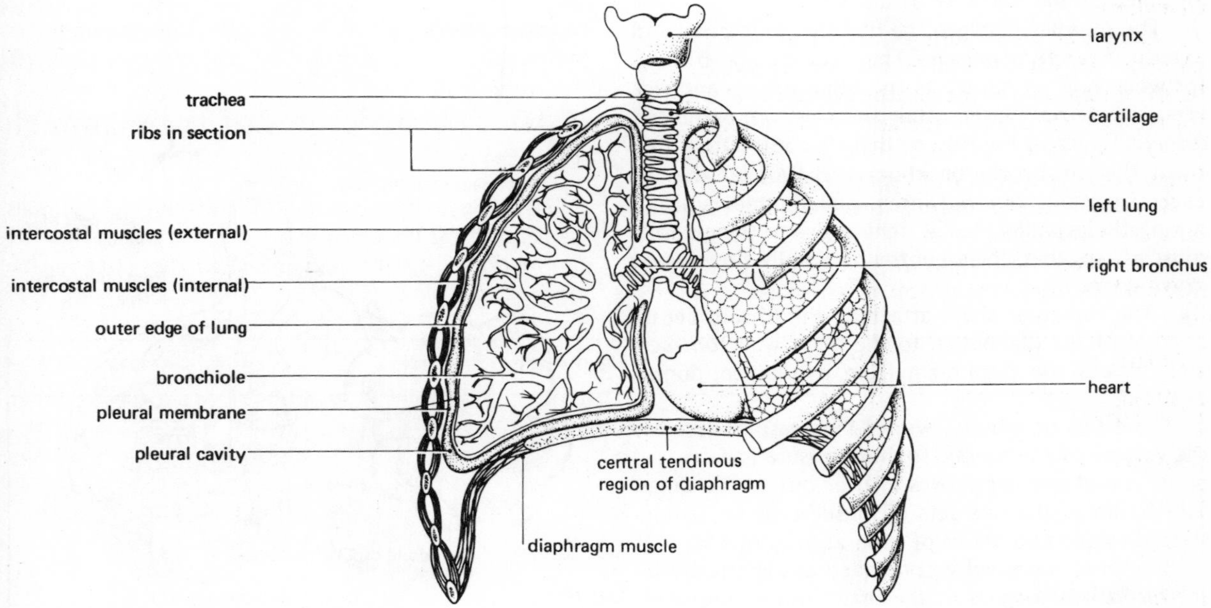

fig 4.8
(a) The lungs, heart and associated organs

Experiment

To examine the breathing organs of a mammal

Procedure

1 Obtain the trachea, bronchi and lungs of a mammal either by dissection of a rabbit or rat, or by purchasing them from an abbatoir (where animals are slaughtered).

2 Put a glass tube into the trachea and blow hard into the tube. What happens when you are blowing, and when you stop blowing? How do you account for this?

3 Observe the size and shape of the larynx and trachea. Note the rings of cartilage.

4 Observe the two bronchi and their points of insertion into the lungs.

5 Observe the thorax and the bones in the thoracic walls. Observe the *diaphragm* which separates the thoracic cavity from the abdominal cavity.

6 Observe the inner walls of the thorax and the surface of the lungs.

7 Cut open the lungs and observe the structure with a hand lens.

Questions

1 Are the surfaces of the lungs and inner walls of the thorax dull or shiny? Are they dry or moist?

2 Notice the large blood vessels. Trace their path. Do they supply the lungs?

3 Can you describe the structure of the diaphragm? Has it a uniform structure throughout?

4.11 Mechanism of breathing

Two groups of muscles are used in breathing, and these are aided by the elasticity of the lungs.

i) The muscles between the ribs are the internal and external *intercostal muscles.* These act antagonistically, for when one set contracts the other relaxes and vice versa. *External* intercostal muscles are arranged diagonally across the ribs, so that by contraction they pivot the ribs on the backbone and thus *lift* the ribcage (in Man). *Internal* intercostal muscles are also arranged diagonally, but at right angles to the external muscles, so that their contraction pulls the ribcage *downwards.*

ii) The muscular sheet attaching the fibrous centre portion of the diaphragm to the body wall *contracts* and *flattens* the diaphragm from its relaxed, domed position.

These sets of muscles work to increase or decrease the volume of the thorax, so that pressure is decreased or increased and air moves into or out of the lungs. The action of the two sets of muscles can be demonstrated by the two pieces of apparatus in fig 4.9.

Breathing is normally controlled automatically by the *medulla oblongata* of the brain, but of course we can control our own breathing when singing and talking; and even stop it for a while when we are swimming under water. The medulla is very sensitive

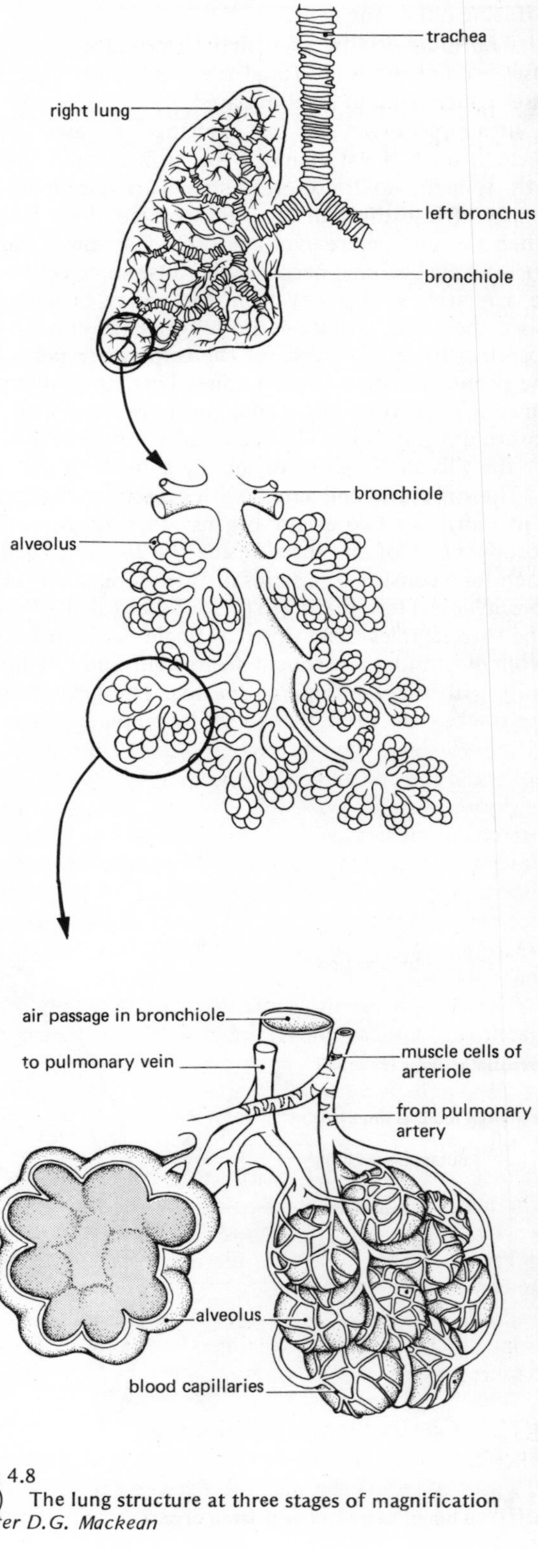

fig 4.8
(b) The lung structure at three stages of magnification
after D.G. Mackean

Inhalation	Exhalation
1 External intercostal muscles contract.	External intercostal muscles relax.
2 Internal intercostal muscles relax.	Internal intercostal muscles contract.
3 Ribs raised in Man (forwards in the rabbit).	Ribs lowered in Man (backwards in the rabbit).
4 Diaphragm contracts.	Diaphragm relaxes.
5 Diaphragm flattens.	Diaphragm arches upwards.
6 (By 3 and 5) volume of thorax increases.	(By 3 and 5) volume of thorax decreases.
7 Air pressure decreases	Air pressure increases.
8 Air moves into the lungs.	Air forced out of the lungs.

Table 4.3 Comparison of mechanism of inhalation and exhalation

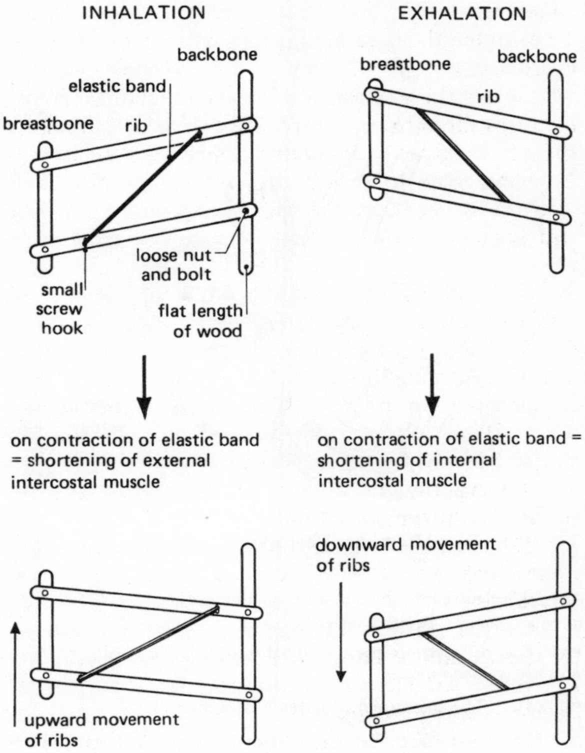

a simple parallelogram wooden model showing the action of antagonistic intercostal muscles

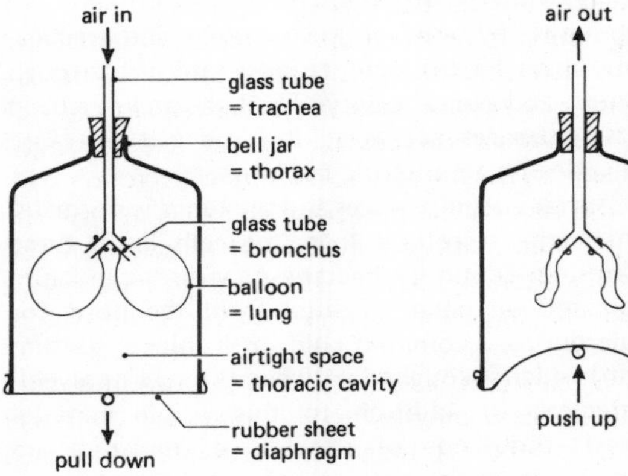

a simple model showing action of the diaphragm

fig 4.9 Models to demonstrate breathing mechanisms

to the *concentration of carbon dioxide* in the blood and when the concentration rises (e.g. in a closed room or during exercise), the brain sends messages to the intercostal muscles and the diaphragm. The breathing movements become faster and this causes the carbon dioxide to be removed more quickly. As the carbon dioxide concentration in the blood goes down the breathing rate returns to normal. This automatic control is called a *homeostatic mechanism*.

There are other movements of the thorax and the diaphragm which are variations of the breathing mechanism.

1 *Hiccough* – caused by a spasmodic contraction of the diaphragm forcing air into the lungs.
2 *Cough* – a sudden forced exhalation to clear the trachea and larynx of mucus that contains bacteria and dust.
3 *Yawn* – a long inhalation of air due to a variety of causes which may have increased carbon dioxide concentration in the blood.
4 *Sigh* – a longer inhalation and exhalation than is usual in breathing, involving the thorax and the diaphragm.

4.12 Gas exchange in the lungs

When the alveoli are filled with air there is an exchange of gases between the air and the capillaries (see fig 4.10). The blood entering the lungs through the pulmonary artery reaches the capillaries. Oxygen and carbon dioxide are present at the levels shown in the diagram. Diffusion of gases takes place between the alveolar cavity and the blood. When the blood leaves the capillary to be collected up by the venules and pulmonary veins, the oxygen concentration has in-

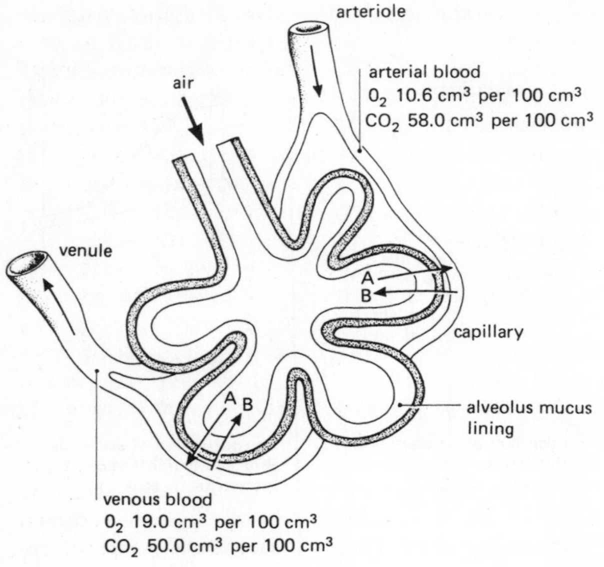

At A, O_2 diffuses into the blood through mucus, alveolar wall and capillary wall

At B, CO_2 diffuses into the alveolus through capillary wall, alveolar wall and mucus

fig 4.10 The exchange of gases in the alveoli

creased and carbon dioxide concentration has decreased.

4.13 Hygiene of the respiratory system

Breathing, which results in the flow of air into and out of the lungs, by way of the pharynx and trachea, should always be through the nose and not through the mouth and buccal cavity. The hairs and mucus in the nasal passages act as a filter and can therefore partially cleanse air entering the lungs.

The surface of air passages and the lungs is normally kept moist by secretions from the cells lining these structures. Infection by bacteria or viruses can cause an increased discharge of fluid from the nose, for example during a common cold, or a thicker sputum (phlegm) which is coughed up when bronchi or alveoli are infected. In addition to this visible material produced, thousands of droplets of moisture are sprayed out from the nose and throat during talking, laughing, singing, sneezing and coughing. The droplets are projected several metres from the mouth and can remain suspended, being moved around by air currents, finally falling and drying as dust. The droplets carry disease organisms, i.e. bacteria and viruses, and these can be inhaled by another person, or they may settle on food or eating utensils. Such diseases as influenza, the common cold, poliomyelitis, pneumonia, whooping cough, smallpox and scarlet fever are all spread in this fashion.

Classrooms, cinemas, theatres and any crowded living-quarters can all be sources of infection in this way, therefore anyone suffering from an infectious disease should stay away from these places. The proper use of a handkerchief when sneezing or coughing prevents the forceful projection of droplets. The elimination of dust containing dried micro-organisms is clearly of great importance. The best measure is plenty of air movement through good ventilation which will disperse the exhaled droplets. (See chapter 10 for details of infectious diseases.)

4.20 Air pollution and health

During a period of twenty four hours, each one of us breathes in about 15 000 litres of air. Many people are concerned that in large cities and industrial areas the air is not as clean as it should be in order to maintain the health of the population. Mainly due to Man's activities a number of 'foreign' substances find their way into the air, causing pollution of the atmosphere. There are two main sources of atmospheric pollution:

i) *Industrial sources* which include the iron and steel industry, chemical industry, refineries of petroleum spirits, oils and coal-fired generating stations.

ii) The *internal combustion engines* of cars and lorries which emit vast quantities of pollutants from their exhaust systems.

The primary pollutants include:

i) *Soot* from unburned fuel used in domestic fires, industrial furnaces and power stations. The older industrial cities of Europe have their buildings blackened by years of exposure to a soot-laden atmosphere.

ii) *Gases,* such as *sulphur dioxide* from the oxidation of sulphur compounds in fuels such as coal and oil; *carbon monoxide, nitrogen oxides* and other gases are poured into the atmosphere from industrial chimneys and car exhausts.

iii) *Hydrocarbons* produced by the combustion of petrol and engine oil.

In the United States more than 140 million tons of these pollutants are poured into the atmosphere every year. For many human beings air pollution has already proved lethal. Secondary pollutants, produced by the *action of sunlight* on primary pollutants are exceedingly damaging to plant life, and lead to the *formation of photochemical smog,* a persistant and dangerous form of fog. Under these atmospheric conditions, mortality of the very young, the old and those with respiratory ailments is greatly increased.

4.21 Smoking

Another major cause of pollution is cigarette smoke which contains a number of hydrocarbons, one of which, benzopyrene, is a cancer-producing agent. The inhalation of cigarette smoke has been proved to be a major cause of lung cancer. This is a rapidly-increasing cause of death in modern society.

Experiment

What are the products of cigarette smoking?

Procedure

1 Set up the apparatus shown in fig 4.11. Light the cigarette and turn on the water tap, adjusting the flow to obtain a steady rate of burning. This rate is increased by placing the fingertip on the glass tube as shown.
2 Place the finger on the tube for about one minute and observe the bicarbonate indicator.
3 Remove the finger and allow the apparatus to run. Observe the bicarbonate indicator.
4 Repeat sections 2 and 3 above.
5 Record the temperature when the finger is placed on the glass tube. Compare with previous readings.
6 Burn three or four cigarettes in the apparatus. Remove the rubber bung and smell the glass wool. Observe its appearance.

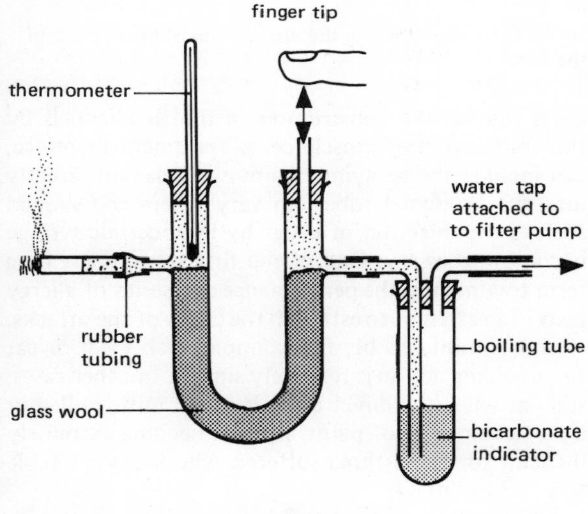

fig 4.11 A cigarette-smoking machine

Questions

1 What do you observe about the bicarbonate indicator when the finger is placed on the glass tube?
2 What do you observe when the apparatus is allowed to run without the finger on the tube?
3 Account for your observations.
4 What do you observe about the temperature of the smoke?
5 What do you notice about the smell and appearance of the glass wool? Account for your observations.

Successive reports by the Royal College of Physicians in the United Kingdom and by the Surgeon General in the U.S.A. have indicated that 'cigarette smoking is now as important a cause of death as were the great epidemic diseases such as typhoid, cholera and tuberculosis that affected previous generations'. Reports published in 1962, 1964, 1971 and most recently in January 1979, by the U.S. Health, Education and Welfare Secretary, have all shown quite clearly that smoking is a definite cause of disease. Points raised in the report include the following:

i) Cigarette smokers generally have a shorter life-span than non-smokers and the shortening of their lives is in proportion to the number of cigarettes smoked. The earlier in life smoking begins the greater is the risk, and when it is begun by young boys and girls in their early teens it will contribute greatly to the shortening of their lives.

ii) Cancer of the lung in the last fifty years has become common in men and has increased at a time when deaths from other forms of cancer have shown little change. In 1974 there were 29 500 deaths from lung cancer in men and 7 500 in women. It has been estimated that by the early 1980s some 35 000 men will die of this disease. By this time the number of female deaths will have reached 10 000 to 15 000 per year.

The January 1979 report strengthens the case against women smoking. Any feelings that women are less vulnerable than men to smoking-linked diseases are clearly dispelled by this report. It gives confident predictions that in a few years more women will be smoking than men and states bluntly that deaths from lung cancer will exceed deaths from breast cancer.

The effects of smoking on the foetus are also well-established and the latest evidence shows that there is physical damage to the blood vessels of the baby in the womb and changes to the maternal blood vessels that supply nutrients. The effects of lack of growth in the womb and immediately after birth may be irreversible. The message must be made clear to every pregnant woman that she is endangering the health, and even the life, of her unborn child if she continues to smoke.

iii) Chronic bronchitis and emphysema kill 40 000 men and women every year in the United Kingdom and in many cases these diseases are brought on by smoking large numbers of cigarettes.

iv) There is a higher incidence of diseases of the heart and blood vessels among smokers. Coronary artery disease is a common cause of death in developed countries and in 1974, 170 000 men and women died in this way in the United Kingdom. The risk of dying of this disease is two or three times greater for smokers than for non-smokers.

v) Smoking may effect health in a variety of other ways. Incidences of cancer of the mouth, pharynx, oesophagus and bladder have been linked with cigarette smoking.

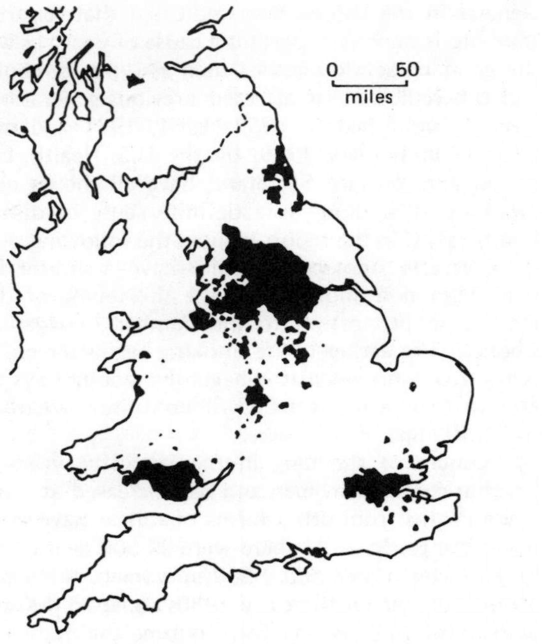

fig 4.12 A map showing the distribution of bronchitis in the U.K.

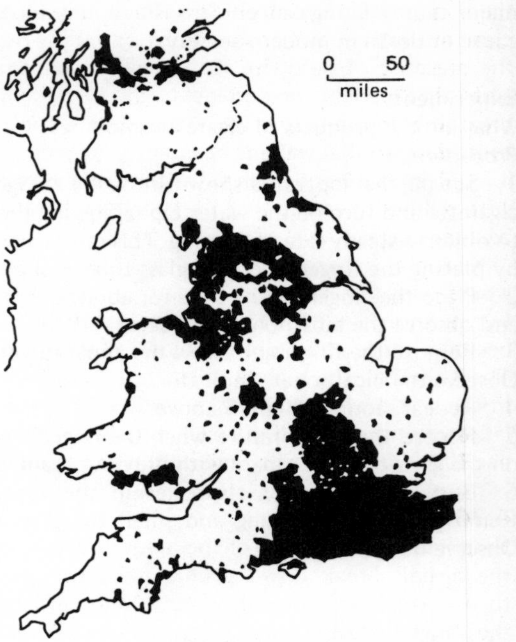

fig 4.13 A map showing the distribution of lung cancer in the U.K.

Questions

Examine the two maps shown in figs 4.12 and 4.13.

6 What similarities do they show?

7 What differences are apparent?

8 Can you suggest reasons for the distribution of bronchitis as shown in fig 4.12?

9 Can you suggest reasons for the distribution of lung cancer as shown in fig 4.13?

We must ask ourselves if the protective mechanisms in the air passages and lungs can guard us adequately against pollution. The approach to this has been to study the incidence of human illness in populations differing in their exposure to air pollutants. In Great Britain the rate of lung cancer is much higher in areas where there are large towns and a great deal of heavy industry. Since men and women in these areas tend to be heavy cigarette smokers, it is not always possible to point to one particular cause of this disease.

4.22 Asthma

More than half of the cases of this very common respiratory disorder are caused by sensitivity to pollens, animal fur or dust in the atmosphere, or to certain foods or drugs. Other cases are associated with mental stress or result from infections of the lungs and bronchi where the patient has become sensitive to the infective bacteria or viruses.

The attacks can be mild or severe but they are always accompanied by a characteristic 'wheeze' when breathing out. The difficulty in inhaling or exhaling air is due to the constriction of the air channels by the contraction of muscle cells. Treatment therefore, can include the spraying of a muscle-relaxant directly into the bronchial tubes. In very severe and sudden attacks the injection of drugs by hypodermic syringe becomes necessary. One of the first steps in any long term treatment is the performance of a series of allergy tests in an attempt to establish the cause of the attacks. If this is found to be, for example, crab meat or cat fur, avoiding action is relatively simple. In other cases, such as when the direct cause is a dust mite, pollen or a particular type of paint, life can become extremely difficult for the asthma sufferer, who is always liable to a sudden attack.

4.23 Bronchitis

Acute bronchitis is a widespread illness, especially among children and frail adults. It often develops as a complication of the common cold, when exposure to low temperature, with the resultant chilling of the body, gives bacteria the chance to break through protective tissues. The disease can also be a complication of measles, scarlet fever, whooping cough, influenza and other infections. The symptoms are headache, mild fever and a cough, with an uncomfortable feeling behind the breastbone. The disease may clear, in a few days, but it can persist for several months.

Chronic bronchitis may develop after a number of attacks of acute bronchitis. During chronic bronchitis there is production of phlegm which at first may be clear but later may become green or yellow owing to the presence of pus. The air passages become narrow and the patient experiences great difficulty in breathing. In the final stages the patient is continually breathless, so that walking becomes practically impossible. Sleep is difficult unless the patient is propped up with pillows to ensure that the bronchial tubes do not become clogged.

Chronic bronchitis kills over 30 000 men and women in the United Kingdom in any one year. Furthermore in the years before death the sufferer becomes more and more disabled and unfit for work. The U.K. has the highest death rate from bronchitis in the world with some 210 per 100 000 of the population dying of the disease, compared with only 23 per 100 000 in Germany and only 9 in France.

4.24 Emphysema

Emphysema is a disease of the lungs in which some of the alveoli break down producing larger air spaces than normal. This condition reduces the efficiency of the lungs since the total absorptive area is less. Thus the patient must take in larger quantities of air to obtain sufficient oxygen and also to release carbon dioxide. This condition is often associated with chronic bronchitis. Both of these diseases have increased greatly in recent years and it is now recognised that they occur much more commonly in smokers. The disease is often accompanied by a gradual change in shape of the chest, caused by the excessive activity of the thoracic muscles and resulting in the so-called 'barrel chest'. This increased breathing activity also causes damage to the blood vessels in the lungs and to the over-worked heart. Thus many emphysema sufferers eventually die from heart failure.

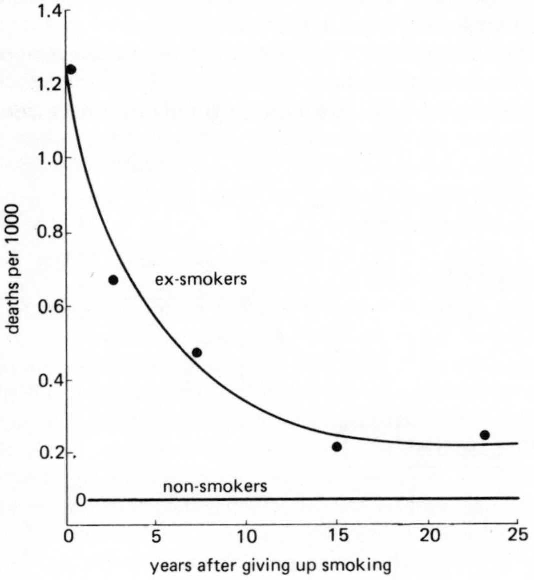

fig 4.14 The death rate from lung cancer after giving up smoking, compared with the death rate from lung cancer in non-smokers

As more and more data become available, it is apparent that a wide variety of human diseases are more common amongst urban populations than in rural communities, especially where the town dwellers are exposed to a good deal of air pollution. These diseases include *chronic bronchitis, heart diseases* and other circulatory problems and lung cancer. Also it has been found that death rates are directly proportional to the number of cigarettes smoked each day, extent of inhaling and the duration of the habit (see fig 4.14).

Hot cigarette smoke destroys the cilia lining the respiratory tract. This interferes with the filter mechanism which protects against microbes. Smokers are therefore prone to respiratory infections.

It is clear that the human respiratory system is not capable of overcoming the extremes of air pollution to which society and the individual subject it.

4.30 Obtaining energy

The work done by a living animal can be compared with that done by an internal combustion engine, such as that of a car. The same is true for plants, but their activities are not always so easily observable as those of an animal. Their energy for living processes is obtained in a similar way, but it is not used for movement and increasing body heat as in animals. Oxygen is not actively breathed in, but diffuses in through pores.

By using *radioactive tracers* in pure foods fed to laboratory animals, it can be shown that the carbon dioxide produced during breathing has come from the foods. On feeding glucose with radioactive carbon ($^{14}C_6H_{12}O_6$) to rats and collecting the carbon dioxide breathed out, it has been discovered that it is the carbon element of this gas that is radioactive ($^{14}CO_2$). The carbon from the sugar has been transformed to the carbon dioxide. Similarly the hydrogen of the sugar can be labelled, and the water present in the urine and the breath shown to contain the radioactive hydrogen.

We may ask, 'but why is the sugar broken down?'. Chemically, it can be shown as follows:

$$C_6H_{12}O_6 + 6O_2 \rightarrow 6CO_2 + 6H_2O$$

Clearly this is not the whole answer, since sugar is not broken down solely to produce waste products of carbon dioxide and water. The water could be very useful to the organism, but most animals obtain the water they require by ingestion. Pursuing our analogy with the internal combustion engine, the most

	Engine	Animal
Preparation of fuel	Crude oil purified	Food digested
Fuel utilised	Carbon compounds of C, H and O – petrol	Carbon compounds of C, H and O – carbohydrates and fats
Supply of oxygen	From air through the carburettor	From air through body surface, lungs or gills into blood stream
Combustion	Oxidised fuel in a chamber	Oxidised food in a cell
Waste products	Carbon dioxide, water and other gases	Carbon dioxide and water
Energy production	Heat and mechanical energy	Heat and mechanical energy
Transfer of energy	By a system of cranks and gears	By a system of muscles, bones and joints (in mammals)

Table 4.4 Comparison of an internal combustion engine and an animal

important result of the breakdown of sugar (fuel) is the *release of energy.*

4.31 Types of energy

In the steam engine the *chemical bond energy* of the fuel, which may be coal, oil or wood, is converted to heat energy which can turn a turbine (*mechanical energy*), rotate a dynamo (*electrical energy*), and eventually light a whole town (*light energy*).

The chemical bond energy of the sugar molecules can be converted eventually to other chemical energy or mechanical energy. These different forms of energy are used to work, in the same way that the chemical energy of oil is used to do the work of moving a car up a hill. The chemical bond energy of the sugar and fuel oils has been obtained as a result of the conversion of the *sun's light energy* by plants. In the case of sugars this conversion was carried out recently, but in the case of fuel oils, coal etc. it was carried out millions of years ago.

Experiment
To make a simple steam engine
Procedure
1 Cut a circular piece of tin from the bottom of a tin can. Punch a hole in the centre of the tin. Fix small pieces of material (e.g. cover slips or pieces of tin) on the rim, at angles of 60 degrees to each other.
2 Hold a nail horizontally in the clamp of a retort stand, and put the nail through the hole in the tin wheel. Wrap a rubber band around the end of the nail to prevent the wheel from falling off the nail.
3 Fit a flask with a glass tube and a rubber bung as shown in fig 4.15. The glass tube should be heated and drawn out into a fine point.
4 Put water in the flask and bring it to the boil over a flame. When the water boils, the steam will come out of the glass tube which should be directed onto the vanes of the wheel.
Questions
1 Where does the energy come from to increase the temperature of the water and make it boil?
2 Record the series of changes in energy from the heating of the water to the turning of the wheel.

Living organisms can use and produce all of these types of energy. The following are examples:
Heat energy – used to keep the body temperature constant in birds and mammals.
Mechanical energy – produced during the movement of animals by their limbs.
Light energy – used to produce light in phosphorescent organisms.

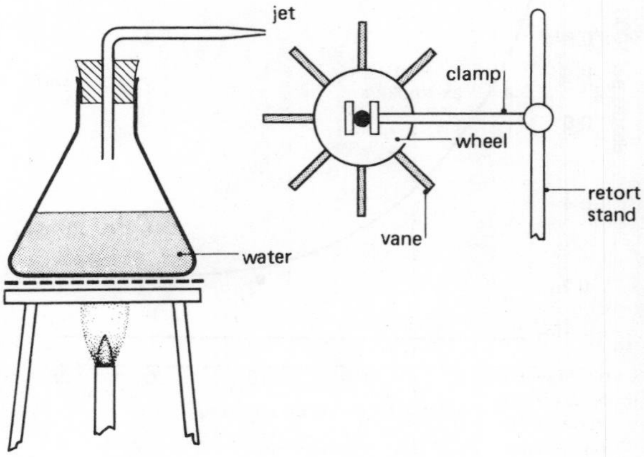

fig 4.15 A simple machine used to demonstrate the transference of energy

Electrical energy – the production of electricity in the electric catfish (*Malapterurus electricus*), and electric eels.
Sound energy – the production of sound in toads and grasshoppers.
Chemical energy – used to bring about muscle contraction.

The heat produced by the bodies of birds and mammals is necessary to maintain the body temperature at a constant level. These animals are *homoiothermic* (warm-blooded). If one places a hand on these creatures one can feel the warmth of their bodies, especially that of a bird which has a body temperature 2 to 3 degrees C above that of Man. It is not so obvious, however, that other animals or plants give off heat. Animals other than birds and mammals are *poikilothermic* (cold-blooded) so that their body temperature *fluctuates* with that of the environment, nevertheless, their bodies do produce heat energy as a result of muscle action and other physiological processes.

Experiment
Do small cold-blooded animals produce heat?
Procedure
1 Set up the apparatus as shown in fig 4.16. (Since the amount of heat produced is very small the vacuum flask is necessary to prevent a large proportion leaking away.)
2 Introduce into the vacuum flask a number of small organisms such as caterpillars or fly larvae.
3 Replace the test-tube and the screw clip; open the clip to equalise the pressure. Close the clip; the liquid levels in the manometer should then be equal. Mark the levels.

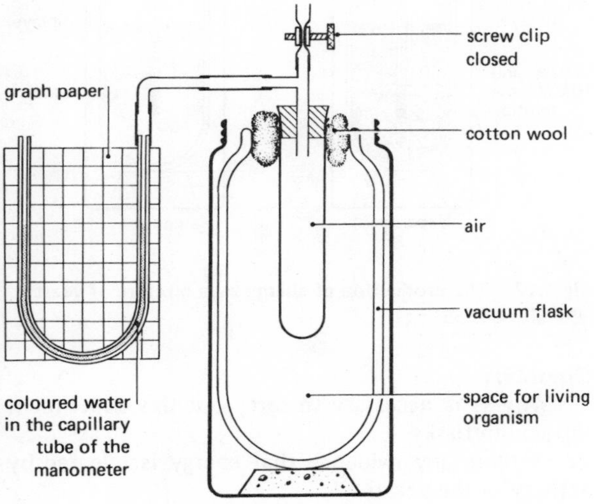

fig 4.16 The production of heat by small animals

4 Leave the apparatus and inspect it at 5 minute intervals.
Questions
3 What happens to the liquid levels in the manometer?
4 Give an explanation of your observations of the change in liquid levels in the manometer. It might help if you were to consider what happens to air when its temperature is increased.
5 What control should be set up for this experiment?

Heat is *produced* by living organisms when they are *respiring*. The faster the rate of respiration, the greater the amount of heat produced. When animals are moving rapidly or when plants are growing quickly then their bodies are producing large amounts of heat energy. Plants grow quickly in the early stages of germination, so that seeds would be particularly suitable for showing heat production. Similarly, fungal spores when germinating, produce heat.

4.32 Cellular respiration

In section 4.30 it was suggested that the important *breakdown product* of glucose is *energy*. In the presence of oxygen the sugar molecule is broken down completely to the waste products, carbon dioxide and water. This goes on slowly, controlled by many different enzymes, in the cells of living organisms. There is a step by step breakdown of the fuel molecule such that energy is released in small amounts.

Each small quantity of energy is stored in a chemical compound known as ATP (*adenosine triphosphate*). This is formed by the joining of a related molecule ADP (*adenosine diphosphate*) with *phosphate*. The binding of chemical energy into these molecules can be shown as:

ADP + phosphate + energy → ATP

Energy in the ATP molecule can be stored in the form of chemical energy and then released restoring the ADP and phosphate.

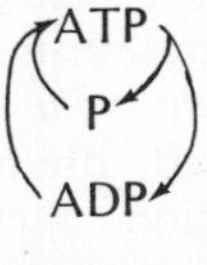

The equation for respiration written in section 4.30 can now be extended as follows:

Aerobic respiration
$$6O_2 + C_6H_{12}O_6 \rightarrow 6\ CO_2 + 6H_2O + \text{Energy}$$

$$\text{Energy} + \text{ADP} + \text{phosphate} \rightarrow \text{ATP}\ (\times\ 36)$$

For the oxidation of each glucose molecule *36 ATP molecules* are produced.

In the rapid, uncontrolled oxidation of burning, the sugar molecule releases all of its energy as heat and light. The controlled, step by step, oxidation of glucose in cellular respiration permits most of the energy to be conserved in the form of chemical energy. The high-energy phosphate bonds of ATP can release this energy for the many activities of the cell.

The 36 ATP molecules represent about *39% of the energy available* in the glucose molecule. The conversion figure from the chemical energy of the glucose molecule to the chemical energy of ATP compares very *favourably* with the internal combustion engine. This converts chemical energy in petrol into mechanical energy, and has an efficiency of only 15–30%.

The living cell breaks down complex, energy-rich molecules into simpler energy-poor molecules and this breakdown process is called *catabolism.* In contrast, in Chapter 1, it was seen that small organic molecules enter the cytoplasm and can be used as building blocks to assemble large molecules such as polysaccharides, fats and proteins. This phase of metabolism is called *anabolism.*

The energy released from the ATP can be converted to the different types of energy given in 4.31.

Cellular respiration in plants and animals results in the production of carbon dioxide and water. The carbon dioxide liberated by green plants in the presence of light is quickly used up by photosynthesis, and therefore the gaseous exhange of plant respiration can only be investigated in the dark. Photosynthesis is the faster process of the two.

4.40 Is energy released without oxygen?

As we have seen in section 4.32, energy can be obtained from complex organic molecules by breaking them into smaller molecules in the presence of oxygen. Where oxygen is not available glucose can be broken down into *ethyl alcohol* (ethanol) and carbon dioxide. This process is called *alcoholic fermentation* and since oxygen is not required, the process is described as *anaerobic.* Fermentation, therefore, is a form of *anaerobic respiration.*

Glucose $\longrightarrow$ ethyl alcohol + carbon dioxide + energy

Although the process of alcoholic fermentation has been used by Man throughout history, it was only about 100 years ago that its true nature was first understood. The French scientist Louis Pasteur found that the process was always associated with the presence of a simple fungus, yeast. He concluded that fermentation was an energy-producing mechanism used by organisms starved of air, for even yeast, given oxygen, will break down glucose completely into water and carbon dioxide.

Experiment

Is energy released by yeast and glucose solution when oxygen is absent?

Procedure

1 Boil 50 cm^3 of 10% glucose solution in order to drive out any dissolved O_2 present. Allow it to cool.

2 Pour the solution into a vacuum flask, and add about 5–10 cm^3 of fine oil (e.g. olive oil, corn oil, groundnut oil or paraffin oil).

3 Mix up a 10% yeast solution by adding dried yeast to 50 cm^3 of cooled boiled water (free of dissolved oxygen). By means of a pipette add the yeast solution, below the oil layer, to the sugar solution.

4 Place a thermometer into the flask and plug the opening with cotton wool. The thermometer should read to one-fifth of a degree between 0 and 50°C (see fig 4.17).

5 Set up a control flask with 100 cm^3 of cooled boiled water and an oil layer. Insert a thermometer into the flask, and plug with a cotton wool bung.

6 Record the temperature of each flask at the beginning of the experiment. Record the flask temperatures every hour for one day.

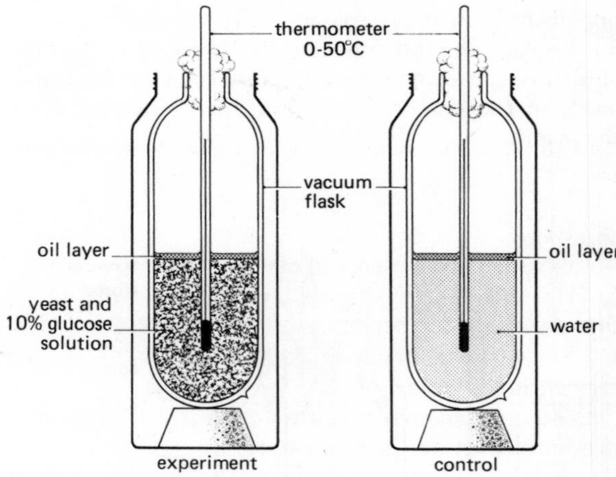

fig 4.17 **The production of energy by a mixture of yeast and glucose**

Questions

1 Why is it necessary to carry out this experiment in vacuum flasks?

2 Is there any evidence that energy is released by activity of the yeast?

3 Why is the surface of the liquid in each flask covered with a light oil?

Experiment
Does the anaerobic respiration of yeast produce carbon dioxide?

Procedure

1 Boil 20 cm^3 of glucose solution to drive off any dissolved O_2. Allow it to cool.
2 Pour the solution into a large test-tube and add a small quantity of light oil.
3 By means of a pipette add 10 cm^3 of 10% yeast suspension below the oil layer.
4 Connect the test-tube by means of a glass delivery tube to a smaller test-tube containing bicarbonate indicator. The large test-tube must be completely sealed by a rubber bung (see fig 4.18).
5 Leave the apparatus for 1 to 2 hours, and then observe the two test-tubes.

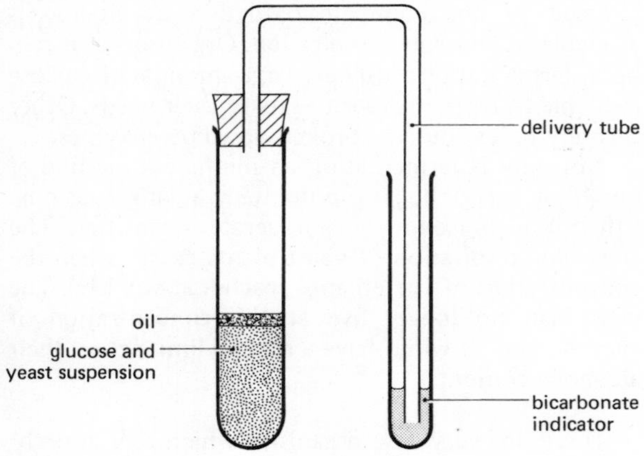

fig 4.18 The production of CO_2 by a mixture of yeast and glucose

Questions

4 What do you observe in each test-tube?
5 Has any colour change taken place in the bicarbonate indicator? What does this indicate?
6 What other product could be present in the fermenting glucose?

Yeast is used commercially for *both* of the chemical products of anaerobic respiration.

i) **Alcohol production** – The alcohol present in beers and wines is produced with the aid of yeast. This small fungus occurs naturally on leaves and fruits. Any crushed fruits containing sugars or starches, if left to stand in a warm atmosphere, will ferment. In the commercial production of alcoholic drinks, yeast is added to the cereal or fruit, and the fermentation is carefully controlled.

ii) **Bread production** – The carbon dioxide produced by yeast added to a dough mixture, (made from starch – containing flour), causes the mixture to rise. The bubbles of carbon dioxide lighten the heavy starch mixture, and when the dough is baked the yeast dies, but the gas bubbles remain as small holes in the bread produced.

Green plants normally have plenty of oxygen for respiratory processes, but germinating seeds in the soil may be flooded with stagnant water or sealed by clay particles so that oxygen is not readily available, and germination will be delayed or prevented.

4.41 Do animals release energy without using oxygen?

Another type of anaerobic respiration is known in animals. When we take vigorous exercise the activity of the muscles *outstrips* the supply of oxygen available for cellular respiration. Under these conditions, glucose is broken down in the muscles into *lactic acid* with the release of a very limited amount of energy. The glucose is derived from glycogen, which is stored in the skeletal muscles.

Glucose → lactic acid + energy

The lactic acid must be removed to avoid harming the body. The removal is slow, but steady, as the lactic acid is rebuilt into glucose and glycogen, or broken down completely to water and carbon dioxide. In the latter case the breakdown uses oxygen which becomes available after the exercise is finished. Examine fig 4.19 which shows the amounts of lactic acid in the

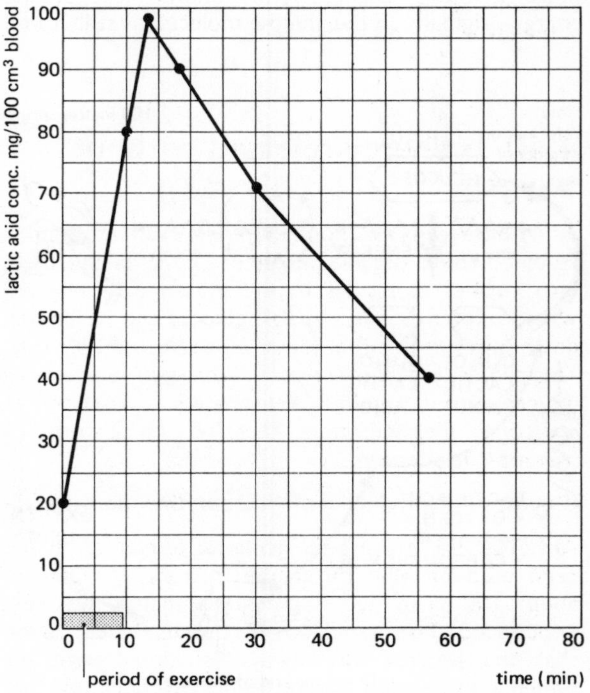

fig 4.19 A graph showing concentration of lactic acid in the blood after exercise *From Nuffield Biology Text III*

blood of a man who has carried on some vigorous exercise for a period of nine minutes.

Questions

1 By how much did the lactic acid increase during the period of exercise?

2 What was the concentration of lactic acid 57 minutes after commencing the exercise?

3 How could you find out the time it takes for the blood concentration of lactic acid to reach its original value of 20 mg/100 cm^3 of blood?

During vigorous use of muscles, as in a 100 metres sprint, it is not possible to provide the energy required for contraction by means of aerobic respiration. This is because the muscular work requires more oxygen than the lungs can take in, and more than the blood can carry, during the exercise. The body is then said to be in *oxygen debt.* In this condition we are breathing very deeply and quickly, with the lungs taking in great gulps of air. After the exercise is finished the breathing rate gradually slows, but the extra oxygen needed to oxidise the lactic acid is taken in over a long period (see fig 4.20).

4.50 Amounts of energy released in respiration

Both alcohol and lactic acid production during anaerobic respiration are *inefficient* processes. Only a small proportion of the energy stored in glucose is made available during fermentation. Most of the energy originally in the glucose molecule is still stored in the end products of ethanol or lactic acid. The high energy content of ethanol is demonstrated by the fact that it may be used as a fuel component in racing cars and rocket engines. The amounts of energy released are as follows (see Chapter 2, section 2.80):

Anaerobic respiration

$$C_6H_{12}O_6 \rightarrow \text{ethanol} + CO_2 + 0.22 \times 10^6\ J$$

Aerobic respiration

$$6O_2 + C_6H_{12}O_6 \rightarrow 6CO_2 + 6H_2O + 2.81 \times 10^6\ J$$

$$\text{Thus only } \frac{0.22 \times 10^6\ J}{2.81 \times 10^6\ J} \times 100$$

= 7% approximately of the total energy released by the complete oxidation of glucose is available in anaerobic respiration. Organisms that rely upon fermentation must have large amounts of glucose available to them in order to satisfy their needs. Other carbohydrates could be broken down by enzymes.

Not only is fermentation an inefficient method of obtaining energy but it is potentially a dangerous one. Ethanol is poisonous in moderate quantities. The anaerobic respiration of yeast plants ceases when the concentration of the ethanol reaches about 14%. The yeast can no longer live at this concentration of alcohol, and so wines have a natural limit set on their alcoholic content.

There are very few organisms which rely entirely on anaerobic respiration for their energy requirements. These organisms that can respire in no other way, are very simple in structure and the *tetanus bacterium* which lives in the soil is one such organism.

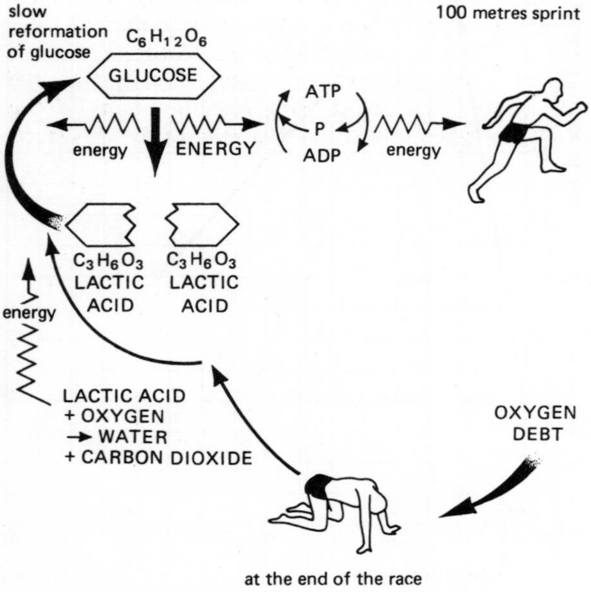

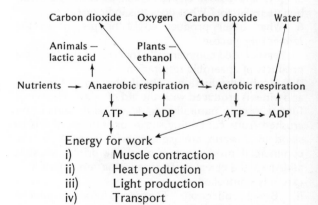

fig 4.20 A diagram of the release of energy in the muscles of a man running a 100 metres sprint

Summary of energy flow in respiration

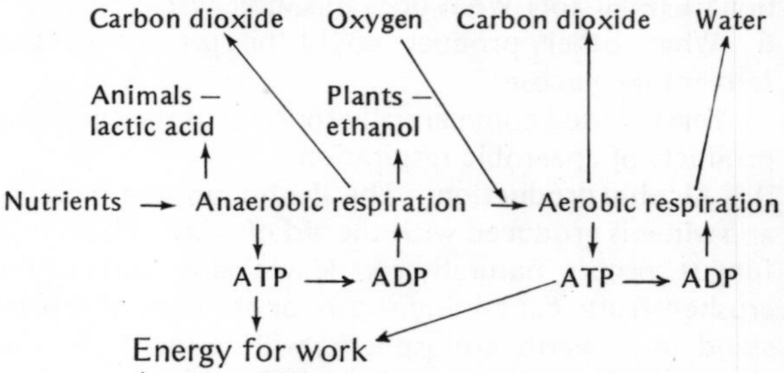

Energy for work

i) Muscle contraction
ii) Heat production
iii) Light production
iv) Transport
v) Synthesis of complex compounds – anabolism

Respiration	Photosynthesis
1 Occurs in all living cells of plants and animals.	Occurs only in plants containing the green pigment chlorophyll.
2 Goes on at all times.	Only occurs in the light.
3 Uses oxygen, but the process can occur without this gas.	Carbon dioxide is needed as a raw material.
4 Carbon dioxide is produced.	Oxygen is produced.
5 Water is produced.	More water is used up than is produced. (There is a net gain of water.)
6 Energy is produced.	Energy of sunlight is absorbed by the chlorophyll and stored in complex organic molecules.
7	Proceeds at a much faster rate than respiration in green plants, in terms of gaseous exchange.

Table 4.5 Comparison of respiration and photosynthesis

Questions requiring an extended essay-type answer

1 a) Describe an experiment by which you could prove that an animal produces carbon dioxide during respiration.
b) Why is there not a continuous increase in carbon dioxide in the atmosphere?
c) What processes in plants and animals balance the quantity of oxygen in the atmosphere?

2 a) Describe, with the aid of diagrams, the structure and position of the trachea, bronchi, lungs and diaphragm in Man.
b) How would you confirm experimentally the truth of the following statements
i) exhaled air is warmer than inhaled air, and
ii) exhaled air is moister than inhaled air?

3 a) Give an equation to show that respiration involves the oxidation of a foodstuff.
b) Describe an experiment to show that this process releases energy in the form of heat. Your experiment can use either plant or animal material.

4 a) Define anaerobic respiration.
b) Describe briefly how and where anaerobic respiration occurs in
i) plants, and
ii) animals.
c) Describe briefly an experiment to show that anaerobic respiration in a plant produces carbon dioxide and energy.
d) What is the immediate source of energy in muscles? How are energy-rich molecules rebuilt after muscular contraction?

5 a) Make a large, labelled diagram of a section of the thorax as seen from the front, to show the organs of gaseous exchange and associated structures.
b) Explain the sequence of events that occurs when air is breathed into the lungs.

6 a) Describe how infection is spread in the case of diseases such as common cold, influenza and tuberculosis.
b) What do you understand by the term *bronchitis*? How is it related to smoking?

7 a) What causes atmospheric pollution?
b) Tobacco smoke has been described as 'the greatest and most dangerous pollutant of the atmosphere'. What are the reasons for this statement?

5 Excretion and regulation — homeostasis

5.00 Regulation of body functions

The ability of complex living organisms to maintain a stable or *constant internal environment* is called *homeostasis.* The concept of the internal environment was first developed in about 1850 by a French physiologist, Claud Bernard. He realised that the ability to maintain a constant internal environment makes animals more independent of the external environment. In order to attain this internal stability, however, the animal must be able to react to any changes in the external environment which could change the internal one. Food, water and oxygen are taken into the body to repair tissues and to provide energy for growth. As a consequence of their intake, regulatory mechanisms come into action which maintain the correct balance of nutrients in the tissues. Waste materials are produced by all metabolic activities and the levels of these substances must also be controlled.

Factors in the external environment also exert influences to which adjustments are required. For example, the animal may be exposed to extremes of heat, cold and light, etc. As a result of adjustments made by the body, its cells remain surrounded by a relatively steady internal environment. The bodies of Man and other mammals only function within *narrow physiological limits.* For example, the range of 'normal' body temperature in Man is quite small (see fig 5.2). Small fluctuations outside of this range can result in death.

Let us consider three steady state mechanisms which affect the internal environment of the body.

1 Breathing mechanisms and gaseous exchange (Chapter 4).
2 Temperature control.
3 Water and ionic control – *excretion.*

Each of these body functions are kept at a constant level by a homeostatic system of the kind shown

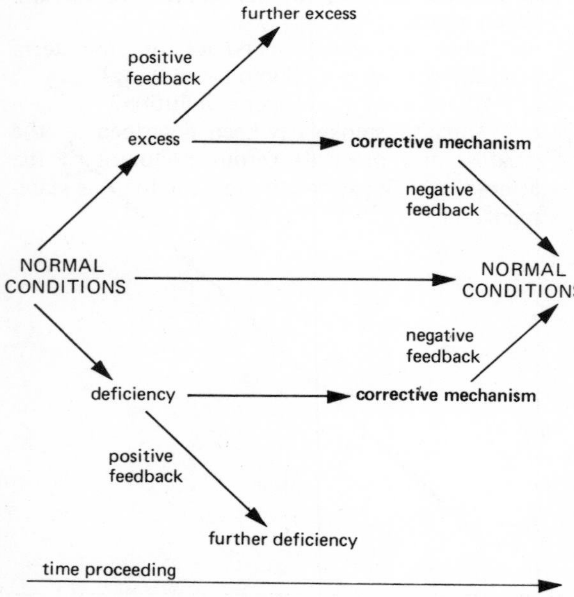

fig 5.1 Homeostasis in its simplest form

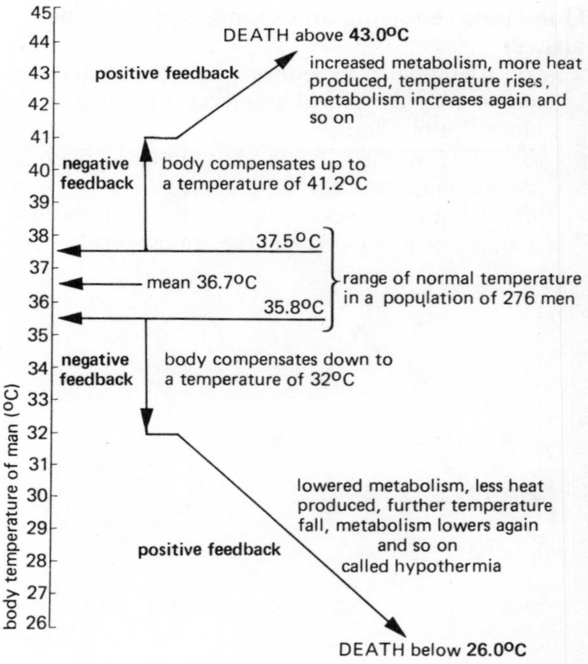

fig 5.2 Normal and abnormal ranges of temperature in Man

diagrammatically in fig 5.1. Any change from the normal operating levels by increase or decrease brings into operation a compensatory change to return the levels to normal. This change is called *negative feed-back*, since it must lessen the effect of the original change. Positive feed-back is rare in living organisms under normal conditions, but when it does occur it accentuates the change and is very difficult to reverse. In fig 5.2 the body of Man, at its danger limits of temperature, is unable to compensate for higher or lower temperatures, and death results.

5.10 Temperature control

A constant warm internal temperature is an advantage to an animal since it permits efficient functioning of the body. In warm-blooded animals with a constant body temperature a heat balance is maintained. Cold-blooded animals have their body temperature determined by the enviroment. The terms 'cold-blooded' and 'warm-blooded' are not really appropriate because a cold-blooded lizard sitting on a rock in the sun may have a higher body temperature than a warm-blooded man. When night comes, however, the lizard's temperature will fall, but the man's will remain much the same.

The best means of distinguishing between the two types of animal is to use the terms:
homoiothermic – indicating a constant body temperature as in mammals and birds, and
poikilothermic – indicating a variable body temperature as in reptiles, amphibia, fish and invertebrates.

Fig 5.3 indicates the difference between the two states. The graphs show the body temperature of students (homoiotherms) and a frog (poikilotherm) plotted against environmental temperature. Examine the graphs.

Questions

1 What happens to the body temperature of the frog as the external temperature increases?
2 What happens to the body temperature of the students as the external temperature increases?
3 From the graph, what can you deduce about the activity of the students and the frog at night time? What advantage does this give one animal over the other in respect of night activities?
4 Man is a homoiotherm, and all men belong to a single species. What relevance have these facts with regard to the distribution of Man on the Earth's surface?

Examine the graph in fig 5.4.

5 What happens to the body temperature of the insect larvae as the external temperature increases?

5.11 Poikilotherms

Fig 5.5 shows that the sand lizard takes its heat from the environment. Homoiothermic animals work most efficiently at a body temperature of 37°C. This is also true of poikilotherms. In cold conditions, where the external temperature is below 20°C, poikilotherms are very sluggish and inactive. At night and in the early morning the temperature may be too low for the lizard to be active. As soon as the sun appears the animal moves into the sunlight, so that heat flows into the

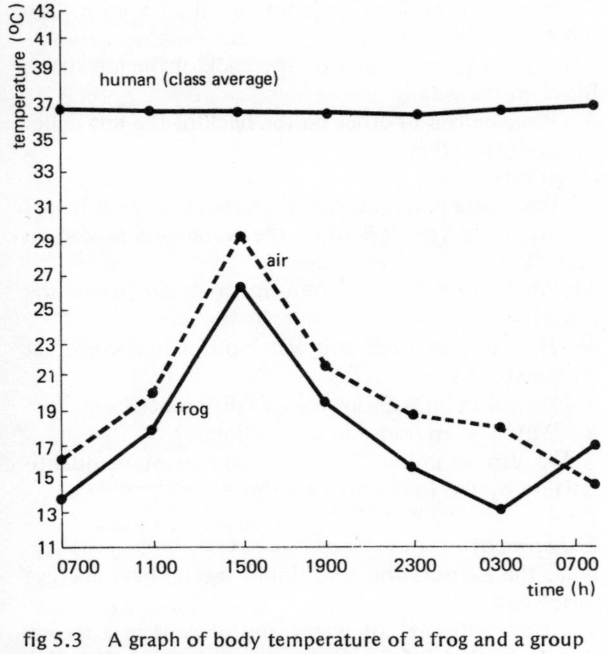

fig 5.3 A graph of body temperature of a frog and a group of students, measured over a 24-hour period

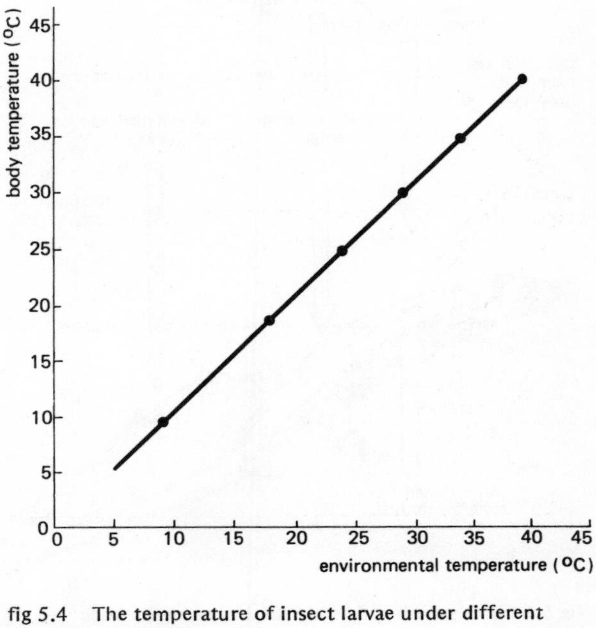

fig 5.4 The temperature of insect larvae under different temperature conditions

body directly from the sun, and also indirectly from the heated rocks or walls surrounding the animal. The body warms up and the lizard becomes very active.

It could not remain for long exposed to the sun because its temperature would rise, there would be positive feed-back (see fig 5.1) and it would eventually reach a lethal temperature. In the hottest part of the day therefore, the lizard hides in the shade, and heat flows from its body to the cooler rocks.

Fig 5.4 shows that insects also adopt the temperature of the environment. Thus in temperate climates such as Europe and North America with a well-defined winter, few insects are seen at this time. Poikilotherms under such conditions must *hibernate or die;* insects often survive in a developmental stage, the *pupa.* The adult insect hatches from this stage on the return of warmer conditions. In the tropics, butterflies flutter their wings in the morning in order to increase their body temperature sufficiently to fly at an earlier time in the day than would be otherwise possible at that environmental temperature. This could be described as a 'warming-up' exercise.

Crocodiles are often seen to lie on sand banks in the hot sun with their mouths open. This is a method of cooling the body which is also shown by some lizards. The metabolism of the butterfly and the crocodile, although poikilothermic, is not entirely conditioned to environmental temperatures. Both animals are able to maintain some control over their internal environments and thus show a link between poikilothermy and homoiothermy.

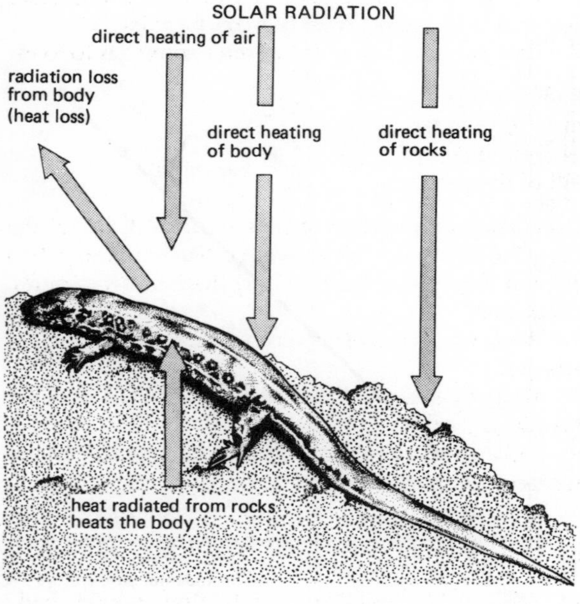

fig 5.5 The sand lizard in the sun increasing its body temperature

5.12 Homoiotherms (including Man)

Body temperature in a mammal is controlled by certain cells in the *hypothalamus* of the brain. Very slight rises in temperature result in this part of the brain discharging nerve impulses that set in action mechanisms to *cool the body.* The result is that the blood is cooled and the feed-back 'switches' off the hypothalamus. Examine fig 5.6 and you will see that as the ice is swallowed the internal body temperature falls from 37.6°C to 37°C but regains its value after 15 minutes. At the same time the skin temperature rises steadily to 37.6°C. Notice, however, that the rate of sweating follows the internal temperature, first dropping and then rising. The explanation of these events must be that the rate of sweating falls to compensate for the falling internal temperature, but the skin temperature rises at the same time because of heat absorption from the hot atmosphere (45°C). These facts indicate that the thermostat controlling sweating is not situated in the skin.

The internal temperature measurements for this experiment were taken at the ear drum. Since they parallel exactly the drop and rise in sweating rate, this indicates that the thermostat control is located near the ear drum. Although the ice is taken into the stomach the drop in temperature of the blood is quickly conveyed by the circulation to the hypothalamus of the brain.

Experiment

Does the skin have temperature detectors?

Procedure

1 Wash the back of your left hand with soap. Dry the skin.

2 Place a drop of water on the back of the left hand. Blow on the water.

3 Place a drop of ether on the back of the left hand. Blow on the ether.

Questions

1 Why were you told to wash the back of your hand?

2 What did you feel when the water was placed on your skin?

3 What effect has the movement of air across the water?

4 How do the ether and water differ in their effect on the skin?

5 How do you account for this difference?

6 Why is a fan useful in a hot climate?

We can examine this phenomenon more quantitatively by the following experiment.

Experiment

Does the evaporation of a liquid use up heat energy?

Procedure

1 Boil 500 cm^3 of water and place it in a round-bottomed flask, clamped onto a retort stand. The flask

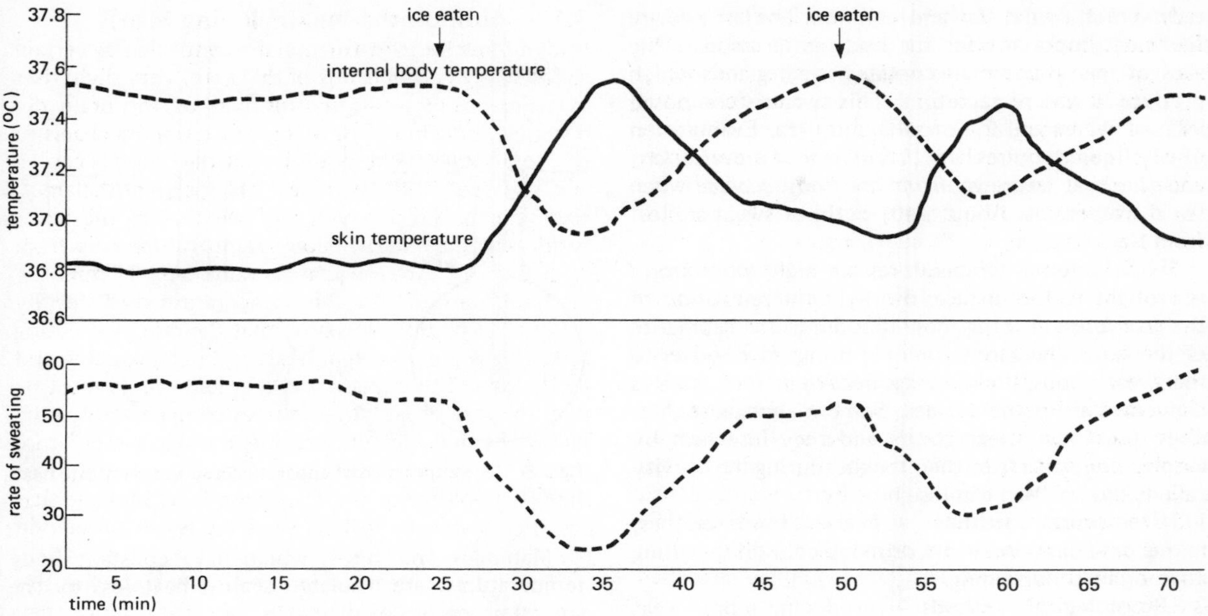

fig 5.6 Graphs of internal body temperature, skin temperature and rate of sweating of a man in a temperature controlled at 45°C

should be fitted with a rubber bung through which passes a thermometer (0°C to 110°C). The bulb of the thermometer should be in the centre of the flask. See fig 5.7.

2 Record the temperature of the water at the time the thermometer is fixed in the flask.

3 Repeat the temperature readings every minute for 10 minutes, recording the time and temperature on each occasion.

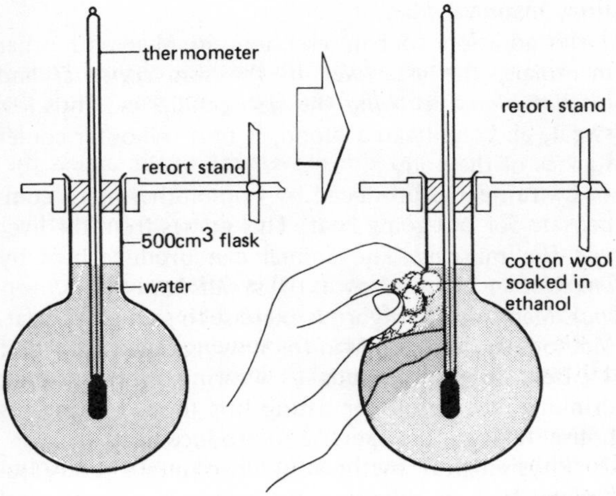

fig 5.7 The experimental apparatus used to investigate the effect of evaporating liquids on cooling water

4 Now wipe the surface of the flask with ethanol continually for 5 minutes. Record the time and temperature at one minute intervals during this operation.

5 When the wiping has stopped, continue recording the temperature every minute for another 10 minutes.

6 Using your results, plot a graph of temperature (vertical axis) against time (horizontal axis).

7 Calculate the rate of fall of temperature as follows:

Fall of temperature over 10 minutes $= A°C$

Therefore rate of fall of temperature $= \frac{A°C}{10}$ per minute

Calculate the three rates of temperature fall for (a) the first 10 minutes, (b) the next five minutes when alcohol was on the flask, and (c) the final 10 minutes.

Questions

7 What was the rate of temperature fall for (a), (b) and (c)?

8 What causes any differences between the rates for (a), (b) and (c)?

9 How could you slow down the rate of fall of temperature in the flask?

Since the control of body temperature of the homoiotherm is the result of a *balance* between heat loss and heat gain, we can see that heat loss in particular can be brought about by the evaporation of liquid from the skin. Heat is lost from the skin by the normal processes of heat transfer from any surface, namely *conduction*, *convection* and *radiation*. The last two are the most important for the body of an animal. The skin of many mammals contains sweat glands which produce a watery secretion. This sweat is composed 98% of water and 2% of salts and urea. Evaporation of any liquid requires heat (*latent heat of vaporisation*) and the heat is drawn *from the body surface* when sweat evaporates. About 140J per g of sweat are lost from the body.

When external temperatures are high, *more blood* is brought to the surface; owing to the *relaxation* of the *arterioles* allowing more blood into the capillaries of the skin. This blood not only brings more water to the sweat glands, it also brings heat to be radiated and convected from the surface. Some mammals such as dogs, have few sweat glands and they lose heat by evaporating saliva from their tongues during the activity called *panting*. Man can lose heat by:

1 **Behavioural methods** – wearing fewer clothes, taking cold baths or swims, drinking cold drinks, using a fan or air conditioning.

2 **Physiological methods** – producing more sweat which evaporates freely (latent heat loss), and the body can bring more blood to the skin (heat loss by radiation and convection).

5.13 Retaining heat in homoiotherms

We have considered the loss of heat in homoiotherms, but their major problem in conserving a heat balance is to prevent heat passing to the environment. Their surroundings are seldom hotter than body temperature, so that there will be a tendency for heat to flow from the external surfaces of the body.

Experiment

How can heat flow from a surface be slowed down?

Procedure

1 Take two round-bottomed 500 cm^3 flasks and cover one (flask B) with an insulating material (e.g. cotton wool, a piece of old blanket or fur).

2 Into each flask, fit a rubber bung and a thermometer as in the experiment in section 5.12 (see fig 5.8).

3 Pour 500 cm^3 of boiling water into each flask. Replace the bung and thermometers, and record the initial temperature of the water.

4 Record the temperature of each flask in the form of a table every minute for twenty minutes. Plot a graph of temperature against time for each flask.

5 Calculate the rate of fall of temperature for the liquid in each flask.

Questions

1 What was the rate of fall of temperature for flask A and flask B in fig 5.8?

2 What deductions can you draw from the graph and your calculations?

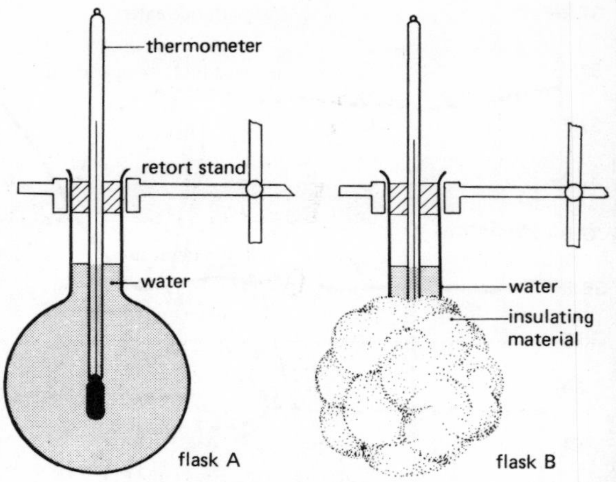

fig 5.8 The experimental apparatus used to investigate heat flow from a surface

Mammals and birds, which have constant body temperatures, are insulated against heat loss by the special properties of their skin.

i) Mammals are covered with *hair*.

ii) Birds are covered with *feathers*.

iii) Mammals and birds have an *insulating layer of fat* beneath the skin which is very thick in homoiotherms living in cold climates e.g. whales, seals, penguins.

Mammals and birds are able to increase the thickness of their covering by *erecting* the hairs and feathers respectively. Each of these structures is attached to a small muscle just below the surface of the skin. When the muscle contracts it causes the hair or feather to stand out from the skin surface. This prevents air flow over the surface and also traps a thicker layer of air that slows down evaporation and radiation. A blanket has a similar structure and function, for it is an interlacing web of hair used to prevent heat loss in a non-furry mammal, Man.

In addition to hair erection, in Man and other mammals, the *arterioles* in the skin *constrict* and *restrict blood reaching* the skin capillaries. Thus less sweat, and less heated blood, is near the outer cooler surface of the body.

Extra heat is produced by homoiotherms to compensate for out-going heat. This comes from the liver and the muscles. The animal can produce heat by *involuntary contraction* of the *skeletal* muscles (*shivering*). Man can take vigorous exercise to gain extra heat. Methods available to Man for slowing down heat loss:

1 **Behavioural methods** – wearing more clothes, drinking hot drinks or eating hot food, heating his house, taking more exercise to produce heat.

2 **Physiological methods** – involuntary body reactions such as shivering, producing less sweat, and bringing less blood to the surface, thus conserving the heat in the inner regions of the organism.

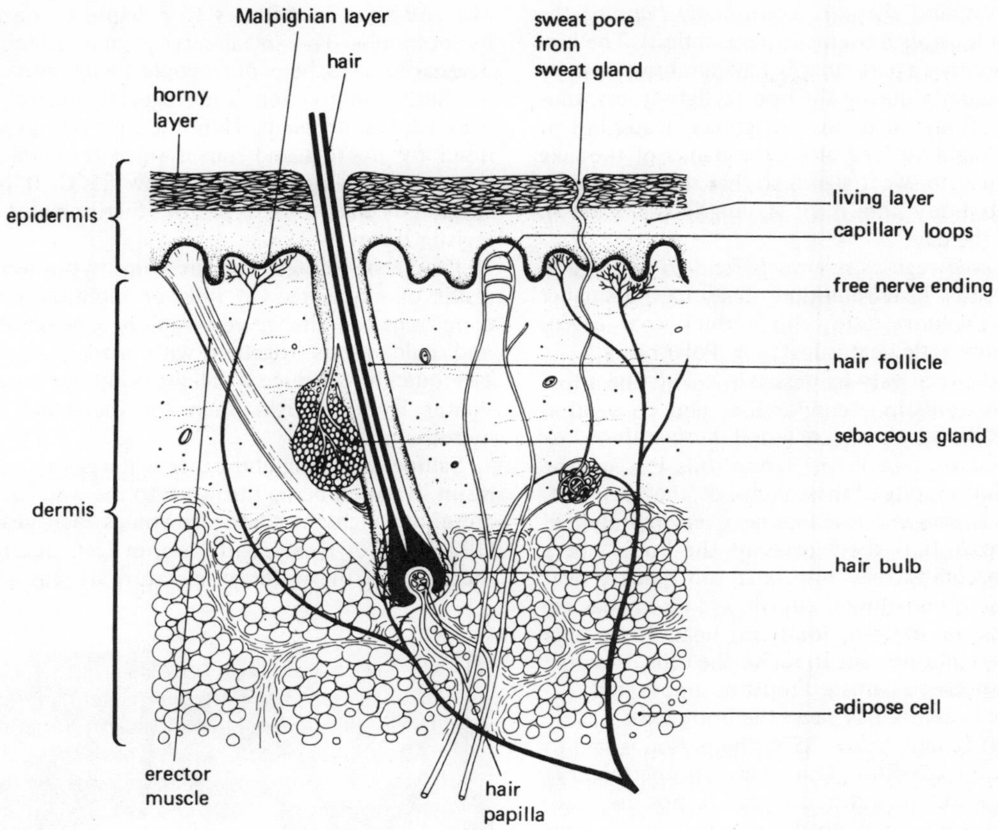

fig 5.9 A vertical section through the skin of Man

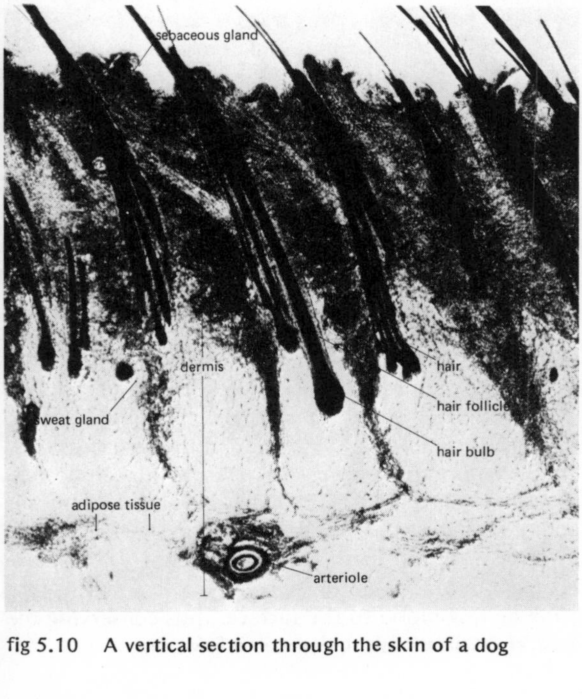

fig 5.10 A vertical section through the skin of a dog

5.14 Heat loss and body size in Man

In Chapter 3 you have discovered that the larger the organism the smaller the surface area/volume ratio. Since heat loss is through the body surface, and heat production is the result of respiration and energy production from the cells in the total volume of the body, this ratio is of vital importance to the animal. Very small animals have a relatively large surface area (high S.A./V. ratio) and find heat conservation a major problem. *Very large animals* in hot climates have the opposite difficulty. They need to *encourage heat loss* in order to keep cool.

Small mammals and birds have a *high metabolic rate* in order to *produce heat* to maintain their body temperatures. Humming birds and small insectivorous mammals have heart beats up to 1 000 times per minute, and need to eat more than their own body weight in twenty-four hours.

Large mammals have a relatively smaller surface area, but a larger volume of body for heat production. The elephant and the hippopotamus have few hairs. This allows for greater heat loss from their naked bodies. The elephant has large ears with a *copious blood supply,* and they are continually *flapping* the ears to encourage air movement (convection). The hippopotamus shows a particular behavioural response by remaining aquatic during the hot daylight hours, thus conducting body heat to the cooler water, but it emerges at night to feed along the banks of the lake or river. It has no sweat glands so that the only way it can keep its body cool is by staying in the water in the heat of the day.

In *very cold* regions mammals tend to be large in size, for greater heat retention. They have a smaller surface area/volume ratio, and a thick coat of fur through which little heat is lost, e.g. Polar bears.

In Man the relatively hairless skin enables heat loss to occur by radiation, conduction, and convection under conditions of low external temperature. We have detailed in section 5.13 how this loss can be decreased, but in spite of these methods, the toes, feet, fingers, hands, ears and nose lose heat very rapidly and thus feel cold. It is these parts of the body where 'frostbite' occurs if they lose heat too quickly over long periods. Blood flow is cut off and the tissues die through lack of oxygen, food and heat. If suitable conditions are not present to revive the cells, then the parts of the body so damaged must be amputated.

Excessive heat loss may cause the body temperature to fall and if it falls below 35°C the person is said to suffer from the condition known as hypothermia. Fig 5.2 shows that this condition is due to positive feedback, with lowered metabolism resulting in less heat being produced, which in turn leads to temperature fall, metabolism lowered still further, and so on. Young babies and old people are particularly susceptible to this condition.

Babies lose heat quickly because of their comparatively large surface area in relation to volume. They move little so they do not generate much heat and as a consequence quickly take on the temperature of their surroundings. A small child sleeping in a cold bedroom on a winter night, in a temperate or continental climate, will suffer a slow downward movement of his body temperature and will quickly pass into a hypothermic coma. This state is often confused with sleep. In spite of guidance given to mothers by the medical profession regarding the dangers of a cold room, parents do not realise that the bedroom must be continuously warmed in order to maintain a temperature of about 29°C and thus ensure normal body temperature.

Old people can also suffer from the same problem and it is now recognised that many elderly men and women living on their own die at home from hypothermia. If an elderly person falls and is unable to rise, then only a few hours on a cold bedroom floor may be fatal. The prohibitive expense of providing adequate heating often results in temperatures that are far too low and thus contributes to collapse or death from hypothermia. The social services and neighbours all have a duty to help old people living alone. Advice on heat conservation and regular visiting help to alleviate the problem. Hypothermia can go unrecognised by doctors and nurses since the clinical thermometer does not register below 35°C. It has been suggested that this type of thermometer should register down to 24°C.

People of all ages can die from hypothermia as a result of exposure. On hills or mountains weather conditions can change very rapidly. The onset of rain and cold winds together with inadequate clothing can quickly produce a fall in body temperature. If shelter is not sought then the inevitable result is hypothermia.

Sailing, canoeing and other water sports can often result in people being thrown into the water and there are about eight hundred drownings each year in the U.K. Many of these deaths are, in fact, due to hypothermia, for the water temperature of seas and lakes

fig 5.11 Climatic adaptation — The Eskimo

in temperate climates is seldom above 14–15°C and often is much lower. If the water temperature is below 10°C the body temperature can drop to a dangerous level in about 15 minutes (see fig 5.2).

The adaptation of animals to differing external temperatures can be seen in many species. Examine photographs fig 5.11 and fig 5.12 showing two races of Man adapted to different environments. The Eskimo weighs about 77kg and his height is 1.64m. He lives in a very cold climate and his body shows the typical thickset figure, flat features and short limbs of men *adapted to these conditions.* The Sudan negro, however, weighs 58.8kg and his height is 1.885m. He is much taller but weighs less than the Eskimo. His thin body and relatively long limbs are an advantageous adaptation to *tropical climates* permitting rapid heat loss from the body.

Heat exhaustion is caused by the loss of salt from the body, due to excessive sweating. It develops in persons exposed to very hot climatic or working conditions where very high temperatures are present, such as in ships' boiler rooms or blast furnaces. The symptoms are profuse sweating and faintness followed by loss of consciousness, weak pulse and shallow breathing. It can be prevented by taking adequate amounts of water, and salt in the form of salt tablets.

Heat stroke is caused by a basic fault in the body's heat regulating mechanism. Sweating ceases and as a result the body temperature rises to 41°C or higher, followed by collapse. The skin of heat-stroke victims is dry and this state must be considered as a serious medical emergency, requiring immediate treatment.

Examine fig 5.13 which shows three species of fox, one each from a cold, a temperate and a hot climate.

Questions

1 Which structure, related to heat loss, clearly shows a difference in each species?

2 Which species has the greatest amount of fur? What advantage could this have?

3 Which species has the smallest head and which has the largest? What advantage does this confer upon each animal?

5.15 Other functions of the skin

1 *Protection* The cornified layer of the epidermis (see fig 5.9) consists of dead cells which provide an outer protective layer. This can perform the following functions.

a) Prevention of water loss.

b) Prevention of frictional damage. This is particularly noticeable on the soles of the feet and palms of the hand in Man, where the cornified layer becomes considerably thicker than in the rest of the body.

c) Prevention of entry of bacteria. Micro-organisms can still enter through the skin through sweat pores, hair follicles or skin wounds.

fig 5.12 Climatic adaptation — The Sudan negro

GATEWAY SCHOOL LEICESTER

Methods by which Man loses heat	Methods by which Man gains heat
1 Production of sweat increases. The water in the sweat evaporates, drawing latent heat from the body.	Production of sweat decreases so that heat lost by evaporation is much less.
2 Arterioles relax (*vasodilation*) and more blood enters the capillary network. Extra heat is lost by radiation and convection from the skin.	Arterioles constrict (*vasoconstriction*) and less blood enters the capillary network. Less heat is lost by radiation and convection.
3 The rate of metabolism decreases so that less heat is produced, thus Man's activity is less in hot weather.	The rate of metabolism increases producing more heat. Loss of heat can cause involuntary muscular action called shivering.
4 Behavioural methods:- wearing fewer clothes, taking cold baths or swimming, drinking cold drinks, using a fan or air conditioning.	Behavioural methods:- wearing more clothes, drinking hot drinks or eating hot food, heating houses, taking exercise.
5 The hair is lowered in other mammals making a thinner coat so that the heat can escape more easily. Man can get no benefit from this since his skin is largely naked.	The hair is raised in other mammals making a thicker coat, thus trapping more air as an insulating layer. In Man the naked skin with its few hairs shows traces of this function. The contracted hair muscles appear as 'goose pimples'.

Table 5.1 Methods by which Man loses and gains heat

d) Protection against radiation from the sun by the development of a black or dark-brown pigment, melanin. This is hereditary in dark-skinned races, but acquired as a result of exposure by light-skinned races.
e) Insulation. Fat layers under the skin prevent heat loss.

2 *Vitamin D production* Ultra-violet rays of the sun penetrating the skin produce sterols which become converted into vitamin D. Thus in sunny weather some of the requirements of the body can be met in this way, but in dull weather or in industrial regions covered by smoke haze the body has to rely on vitamin D in the diet.

3 *Energy storage* The lower layers of the dermis can store fat. As stated above, this can act as an insulating layer; but in many mammals, especially those that hibernate, the subcutaneous layers of fat are broken down to provide energy.

5.16 Skin hygiene

The skin serves as a protective layer and plays an important part in the maintenance of a healthy body. However, the sebum and sweat produced provide an ideal breeding ground for bacteria, so if the skin is not kept clean, infection can result. A healthy skin has considerable resistance to infecting organisms but if this resistance breaks down pimples, boils, carbuncles and other types of sepsis occur. Infecting organisms which gain entry in this way can also spread to other

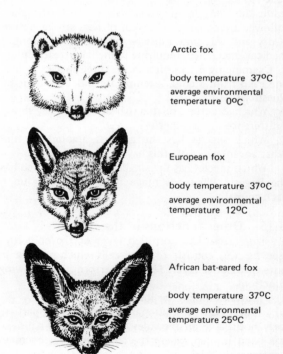

fig 5.13 The differences in appearance of three foxes: from the Arctic, Europe and Africa

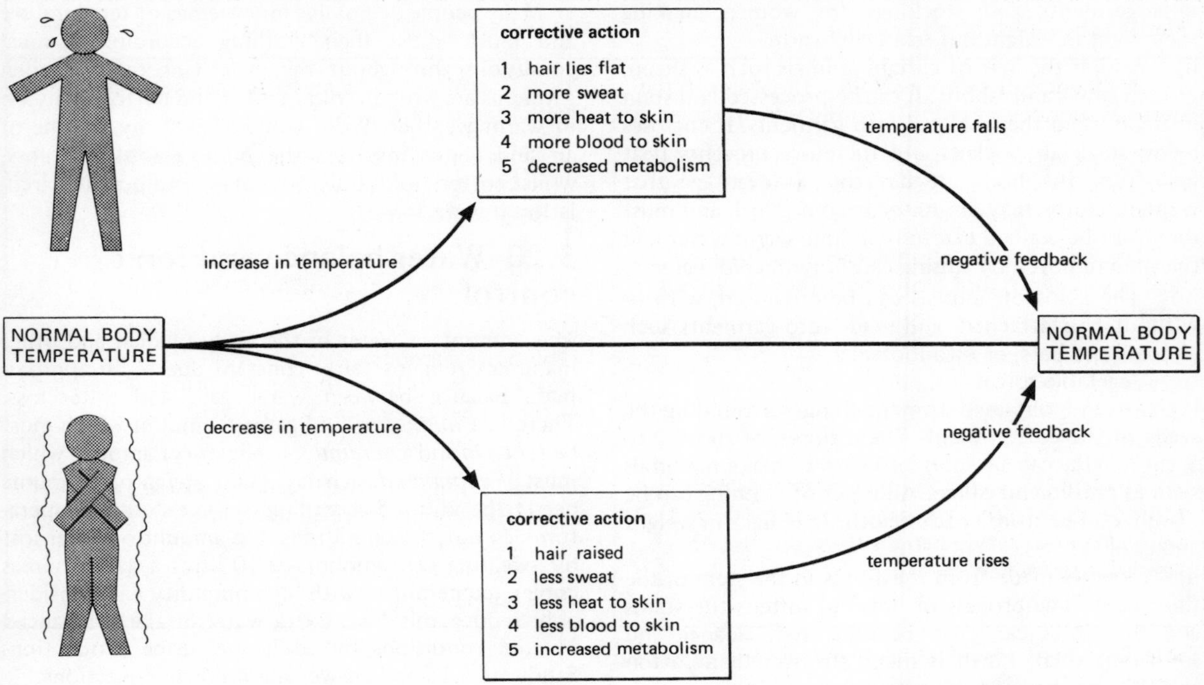

fig 5.14 Homeostatic temperature control

parts of the body. Soap and warm water remove natural oils, sweat and dirt and washing the body, particularly the hands, head and neck, can remove the possibility of infection. Ideally a warm bath or shower should be taken daily, but certainly the routine must include regular washing of the hands and face. Hands must always be cleansed after going to the lavatory, before meals and after performing any task likely to place bacteria on the hands, e.g. cleaning drains, attending to the sick and so on. The skin must be dried well after washing, particularly between the toes, to avoid fungal infection.

The first evidence of sickness often appears on the skin, e.g.

i) the whiteness of the skin due to fatigue or faintness,

ii) the flushing of the surface blood vessels during fever,

iii) changes in the skin due to malnutrition, particularly vitamin deficiency,

iv) redness of the skin due to inflammation at the site of cuts, abrasions, boils and pimples.

Parasites may live on the surface of the body or the hair. These include fungal parasites causing ringworm and parasites such as fleas, bed bugs, head-lice, body-lice and itch-mites. Cleanliness of skin, hair and clothing are all important in avoiding these parasites (see Chapter 10).

Chemicals used in everyday life in the home or in industry can contain substances injurious to the skin. Gloves can be worn as protection when these substances are being handled, but gloves are not always suitable, so special protective creams are often used, called *barrier creams*. These can protect the hands of the housewife against washing-powder detergents or those engaged in industry against immersion of the hands in chemicals, grease or oil.

5.17 Clothing and its functions

By the use of clothing the skin can also be protected against the effects of wind and weather and against injury, e.g. feet and legs. The purely functional value of clothing has often been superseded by a third function, that of adornment.

In cold climates clothing is most important to prevent heat loss from the body. The unprotected, naked skin has very limited powers of dealing with extremes of temperature. In hot and humid climates the excess of moisture in the air may prevent the evaporation of sweat and thus prevent the cooling of the body. The wearing of clothing is a necessity, but clearly the type of clothing and the materials of which it is made must vary according to the climate.

The materials of which clothing is made have the following characteristics:

a) Animal origin

i) Silk is made from silk-worm cocoons and can be woven into different materials such as brocade and crepe. It is used also in the manufacture of various

undergarments and stockings for women, making them light in weight but relatively warm.
ii) Wool is the hair of certain animals such as sheep, goats, camels and rabbits. It can be processed and spun into yarn and then knitted into garments. It encloses many small air pockets and therefore prevents heat loss from the body. It has the disadvantage that woollen cloths may shrink or become hard, and must therefore be washed carefully in luke-warm water and the soap removed by rinsing carefully in cold water.
iii) The skins of animals can be processed, with or without hair attached, and made into garments such as coats and hats, or into footwear.
b) Vegetable origin
i) Cotton is obtained from the fibres surrounding the seeds of the cotton plant. These fibres, of about 2 to 5 cm length, can be spun into yarn to make materials such as muslin and calico. All types of clothing can be manufactured from cotton cloth. It is light in weight and cool.
ii) Linen is made from the fibres in the stem of the flax plant. The process of 'retting' softens the stems and the fibres can then be separated, cleaned and spun into yarn. Linen is much stronger than cotton and thus can be used to make jackets and suits.
c) Articifical and synthetic fibres
These include rayon, which is spun from cellulose, and completely synthetic fibres such as nylon, terylene and dacron. The latter are synthesised from carbon compounds and vast quantities of these materials are now used in clothing manufacture.

Clothing generally should be light in weight. It should be loose, since tight-fitting clothing is uncomfortable and in certain cases positively harmful, e.g. tight-fitting corsets, garters and shoes. All clothing should be porous to enable sweat to evaporate from the skin surface. Flowing clothing should be non-inflammable to prevent the possibility of it swinging into a fire and catching light. Clothing must be kept clean. Underwear needs to be changed and washed frequently and outer clothing also washed or dry-cleaned periodically.

Considering the above characteristics of clothing and the requirements of the body we can now consider the use of clothing in different climates. In a hot climate the minimum of clothing is required and should be of the type to absorb sweat and allow it to evaporate. This is aided if the clothing is loose and allows air to circulate. The colour should be light to reflect heat. In intense cold several layers of clothing should be worn so that air is trapped between the layers and the outer layer should be wind-proof. If hard work is performed in cold weather the body heats up and produces more sweat, so to allow evaporation the clothing is opened up, generally at the neck, but immediately closed up again after work stops.

Many people do not live in extremes of temperature and must adjust their clothing according to local conditions throughout the year. Generally woollen garments are worn in cold weather and cotton or nylon in warm weather. Wool contains 90% by volume of air and hence insulates the body against heat loss, whilst cotton holds only 50% of air and permits circulation if worn loose.

5.20 Water balance and ionic control

The amount of water in the body of Man and other mammals remains fairly constant due to an approximate balance between water gain and water loss. There is a *minimum* (obligatory) amount which must be *taken in* and a *minimum* (obligatory) amount which must be *eliminated* each day. These obligatory amounts need to be adjusted according to the external temperatures. In hot, dry conditions, the amount of water lost by sweating can amount to 10 litres a day, whereas cooler temperatures with high humidity will considerably reduce this loss. Extra water intake is balanced in cool conditions by additional urine production. Study fig 5.15 and answer the following questions.

Questions

1 If we sweat a great deal, how do we generally replace this water?

2 Under what conditions would the loss of water in urine increase considerably?

3 Why do you think dehydration rapidly occurs in a man suffering from cholera, a disease involving continuous diarrhoea?

5.21 The kidneys and associated structures

The kidneys are the most important organs involved in regulating the amount of water in the body in addition to their role as excretory organs, so let us try and understand their structure.

Investigation
The kidney and associated structures
Procedure

1 Examine the demonstration dissection of a small mammal. (The gut will have been removed in order to show the *kidneys, ureters, bladder* and *urethra.* See fig 5.16).

2 Examine a large kidney removed completely from a sheep, goat or cow. Note any tubes attached to the structure. See fig 5.17.

3 Place the kidney flat on the bench, and by means of a sharp blade, cut it into two equal halves by means of a horizontal cut parallel to the bench top. This is a longitudinal section.

4 Separate the two halves and examine the internal surfaces and the relationship of the tubes to the internal structure.

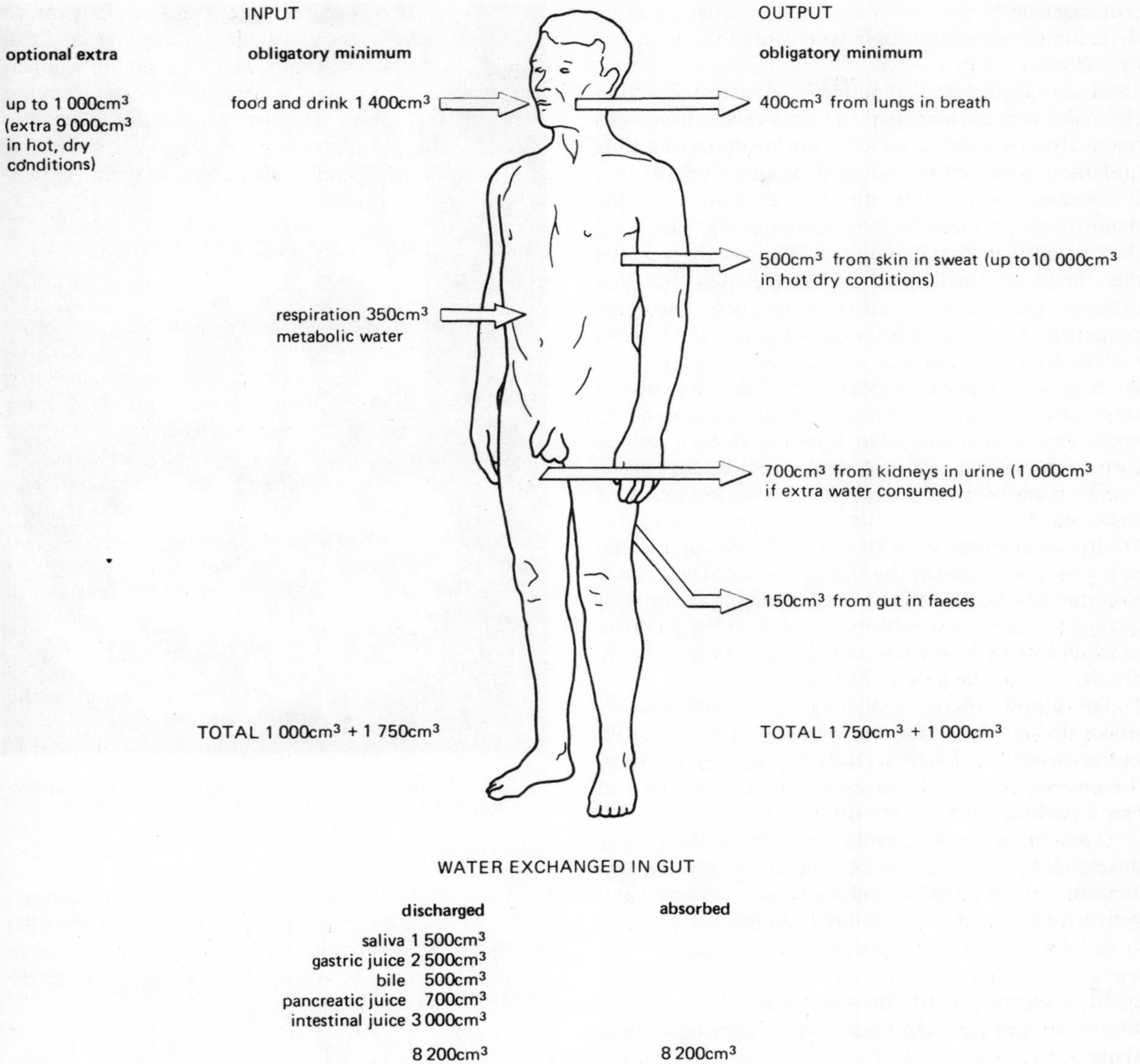

fig 5.15 Water gain and loss in Man

5 Make a diagram of the internal surface and the tubes.

Questions

1 What are the three tubes attached to the kidney?
2 How can you distinguish between them.
3 How would you describe the structure of the internal face of the kidney?
4 From what part of the kidney does the ureter directly receive its urine?

The *urinary system* can be seen in fig 5.17. It consists of two kidneys lying in the dorsal wall of the abdominal cavity and supplied with blood from *renal arteries.* The blood leaves the kidneys through the *renal veins,* and the urine produced, passes from each kidney into a *ureter.* The left and right ureters join with a muscular sac, the *bladder.* The mouth of this sac connects with a short tube, the *urethra,* which leads to the exterior. There are two small rings of muscle closing the exit of the bladder.

Internally the kidney has an outer portion, the *cortex,* and an inner portion, the *medulla.* The urine leaves the medulla and enters the pelvis, it then enters the ureters.

Investigation

To examine the microscopic structure of the kidney

Procedure

1 Take a prepared slide of a transverse section through a kidney, in which the blood vessels have been injected with a coloured dye and examine the slide under low power of the microscope (see fig 5.18).

2 Examine the slide under high power of the microscope, particularly the cortex (see fig 5.18).

You will note that under the low power the injected dye shows up small knots of blood vessels, but that most of the kidney consists of *tubules.* These are present particularly in the medulla where the tubules radiate from the *pelvis* of the kidney. The interpretation of this complicated structure is difficult from a single section, but by means of serial sections and by microdissection it has been found that each kidney consists of millions of such tubules, called *nephrons.*

The nephron begins in the cortex as a cup-shaped structure, the *Bowman's capsule,* surrounding a small knot of capillaries, the *glomerulus.* These can be seen in the cortical region of the kidney section. The capsule is connected to the tubule of the nephron which is divided into a coiled section, a loop into the medulla, a second coiled section and a duct which finally discharges into the pelvis of the kidney.

The blood enters the kidney by the renal artery which divides into arterioles supplying the capillaries of the glomeruli. Further capillaries surround each of the coiled sections of the tubule. A large amount of blood reaches each kidney (in Man this is about 500 cm^3 per minute) and slightly less leaves the kidney through the renal vein. The difference between the two amounts is due to the excretory substances and water which is extracted by the kidney tubules.

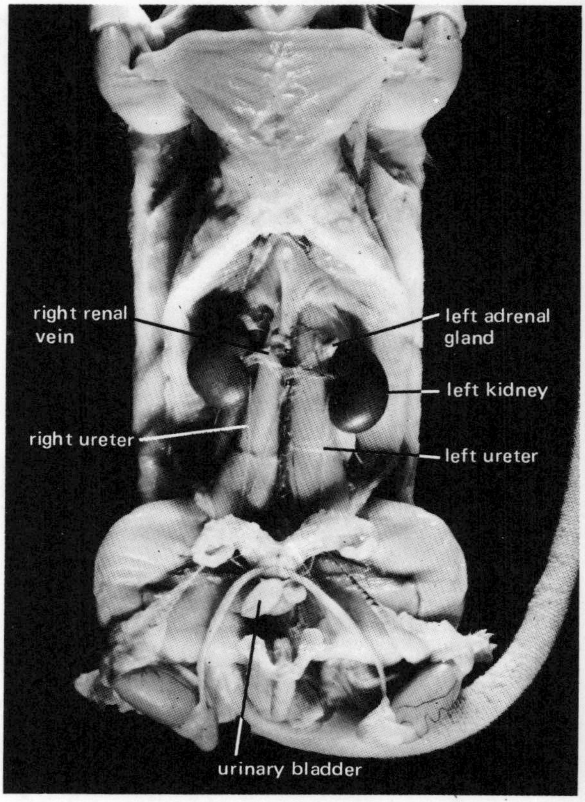

ig 5.16 A dissection of the rat to show kidneys and associated structures

5.22 The work of the nephron

In the preceding chapters we have seen how living things

i) take up materials and energy from their environment (Chapter 2 and 4),

ii) transport materials through their bodies (Chapter 3), and

iii) transform and store these materials within their cells (Chapter 2).

All of these activities are *anabolic,* involving a building-up process, but to complete our understanding of metabolism we must now examine the ways in which living things return their waste products to the environment. These are produced by the breaking down activities of living cells, activities described as *catabolic.* The most abundant waste products are water, carbon dioxide and ammonia, but they should not be regarded simply as wastes. Remember the importance of carbon dioxide in the regulation of

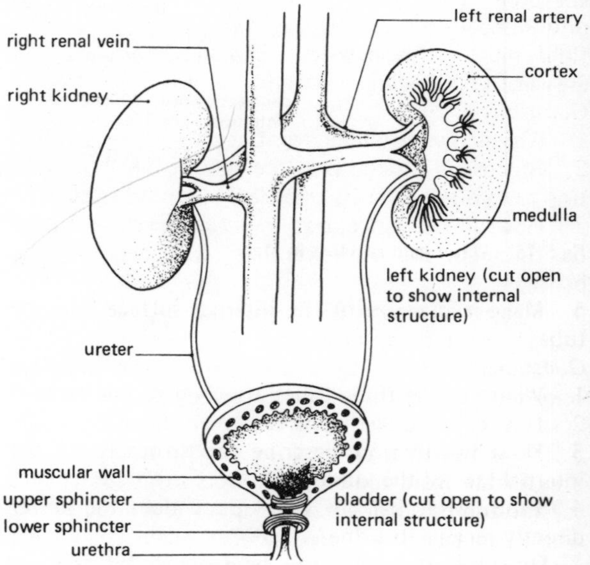

fig 5.17 General structure of the kidneys and bladder in Man

breathing, and even ammonia, a toxic substance, is essential for some amino acid metabolism.

These substances are dangerous because they can *accumulate* to levels in excess of the organism's needs. The *elimination* of these *products of metabolism* from the body is called *excretion*. The most important organ concerned in this function is the kidney and within this organ the nephron is the working unit. A single human kidney is made up of about one million of these units. How does the nephron function? Many answers to this question were produced by a technical operation which involved sampling, by means of a micropipette, the fluid passing along the tubule. Extremely fine-drawn glass tubes were pushed into the tubule at different points, samples of the fluid withdrawn and then chemically analysed.

The nephron functions as follows. The *exit* capillary of the glomerulus has a *narrower bore* than the capillary which delivers blood (see fig 5.19). The blood is under higher pressure as a consequence and so much of the blood is forced out in the Bowman's capsule. The capsular space and the capillaries are separated only by two thin layers of cells, which act as a filter allowing only the liquid of the plasma and small molecules to pass. Since the filtration is a result of increased blood pressure, it is called pressure filtration or *ultrafiltration*. The fluid which is forced into the capsule is very similar to tissue fluid and lymph (see Chapter 3).

The rate of production of fluid into the tubules is about 130 cm^3 per minute (about 100 litres per day). Over a period of twenty-four hours this is equivalent to a volume of fluid more than twice the weight of the body. The final production of urine, however, is only some 1 500 cm^3 per day. Thus about 98% of the fluid must be reabsorbed. Examine table 5.2 and answer the following questions.

Questions

1 Why is the protein concentration in urine zero?

2 Glucose is filtered into the tubule. What explanation can you give for its complete absence in urine?

3 How do the concentrations of urea, ammonia and creatinine in the urine compare with those in the plasma?

4 Suggest a reason for any differences in concentration of the three substances.

The values in the table are approximate. There are considerable variations even in healthy people.

5.23 Chemical constancy of the body

The filtrate from the blood collects in the Bowman's capsule and passes down to the first convoluted tubule where considerable *reabsorption* occurs. *Glucose, amino acids, vitamins, hormones* and a large number of *inorganic ions* (Na^+, K^+, Ca^{2+}, HCO_3^-, Cl^-, etc.) are reabsorbed into the blood stream. This reabsorption

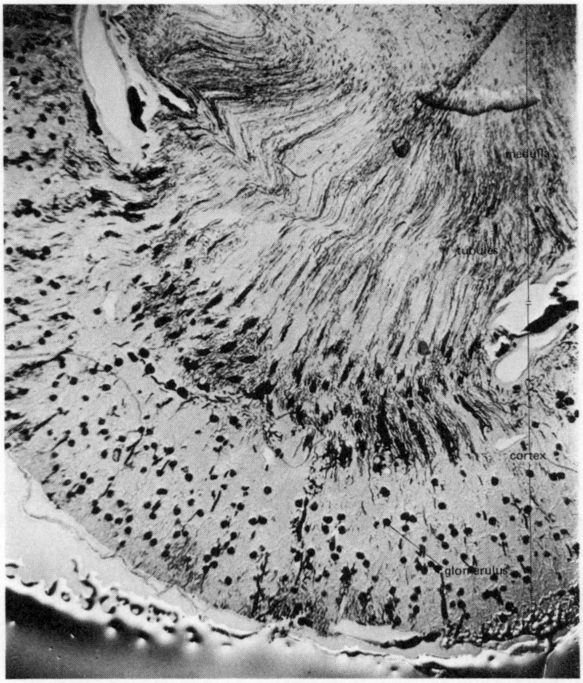

fig 5.18 A photomicrograph of kidney tubules

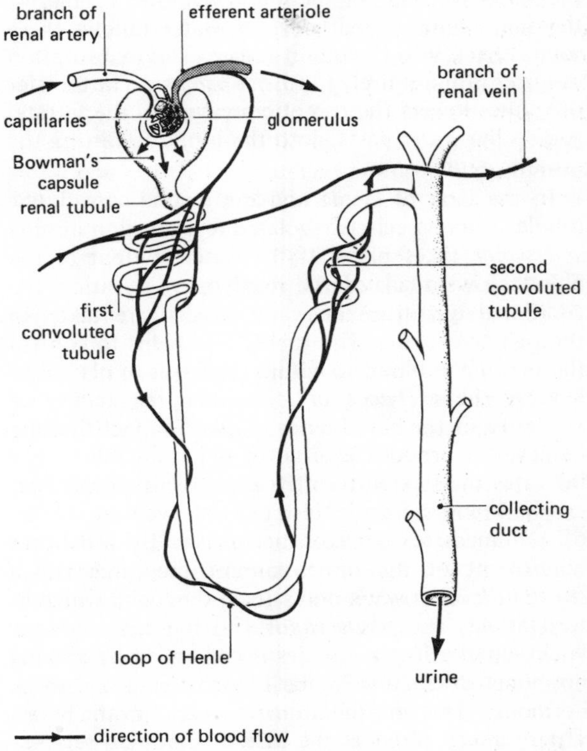

fig 5.19 A diagram of one renal tubule showing the blood supply and the main areas of reabsorption

Substance	% in plasma	% filtrate into nephron	% in urine	Concentration factor
Water	90–93	90–93	95.0	
Protein	7.0	0	0	
Glucose	0.1	0.1	0	
Sodium	0.3	0.3	0.35	x 1.0
Chloride	0.4	0.4	0.6	x 1.5
Urea	0.03	0.03	2.0	x 60.0
Uric acid	0.004	0.004	0.05	x 12.0
Creatinine	0.001	0.001	0.075	x 75.0
Ammonia	0.001	0.001	0.04	x 40.0

Table 5.2 Concentrations of substances in blood plasma and urine

requires energy for active transport so that the tubule cells have a high rate of metabolism with much ATP being formed. The inner surface of the cells lining the tubule have *microvilli* to increase greatly the area for absorption.

The solutes are returned to the blood in the capillaries surrounding the first convoluted tubule. At the same time a great deal of water follows these solutes back into the blood stream. This reabsorption of *water* is due entirely to *passive osmosis.* The transfer of solutes lowers the osmotic pressure of the filtrate, and so the water passes into the blood to restore the osmotic equilibrium.

In the loop of Henle and the second convoluted tubule, a more precisely regulated reabsorption of *salts and water* takes place. If the water content of the blood is lower than usual (through heavy sweating) the additional quantities of water needed are taken in through the tubule. The tubule also helps to control the pH of the blood to within the limits of pH 7.3 to 7.4, by the *exchange of ions* when the acidity or alkalinity of the blood tends to rise. The fact that the kidney can produce a urine of pH from 4.5 to 8.5 indicates the flexibility of its adjustment mechanism.

The kidneys are the third example we have chosen of a homeostatic control mechanism. By a delicate adjustment of the urine content they maintain a *constant level* of water and salts in the *blood* which in turn affects the *tissue fluids* and the *cell contents.* Water uptake in the tubules is controlled by various hormones circulating in the blood stream and these hormones are under the ultimate control of the brain. The *pituitary gland* at the base of the brain secretes the hormone *vasopressin* (ADH). Dehydration of the body is detected as a drop in water content of the circulating blood in the brain. The pituitary gland is stimulated to release vasopressin, which is carried by the blood to the kidneys. Additional water is reabsorbed by the tubules and the water content of the blood is restored to normal levels. Urine production drops and the urine itself becomes more concentrated. If blood becomes too diluted then the secretion of vasopressin is inhibited. The tubules fail to absorb water and large quantities of watery urine are produced (see fig 5.20).

Examine fig 5.20.

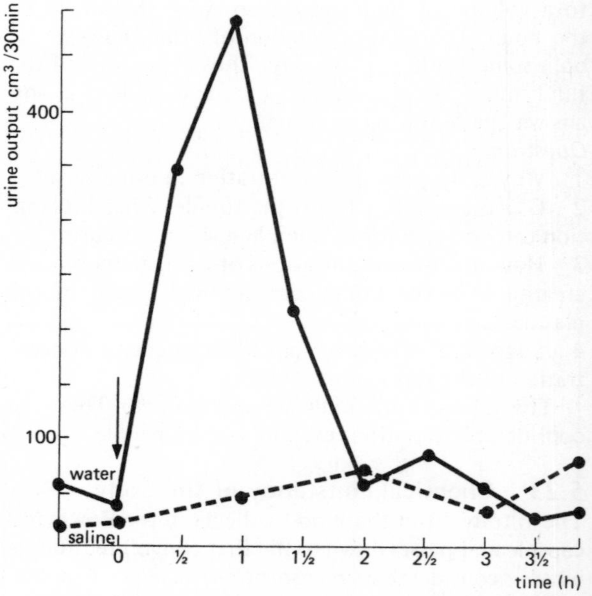

fig 5.20 A graph of response to drinking one litre of water, and on the following day, one litre of 0.96% sodium chloride solution, by a healthy man

Questions

1 What period of time passed before the entire litre of water was eliminated from the body?

2 Why do you think that drinking 1 litre of 0.96% sodium chloride solution made little difference to urine output?

If a human is in conditions where no drinking water is available and instead drinks *sea water,* his tissues will dehydrate and he will die. The salt content of the blood is about 1% whereas that of sea water is 3%. Thus when sea water is drunk, the salt content of the blood will rise towards 3%, but the kidney can only eliminate at best about 2% salt in urine. In order to do this, however, the blood drains water even more quickly from the tissues and there is a *net loss of tissue fluid.* As a result, tissue dehydration occurs and this leads quickly to death.

The analysis of urine can give an indication of certain malfunctions of the metabolism of the body. For example, glucose is not normally present in urine, for it is reabsorbed by the tubule after filtration. Evidence of *glucose in the urine* indicates a disease called *diabetes mellitus,* in which the pancreas fails to secrete the hormone *insulin.* Occasionally people lose the ability to secrete vasopressin and they then excrete an enormous volume of urine. This is accompanied by a terrible thirst and the person *drinks large quantities of water* to replace that lost in the urine (up to 20 litres daily). This disease is called *diabetes insipidus,* or *water* diabetes as opposed to *sugar* diabetes.

Kidney failure may occur and, although we can live with only one kidney, the failure of both is fatal. The simple phenomena of diffusion and permeability of membranes are used in *artificial kidneys* in order to help patients over a period of danger. Blood is passed through cellulose tubing (visking tubing) which is immersed in a bath of carefully prepared fluid. The fluid can be made of a precise composition to eliminate any particular ion from the blood. The efficiency is very great, removing urea, for example, much more quickly than normal kidneys.

Lungs	Skin	Kidneys
Water	Water	Water
Carbon dioxide	Sodium chloride	Sodium chloride
	Urea	Urea
		Uric acid
		Creatinine
		Ammonia

Table 5.3 Summary of substances excreted from body organs

5.30 The liver as a regulator

The body of a mammal is unable to store proteins or amino acids. *Excess amino acids* in the blood are taken to the *liver* where they are *deaminated.* The nitrogen-containing amino ($-NH_2$) portion of the molecule is removed and converted to nitrogenous waste, *urea.* The non-nitrogen containing residue from the amino acid can be converted to glycogen for storage in the liver, or it can enter the blood as glucose and be used in cellular respiration.

In this connection the liver is concerned with the control of *blood sugar levels.* As these levels *rise* in the blood, the pancreas secretes *insulin* which changes the *glucose to glycogen* (an insoluble polysaccharide), for storage in the liver. During heavy exercise, blood sugar levels *fall* and the liver compensates by changing back *glycogen to glucose.*

Fat is stored in the liver. As fat stores elsewhere in the body decrease during starvation, stored fat is released from the liver. The fat in the liver is very active metabolically for it can be changed to glycerol and fatty acids.

All the blood from the intestine, containing absorbed food materials, reaches the liver via the hepatic portal vein. The liver screens the blood passing through it so that the composition of the blood leaving the liver will be that which is correct for the organism. For example, even after a meal rich in carbohydrate, the blood leaves the liver with its blood sugar level at the normal 0.1%. In the reverse process, blood sugar levels below 0.1% will be restored by the release of glucose obtained from glycogen. Thus we can see that the liver is an organ of prime importance in the maintenance of a constant internal environment.

Questions requiring an extended essay-type answer

1 a) Define the process of excretion in living organisms.
b) The water content of the body of Man remains constant. Show briefly how this balance is achieved.

2 a) What is meant by *homeostasis?* Give three examples in mammals.
b) Discuss the phenomenon of *homoiothermy.*

3 a) Compare the methods by which mammals gain heat and lose heat.
b) How are the processes mentioned in part (a) related to water loss and ionic balance in a mammal?

4 a) Describe a *controlled* experiment to investigate heat loss through evaporation of a liquid from a hot surface.
b) How are your findings in part (a) relevant to control of temperature in a mammal?

5 a) What substances are excreted by
i) the lungs,
ii) the skin, and
iii) the kidneys?

b) Describe the part played by the liver and kidneys in the process of excretion in a *named* mammal.

6 a) Make a large, labelled diagram of a single kidney tubule (nephron) with its blood supply.

b) Describe the composition and fate of the liquid contents of the tubule, from its extraction to its entry into the bladder.

7 a) Draw a simple diagram of a vertical section through the skin of Man.

b) Describe methods by which Man preserves heat in his body.

8 a) What must we do to maintain the skin in its normal, healthy state?

b) How may
(i) the type of clothing material and
(ii) the style of clothing affect Man's ability to lose heat from the body?

6 The skeleton and movement

6.00 Introduction

One of the features that enables us to distinguish animals from plants is the ability of animals to move from place to place. This ability – locomotion – is necessary because animals, unlike plants, cannot make their own food but have to move around to obtain it.

Three essential elements occur in almost all animal locomotory systems: a *contractile tissue*, usually muscle, acting upon a *rigid skeleton* (made of bone in the case of vertebrates) which in turn acts on a supporting and *resisting medium* (earth in the case of most mammals, water in the case of fish and air in the case of birds).

Not all skeletons are internal, and not all are made of bone. Insects, for example, have an *external* skeleton made of a horny material, and annelid worms have a rigid *hydrostatic* skeleton consisting of fluid-filled chambers.

Of course it is neither possible nor desirable to consider the skeleton in isolation from the other organs and tissues of the body. Most movements occur in response to a particular stimulus, for example the sight of food. In this case, the stimulus is received by the eye and the resulting nerve impulse is transmitted to the brain by the sensory neurons of the optic nerve. Impulses from the brain are conducted via motor neurons to certain muscles, which *contract* to produce movement of parts of the skeleton and hence locomotion of the body towards the food. Thus, the functioning of sense organs, nervous system, muscles and skeleton are closely interlinked. In particular, the skeleton and its associated muscles form a co-ordinated *effector system.*

Investigation

To examine the arm

Procedure

1 Stretch out your right arm horizontally in front of you, palm upwards.

2 Observe the muscles of the upper part of the arm, and feel them carefully with your left hand.

3 Bend your right arm at the elbow, so that the fingers of your right hand are touching your right shoulder.

4 Again, observe the muscles of the upper arm and feel them carefully with your left hand.

Questions

1 What changes have taken place in the muscle on top of the upper arm (the *biceps*)?

2 What changes have taken place in the muscle underneath the upper arm (the *triceps*)?

6.10 The mammalian skeleton

Throughout the class *Mammalia*, in spite of great variation in form and size, the pattern of the skeleton is remarkably constant. It may be divided into two main parts: the *axial* skeleton and the *appendicular* skeleton. The axial skeleton consists of the skull and the vertebral column, together with the ribs and sternum; the appendicular skeleton consists of the limbs and limb girdles (see figs, 6.1, 6.7 and 6.11).

6.11 Functions of the skeleton

In addition to locomotion, skeletons have two other main structural functions: support and protection (bones also produce blood cells in the marrow and provide reserves of calcium and phosphorus, but these non-structural functions are dealt with elsewhere).

The *supporting* function of the skeleton is more obvious in large, terrestrial animals that are surrounded by air than in those that live in the more dense medium of water. As well as supporting the body as a whole, the skeleton provides a framework which keeps the major internal organs of the body in constant positions relative to each other.

Certain parts of the skeleton are adapted for the *protection* of delicate structures. The brain is completely surrounded by the skull, and the ribs form an

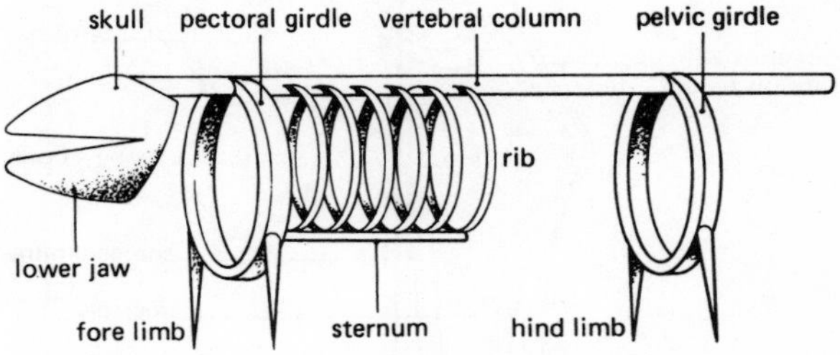

fig 6.1 A diagram showing the general plan of a mammalian skeleton

expandable protective framework around the heart and lungs. In addition to its primarily supportive function, the vertebral column protects the spinal cord and its nerves.

6.12 The tissues of the skeleton

Before we go any further, we should make a distinction in our minds between the two uses of the term 'bone'. Its first and most familiar use is to describe the individual structures of which the skeleton is composed: the thigh-bone, shin-bone, etc. However, the term 'bone' is also used to describe a particular *type of tissue* which forms only a part, albeit a major part, of each of these structures. The other major type of tissue found in the skeleton is *cartilage* (see fig 6.2).

Typically, a bone is covered by a thin layer of dense, hard bone tissue. The hardness is caused by the presence of large quantities of calcium salts (mainly *calcium phosphate*). However, it should not be assumed that bone is non-living. The calcium phosphate forms part of a 'matrix' which makes up only about 70% of bone tissue. This matrix is secreted by bone cells and is perforated by a series of canals (*Haversian canals*) running along the axis of the bone. The canals contain blood vessels, and the mineral-secreting bone cells are arranged around them in concentric circles (see fig 6.3).

In the centre of most limb bones is a soft, fatty substance containing blood vessels. This is the *bone marrow,* where the red blood cells (and certain white cells) are produced. The marrow cavity is continuous with spaces inside the bone tissue at the ends of limb bones. This results in the 'spongy' appearance of the bone in this area.

Where one bone joins another, as in the hip joint, the touching surfaces consist of a smooth layer of clearer, less rigid material known as cartilage or 'gristle'. This tissue owes its elasticity to the protein fibres secreted by its cells and, since it does not contain the hard calcium salts of bone, it acts partly as a *shock absorber* and partly to *reduce friction* between the bones at a joint. In the mammalian foetus, the skeleton is composed of cartilage, which is largely replaced by bone tissue by the time of birth. Some cartilage remains in the shafts of long bones, until growth is completed at the end of puberty.

At the joints, bones are attached to each other by *ligaments* consisting of tough, fibrous and more or less *elastic* tissue. Another type of structure found associated with the skeleton is the *tendon.* Tendons, like ligaments, consist mainly of fibres, but in their case

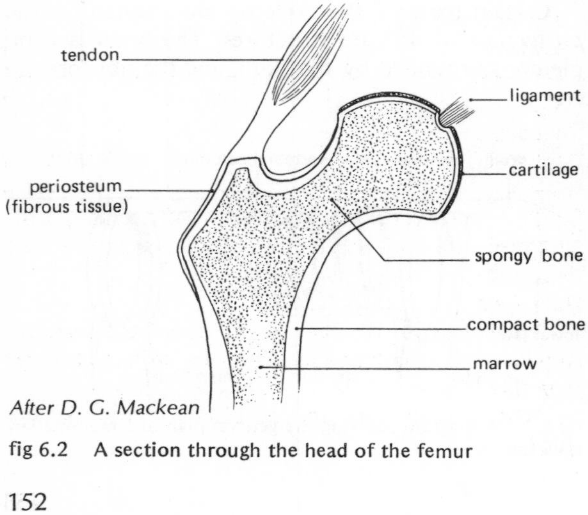

fig 6.2 A section through the head of the femur

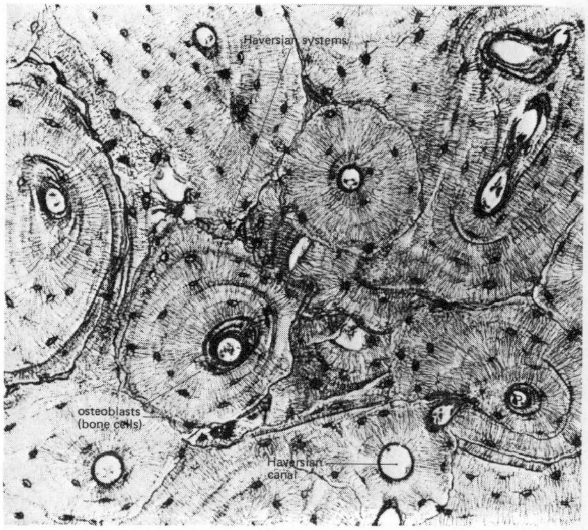

fig 6.3 Microscopic structure of bone

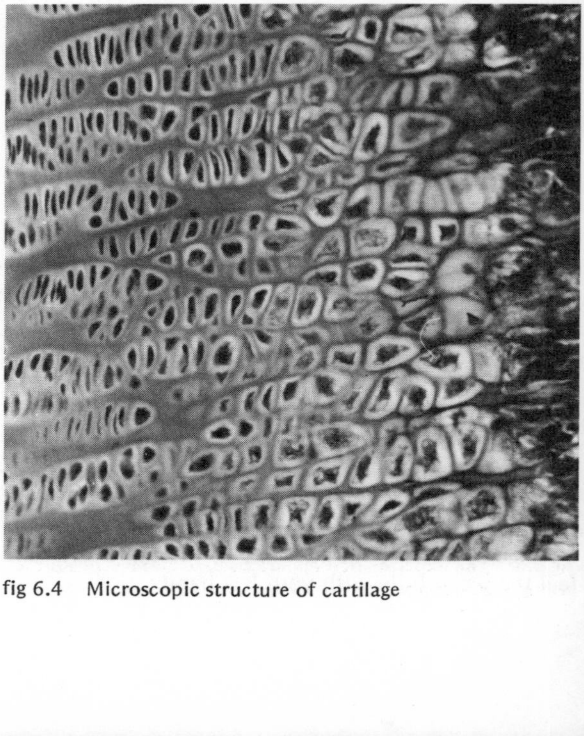

fig 6.4 Microscopic structure of cartilage

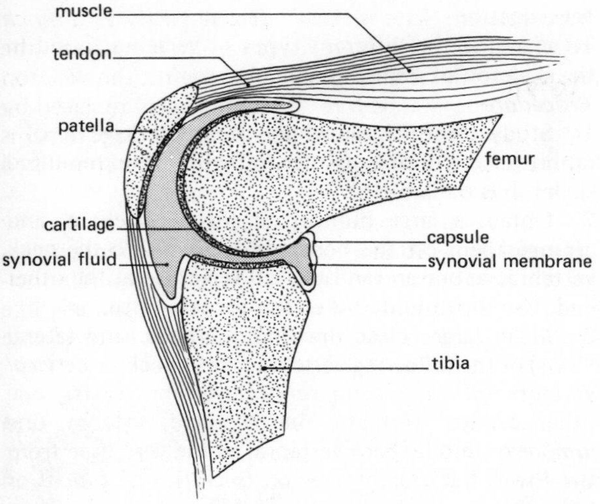

fig 6.5 The hinge joint of the knee

the fibres are *inelastic*. Tendons occur at the ends of the muscles and serve to join muscles to bones. Their toughness is particularly important where they pass over a bony structure, since they are much less liable than muscles to rupture as a result of vigorous movement.

Experiment

What is the relationship between structure and strength?

Procedure

1 Obtain a glass rod and a piece of glass tubing, each about 20 cm in length.

2 Weigh the rod and the tube.

3 Place two identical chairs (or stools) facing each other, about 18 cm apart.

4 Place the rod across the gap between the chairs, with exactly 1 cm of rod resting on each chair.

5 Suspend from the exact centre of the rod a small polythene beaker on a piece of string (see fig 6.6).

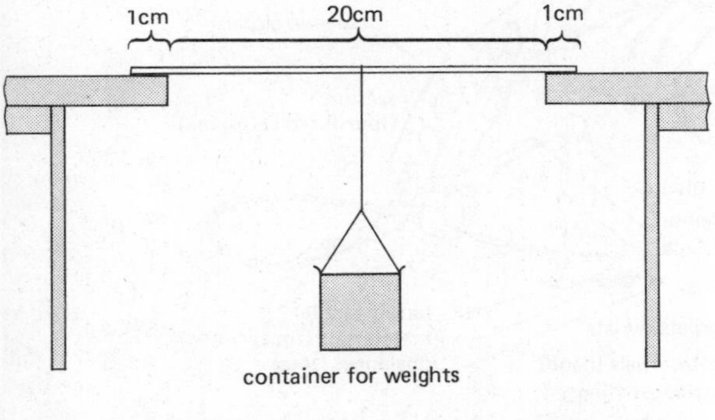

fig 6.6 The experimental apparatus used to investigate the strength of a supporting structure

6 Very gradually, add sand to the beaker until the rod breaks. Take great care to keep your face well away from the rod at this time: if possible wear a face mask or sunglasses to protect yourself against glass splinters. Weigh the sand used. Repeat the operation, substituting the glass tube for the rod. Weigh the amount of sand required to break the tube.

Questions

1 Divide the weight of sand required to break the tube by the weight of the tube.

Divide the weight of sand required to break the rod by the weight of the rod.

Which structure supports the greatest weight of sand relative to its own weight?

2 What do your results tell you about the relationship between strength and structure?

3 What do your results tell you about the design of limb bones?

Experiment

What is the relationship between length and strength?

Procedure

1 Set up the chairs (or stools) as in the previous experiment.

2 Take a few standard 22 cm, drinking straws. Cut one to 18 cm length, one to 14 cm, one to 10 cm and one to 6 cm.

3 Place a 22 cm straw between the two chairs as in the previous experiment, leaving 1 cm lengths resting on each chair.

4 Suspend a polythene beaker (or polythene bag if the beaker is too heavy) from the centre of the straw.

5 Pour sand gradually into the beaker until the straw collapses. Weigh the sand used.

6 Repeat the operation for the other four lengths of straw.

7 Record your results in the form of a table (see below).

Length of unsupported straw	4 cm	8 cm	12 cm	16 cm	20 cm
Mass of sand needed to collapse straw					

9 Plot your results in the form of a graph, with length on the horizontal axis and mass on the vertical axis.

Questions

4 What do your results tell you about the relationship between length and strength of a uniform tubular structure?

5 What explanation can you offer for this relationship?

6 What do your results tell you about the nature of bones of very tall animals?

6.20 The axial skeleton

The axial skeleton consists principally of the *skull*, together with the *vertebral column* (spine or backbone). For most purposes, the *ribs* and the *sternum* or breastbone are also included under this heading. Its functions are concerned mainly with support and protection.

6.21 The vertebral column

The backbone is made up of a number of small bones, the *vertebrae*, placed end to end and separated from each other by small pads of cartilage termed *intervertebral discs*. Although the detailed structure of the vertebrae varies betweeen the base of the skull and the tip of tail, they all conform to the same basic plan (see fig 6.14).

Each vertebra consists of a central disc of bone, the *centrum*, to the dorsal surface of which is attached the bony *neural arch*. The tunnel formed by placing the neural arches of all the vertebrae end to end provides protection for the spinal cord in the *neural canal*.

Typically, seven bony projections arise from the neural arch, these projections serving for the attachment of muscles and/or articulation with other bones. The *neural spine* is situated in the mid-dorsal line of the neural arch, and two *transverse processes* arise from the junction of the arch with the centrum. Four *articulating facets*, two anterior and two posterior, are also present. The downward-facing articulating facets on the posterior surface fit into the upward-facing anterior facets of the vertebra immediately behind (see fig 6.14).

Investigation

To examine the different types of vertebra found in the skeleton of a small mammal

Procedure

1 Study carefully the whole, mounted skeleton of a rabbit (or rat, or dog). (Refer to fig 6.7, if no mounted skeleton is available.)

2 Obtain a large number of loose vertebrae, and arrange them on the bench in a line, with the neck vertebrae at one end and the tail vertebrae at the other end. Use the mounted skeleton for reference.

3 Make large, clear drawings (anterior and lateral views) of the following vertebrae: first neck or *cervical* vertebra (*atlas*), second cervical vertebra (*axis*), one other cervical vertebra, one *thoracic* vertebra, one *lumbar* or middle back vertebra, all the vertebrae from the lower back or hip region (*sacral*), one *caudal* or tail vertebra.

Questions

1 Which vertebra has a very wide neural canal, a small centrum and broad, flat transverse processes?

2 What kind of movement of the head will this arrangement allow?

3 Note the *odontoid* process of the axis vertebra, which slots into the atlas. What kind of movement of the head does this allow?

4 Examine a thoracic vertebra carefully. Note the large neural spine. What is its function?

5 Note the small depressions at each end of the centrum and on the ends of the transverse processes. What is their function?

6 What differences can you observe between the thoracic and lumbar vertebrae? Present your answer in the form of a table.

7 Count the sacral vertebrae. Are any of them fused?

8 At what points does the sacrum articulate with the pelvic (hip) girdle?

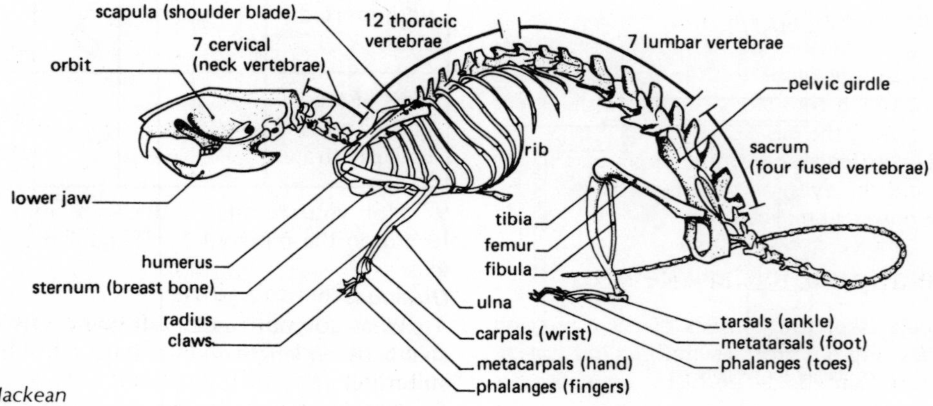

After D. G. Mackean

fig 6.7 The skeleton of a rat

9 How do the more posterior caudal vertebrae compare with the more anterior ones?
10 Examine the mounted skeleton again. How many pairs of ribs can you count?
11 How many of these are connected to the breastbone (sternum)?
12 Note the pieces of cartilage between the first seven pairs of ribs and the sternum. How many pairs of ribs are attached to those in front by cartilage?
13 How many 'floating' ribs, with no ventral attachment are there?

Vertebra	Human	Rat
cervical	7	7
thoracic	12	13
lumbar	5	6
sacral	5	4
caudal	4	30 (approx)
Total	33	60 (approx)

Table 6.1 Comparison of human and rat vertebrae

6.22 The skull

Examination of the skull bones of any mammal reveals that the skull is made up of two separate parts. The upper part consists of a brain-box or cranium, and fused upper jaw. Articulating with this is the lower jaw.

The cranium consists of a number of flattened bones, joined to form a rounded case protecting the brain. In newly-born babies, the bones on the top of the cranium may not have joined up completely, resulting in a delicate, unprotected area known as the *fontanelle*.

The cranium also *protects* the organs of special sense. The eyes are partially enclosed in bony sockets and the delicate structures of the inner ear are surrounded by a capsule of bone. The hard palate is a bony structure which runs along the top of the upper jaw, partly enclosing the nasal cavity.

The lower jaw articulates with the upper part of the skull just behind the eyes, and the two structures are connected by powerful muscles.

6.30 The appendicular skeleton

The appendicular skeleton consists of the *limbs* and the *limb girdles.* Fig 6.9 gives a simplified illustration of these structures, and the details of the arrangements of the bones in Man and in the rat are shown in figs 6.11 and 6.7 respectively.

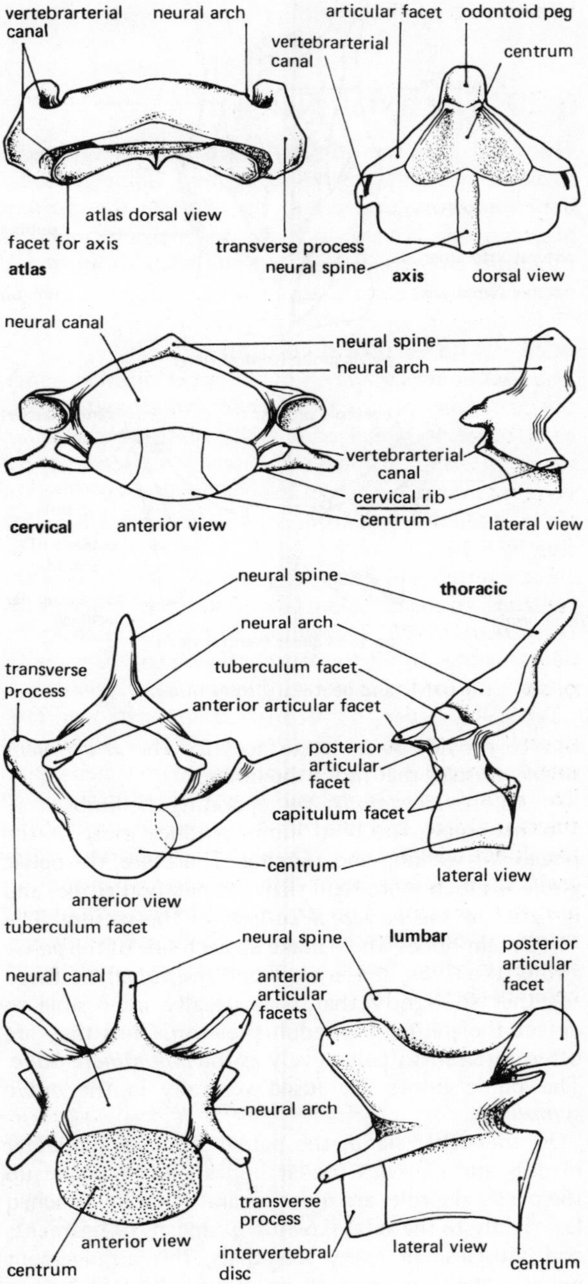

fig 6.8 The different types of vertebrae in Man

6.31 The limb girdles

There are two limb girdles; the *pectoral* or shoulder girdle and the *pelvic* or hip girdle. They have the following three main functions:

i) to form a more or less rigid *connection* between the axial skeleton and the limbs;

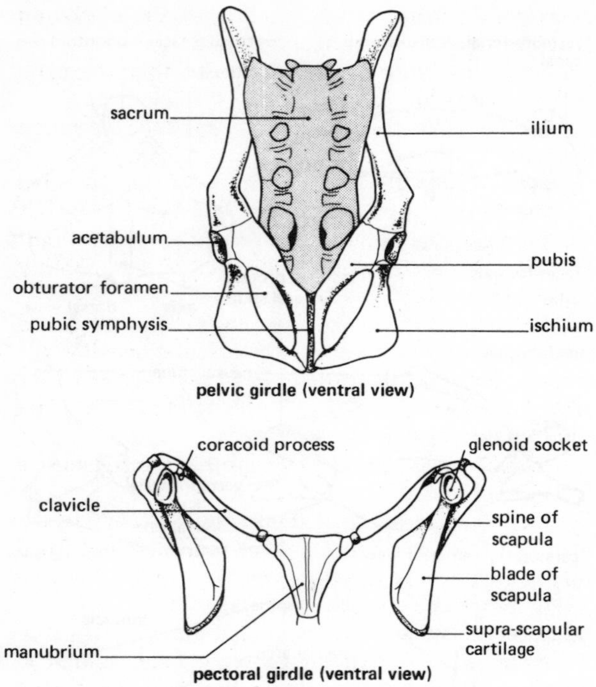

fig 6.9 The pelvic and pectoral girdles of a rat

ii) to provide suitable surfaces for the *attachment* of the *muscles* that move the limbs;
iii) to provide *stability* by separating the limbs.

In mammals, the hind limbs produce most of the power for walking and running. Therefore, the pelvic girdle is much more rigid than the pectoral girdle, and is *fused to the vertebral column* at the sacrum. The three main bones which make up each side of the pelvic girdle (the *ilium*, the *ischium* and the *pubis*) are fused together so tightly that it is usually impossible to detect the joints in an adult skeleton. Thus they are often referred to collectively as the *innominate bone*. The pubic bones are fused ventrally in the *pubic symphysis*.

On the other hand, the paired *scapulae* (shoulder blades) and *clavicles* (collar bones), which make up the pectoral girdle, are quite separate and are attached less rigidly to the axial skeleton by means of ligaments and muscles. In many mammals, this arrangement enables the forelimbs to be moved through a wide variety of planes and angles. Unlike the pelvic girdle, the pectoral girdle is not a complete girdle.

Investigation
The pectoral girdle of a rat
Procedure
1 In your specimen, the long, thin clavicles have probably become separated from the scapulae. Place one scapula flat on the bench, with the 'spine' (the long ridge, projecting at right angles from the 'blade') facing upwards.
Question 1 Which surface can you now see?
2 Make a careful drawing of the scapula, labelling the following:
i) anterior edge;
ii) dorsal edge;
iii) posterior edge;
iv) point of attachment to the clavicle;
v) point of attachment to the forelimb.
3 At (iv) above, note the remains of a piece of cartilage which was attached to the clavicle.
Question 2 To which part of the skeleton was the other end of the clavicle attached?
4 Where two bones are attached together so that they can still be moved in relation to each other, they are said to articulate with each other.
Question 3 What do you notice about the shape and texture of the part of the scapula that articulates with the forelimb (the *glenoid* socket)?
Question 4 What is the likely function of the spine of the scapula?

Investigation
The pelvic girdle of a rat
Procedure
1 Use the mounted skeleton of a rabbit to enable you to orientate the rat pelvic girdle correctly.
2 Your specimen will consist of two innominate bones, probably separated because of the breakdown of the cartilage which joins them in life at the pubic symphysis (see fig 6.9).
Question 5 What is the likely function of this cartilage in the female?
3 Make a careful drawing of the lateral view of one innominate bone. Identify and label the following features:
i) the ilium;
ii) the ischium;
iii) the pubis;
iv) the *obturator foramen* (the large 'hole' between the bones);
v) the *acetabulum*, where the pelvic girdle articulates with the hind limb.

Remember that, execept in young specimens, you will be unable to see the *sutures* between the three main bones.
4 Examine the inner surface of the ilium.
Question 6 What is the function of the roughened area just above the 'notch'?

6.32 The pentadactyl limb

The limbs of all the major groups of vertebrates are based on the same pattern. This pattern is shown diagrammatically in fig 6.10.

Because the basic pattern includes five digits (fingers

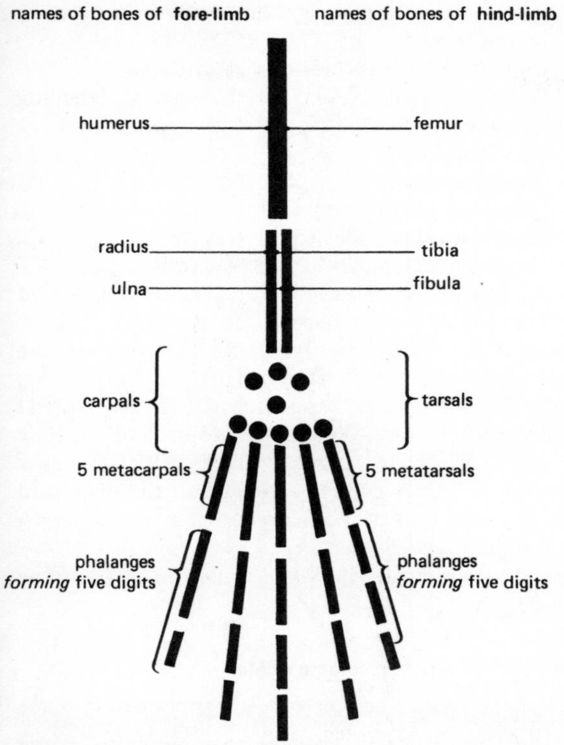

fig 6.10 A plan of the pentadactyl limb

and toes), the structure is known as a *pentadactyl limb.* In many cases, e.g. horses, pigs and birds, the number of digits has been reduced during evolution, and the limbs of the snakes have disappeared completely.

The typical pentadactyl limb consists of one upper bone (articulating with the girdle), two parallel lower bones, three rows of small ankle or wrist bones (9 in all), five parallel sole or palm bones and fourteen digit bones (four sets of three and one of two). The names of these bones, as applied to both fore- and hind limb are given in fig 6.10.

Investigation
To examine the forelimb of a rat
Procedure
1 Obtain the individual bones of a rat forelimb and place them flat on the bench, in correct relationship to each other. Use the mounted skeleton as a model.
2 Make a careful drawing (lateral view) of the assembled bones, labelling the *humerus, ulna, radius, carpals, metacarpals,* and *phalanges.* The ulna has a projection (the *olecranon process*), at the elbow end.
Question 1 What is the function of this projection?
3 Hold out your own arm in front of you, with the palm facing downwards (this is known as the *prone* position).
Question 2 What is the position of the olecranon process (use your other hand to locate it)?
4 Now turn your extended arm so that the palm faces upwards (*supine* position).
Question 3 What is the new position of the olecranon process?
Question 4 Is the same mobility of the forelimb possible in the rat or rabbit?
Question 5 What is the position of the rat forelimb?
Question 6 What is the importance of the mobility of the arm in Man?
5 Examine the rat humerus. Note that the head (*proximal*) end is rounded for articulation with the scapula, and that the other (*distal*) end is grooved for articulation with the ulna. Both ends bear a number of processes for muscle attachment.
Question 7 Can you identify another site for muscle attachment?
6 Examine the bones of the forelimb. Note that the ulna and the much more slender radius are not fused but lie in very close contact.
7 Examine the carpals.
Question 8 How many can you count?
8 Note that the first digit is very small compared with the other four; much smaller than the human thumb is in comparison with the fingers.

Investigation
To examine the hind limb of a rat
Procedure
1 Obtain the individual bones of a rat hind limb and place them flat on the bench, in the correct relationship to each other. Use the mounted skeleton as a model.
2 Make a careful drawing (lateral view) of the assembled bones, labelling the *femur, tibia, fibula* (partly fused to the tibia), *tarsals, metatarsals* and *phalanges.* The patella or knee cap may have been lost if the bones have been separated for some time.
3 Examine the tarsals.
Question 9 How many can you see?
4 Note that one of the tarsals projects backwards to form a heel-bone for the attachment of muscles.
5 Examine the metatarsals.
Question 10 What does their length tell you about the way in which the rat moves?
6 Note the tip of the phalange at the end of each digit is pointed, for insertion into a *strong claw.*

6.40 Human skeleton

The human skeleton has the same basic structure as the mammalian skeletons discussed in sections 6.10 to 6.30.

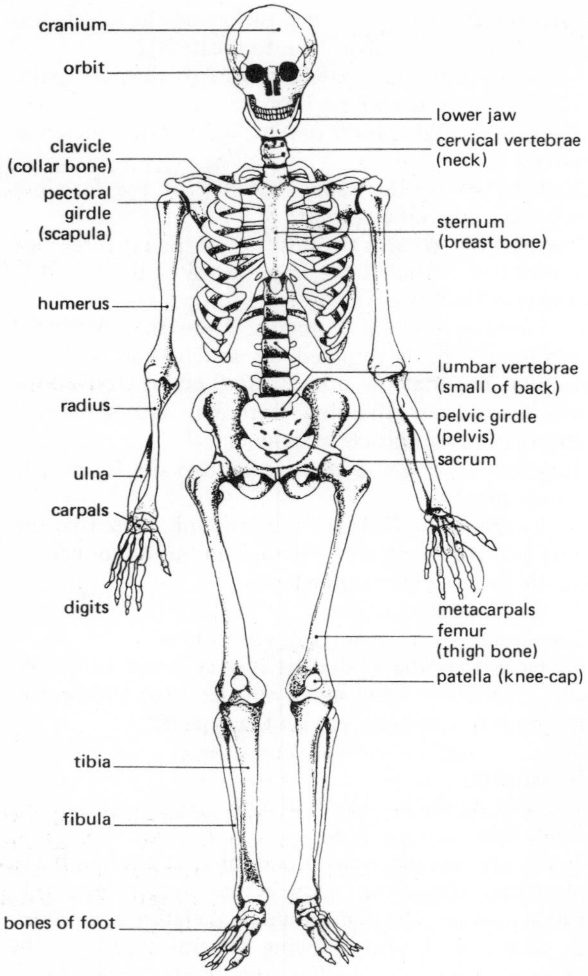

fig 6.11 The human skeleton

6.41 The axial skeleton

The vertebral column has the following regions and vertebrae.
Cervical region (neck) – 7 vertebrae
Thoracic region (chest) – 12 vertebrae
Lumbar region (small of the back) – 5 vertebrae
Sacral region (attached to the pelvic girdle) – 5 fused vertebrae (sacrum)
Coccygeal region ('tail') – 4 fused vertebrae (coccyx)
Total – 33 vertebrae
Attached to the thoracic vertebrae are twelve pairs of ribs connected to a breastbone or sternum.

6.42 The skull

The skull has a cranium or brain box of eight bones, a face of fourteen bones and auditory ossicles of six bones. The dentition of the skull is dealt with in Chapter 2, section 2.

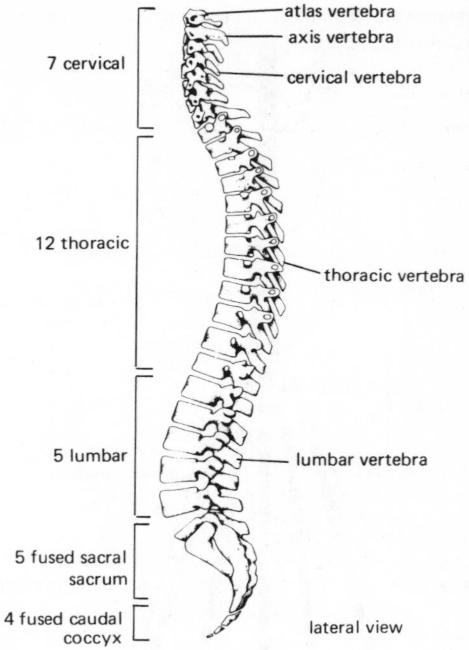

fig 6.12 The vertebral column of Man

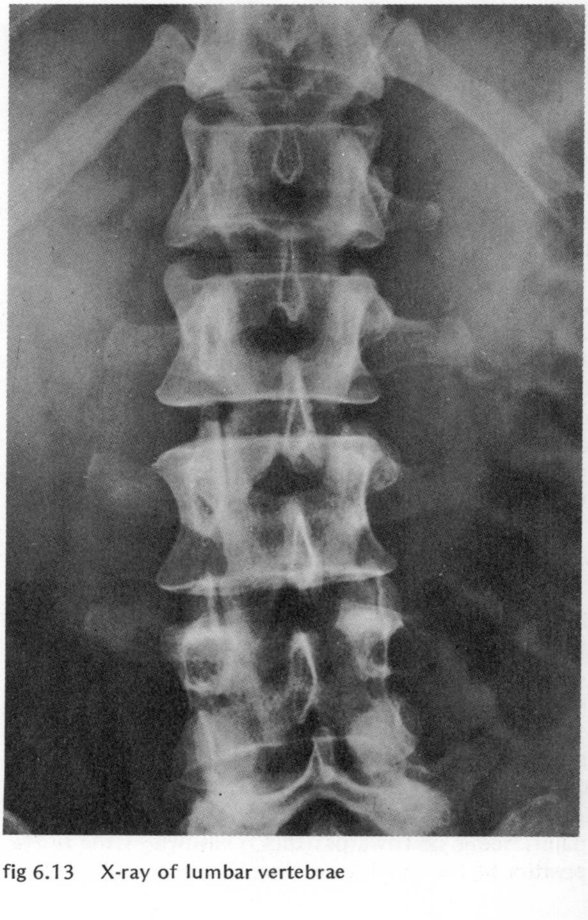

fig 6.13 X-ray of lumbar vertebrae

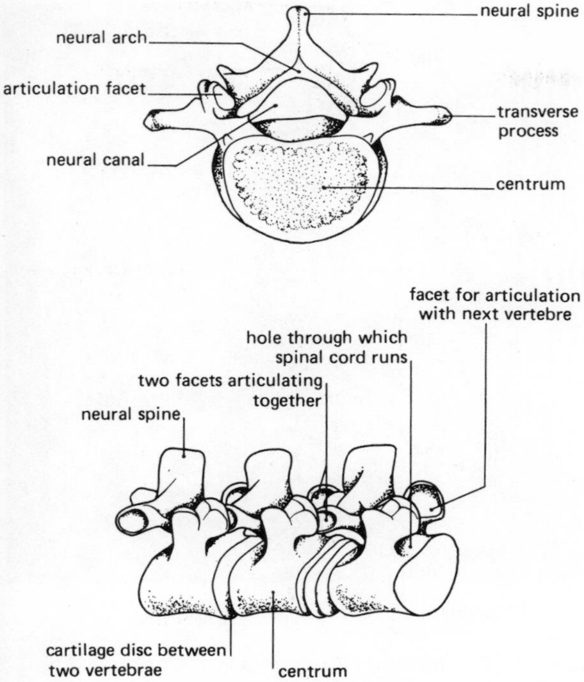

fig 6.14 The structure of a lumbar vertebra and the articulation between several vertebrae *After D.G. Mackean*

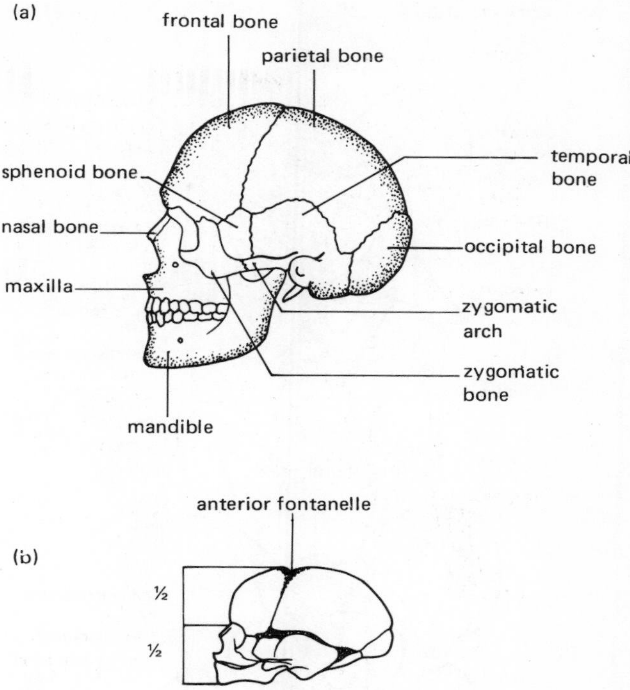

fig 6.15 (a) Lateral view of adult skull
(b) Lateral view of infant skull

6.43 The appendicular skeleton

a) *Limb girdles*

The shoulder girdle has two scapulae (shoulder blades) and two clavicles (collar bones). The clavicles articulate with the sternum but since the scapulae do not meet at the dorsal surface the girdle is not a complete circle.

The hip girdle has two large innominate bones each composed of pubis, ischium and ilium. These form a complete ring of bone meeting ventrally at the pubic symphysis and dorsally at either side of the sacrum. The pelvic girdle thus completely surrounds the pelvic cavity. The pelvic girdle of a woman is wider and shallower than that of a man. At the junction of the three bones in each innominate is a socket, the acetabulum, into which the head of the femur fits (see fig 6.16).

b) *The limbs*

The upper arm is formed from the humerus and the forearm from the radius and ulna. There are eight small carpal bones in the wrist, arranged in two rows of four, and five metacarpal bones form the palm of the hand. There are two phalanges in the thumb and three in each of the four fingers (fourteen in all).

The femur is the long bone of the thigh, with the tibia (shin bone) and fibula completing the lower limb. The knee joint has a patella or knee-cap lying in the tendon of the quadriceps femoris muscle. In the foot there are seven tarsal bones forming the ankle, together with five metatarsal bones forming the arch of the foot and fourteen phalanges distributed in the toes in a similar pattern to those in the fingers. The distal phalanges of the little toe are often fused.

6.50 How do muscles move the skeleton?

Investigation

The muscles of a rat's hind limb

Procedure

1 Obtain a rat carcass which has been preserved after a previous dissection. It will probably have been opened ventrally for investigation of the abdominal organs. Pull the skin at the base of the abdomen clear of the body wall and make a cut along the edge to as far as possible below the knee.

The lower you cut, the more difficult it will become to separate the skin from the underlying muscle.

2 Make a second skin cut completely around the top of the leg.

3 Peel back the skin from the top of the leg downwards, using the back of your scalpel to remove the connective tissue between skin and muscle. Expose the heel.

Question 1 What is the general appearance of the calf muscle (*gastrocnemius*)?

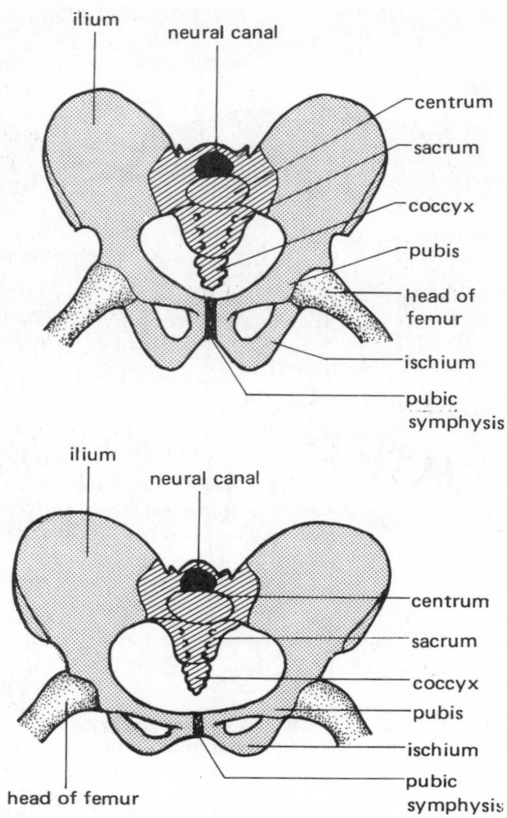

fig 6.16 (a) Pelvic girdle of male
(b) Pelvic girdle of female

Question 2 How does this appearance differ from that near the heel-bone?

4 Use the scalpel (blunt side) to separate the longitudinal bundles of muscle fibres. Note that they are surrounded by a sheath of connective tissue.

Muscles are unable actively to increase their length: they can only *contract and relax.* When a muscle relaxes, it can be lengthened only by the action of another muscle pulling indirectly upon it. Therefore, muscles tend to *work together* in *antagonistic pairs.* Such a pair are the biceps and triceps muscles of the arm (see section 6.00). Muscles that bend limbs are termed *flexors* and those that straighten limbs are termed *extensors.* Thus the biceps is the flexor of the arm and the triceps is the extensor of the arm.

Fig 6.19 shows the action of these muscles in raising the forearm.

For many pairs of antagonistic muscles, one member of the pair is more powerful than the other. This is because one member may be used for the principal function (e.g. *lifting* of a weight by the arm) and the other member is used simply to return the limb to its original position.

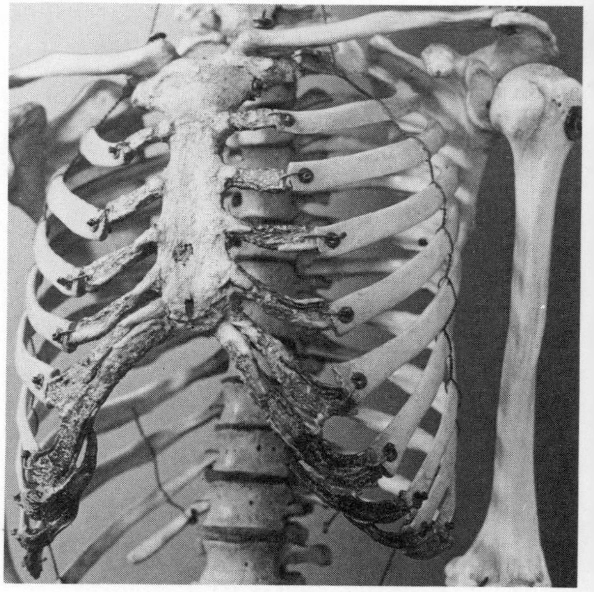

fig 6.17 Upper part of the human skeleton

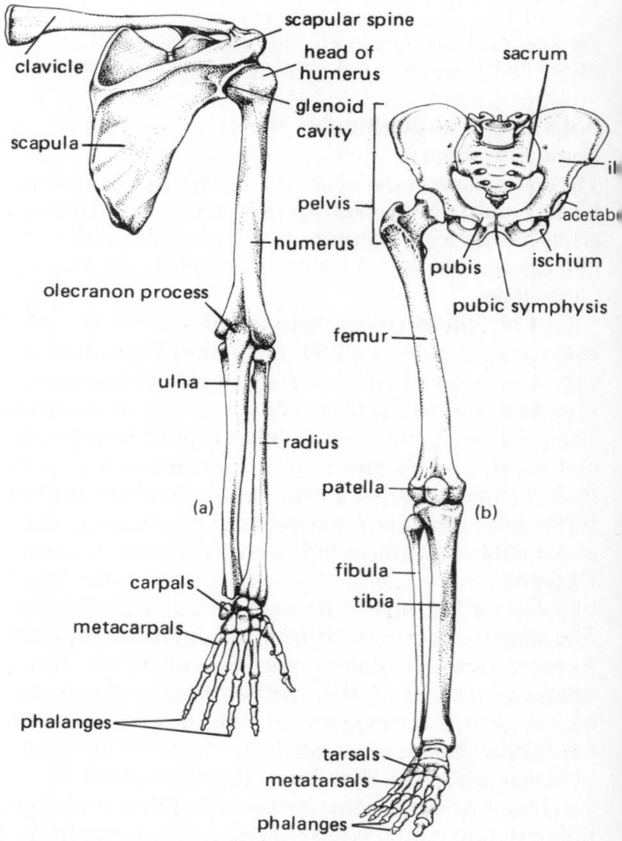

fig 6.18 Bones of (a) right arm, hand and pectoral girdle
(b) right leg, foot and pelvic girdle

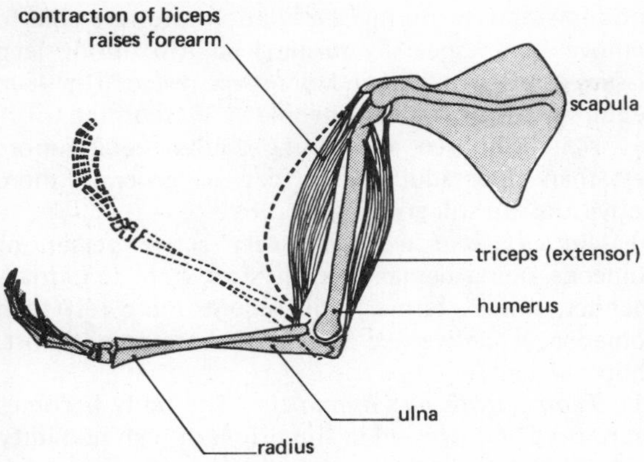

fig 6.19 The antagonistic muscles moving the forearm
After D.G. Mackean

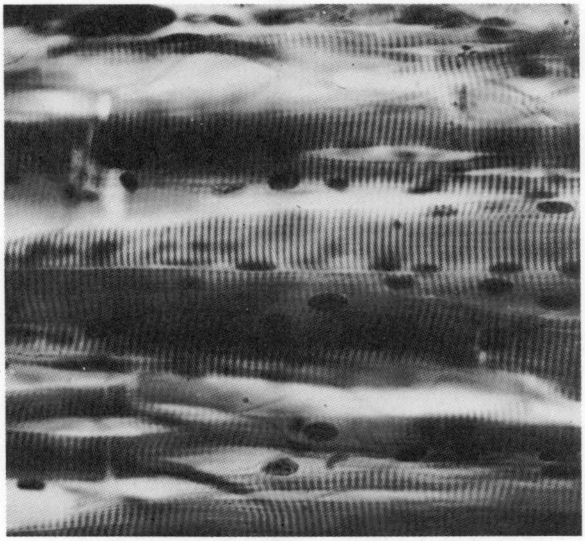

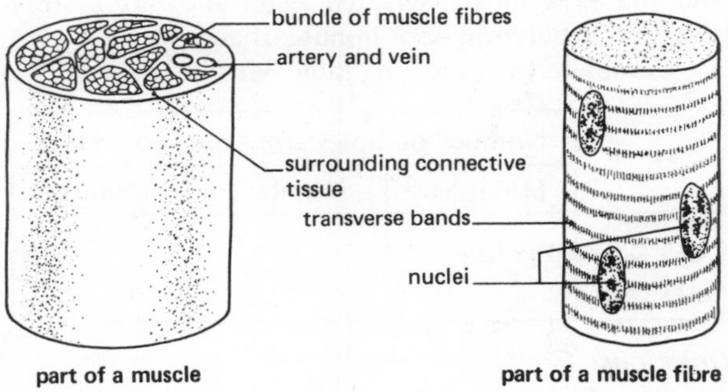

fig 6.20
(a) Microscopic structure of skeletal muscle
(b) Stereograms of muscle structure

Limbs work as a series of levers and in the case of the arm, the elbow joint acts as the fulcrum and the effort is produced by the biceps (see section 6.60). Thus the flexor of the arm is more powerful than the extensor.

An example of the opposite situation can be seen in the running action performed by a rat. This is brought about in part by the extension of the lower hind limb, which pushes downwards and backwards on the ground, causing the rat to move forwards and upwards. The gastrocnemius or calf muscle (the extensor) contracts, pulling the heel-bone upwards and bringing the metatarsals into line with the tibia-fibula. The muscle on the anterior side of the leg (the flexor) is less powerful, being used only to return the foot to its original position.

6.51 The structure of muscle

As with the term *bone* (see section 6.12), we use the term *muscle* in two different ways. In one sense it is used to refer to *specific organs* (the biceps muscle, the deltoid muscle, etc); in the other sense it refers to one of the *tissues* of which these organs are made up.

The unit of muscle tissue is the muscle cell or fibre. This is an elongated, approximately cylindrical structure, which under the light microscope can be seen to have several nuclei and a series of transverse bands (*striations*) parallel to its short axis. These cells occur in *bundles*, each bundle being surrounded by a sheath of connective tissue. A muscle such as the biceps consists of a *number of such bundles*, together with nerve and blood supplies, all surrounded by a further sheath of connective tissue.

The above description and diagram applies only to the kind of muscle that moves the arms and the legs. This type of muscle is termed *skeletal* muscle, *voluntary* muscle (because it is under the control of the will) or *striated* muscle.

Two other types of muscle tissue are found in mammals, *cardiac* muscle, where the *transversely banded* fibres are *branching*, and *smooth* muscle, where the fibres are *spindle shaped* and *striations are absent*. Cardiac muscle is found in the heart and smooth muscle is found in the bladder and certain other non-skeletal organs.

6.52 Energy for muscle function

In order to contract, muscles require energy. Some of this energy is stored in the muscle in the form of the carbohydrate glycogen (see Chapter 4, section 4.32). This is oxidised to carbon dioxide and water, with the production of ATP for 'contraction energy' during cellular respiration in the muscle fibres.

However, since the amount of glycogen stored in a muscle seldom exceeds one or two per cent of its weight, this is soon used up during vigorous exercise. Energy is then obtained by the oxidation of glucose transported to the muscle through its rich blood supply (the extent of this supply is obvious from the appearance of fresh meat which is skeletal muscle).

Experiment

Can muscles continue to work indefinitely?

Procedure

1 Place your elbow and forearm flat on a bench or table top and grasp a large stone or weight of about 2 kg in your hand.

2 Raise your forearm, grasping the weight firmly, to a vertical position as many times as possible in a period of three minutes. Each time return the forearm to the bench top, ensuring that the knuckles touch the bench.

A second person should act as timekeeper to count the number of times the weight is lifted to the vertical in each minute of the three consecutive minutes.

3 The arm should then be rested for five minutes and the experiment repeated but this time allow a thirty second rest between each minute of activity.

4 The arm should then be rested for five minutes and the experiment repeated again allowing a sixty second rest between each minute of activity.

5 Complete the following table with your results.

	Number of times arm raised to vertical		
	Minute 1	Minute 2	Minute 3
no rest			
30 s rest			
60 s rest			

Questions

1 Describe your results.

2 Did the rest periods make any difference?

3 Why were you unable to maintain the same rate of lifting in the three successive minutes? (See Chapter 4, section 4.41.)

Fatigue

Skeletal muscles working unceasingly become fatigued. The muscles are no longer able to contract and therefore the arm cannot be moved. Given a period of rest, however, the muscle is able to recover and resume working for another period. This fatigue is due to the build-up in the muscles of waste products which must be removed before the muscle can work normally again.

Smooth or involuntary muscle works more slowly and infrequently, with less energy consumption, so no fatigue is experienced.

Heart muscle has the unusual capacity of never developing fatigue at its normal rate, in spite of beating non-stop from birth to death.

In order for muscles to recover from fatigue, rest is necessary so that blood vessels can remove the waste products and renew the supply of energy-rich substances and oxygen. Muscles at rest produce few waste products and so normal circulation rates are able to remove them, especially during sleep. Also during sleep tissues can be repaired and all organs rested. The sleep requirements of different people vary according to:

a) *Age* Children and young adults require more rest than older adults since they are generally more active and are still growing.

b) *Activity and work* A very active person or someone doing demanding physical work (e.g. road-mender, miner, farmer) will require more rest than someone less active (e.g. bank clerk, hotel receptionist, shop assistant).

c) *Temperature and humidity* The body becomes more tired and stressed in conditions of high humidity and temperature than in cool, dry situations, e.g. tropical jungle or blast furnace compared with Northern Temperate forest or air-conditioned office. Therefore, under the former conditions, more sleep is needed.

Muscle cramp usually involves muscles of the legs and is found more commonly in elderly people than in the young. It is caused by lack of nutrients or oxygen arriving at the muscle or by an accumulation of waste products that are not carried away. The reason for this state of affairs is usually an interference with the blood flow. This can be brought about by varicose veins, overweight or disease of the heart and arteries. Exercise of the muscles during the day can help, and hot and cold baths can improve the circulation. At night, heating pads in the small of the back or on the legs also help to maintain blood flow in the affected muscle.

6.60 The action of muscle through levers and joints

A *lever* is a simple *machine* which enables a certain amount of work to be done. It involves the expenditure of energy in producing an *effort* to move a *load*. All levers have a *fulcrum* or pivot, which is their fixed point of support.

There are three orders of levers, classified according to the relationship between the positions of the load, the fulcrum and the point of application of the effort.

In a *first order lever*, the fulcrum is between the effort and the load. A crowbar and a pair of scissors are two examples of this order.

A wheelbarrow is an example of the *second order of levers*, where the load is between the fulcrum and the effort.

Finally, there are the *third order levers*, where the effort is applied between the fulcrum and the load. This order is used when a ladder is lifted away from a wall.

Examples of all of these orders can be found in the joints of Man (see fig 6.23).

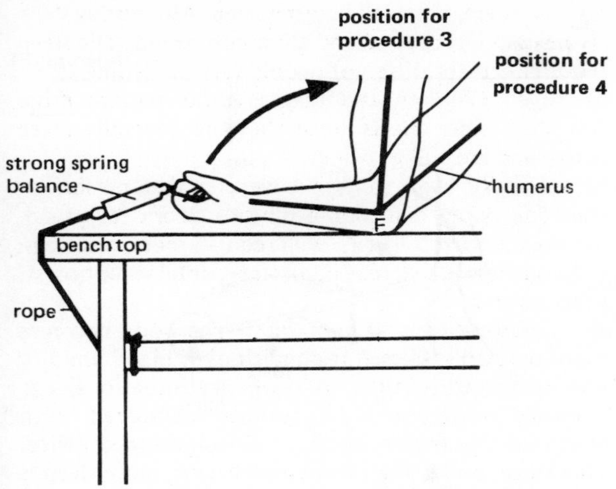

fig 6.21 Position of arm for application of effort

Experiment
Does the application of effort in a lever system affect the efficiency of this simple machine?

Procedure

1 Attach a strong spring-balance (max. 20 kg or 200N) to the end of a bench as shown in fig 6.21. Tie the strong rope around the end of the bench and attach the spring-balance to the rope. A rope hand-grip should then be attached to the hook of the balance.

2 Place the forearm horizontally, with the elbow resting on the bench top and the hand grasping the rope grip.

3 Keep the upper arm vertical, the elbow firmly on the bench and lift the forearm as far as possible whilst pulling steadily on the hand-grip. Record the final force exerted.

4 The arm should now be rested for five minutes. Repeat the action with the upper arm at 45° to the bench top as shown in fig 6.21. Record the final force exerted.

Questions

1 In which position was the arm able to exert the maximum force?

2 Can you suggest any reason for this?

You will have discovered that less effort can be exerted with the upper arm inclined to the vertical. The work produced by a force acting on a lever is the product of force and distance.

Force x Distance (from the fulcrum) = Work

The greatest amount of work is produced if the force is applied at right angles to the lever. A similar principle applies in opening a large and heavy gate. The easiest way to pull it open at the end of the gate is if the pull

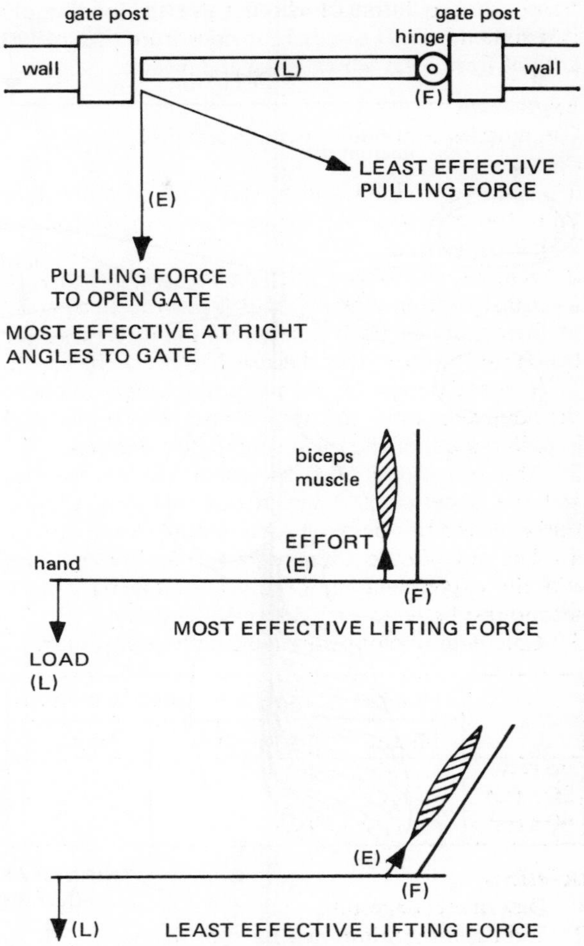

fig 6.22 Effective application of effort

is at right angles (see fig 6.22). At the elbow joint, the angle at which the tendon joins the bone is vital in relation to the work produced by the muscle and thus to the work done in moving the load (the spring balance).

6.61 Joints

In all the above examples, the fulcrum of the lever has been at the position of a joint between two bones. These joints allow varying degrees of movements, and can be classified on this basis into several different groups.

Perhaps the easiest type of joint to understand is the *hinge joint*, such as is found at the *elbow and knee.* This allows movement in one plane only, and so is directly comparable with the simple examples of levers considered in section 6.60.

Ball and socket or *universal joints* are found at the *shoulder* and *hip.* These allow movement in all planes,

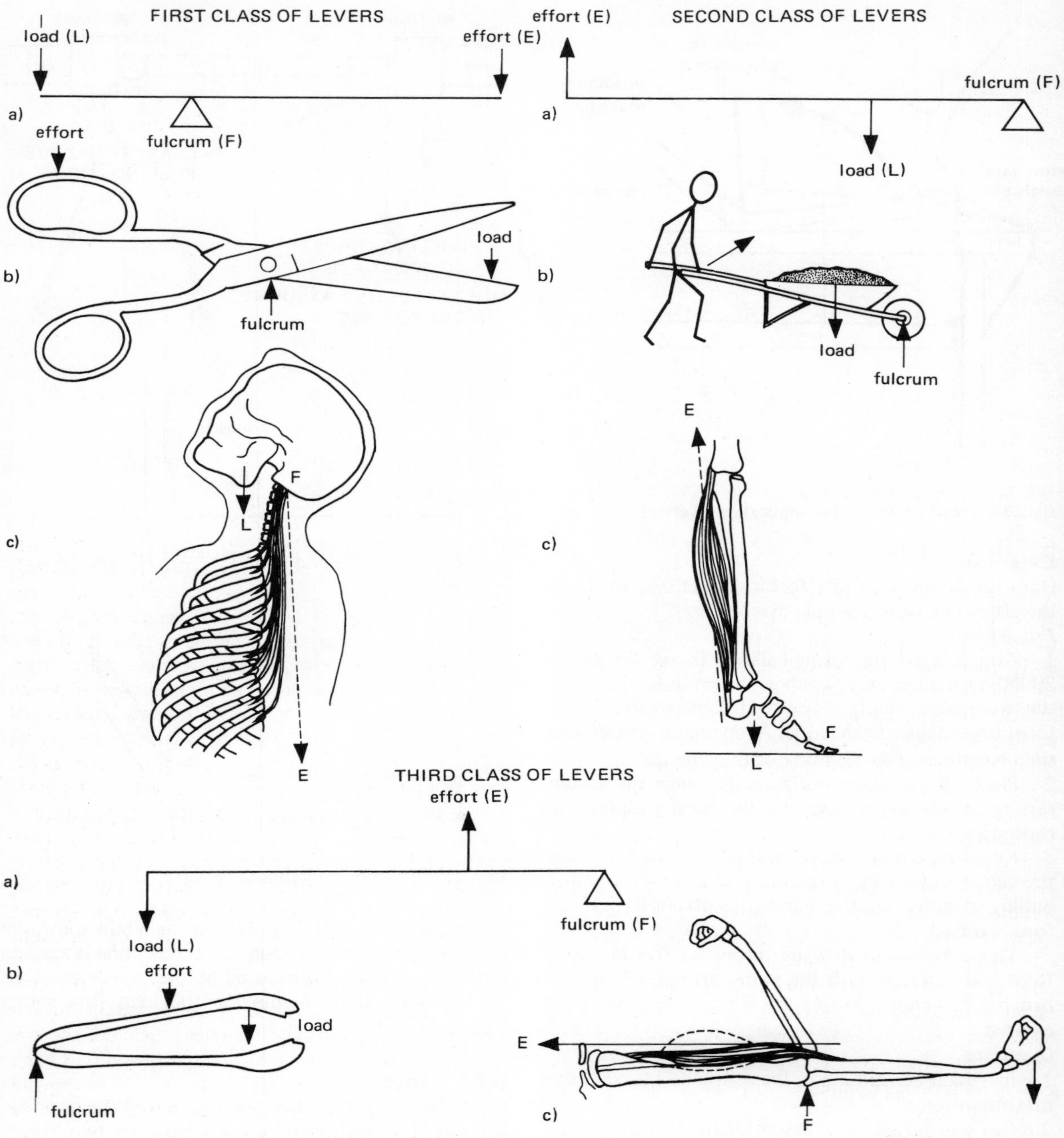

fig 6.23 The three classes of lever (a) diagram (b) simple machine using same principle (c) example in the body of Man

and thus we can move our arms forward, sideways or in a circle.

Examples of *gliding joints* are found between the carpals and between the tarsals, where one bone moves over the surface of another.

A fourth type of joint is the *pivot*. This allows rotation of one bone around another, and the most obvious example is found between the *atlas* and the *axis vertebrae,* which allows rotation of the skull.

Finally, we should consider the *fixed joint or suture*. We have seen examples of this in the skull and in the innominate bone, where the ilium, ischium and pubis have become fused so that they are not moveable in relation to each other.

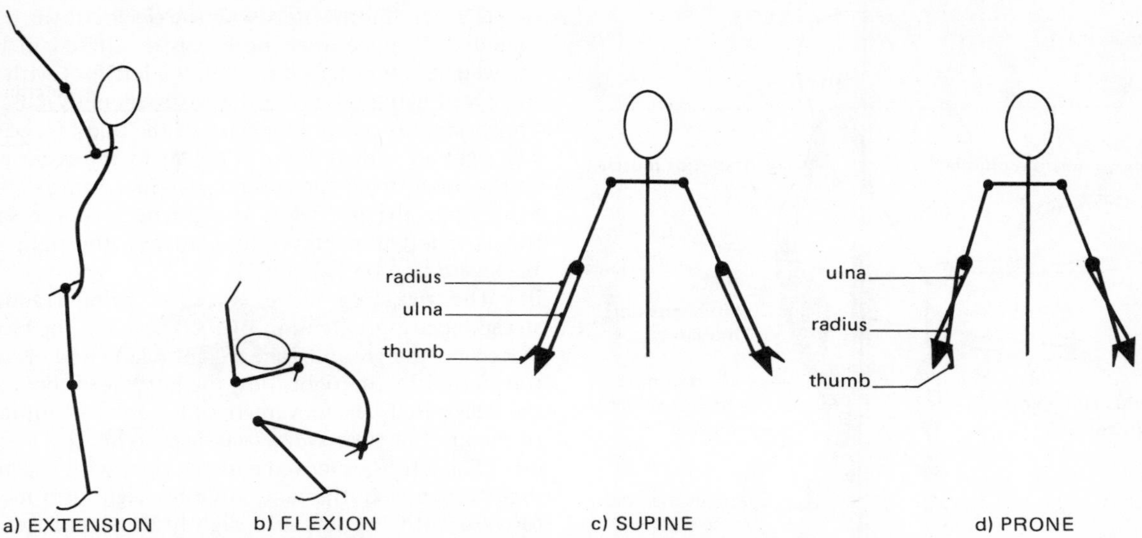

fig 6.24 Movement of bones about joints (a) extension (b) flexion (c) supine (d) prone

6.70 Posture

Maintenance of posture is effected by a whole series of muscles in the neck, trunk and legs acting on the framework provided by the skeleton. Throughout the body there are antagonistic muscles: those tending to pull the body forwards are equally opposed by muscles tending to pull it backwards, likewise those pulling it to one side are opposed by a set which pull it equally in the opposite direction. When we stand upright, our muscles have just the right degree of tension to maintain the body in the erect position without movement. If a person standing upright should faint, as sometimes happens, for example, when soldiers are standing to attention for a long time in hot weather, the nervous control of the muscles ceases, muscular tone is lost and they collapse.

The problem of standing erect is unique to Man (and certain apes) since during his evolution he has taken to walking on his two hind legs instead of all four. Consequently the body weight is distributed over two rather small areas (the feet) and Man can quite easily overbalance compared with an animal standing on all four legs.

Not only when standing, but when sitting also, the muscles of the trunk and the head play an important part in maintaining posture. Obviously all muscles must be kept in good working order: muscles that are not used become weak, whereas those that are used become strong. Hence the importance of exercise for general well-being and of specialist training for athletic events. The particular muscles that are required, for example, in running, rowing or weight-lifting are strengthened in this way and thus produce the required movements without any problems of strain.

Fig 6.25 shows mucles involved in the erect posture. Particularly important are the quadriceps and gluteals in the thigh, and the tibial, gastrocnemius and soleus muscles in the lower leg. Good posture involves the natural and comfortable bearing of the body and should be achieved with a minimum of muscular effort. The erect posture is maintained by minimum expenditure of energy partly because only a small number of muscle fibres are contracted at any one time, partly because ligaments hold the joints in position at the knee and hip and partly because the centres of gravity of the head and trunk are balanced over the centre of gravity of the body as a whole. The latter is approximately in the centre of the hip girdle and, for a correct standing position, it should be located immediately above the middle of the area covered by the two feet. Each foot has an arch which is formed by the metatarsal bones and the arch has three supports, at the heel, the base of the big toe and the bones of the little toe. As shown in section 6.71, immediately the line through the centre of gravity moves outside of the area covered by the feet and between the feet, either backwards or forwards, then the body must start moving. A foot must be lifted from the ground and the initial movements to regain balance begin.

Incorrect posture can result in malfunction of digestive, lymphatic, muscular and skeletal systems. Continued sitting in a slumped position whilst reading or writing can produce a hunched back. Tight clothing and incorrect footwear can also lead to poor posture. Good postural habits are important from the beginning, since the bones of young children are still growing and thus able to change their form. A conscious effort by parents, teachers and children must be made to improve posture.

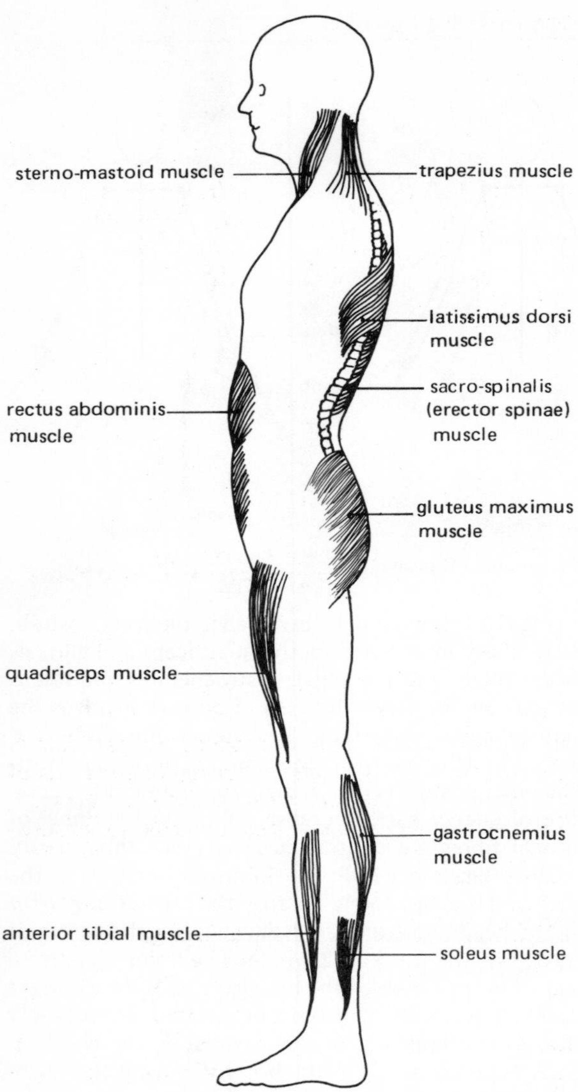

fig 6.25 The principal muscles used to maintain posture

i) The standing position, with the weight of the body shared equally between both legs, is altered in that the weight is transferred to, say, the left foot with the right heel being raised by the gastrocnemius muscles. This brings the centre of gravity of the body forwards. The right leg is then flexed at the hip by the quadriceps and is raised from the ground, the knee being slightly bent while the left leg is straightened. At the same time the left arm moves forward and the right arm backward (see fig 6.26 no.1).

ii) The right leg moves forward, being gradually straightened by extension of the knee, and the heel is placed on the ground. The weight of the body is now transferred to the right foot, the left knee is bent and the heel raised. The movement of the arms is continued to the limit of their swing (see fig 6.26 no.2).

iii) The left leg is moved past the right with the knee slightly bent. At the same time the right arm moves forward with the left arm slightly backward (see fig 6.26 no.3).

iv) The left heel reaches the ground, the foot is flattened and the right heel raised while the left arm swings forward again. The whole movement is repeated (see fig 6.26 no.4).

Walking upstairs requires a great deal more energy than walking on level ground. The whole weight of the body has to be lifted against gravitational forces and thus a much greater strain is placed on the muscles, the heart and the breathing mechanism. The movements of walking upstairs are modified as follows:

The right leg flexes at the knee, with the body weight acting through the left leg. The flexion at the hip raises the thigh, knee and lower leg. The right foot is placed on the step and the body is raised by the muscles on the leg and trunk. The left foot is raised, the left knee and hip are flexed and the right knee is straightened.

During these movements of walking or climbing stairs the muscles of the neck and back regions all play their parts in keeping the head and trunk in the correct position.

6.71 Walking, running and jumping

Man does not often stand still and from the standing position just mentioned he must be able to walk forwards or backwards, to run and occasionally to jump. These actions are part of everyday life but are emphasised in many sporting or athletic activities. During walking one foot is always on the ground and the heels of the right and left feet are brought to the ground alternately, whereas in running the weight of the body is on the toes, and both feet are off the ground for a period of time. The sequence of movements of the legs and arms during walking are as follows:

6.72 Lifting heavy objects

As we have pointed out in the discussion of posture (section 6.70), Man has evolved a method of standing on two legs instead of four and as a result has become subject to certain problems of back and feet. Varicose veins and phlebitis (see Chapter 3) are partly due to the long column of blood in the veins of the legs. Backache is one of the commonest causes of absence from work in developed countries. Some causes for backache originate in the skeletal and muscular organs of the back. Many men and women have to sit in chairs for long hours in offices, or follow other work patterns

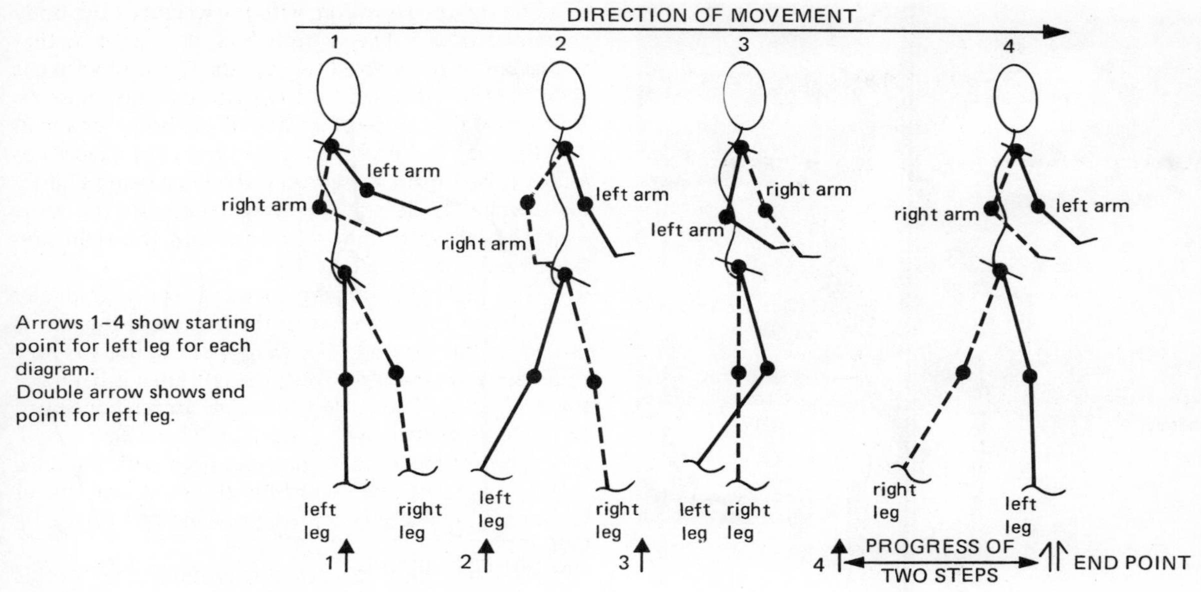

fig 6.26 Movements of arms and legs during walking

which involve bad posture and give no exercise to back muscles. The result is that when strain is placed on these weakened back muscles during work or leisure activities the individual experiences back-pain which can be very acute. This is brought on particularly when lifting heavy objects or during long periods of activity involving the back muscles, such as digging.

The vertebral column is often subjected to consider-

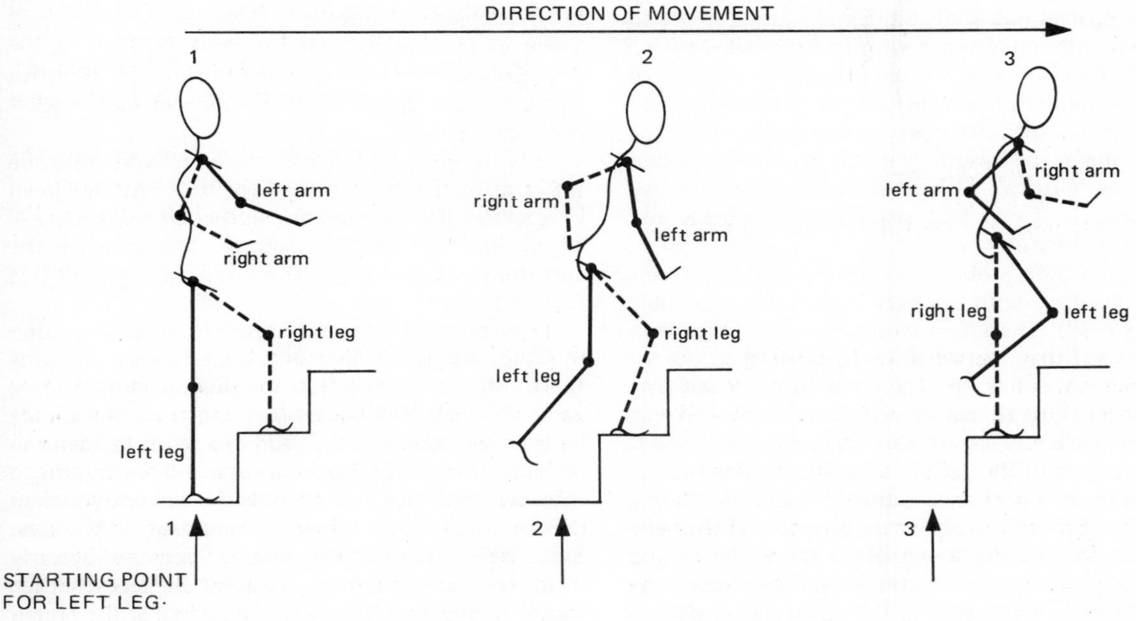

fig 6.27 Movements of arms and legs during stair-climbing

fig 6.28 A sprinter – note the centre of gravity well forward from the feet and the muscular action of the legs

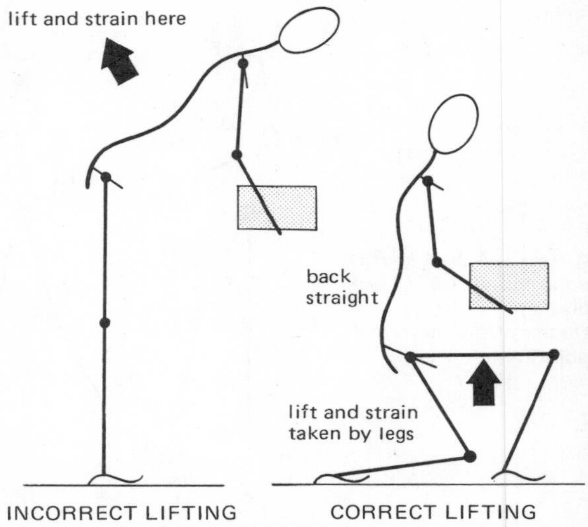

fig 6.29 Lifting heavy objects

able mechanical forces, which can result in damage to the intervertebral discs. These are cushion-like structures formed of fibro-cartilage and situated between the vertebrae. By their resilience they permit the bending and twisting movements of which the vertebral column is capable. A pair of spinal nerves emerge from between each pair of vertebrae. As a result of pressure the disc may deteriorate and a portion of it may be pushed out against a nerve, causing injury to the nerve and interference with its function. This is a painful condition and is often referred to as 'slipped disc'. It commonly occurs in the lower part of the back, thus affecting the sciatic nerve passing down the back of the thigh and resulting in the painful condition known as 'sciatica'. This pressure on the nerve may cause muscles to go into spasm and skin areas lose their normal sensitivity.

To avoid these problems any lifting of heavy objects should be done with the back as straight as possible (see fig 6.29). The knees should be bent rather than the back so that the weight is transferred to the leg muscles on standing up. The curve in the lower part of the back must be maintained when lifting, carrying children or merely sitting, and this can be achieved if the upper part of the back is always kept straight.

Obesity is a common cause of backache. Heavy layers of fat on the abdomen pull forward on the vertebral column and thus there is strain and fatigue on the back muscles. This condition is more common in developed countries and is nearly always due to overeating, although in some cases it may result from a glandular imbalance. Obesity can cause a number of other complications in body function, including strain on the heart and diabetes.

Pregnancy in a woman shifts the centre of gravity forwards. The extra weight of the foetus has to be held up by the back muscles, thus fatigue and backache can be a painful problem even if she does no heavy lifting.

6.73 Postural defects

There are three main types:

1 Kyphosis or 'round shoulders'. This is an excessive curve of the spine between the shoulder blades, producing a hump.

2 Lordosis or hollow back. This is an excessive curve forwards of the lower part of the back resulting in an exaggerated hollow curvature in the lumbar region.

3 Scoliosis or lateral curvature. This is when the curvature of the spine is either to the left or right (see fig 6.31).

These postural defects can result from malformation of bones during development, some deficiency during foetal life or disease (such as poliomyelitis) during early childhood. Poor postural habits carried on for long periods during childhood can result in kyphosis or lordosis, while scoliosis can result from carrying a heavy weight (such as a school bag) from one shoulder or one hand without regular exercising of the body. Since bad sitting and standing positions during early childhood can result in postural defects children should be supplied with chairs and desks of the proper height, both at home and at school, in order to provide a favourable position for reading and writing.

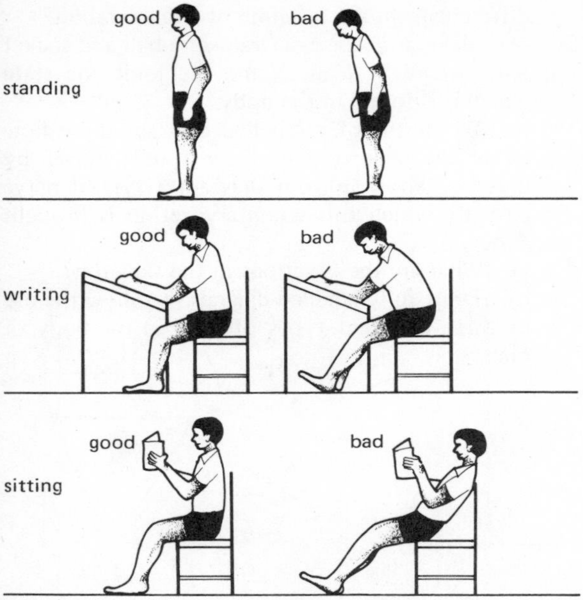

fig 6.30 Good and bad posture

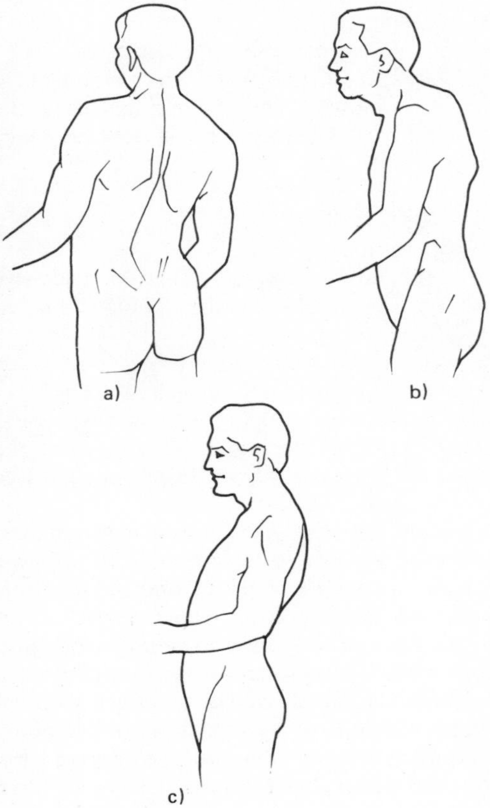

fig 6.31 Spine deformities (a) Scoliosis – lateral curvature (b) Kyphosis – hunchback (c) Lordosis – hollowback

6.74 Exercise and its importance

Children and young people should be encouraged to take part in games and other physical activities. Physical education staff in schools, while supervising organised forms of exercise, keep a watch on growing children to spot any postural defects which might merit referring them to a doctor. As children grow their physical exercise must be suitable for their age, sex and nutritional standard. Young people accustomed to regular physical exercise will grow into healthier adults than those whose bodies have not been regularly exercised.

Exercise of groups of muscles brings about many physiological changes. The muscles will produce heat as a result of cellular respiration (see Chapter 4) and at the same time rapidly change their lengths and widths. These changes will be accompanied by increases in the heart and respiratory rates. Exercise should therefore always be subject to certain conditions.

a) Before taking strenuous exercise the body should be warmed up with gentle exercise. This will prevent sudden strain on the muscles resulting in rupture of muscle fibres ("pulled muscles") or torn tendons. Furthermore, the muscles warmed and prepared for hard work in this way will not fatigue as quickly as they would if plunged straight into violent activity.

b) Exercise should be followed regularly so that muscles never lose their tone and preparedness. If long intervals of inactivity come between bouts of exercise the muscles are unfit and are then more likely to be damaged, also the heart and lungs are not able to cope effectively with the sudden demands placed upon them.

c) Exercise helps to improve the vital capacity of the lungs (see Chapter 4) since they are expanded beyond the normal tidal volume. This increased air flow enables more oxygen to be taken to the tissues and more waste gases to be returned for elimination. Appetite, rate of digestion and the peristaltic action of the gut are all increased by exercise.

Questions requiring an extended essay-type answer

1 a) The vertebral column of Man has five regions. *Name* each region and show the basis for distinguishing each one.
b) What similarities are present in the vertebrae from each of the five regions?
c) Describe the ways in which the first two vertebrae of the vertebral column are specialised for the functions they serve.

2 a) Draw a diagram of generalised mammalian vertebra.
b) Name four types of vertebra in Man's skeleton and state how each one differs from a generalised vertebra.

c) What are the functions of the sacrum?

3 a) Describe (with the aid of diagrams) the structure of

i) a thoracic vertebra, and

ii) a true rib of Man. Show how the rib is attached to the vertebrae and the sternum.

b) Describe the mechanism of breathing in Man with particular reference to the structures mentioned in part (a).

4 a) Draw a diagram to show the general plan of the pentadactyl limb. Label the bones drawn as though they are part of the forelimb.

b) How is the forelimb of Man adapted for use as a tool-holding limb? How are these adaptations different from the forelimb of a rat or rabbit?

5 a) Make a labelled diagram of a ball and socket joint found in Man. Name the joint and state where it is found in the body.

b) By means of a labelled diagram show how either the knee or the elbow joint is moved by muscles. Show in your diagram a typical nerve path by which this voluntary action is brought about.

6 a) What are the functions of the skeleton?

b) Draw fully labelled diagrams to show the use of three different types of lever in the body of Man.

7 Response

7.00 Irritability

Irritability is the ability of an organism to react in response to changes in its environment. It is one of the characteristics of life listed in the Introductory Chapter.

All living organisms show this ability to some degree: a flower may close up at night in response to a fall in the relative humidity of the environment, while a man may run across a road faster when he hears and sees an approaching car. Both of these extreme examples have two factors in common, a change in environment (the *stimulus*) and a reaction to that change (the *response*). In the case of the flower, the stimulus brings about the response directly, but for the man crossing the road the process consists of three distinct stages — (i) the stimulus of the approaching car is detected by his eyes and ears, which (ii) send *impulses* by means of his nervous system to his legs, which (iii) carry out the response of running. The eyes and ears act as *receptors,* while the nervous system serves for *co-ordination* and the legs act as *effectors.*

It is easy to find other examples of irritability in Man, including many that involve responses to changes in elements of the *internal* environment, such as body temperature and blood sugar level. These reactions are going on continuously as part of the *homeostatic mechanism* of the body (see Chapter 5) and we are usually quite unaware of them.

7.10 The eye — a light receptor

It is popular belief that we have five senses: sight, hearing, smell, taste and touch. In reality, the situation is much more complex. For example, we are able to balance and remain upright because of a 'sense of gravity'; smell and taste are closely related chemical senses, depending upon the detection of molecules carried in the air or dissolved in saliva; 'touch' may be regarded as a combination of sensitivity to pressure, heat and pain.

Nevertheless, we do possess certain well-defined *organs of special sense* (the eyes, ears, nose and tongue), which are extremely complex structures, made up of many different kinds of cell. Of these organs, our eyes tell us more than the other receptors about the environment in which we live.

Many organisms have the ability to respond to light. Certain green plants, for example, grow towards bright light, whereas blowfly larvae and woodlice move away from it. There is a world of difference, however, between these simple reactions to changes in light intensity and the mass of information made available to us by our own eyes. For instance,

i) we can swivel our eyes to cover a wide field of vision
ii) we can judge distances
iii) we can see the shape of objects in three dimensions (height, breadth and depth)
iv) we can focus clearly on objects far away and near to us
v) we can see colours
vi) we can see clearly both in bright sunlight and in conditions where the light is quite dim.

7.11 Investigation

The external appearance of the eyes

Procedure

1 Examine your eyes in a mirror. Observe the external appearance of the eye as shown in fig 7.1.

Question

1 How is the eye protected (a) from physical damage, (b) from dust?

7.12 Experiment

How do the eyes react to light and dark?

Procedure

1 Look in a mirror. Place the left hand over the left eye and observe any changes that take place in the

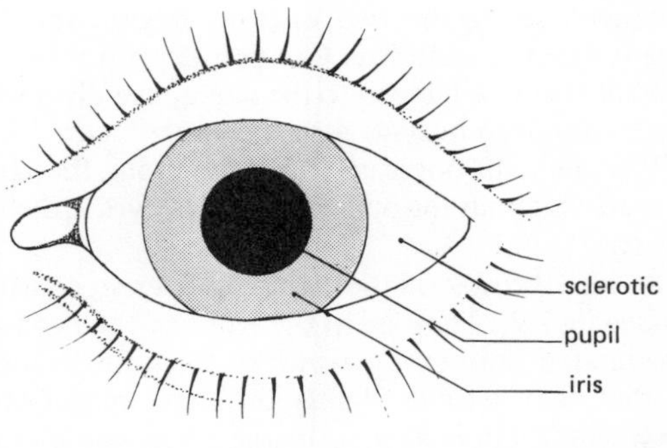

fig 7.1 A diagram of the front view of the eye of Man

• 1 2 3 4 5 6 7 8 9 10 11 12

right eye. Write down any changes that occur.

2 Remove the left hand from the eye and allow both eyes to adjust to light. Continue to look in the mirror, cover the left eye again with the left hand, count slowly up to 10 then uncover the left eye. Observe and record any changes that occur in the left eye immediately after it has been uncovered.

7.13 Experiment

Is there any part of the eye which does not react to light?

Procedure

1 Place the page of this book so that the black dot on the left of the line of figures above is opposite the left eye. The page of the book should be at the normal reading distance of about 25 cm. Close the right eye. Without moving the book, or turning the head, read the numbers from left to right with the left eye. Swivel the eye from left to right as you do when reading a line of print in a book.

Question

1 What happens to the black dot?

7.14 Experiment

Do two eyes work more effectively than one, in seeing the shape of objects?

Procedure

1 Cut two circular discs of card, one with a diameter 2 cm and the other with a diameter 3 cm. Mount each disc on a long pin or a piece of wire.

2 Hold the 2 cm disc in the left hand in front of the eyes, at a distance of about 45 cm. Hold the 3 cm disc in the right hand, at the same distance.

3 Close the right eye and move the right hand disc further from the eye until the two discs appear exactly the same size.

4 Open the right eye and observe the disc sizes.

Questions

1 When viewing the two discs in stage 3 the discs appear to be the same size. Can you explain this?

2 What seems to happen to the size of the discs when the eyes are open in stage 4 of the experiment?

3 What conclusions can you draw from the above observations about the operation of two eyes as distinct from one?

The *visual fields* of the two eyes *overlap* considerably (see fig 7.2). They are in the front of the head and are separated only by the width of the nose, in such a way that they receive slightly different images of an object. The brain is able to merge these two pictures into one, and in this way we can appreciate the depth of an object and its distance from us. This is called *stereoscopic* vision.

In herbivorous mammals, such as ungulates (deer, horses and cattle), the eyes are most important for detecting the presence of predators. The eyes are *large* and *set high* on the sides of the head, with the result that an ungulate such as a horse has a very *wide visual field* (see fig 7.3). The visual fields of the eyes overlap to a small extent only, in front of the head; also the pupil has a *rectangular* shape with its long axis horizontal.

The experiments have shown you some of the functions of the eye, but a knowledge of its internal structure will aid understanding of its working. A sheep, pig, or cow's eye is very similar in structure to that of your own, and therefore can be used for this investigation.

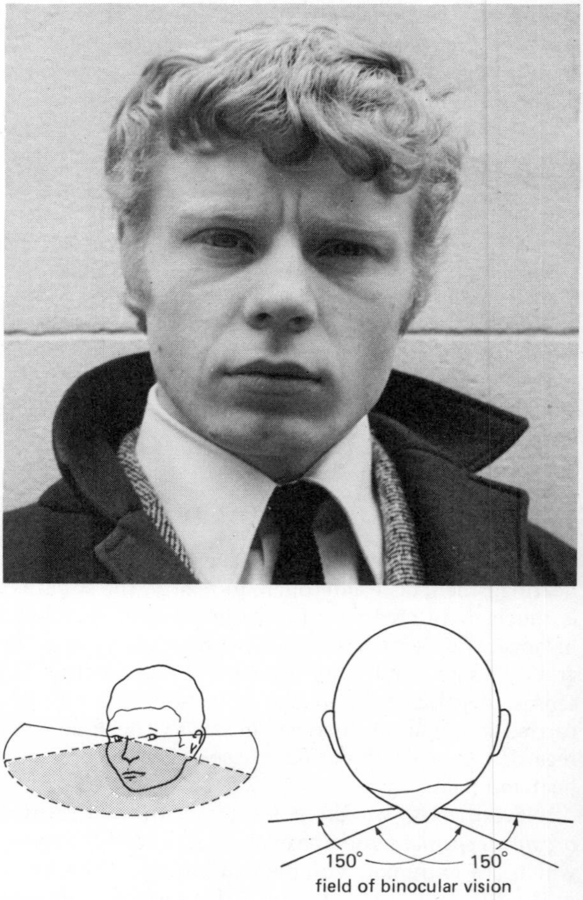

fig 7.2
(a) The front view of the head of a man
(b) A view of the head of a man showing his field of vision

7.15 Investigation

To examine the structure of the eye of a mammal

Procedure

1 You are provided with the eye of a sheep (or bullock). Use forceps and scissors to trim away the yellowish-white fat at the back of the eye. Do not cut away any of the structures that are firmly fixed to the eyeball. The fat cushions the eye in its bony socket or orbit.

2 Removal of fat exposes a number of strips of pink tissue attached to the eyeball (examine fig 7.4).

Questions

1 What do you think is the function of this tissue?
2 How may strips are there?
3 Where would you expect the other end of each strip to be attached?

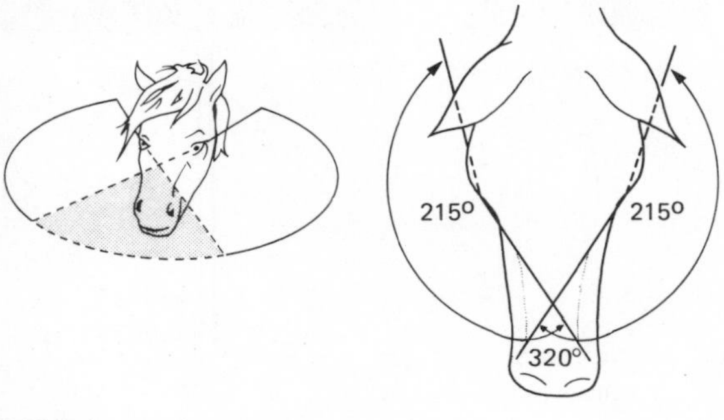

fig 7.3
(a) A front view of the head of a horse
(b) A view of the head of a horse showing its field of vision

3 Remove the strips of tissue and expose the optic nerve. This is a white cord-like structure attached to the back of the eyeball.

Question

4 What do you think is the function of this structure?

4 Using sharp scissors, make a circular cut around the eyeball with the optic nerve as the circle centre. Remove the disc attached to the optic nerve. The outer white layer (the *sclerotic*) maintains the shape of the eye and is very tough to cut. Within the sclerotic is the pigmented *choroid*, while inside this is a thin transparent layer, the light-sensitive *retina*. A jelly-like transparent substance, the *vitreous humor*, fills the main cavity of the eye. It probably began to emerge when you made the cut.

Questions

5 What do you think is the function of the choroid?
6 What do you think is the function of the vitreous humor?

5 In the front half of the eyeball can now be seen the *lens*. Remove it from the eye and place it in a petri

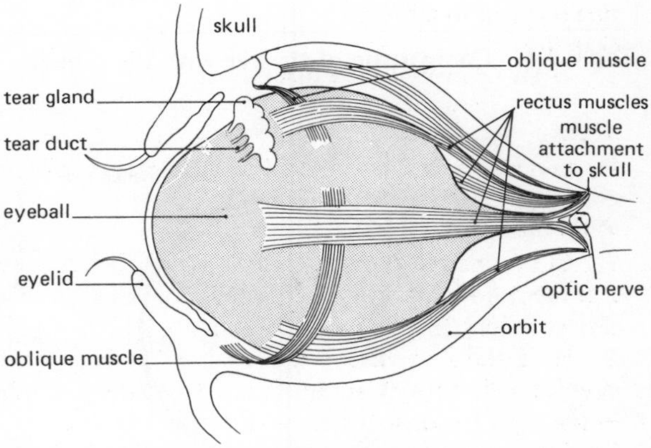

fig 7.4 A diagram of the side view of the eyeball in the socket, showing eye muscles

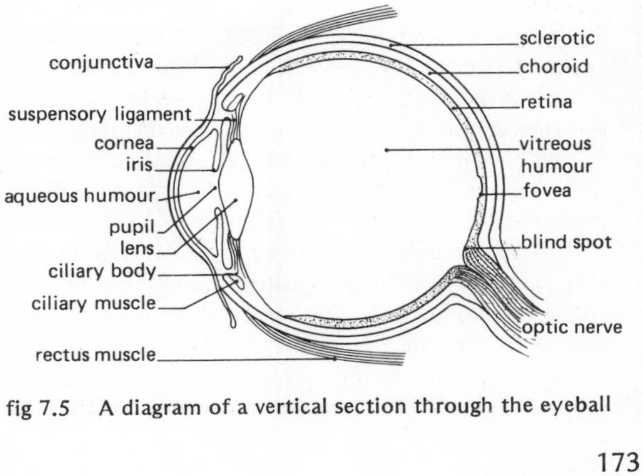

fig 7.5 A diagram of a vertical section through the eyeball

dish. You will note that it is slightly compressible under the pressure of a finger. If the eye is fresh the lens can be used to focus an image of a bench lamp onto paper.

6 Look again into the front portion of the eye. The black choroid continues forward to form the *iris* surrounding the aperture (the *pupil*), through which light enters. Behind the iris is the ciliary muscle and suspensory ligament, to which the lens was attached.

The sclerotic is continued to the front of the eye as a transparent layer, the *cornea*. The jelly-like substance in front of the iris and lens is the *aqueous humor*.

Question

7 What shape is the pupil? Compare it with your own pupil. Why has the pupil this shape in the eye of this dead animal?

7.16 Focusing the eye-accommodation

The eye functions in much the same way as a camera.

Examine table 7.1 and fig 7.6, and you can see how similar they are in structure and in function.

Question

1 State one way in which the eye is fundamentally different in function from the camera.

Mammalian eye	Camera	Function
1 Iris	Iris diaphragm	Adjusts the quantity of light entering
2 Cornea and convex lens	Convex lens	Focuses light
3 Sensitive retina	Sensitive film or plate	Detects light (image is formed here)
4 Change in lens thickness	Lens moves backwards and forwards	Adjusts focus for near and distant objects

Table 7.1 Comparison of the eye with the camera

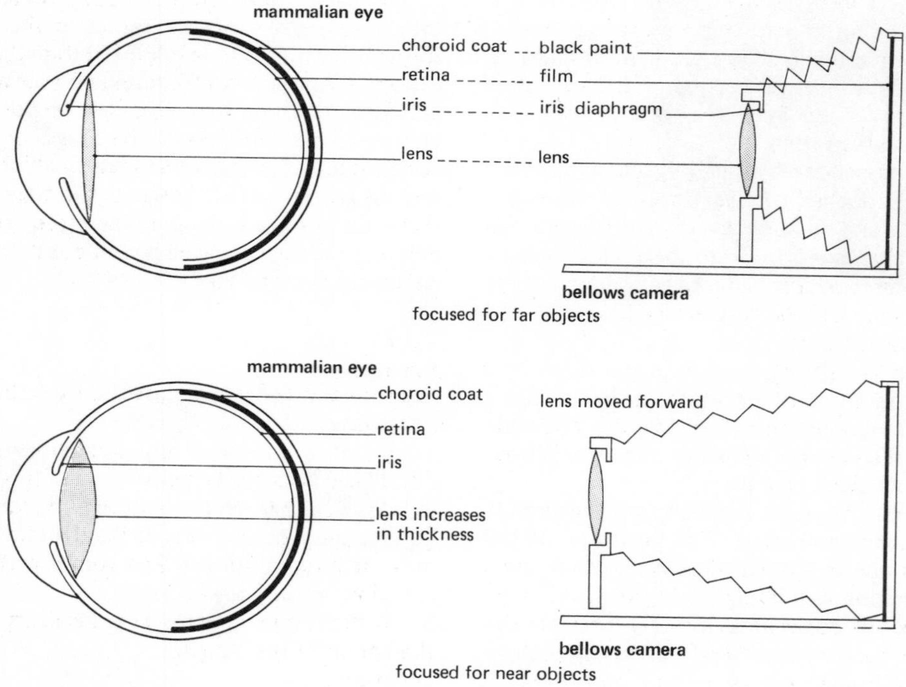

fig 7.6 A comparison between the eye and the camera, focused for near and far objects

Experiment

How does the lens focus an image?

Procedure

1 Obtain an empty 500 cm^3 flask and fill it with water. Fix a sheet of white paper on a drawing board or stiff cardboard and stand it upright on the bench.

2 Place the filled flask between the paper and the window. Move the flask away from the paper, at the same time observing the paper.

Question

2 What do you observe on the paper?

The flask filled with water behaves as a *convex lens.* The light from the window and the outside scene, passing into a more dense medium such as glass and water, is bent or *refracted* and an image is thus formed (*focused*) on the screen of white paper. The image is upside down or *inverted.* In this case, the light is coming from a distant object, but as objects approach closer to the lens the image moves further back behind the screen.

Thus in a camera the lens must be pulled or screwed further away from the film in order to keep the image in focus on the film.

The mammalian eye resembles very closely a camera except in one respect; the focusing of a near object onto the retina. Most refraction takes place as the light passes through the transparent cornea in the front of the eye, but the final adjustment is brought about by the lens so that the image is focused clearly on the retina. Any object *more than 10 metres* from the eye is in focus on the retina with the lens at its *thinnest.* Objects closer than this distance are kept in focus by the eye lens bulging and becoming thicker, thus increasing its power to bend the light rays so that they converge on the retina.

The lens is suspended by small ligaments (*suspensory ligaments*) attached to *ciliary muscles.* When the eye lens is thin and focused for distant objects, the ciliary muscle is *relaxed.* In order to see close objects the ciliary muscle *contracts,* pulling the wall of the eye inwards, *slackening* off the suspensory ligaments and allowing the lens to *bulge.*

Imagine you are sitting outside in the shade of a tree reading a book; you look up from the book to a distant sunlit mountain and then down to the book again. Table 7.2 shows the sequence of events in your eyes as the two actions take place.

Looking up from the book to the distant scene.	Looking back to the book from the distant scene.
1 Circular muscles of the iris contract, radial muscles relax.	Radial muscles of the iris contract, circular muscles relax.
Pupil gets smaller.	Pupil gets larger.
Light intensity adjusted.	Light intensity adjusted.
2 Ciliary muscles relax.	Ciliary muscles contract.
Lens becomes thinner and less refractive.	Lens becomes thicker and more refractive.
Distant scene focused on the retina.	Book print focused on the retina.

Table 7.2 Accommodation of the eye to near and distant objects

The mechanism by which the eye focuses accurately is known as accommodation. The accuracy of the picture seen is due to special sense cells called *cones* which only function well in high light intensity. There are about 3.5×10^5 cones in each retina. They are also responsible for *colour vision.* The most common light receptor in the human eye is the *rod,* and there are about 6.5×10^7 of these in each eye. The rod is extremely sensitive to light and shade and therefore in *nocturnal animals*, e.g. bats, the rod is the most important light receptor. An animal with only rods in the eye can see in daylight but it would not be able to distinguish any great detail since this is the function of the cones.

The nerve fibres from the optic nerve spread over the inner surface of the retina so that light passes through the fibres to reach the rods and cones. Where the optic nerve pierces the retina no rods and cones are present and thus no image can be formed. This is known as the blind spot. The *fovea,* or yellow spot, contains only cones, so that images which fall here are seen in greater detail. Towards the edge of the retina there are mainly rods and, therefore, an object just entering the field of vision is seen but its colour cannot be clearly distinguished.

7.17

Experiment

Are rods and cones distributed differently?

Procedure

1 Sit on a chair and look straight in front of you. (You need a partner to help you with this experiment.)

2 Ask your partner to stand behind you with a pencil (which you have not seen) in his right hand. He should move it forward until it just comes within your field of vision, when you say 'stop'.

3 Without moving the eyes or head, try to determine the colour of the pencil.

Question

1 Can you decide on the colour of the pencil? If not, try to give an explanation of this.

7.18 Defects of the eye

If the eye is to function correctly, there must be a balance between the length of the eyeball and the refractive power of the cornea and the lens. When a child grows, the axis length of the eyeball increases from 16 mm in the newborn to 24 mm in the adult. During this time, refractive power must be adjusted to the increasing axis length.

In extreme cases, the axis length of the eyeball falls outside normal limits, and the eyes are unable to focus by cornea and lens alone. They are aided by *external lenses* (glasses) to focus light rays onto the retina.

Myopia (short sight)

The axis length of the eyeball may be *too long* due to defective nutrition or some genetic reason, and *short sight* or *myopia* results. The term short sight refers to the fact that only objects *near* to the eye can be seen clearly. In children this defect (detectable by their reading small print close to the eyes) can become progressively worse and the condition must be treated. Light rays from a distance are brought to a focus *in front* of the retina and this situation can be corrected by *concave* spectacle lenses (see fig 7.7).

Hypermetropia (long sight)

In some children the axis length of the eyeball remains *too short.* Light rays from a distance are brought to a focus *behind* the retina, the ciliary muscles come into action by the accommodation reflex and thus distant objects come into focus. This means that the nearest point of distinct vision may be well away from the eyes. Near objects cannot be seen clearly. This condition is corrected by a *convex* spectacle lens.

Another type of long sight *(presbyopia)* occurs in older people and is due to a decrease in the refractive power of the lens of the eye. Again this is corrected by spectacles with convex lenses.

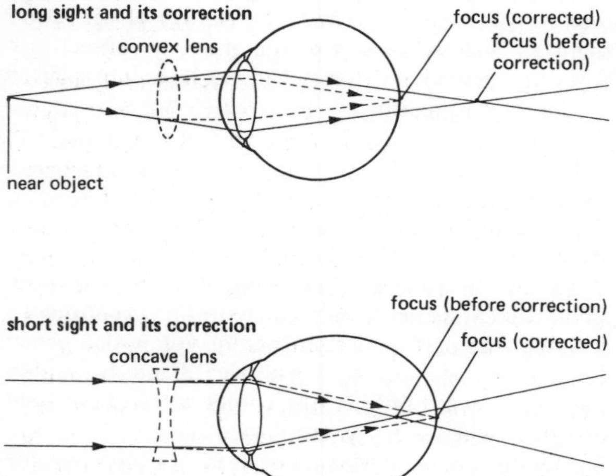

fig 7.7 Diagrams showing long sight and its correction and short sight and its correction

A squint is a very common eye defect, particularly in children. There are many different kinds of squint but basically they all arise from the unequal development of the rectus muscles. One of the most familiar forms is due to hypermetropia. The long-sighted child makes a very great effort to focus near objects clearly. The muscles used in accommodation are in fact supplied by the same nerve as the internal rectus muscle. Thus the harder he or she tried to focus, the more the internal rectus muscles contract, causing the eyes to converge and producing a 'cross-eyed' appearance.

Conjunctivitis (inflammation of the conjunctiva) is a condition that almost everyone experiences at some time in their life. Although not serious in itself, if it is ignored it can lead to serious complications such as ulceration of the cornea. The most obvious sign of conjunctivitis is the reddened or 'bloodshot' appearance of the 'white' of the eye, caused by dilation of the blood vessels in the conjunctiva. In its simplest form conjunctivitis may result from irritation by dust or wind, but it is often associated with infection by micro-organisms. Care should therefore be taken to avoid the sharing of towels and face flannels, particularly in large communities such as boarding schools, where conjunctivitis infections can spread very rapidly.

Cataract is a condition in which the eye lens becomes opaque (loses its transparency). It occurs frequently, as part of the ageing process, in people over the age of fifty whose eyes may previously have been quite healthy. In other cases it may arise as a result of the disease diabetes or because of exposure to intense heat or atomic radiation. In many cases, blindness caused by cararacts may be cured by removing the lenses surgically (much of the refraction of light to form images on the retina is performed by the cornea, which is unaffected). More recently, it has been possible to replace the opaque crystalline lenses with clear acrylic (plastic) ones.

Glaucoma is another condition that occurs mainly in older people. It results from increased pressure inside the eyeball, which damages the optic nerve and may result in both blindness and great pain. Glaucoma arises in the following way. Under normal conditions the fluids of the eye filter out of the eyeball through the sharp angle formed between the iris and the cornea and into the blood vessels of the ciliary body. If the size of the lens should increase with age or because of inflammation, it may push against the iris, closing up this sharp angle. Thus the exit of fluid is prevented or slowed down and pressure builds up. This condition can be relieved surgically by making an incision in the sclerotic and allowing the fluid to filter out through the resulting scar.

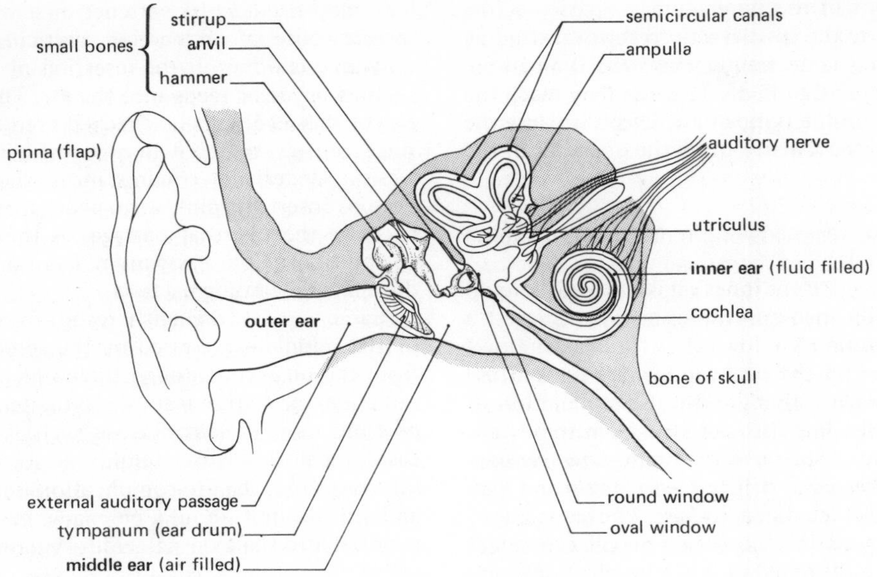

fig 7.8 The structure of the ear

7.20 The ear – a sound and gravity receptor

The *pinna* or external ear flap, together with the outer ear tube, serve to concentrate sound waves. In many animals, but not in Man, the pinna can be turned to the direction from which the sound is coming. The pinna is connected by the *auditory canal* to the *tympanum.*

The middle ear is an air-filled chamber connected by the *eustachian tube* to the back of the throat (the *pharynx*), and thus to the outside air by the *buccal cavity* and mouth.

Question

1 What is the function of the eustachian tube?

The tympanum is connected across the middle ear to the *oval window* by the three tiny, linked bones called ear *ossicles* (hammer, anvil and stirrup). The hammer is attached to the tympanum, and the anvil passes on the movements of this bone to the stirrup which touches a flexible membrane, the oval window. Just below is another membrane, the *round window.*

The inner ear is fluid-filled. It contains two structures.

i) The *cochlea.* This is a tube about 3 cm in length, coiled like a snail shell and filled with lymph. Lying within the cochlea is the *organ of Corti* containing thousands of sensitive cells which are the receptors of the vibrations conducted from the tympanum.

ii) Three *semi-circular canals.* These are fluid-filled tubes each one at right angles to the others so that they lie in the three planes of space. At one end of each canal is a small chamber, the *ampulla,* containing sensory hairs. Both structures are connected to the brain by the auditory nerve.

7.21 Functions of the ear

i) *Detection of sound (hearing).* Sound waves (vibrations of gas molecules in the air), having been concentrated by the pinna, pass down the auditory canal of the outer ear and strike the tympanum causing it to vibrate. To understand this, the following facts about transmission of sound should be known.

a) Sound can be transmitted through a solid, liquid or gas, but not through a vacuum. Light and heat from the sun can travel across outer space but not sound.

b) Transmission of sound in air is due to alternate *compression* and *rarefaction* of the air at the front of the sound wave. These compressions and rarefactions strike the tympanum in succession causing it to move in and out. In a more dense medium, such as wood, the movement of sound is like the rapid transference of energy from one group of molecules to the next.

c) Sound, such as a musical note, has a measurable *frequency,* and this frequency can be transmitted through a medium to cause another object to vibrate at the same frequency. Frequency is expressed in *cycles per sec (c/s)*. The 'pitch' of a sound is its frequency.

d) Sound travels relatively slowly in the air at about 330 m/s. Thus the sound of distant thunder arrives long after the lightning flash.

The vibrations of the tympanum are passed across the middle ear by the ossicles and so the oval window vibrates with the same frequency as the tympanum. The vibrations are magnified x 22 when they reach the oval window from the tympanum. This is because the tympanum has over twenty times the diameter of the oval window.

Questions

1 Why should the perforation of the tympanum cause deafness?

2 Remembering that the inner ear is filled with liquid that is practically incompressible, can you suggest a function of the round window below the oval window?

The vibrations of the oval window are transmitted through the cochlear fluid, and cause stimulation of the sensitive cells. Impulses are sent from these cells through the auditory nerve to the brain. *Low frequencies* stimulate the organ of Corti near its *tip* and *high frequencies* are detected near its *base.* The ear is a very versatile sound receptor and young people can detect frequencies from 20 to 20 000 c/s. As the body ages the ability to hear higher frequencies is lost and the upper limit may fall to as low as 5 000 c/s. Some mammals, such as dogs, can hear higher frequencies than Man. Bats can detect frequencies as high as 100 000 c/s.

The range of volume of audible sounds is also very great since the loudest sound is about one trillion times as loud as the faintest we can detect.

ii) *Detection of gravity and motion.* The inner ear has a sac, the *utriculus,* connecting the bases of the semi-circular canals. This sac is lined with sensory cells connected to nerve fibres. Entangled in the sense cells are tiny spheres of *calcium carbonate.* These are acted upon by gravity, and press against the sense cells. The nerve fibres send impulses to the brain and so the body becomes aware of its position. If we are lying down, the spheres press on certain of these sensory cells, and the body is thus conscious of its position in space.

Motion of the body is detected by the semi-circular canals. When the head is moved, the fluid in the canals moves also. The sense cells in the canals are able to send impulses to the brain regarding the moving fluid, and consequently the muscular control of the body is maintained. The control of the body during athletic activity, particularly for balance, would not be possible without this mechanism.

7.22 Care of the ears

Most common ear problems arise from obstructions of either the external auditory passage or the eustachian tube. In both cases medical help should be sought.

The external passage may be blocked by accumulation of wax, which may lead to bouts of partial deafness, sometimes lasting for years. This can usually be cleared by expert syringing, although hardened wax may need to be softened first with olive oil. A second cause of obstruction, particularly in the case of young children, is the insertion of foreign objects such as beads or seeds into the ear. The nature of the object should always be checked before it is removed, since organic material (e.g. seeds) tends to swell in water. **Under no circumstances** should untrained persons insert hairpins, wire, pencils, etc. in a attempt to clear the external passage, as this could lead to perforation of the tympanum. Even fingers can introduce disease-causing bacteria.

Earache most frequently results from inflammation of the middle ear, caused by the spread of infection from the pharynx through the eustachian tube. Such infection may arise from decaying teeth, from colds or from more serious diseases such as measles. In all cases medical advice should be sought since such infections may lead to complications such as abscesses, and no attempt should be made to pour or insert material into the external auditory passage.

7.30 Reception of chemical stimuli

The presence of certain chemicals in the environment is detected by the senses of taste and smell, which are closely linked with each other.

7.31 Taste

One of the organs of the mammal that detects *chemicals* in the environment is the *tongue.* In order that a chemical may be tasted it must dissolve in the moisture of the buccal cavity, then, when in solution, it can stimulate the taste buds. There are only four primary tastes: sweet, sour, salt and bitter.

Experiment

Are the four primary tastes detected in different regions of the tongue?

Procedure

1 Prepare four solutions giving different tastes, e.g. (a) sugar solution (sweet), (b) very dilute hydrochloric acid, citric acid or vinegar (sour), (c) common salt solution (salt), and (d) quinine sulphate (bitter).

2 Working with the help of a partner, explore the areas of the tongue for sensitivity to the four liquids. Use small glass tubes to place drops of each liquid on different parts of the tongue. The mouth and the tongue should be rinsed well between tests to eliminate the previous tastes.

3 For each liquid, record, on a drawing of the tongue, where it can be tasted (pay particular attention to the tip, sides and back of the tongue). Complete the tests for each liquid before passing on to the next. Leave the bitter substance until last.

Questions

1 Do your tests show any clear pattern of taste areas over the tongue?
2 Are the taste areas separated or do they overlap?
3 Do individuals differ or is there a basic pattern?

7.32 Smell

Chemicals in the air are detected high up in the nasal cavity by means of *olfactory epithelial* cells. These cells are receptors and stimulate nerve endings to produce nerve impulses which are interpreted in the brain as smell.

Man's sense of smell is poor, compared with that of a dog or an antelope but he can detect a wide variety of smells and, in many cases, the molecules are at very low concentrations. It is difficult to analyse all the different smells that we receive. Many smells represent the combined effects of a number of chemicals.

The flavour of food is a combination of taste and smell. When we have a cold in the head, food has little flavour, as the olfactory epithelium is unable to function because of additional mucus covering the sense cells. Smell is probably the least important of our senses, whereas in other animals, such as insects, the detection of chemical stimuli is vital.

Sense cells can become adapted. When we first step into a room where there is a very obvious smell of newly cooked food or new paint we are immediately aware of the smell, but if we stay there long, our perception of the smell diminishes almost to vanishing point. There is a 'conditioning' of the olfactory cells for that particular smell. New smells, however, are detected quickly.

7.40 Receptors in the skin

See Chapter 5 fig 5.9 for the structure of the skin. The skin is an important sensory organ. In the dermis are many nerves which carry information about changes in the external world to the central nervous system. Areas of skin which are hairless, such as the finger tips of Man or the lips of most mammals, tend to be more sensitive than hairy skin. There are receptors at the ends of the nerves in the dermis but it is not always possible to differentiate those responsible for receiving particular stimuli. The following stimuli are detected by skin receptors: (a) heat, (b) cold, (c) touch (roughness, smoothness, etc), and (d) pain.

Experiment

Can the skin distinguish different temperatures?

Procedure

1 Prepare three beakers of water as follows:
i) hot water (about 50°C)
ii) water at room temperature (about 22°C)
iii) ice cold water (about 5°C).

2 Immerse the index fingers of both hands in beaker (ii) for 10 seconds, then immerse the left index finger in beaker (i) and the right index finger in beaker (iii) for about 30 seconds.
3 Transfer both fingers simultaneously into beaker (ii).

Questions

1 Does the temperature of the water in beaker (ii) feel the same to both fingers at the beginning of the experiment? Why were the fingers both placed in the water of beaker (ii) at the beginning?
2 Do you notice any difference in the response of the two fingers after transfer back to beaker (ii) from the other beakers?
3 What conclusions can you make from your observations regarding the response of the skin to temperature?

Experiment

How do different parts of the skin differ in touch sensitivity?

Procedure

1 Push two pins through a cork so that the points are 2 mm apart.
2 Working in pairs, test different surfaces of the body by pressing gently on the skin with the two pins. First, test the upper and lower surfaces of the hand and then move to fingertips, upper arm and back of the neck.
3 Each time the pins are pressed on to the skin the student should indicate whether he/she can feel two points or one.

Questions

4 Are all areas of the skin equally sensitive to the touch stimuli of the pins?
5 What can you conclude about the distribution of receptors under the skin?

7.50 Co-ordination

The quickness of many of our responses to stimuli is due to the extremely rapid conduction of electrochemical impulses by the nervous system. In the sciatic nerve of the leg, for example, impulses travel at a rate of 65 metres per second or 140 miles per hour!

The nervous system is assisted in co-ordinating the body functions by the slower-acting *endocrine system.* This consists of a number of glands that discharge chemicals called *hormones* into the blood circulation. These secretions produce their effects on particular target organs or tissues within the body. The nervous system controls the endocrine glands by sending appropriate messages to increase or decrease secretion either directly or through the release of intermediate hormones.

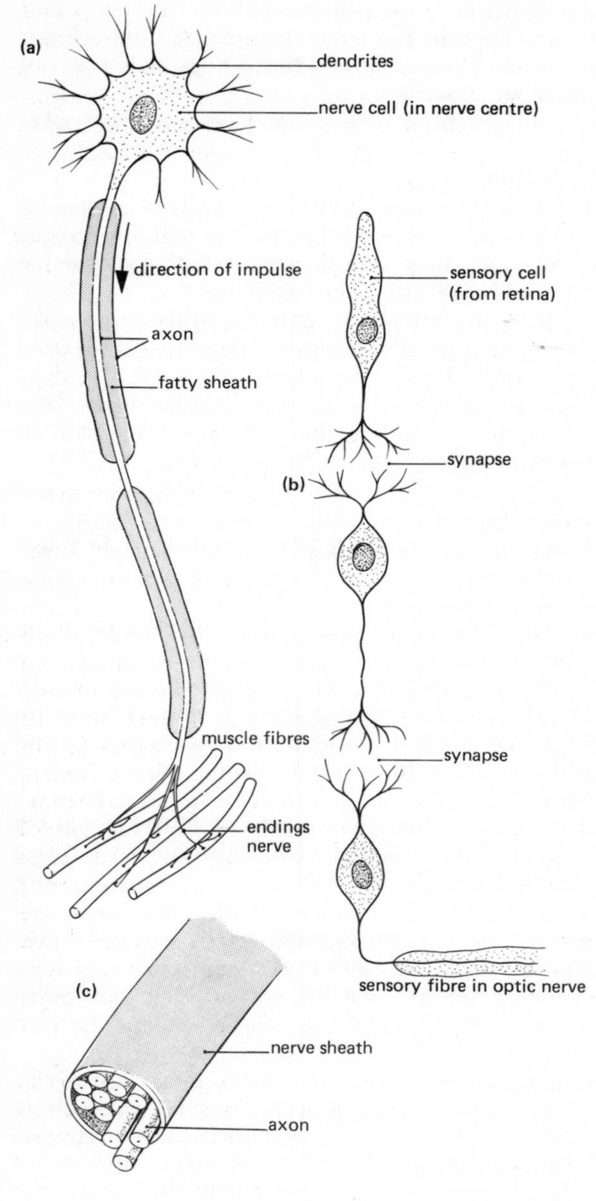

fig 7.9 Diagrams showing (a) a single motor neuron and its connection with muscle fibres (b) synapses between a sensory neuron and a sense cell of the retina of the eye (c) fibres of neurons bound together to form a nerve *After D.G. Mackean*

7.51 How do nerves work?

Nerves are made up of cells called *neurons,* highly specialised for their function of receiving and conducting impulses. A typical neuron has a very long *axon* or fibre covered and *insulated* by a layer of *fat.* Attached to the axon is the *cell body* containing a nucleus and

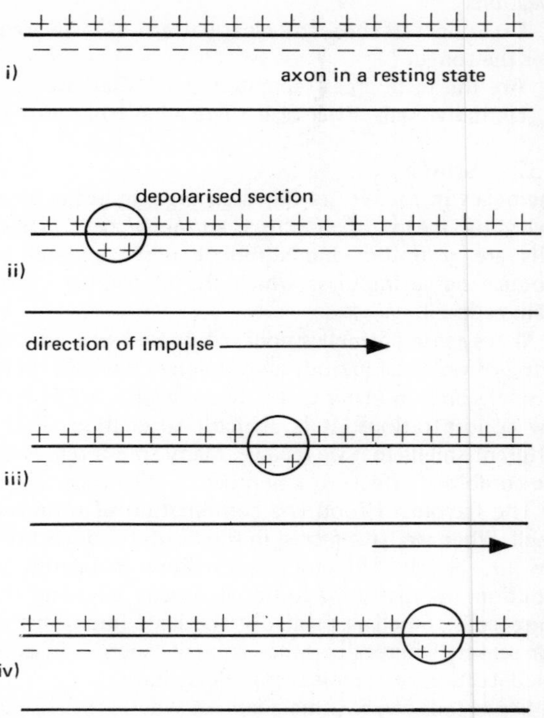

fig 7.10 The conduction of a nerve impulse along an axon as a wave of depolarisation

having many threadlike extensions of the cytoplasm called *dendrites* (see fig 7.9).

The dendrites are in close contact with those from cell bodies of other neurons or with sense cells from which impulses originate. The dendrites do not connect completely with each other but are separated by a minute gap called a *synapse.* The brain and spinal cord have several millions of neurons, each with numerous dendrites: the number of synapses in the nervous system is very great indeed.

Neurons carry nerve impulses which are electrochemical in nature and spread rapidly along the fibres. The impulse is produced by minute electrical charges that normally travel only in one direction, and during this time use up small amounts of energy. In its resting state, the outer surface of the axon is positively charged and the inside is negatively charged, a condition known as *polarisation.* When an impulse is generated this situation is reversed and the axon becomes temporarily *depolarised* over a small area, with the inside positive and the outside negative. Depolarisation lasts for only about a thousandth of a second but, as the resting state of polarisation develops again, it causes the next part of the axon to become depolarised. Thus the impulse travels along the axon as a *wave of depolarisation* (see fig 7.10).

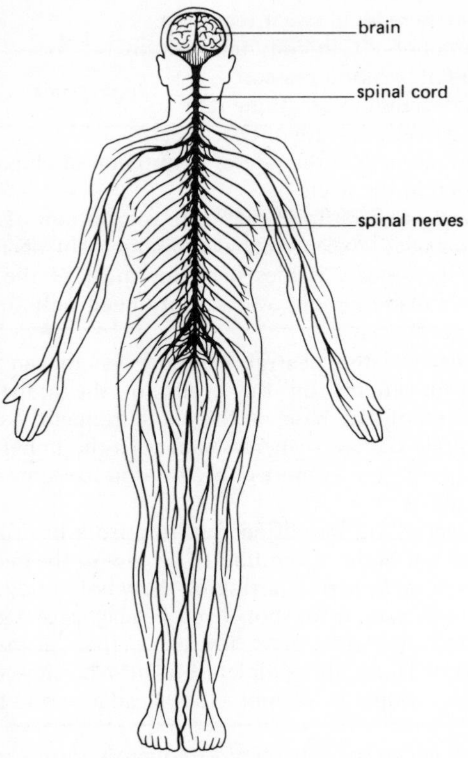

fig 7.11 A diagram of the central nervous system together with the spinal nerves and their major branches

An impulse reaching the end of a fibre causes the secretion of minute amounts of a chemical substance called *acetylcholine.* This chemical moves across the synapse, causing a new electrical impulse to be produced and travel along the next neuron. Acetylcholine is unstable and is quickly broken down by an enzyme to prevent the continuous production of new impulses.

7.52 Organisation of the nervous system

The nervous system has a very complex organisation, which can most conveniently be summarised as follows. There are two main sub-divisions, the *central nervous system* (brain and spinal cord) and the *peripheral nervous system.* The latter consists of the *cranial and spinal nerves* and their branches, together with the *autonomic nervous system.* The autonomic nervous system, which controls the internal activities of the body, is further divided into *sympathetic* and *parasympathetic* systems.

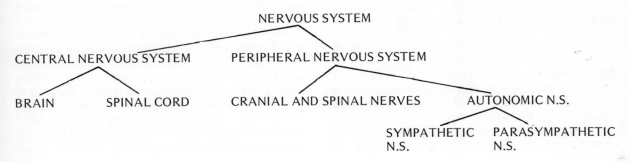

The mammalian brain is much larger in relation to body size than the brain of other animals. This is due to the proportionately greater size of the cerebrum and the cerebellum associated with the increased variation in behavioural patterns and the greater muscular control that goes with them. The brain of Man is particularly notable for the large *cerebrum,* which extends over all other areas of the brain.

The main regions of the brain are as follows:

Olfactory lobes: Connected by sensory neurons to the organ of smell, which also sends sensory neurons to the cerebrum. These lobes are proportionately larger in animals such as fish which depend heavily on a sense of smell for survival.

Cerebrum (cerebral hemispheres): They have a much folded and wrinkled surface giving the outer cortex a very large surface area. The *cortex* is formed of *grey matter* (cell bodies of neurons) as distinct from the inner part which is composed of *white matter* (axons of neurons). Different areas of the cortex control distinct functions. Sensory areas control sight, hearing, smell and skin sensation, while motor areas control muscles of the legs, arms, face, eyes and head. Large areas of the human cerebral cortex are not concerned with sensation or motor control. These areas are missing in all other mammals, and thus it could be that they are concerned with Man's intellectual functions including speech, music, mathematics and other activities requiring the use of symbols and abstract thought.

Hypothalamus: This is the reflex centre concerned with a number of homeostatic mechanisms such as temperature control, water balance and carbon dioxide levels in the blood.

- NERVOUS SYSTEM
  - CENTRAL NERVOUS SYSTEM
    - BRAIN
    - SPINAL CORD
  - PERIPHERAL NERVOUS SYSTEM
    - CRANIAL AND SPINAL NERVES
    - AUTONOMIC N.S.
      - SYMPATHETIC N.S.
      - PARASYMPATHETIC N.S.

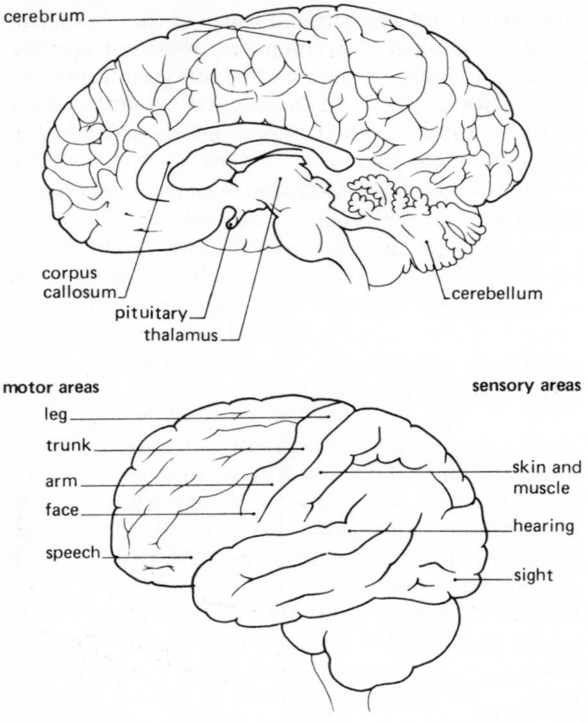

fig 7.12 A longitudinal median section of the brain of Man and areas of control in the cerebrum of the brain of Man

Optic lobes: These receive sensory neurons from the eyes. They are small in Man compared with other animals since the cerebral cortex has taken over much of their role.

Cerebellum: This is a large structure in mammals, as it is concerned with the maintenance of balance, locomotion and positioning of the body, and is thus involved in the co-ordination of muscular activity. It receives sensory impulses from the skeletal muscles and sends motor impulses out to them.

Medulla oblongata: This is a reflex centre of the brain controlling blood pressure, coughing, swallowing, sneezing, yawning and vomiting (see fig 7.12). Inside this region is a centre that is thought to be concerned with 'wakefulness' in that the activity of its neurons enables the cerebral cortex to control all the concious activities of hearing, seeing, moving, talking, etc. When our wakefulness centre becomes inactive, we fall asleep.

Sleep is necessary for the rest and recovery of both body and mind after the activities of the day. As we go to sleep the following physiological changes take place:

a) a reduction in the number of impulses occuring in the cerebral cortex
b) an increase in sweat secretion
c) a fall of 1°C in body temperature
d) a fall in blood pressure
e) a decrease in breathing rate
f) a decrease in rate of heart beat
g) an increase in the rate of conversion of glucose to glycogen in the liver.

There is no definite rule as to the amount of sleep that we need, except that our requirement decreases as we get older. Babies sleep for most of the day, whereas many people over seventy need only four or five hours' sleep.

Injury to the central nervous system can have disastrous effects on the health of the individual. Fractures of the base of the skull frequently cause fatal brain damage, whereas injury to the upper part of the cerebrum is more likely to produce some form of paralysis.

Concussion is a condition resulting from bruising of part of the brain, often due to a blow to the back of the head or to severe jarring of the whole body. The symptoms vary from short-lived headaches and giddiness to unconsciousness for more than a month. Recovery is usually complete but may be slow, with persistent bouts of memory loss, headache and tiredness.

Damage to the spinal cord frequently results in the condition known as **paraplegia**, in which there is paralysis and loss of sensation in the parts of the body below the injury. Often the legs are affected, together with control of the bladder and bowels; but in some cases all four limbs may be paralysed (quadriplegia).

The effects of certain 'mood-influencing' drugs on the health of the individual are discussed in Chapter 10. Briefly, they include *stimulants* such as *amphetamines*, which increase the activity of the central nervous system, and *depressants*, such as *barbiturates*, which have the opposite effect. Contrary to popular opinion, alcohol is a depressant rather than a stimulant.

From the brain, twelve pairs of *cranial nerves* leave through small holes in the cranium. They are concerned with sensory impulses (from organs of smell, sight and hearing) and motor impulses (to jaw muscles, eye muscles and tongue muscles). The cranial nerves cross over in the brain so that nerves serving the left side of the head originate in the right side of the brain and vice versa.

Spinal nerves are pairs of mixed nerve fibres (motor and sensory) leaving the spinal cord at regular intervals from spaces between the vertebrae. Each nerve arises as a *dorsal (sensory)* root and a *ventral (motor) root*, which join just outside the cord (see fig 7.14). The cell bodies of the sensory neurons occur about halfway along the dorsal root in a swelling called the *dorsal root ganglion* (a *ganglion* is a collection of nerve cell bodies). The motor neurons have their cell bodies within the spinal cord itself. The sensory and motor neurons connect with short *internuncial* neurons, or longer secondary neurons passing up and down the spinal cord. These cross over, mainly in the medulla, so that nerves from the left side of the body reach the right side of the brain and vice versa (see fig 7.13).

The *autonomic* nerves control the *internal* activities of which the individual is not normally aware, such as peristalsis of the gut, glandular activity and heart beat. They enter the central nervous system in association with certain of the cranial and spinal nerves. The autonomic nervous system is composed of the sympathetic and parasympathetic systems. The actions of these two systems are usually opposite, or antagonistic, as is shown in Table 7.3. Anatomically, they differ in the positions of their ganglia, and chemically, in the nature of the transmitter substance produced by the ends of their fibres.

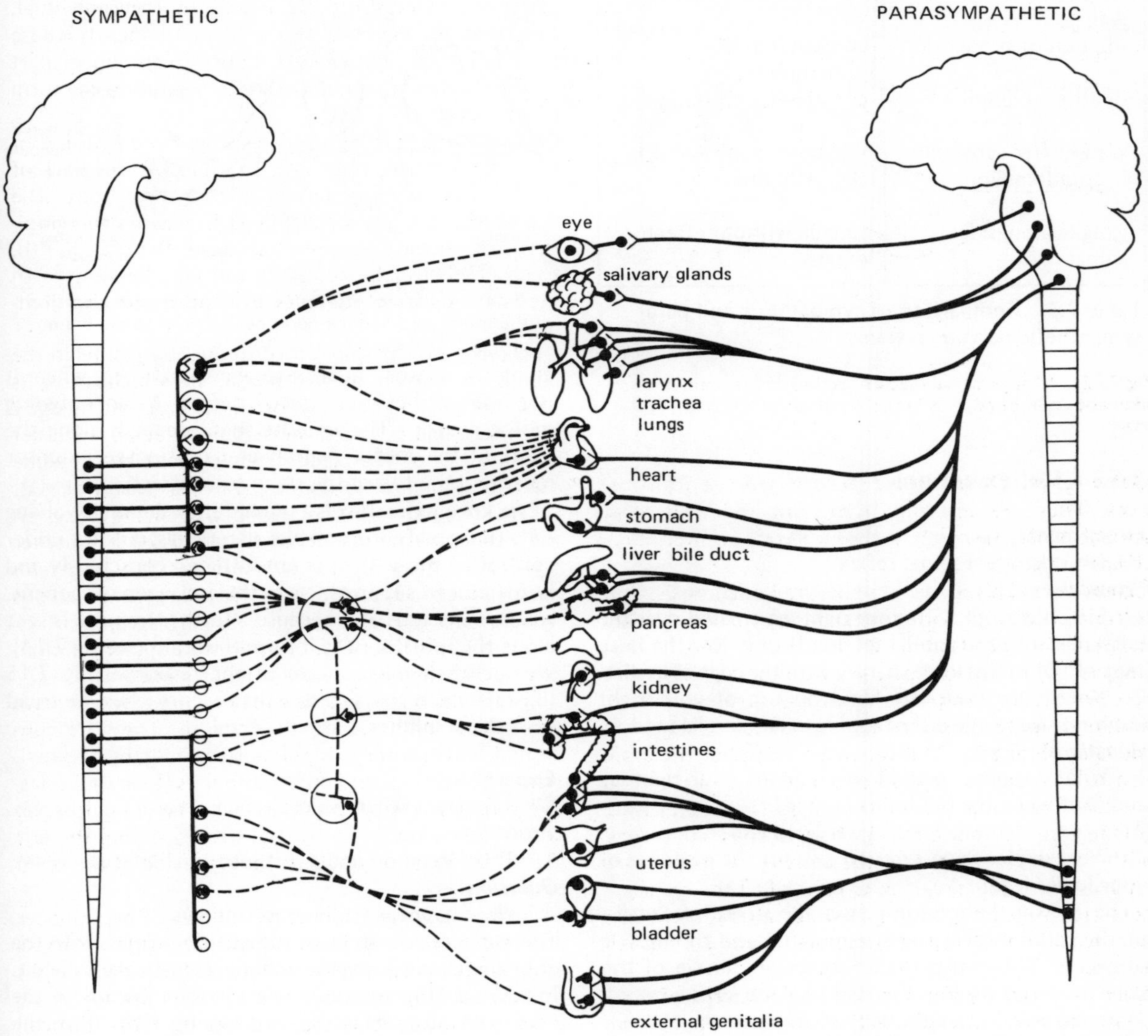

fig 7.13 The autonomic nervous system

Sympathetic	Parasympathetic
causes: increased rate of heart beat decreased rate of gut peristalsis constriction of arteries constriction of anal sphincter constriction of bladder sphincter relaxation of bladder wall dilation of bronchioles dilation of pupil	causes: decreased rate of heart beat increased rate of gut peristalsis dilation of arteries dilation of anal sphincter dilation of bladder sphincter contraction of bladder wall constriction of bronchioles constriction of pupil
transmitter substance is noradrenaline	transmitter substance is acetylcholine
ganglia alongside spinal cord	ganglia without effector organs

Table 7.3 Comparison of sympathetic and parasympathetic nervous systems

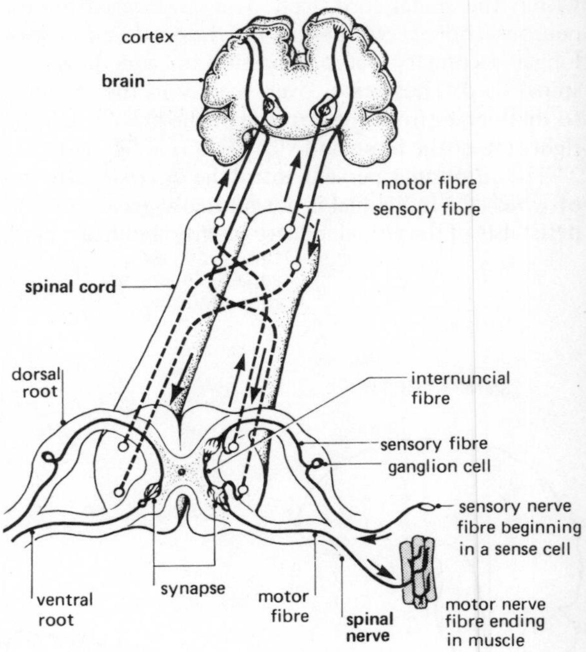

fig 7.14 A diagram of a reflex arc together with a portion of the spinal cord and the nerve connections to the brain

7.53 Reflex actions

Experiment
To demonstrate a spinal reflex
Procedure
1 Sit on a stool with your right leg crossed over the left knee. Tap the patella tendon (just below the knee cap) of your right knee sharply with the edge of a ruler.
2 Now place your left hand on top of your right thigh muscle. Repeat the tapping of the patella tendon.
Questions
1 What happens to the lower part of your right leg and foot when the tendon is tapped?
2 What do you feel with your left hand?
3 What explanation can you give for the response of your lower leg?

The tap on the tendon causes the *stretch receptors* of the thigh muscle to be stimulated and the muscle contracts. This muscle is an *extensor* muscle of the knee joint and the lower part of the limb jerks forward. Thus we have a simple behavioural pattern called a *reflex action.* It results from a sequence of impulses in the nerves and the spinal cord constituting a *reflex arc.* The reflex arc sequence is as follows: the tap on the tendon stimulates an impulse which is transmitted along the sensory neuron to the spinal cord, where it is passed via an internuncial neuron to an outgoing motor neuron. The impulse then proceeds along the axon of the motor neuron and causes the extensor muscle in the top of the thigh to contract.

The knee jerk reflex is a spinal reflex action involving only the spinal cord, and like all reflexes, it is not under control of the will. It is not initiated consciously and is difficult to suppress. Nevertheless, when it happens, we are conscious of it because a further impulse is sent along the spinal cord to inform the brain (see fig 7.14). We are then made aware of the response. Fig 7.15 illustrates the spinal reflex that occurs when we tread on a sharp splinter.

Experiment
What happens when we cut up an onion?
Procedure
1 Take an onion and cut it into small pieces.
Question
4 What happens to the nose and eyes?
The vapours arising from the cut onion pass into the nose and eyes causing the olfactory sense cells to react. In addition, the vapour dissolves in the moisture of the eyes. The latter react by discharging fluid from the tear gland to dilute the dissolved vapour. Here is a clear reflex action, *not controllable by the will* and giving a definite response to the chemical stimulus. In this case, however, although quite automatic, the reflex

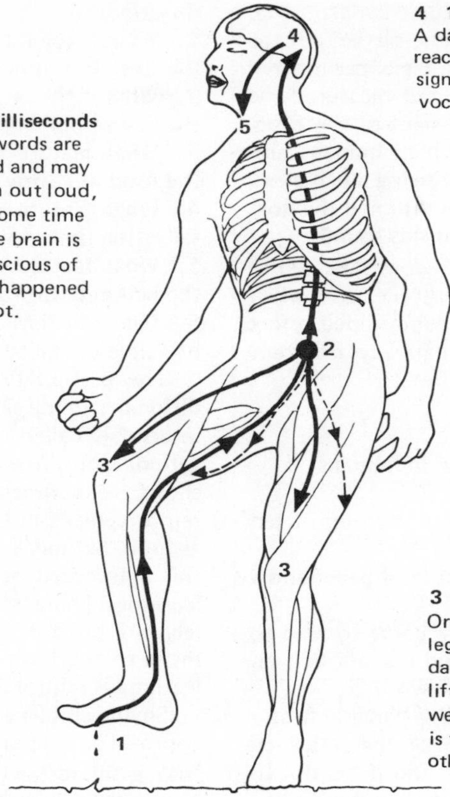

4 100 milliseconds
A damage report reaches the brain. A signal is sent to the vocal cords.
5 115 milliseconds
Suitable words are composed and may be spoken out loud, but it is some time before the brain is fully conscious of what has happened to the foot.
2 60 milliseconds
The message is received by a nerve junction in the spinal column. A message is sent to the brain. Orders are immediately sent to the leg muscles.
3 68 milliseconds
Orders reach the leg muscles. The damaged foot is lifted while the weight of the body is transferred to the other leg.
1 zero milliseconds
A splinter penetrates the foot. A message is flashed to the base of the spine.

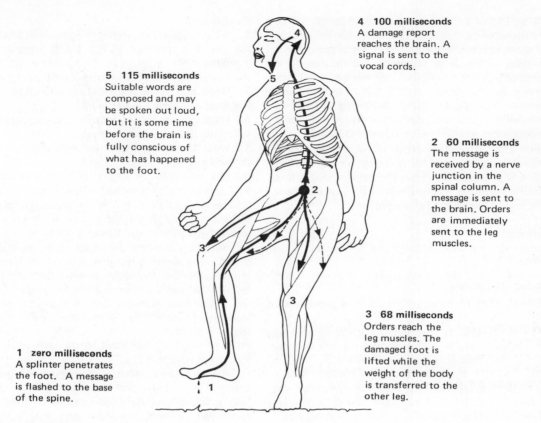

fig 7.15 A spinal reflex

involves the brain rather than the spinal cord, so that it can be called a *cranial reflex action.* The response is a *glandular discharge* as distinct from the muscular action in the case of the knee jerk.

Another example of the cranial reflex action is the *blinking* of the eyelids in response to a threatened blow, while the quick withdrawal of the hand or finger when pricked or burned is a *spinal reflex.* Most of the movements of the limbs and other parts of the skeleton are the result of voluntary action under the control of the brain. The kicking of a football by the action of the leg muscles is very similar to the action of the muscles in a knee jerk reflex. The essential difference, however, is that the brain initiates the impulse and causes muscular action in order to bring about the kick, whereas the knee jerk reflex is an involuntary action. The brain is informed of the contraction of muscles in the knee jerk reflex, but it has no control over the movement. The differences between these two actions are summarised in table 7.4.

Spinal reflex	Voluntary action
Stimulus affects external or internal receptor.	Initiated from the brain at the conscious level.
Spinal cord only involved – not under the control of the will.	Forebrain involved – under the control of the will.
The impulse travels only up or down the spinal cord.	The impulse travels from the brain down the spinal cord.
The path of the nerve impulse is by the shortest route.	The path of the nerve impulse is much longer.
The response is immediate.	The response can be delayed.
The response is in skeletal or involuntary muscle or glands.	The response is in skeletal muscle only.

Table 7.4 Comparison of a spinal reflex action and voluntary action

7.60 More complex behaviour

2
3
3
3

Many of the reflexes of an animal are innate, that is already present when the animal is born. The changing of the pattern of a reflex action is called *conditioning.* The Russian physiologist, Ivan Pavlov, carried out the early work on *conditioned reflexes.* He experimented with dogs, to which he fed meat and measured the amount of saliva produced at the sight of the food. This is a normal reflex action, which humans also experience at the sight of food. By ringing a bell every time food was produced, Pavlov conditioned the dogs to expect meat whenever they heard this sound.

Eventually after many trials, the dogs salivated in response to the ringing of the *bell only* (i.e. even when no food was produced). Thus the conditioned reflex producing salivation involves nerve pathways *different* from those of the unconditioned reflex.

Experiment

To demonstrate a conditioned reflex in humans

Procedure

1 You will need the cooperation of a group of ten people to act as subjects.

2 Provide each of them with a piece of paper and a pencil.

3 Tell your subjects that you are going to read to them a passage and that they should make a tick on the paper each time that you say the word 'a'.

4 Now read to them the following paragraph at a fairly brisk pace, tapping your pencil on the desk each time that you say 'a'. (You might find it helpful to encircle the letters in pencil, so that they stand out.)

'A farm worker was walking down a country road on a chilly, spring morning, when he saw a white shape immediately ahead of him. A closer look revealed that it was the body of a large sea-bird. It had a pair of webbed feet, a dark band on each wing, a long, orange beak and a crest of black feathers on its head. Death had obviously been caused by a blow on its neck (possibly from a passing car), although there was a smear of oil along its left wing.'

5 Now ask your subjects to draw a line under their ticks before continuing.

6 Read out the second paragraph, this time tapping your pencil on the desk when you say the word 'a' and when you say the words printed *in italics.*

'A few days later, he was passing *the* same spot when he saw the body of a second bird. He realised that *it* belonged to a different species (its feet were *not* webbed), but *its* overall size, shape *and* colouring were the same as those of the first bird.'

7 Collect the pieces of paper and count the numbers of ticks above and below the lines. Calculate the mean number of ticks per subject above the line and the mean number of ticks per subject below the line.

Questions

1 What is the ratio of average number of ticks above the line to number of 'a's in the first paragraph (13)?

2 What is the ratio of average number of ticks below the line to number of 'a's in the second paragraph (3)?

3 What feature of the experiment corresponded to the food in Pavlov's experiment?

4 What feature of the experiment corresponded to salivation in Pavlov's experiment?

5 What feature of the experiment corresponded to the bell in Pavlov's experiment?

6 Can you think of a children's game in which the principle of the conditioned reflex is used?

The presentation of the conditional and unconditional stimuli simultaneously or with a short time interval is called *reinforcement.* Many recent investigations into *learning processes* with animals have shown, experimentally, that learning progresses by reinforcement of the response with other stimuli such as food. Skinner's work on rats and pigeons shows how this reinforced stimulus can increase the speed of learning. These studies have been used to help the learning processes in Man, especially in the development of teaching machines where each small step in learning is reinforced.

Small children learn many things through the approval or disapproval of their parents. They recognise that good behaviour brings rewards in the form of encouraging words. As they grow up, this reward becomes the approval of other children or adults.

7.70 Co-ordination by the endocrine system

Homeostatic mechanisms in the mammalian body (see Chapter 5) are co-ordinated in two ways, by the nervous system and by the *endocrine system* of ductless glands. The nervous system has been likened to a telephone exchange with its telephone wires connected to individual telephones, but the endocrine system is less exact and its action could be compared to a public address system. This means that the products liberated into the *blood stream* by the ductless glands have a much wider effect on the body. The response could be temporary changes produced by a secretion from a ductless gland (hormone) at a particular moment, e.g. *adrenalin* (see table 7.5), or they could initiate a complicated cycle of changes, e.g. sex hormones preparing the female for ovulation, pregnancy, birth and lactation.

7.71 Evidence of activity of endocrine glands

It is common practice to remove the testes of certain domestic animals in order to make the males more docile, and also to bring about an increase in weight. This operation, called *castration,* is performed on bulls (bullocks) and stallions (geldings) and male pigs. This operation is relatively simple in mammals, where the testes are external. It is also performed on birds with internal testes. When this is done with young cockerels they grow fat and more suitable for eating. They are known as *capons.*

In 1848, Adolf Berthold performed a series of experiments on the testes of young cockerels (see fig 7.16).

i) He removed the testes, and the bird became a capon.

ii) He disconnected nerve and blood supplies to the testes. A new blood supply developed but not a new nerve supply.

iii) He transfered the testes to another part of the body and the testes developed a new blood supply but not a new nerve supply.

Questions

1 Examine the mature birds in experiments 2 and 3 shown in fig 7.16 below. Do they resemble normal mature cockerels?

2 What controls the development of secondary sexual characters in the cockerel, hormonal secretions or nervous control? Explain the reasons for your decision.

One classical investigation of hormonal activity was made on the pancreas. In 1889, Mering and Minkowski found that on removing the pancreas from dogs, the animals developed a fatal condition called diabetes in which the urine contained sugar. In 1917, Kamimura tied off the pancreatic duct in rabbits, and although the cells secreting pancreatic juice died off, the animals did not discharge sugar in their urine i.e. did not develop the disease. In 1922 Banting and Best injected extracts of pancreas into dogs whose pancreas had been removed. The blood sugar levels of these dogs immediately fell from 0.3% to 0.21%. In their paper, dated 1933, they state:

'. . . in the course of our experiments we have administered over 75 doses of extract from pancreatic tissue to 10 different diabetic animals . . . the extract has always produced a reduction of the percentage of sugar in the blood and of the sugar excreted in the urine.'

Questions

3 Why could the experimenters finally conclude that the chemical secretion of the gland controlling blood sugar levels was secreted into the blood?

4 The pancreas extract does not completely cure diabetes, but has to be given at regular intervals. Why is this?

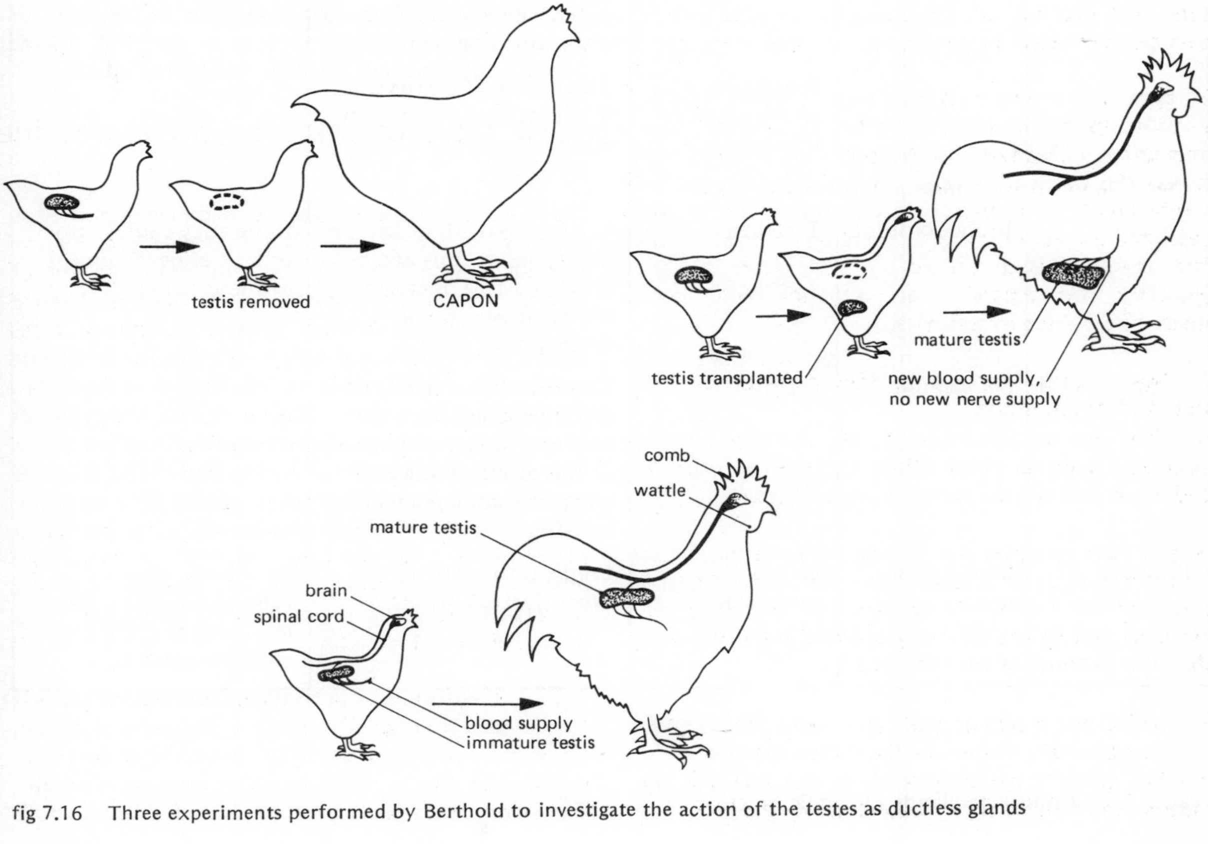

fig 7.16 Three experiments performed by Berthold to investigate the action of the testes as ductless glands

7.72 Endocrine glands and their functions

Gland	Position in body	Hormone secreted	Response of body to hormone	Abnormal functions
Thyroid	Neck	Thyroxin	Controls basic metabolism and growth rate.	Deficiency causes dwarfism and mental retardation in childhood, myxoedema in adult. Over production causes increased metabolism.
Islets of Langerhans	Pancreas (dual function exocrine and endocrine)	Insulin	Controls balance of sugar in the blood.	Deficiency results in diabetes mellitus.
Adrenal gland	Attached to kidneys	Adrenalin (medulla)	Controls response for 'fight or flight', i.e. increased heart beat, increased blood sugar, dilates coronary artery, dilates pupils, etc.	
		Cortisone and other hormones (cortex)	Release glucose from protein in stress. Control salt balance in the body.	
Ovary	Dorsal abdominal wall	Oestrogen	Controls growth of uterus, hip skeleton, underarm hair, pubic hair, breasts, (secondary sexual characteristics).	Deficiency causes delay of appearance of these changes.
		Progesterone (from corpus luteum)	Thickening of uterus wall in preparation for implantation of embryo.	
(See Chapter 8 – hormones in the sexual cycle)				
Testis	In the scrotum	Testosterone	Controls growth of hair on pubis, under arms and on chest and face, increased muscular development, deepening of voice.	Deficiency causes delay or lack of development of these changes.
Intestinal wall	Duodenum	Secretin	Controls the secretion of digestive juices from the pancreas (exocrine function of this gland). Produced when acid food enters the intestine.	
Pituitary	Beneath the brain	Several hormones	Controls the activity of other ductless glands.	Deficiency causes a type of dwarfism, and inactivity of certain endocrine glands.

Table 7.5 Endocrine glands and their functions

7.73 Comparison of nervous and endocrine activity

	Nervous control	Endocrine control
Stimulus	Through receptors, eyes, nose or internal receptors, include light, gravity, sound, temperature, etc.	Through external or internal receptors.
Linking mechanism	Central nervous system and nerves	Blood and circulatory system
Effectors	Muscles and glands	Whole body, organs or organ systems
Speed	Rapid reaction – reflex arc or voluntary nerve paths	Slow for some such as growth; rapid for others such as fight or flight hormone, adrenalin.

Table 7.6 Comparison of nervous and endocrine activity

Questions requiring an extended essay-type answer

1 Describe how the eyes of a man adjust
a) when he looks down at a newspaper in his hand after watching distant walkers on a hillside, and
b) when he steps from a well-lit room into the darkness.

2 a) Compare the structure of the human eye with a camera. In what ways do their workings differ from each other?
b) Describe how the human eye is corrected for long and short sight.

3 a) What do we mean by *irritability* in living organisms?
b) Describe how each of the following illustrates this phenomenon:
i) A deer sniffing the wind with a predator near
ii) A boy dropping a hot plate that he has just picked up
iii) A cook's eyes watering whilst peeling an onion.

4 a) Compare the reactions to stimuli, the speed of response and the effectors of the nervous and endocrine systems.
b) Name the hormones produced by three ductless glands and describe the body's response to them. What abnormal states could arise in the body due to their deficiency?

5 a) Describe the main regions of the human brain and the functions of the body that these control.
b) What part does the sympathetic nervous system play in the autonomic nervous system?

6 a) What do we mean by *a reflex action?*
b) Draw a diagram to show the reflex arc associated with the knee-jerk reflex.

7 a) Draw a labelled diagram to show the structure of the inner ear.
b) Describe a controlled experiment you would carry out to determine the frequency range audible to an old man.
c) How would you expect the auditory range of a boy of fifteen years to differ from that of an old man?

8 Reproduction, development and growth

8.00 Introduction

Every individual living thing has a limited life span. This may range from a few seconds for certain micro-organisms to thousands of years in the case of large trees. However, all organisms have the ability to pass on life by producing new individuals with the same general characteristics as themselves; this is known as *reproduction.*

Basically, there are two different methods by which reproduction may take place; *sexual* and *asexual.* In sexual reproduction, new individuals are formed by the *fusion* (joining) of small pieces of living material, called *gametes,* from two different organisms. In mammals (including Man) a male gamete or *sperm* fuses with a female gamete or *egg.* The sperm and the egg (or ovum) are highly specialised cells, completely different from each other in appearance but complementary in function. The details of sexual reproduction are slightly different for plants and for unicellular organisms, but the principle is exactly the same in all cases.

The process of fusion between male and female gametes is termed *fertilisation;* the male gamete is said to fertilise the female gamete and not vice versa. The fertilised ovum is known as a *zygote,* which will develop to form, eventually, a new adult organism.

Asexual reproduction takes place when a new individual arises from part of a previously-existing organism, *without fertilisation* having taken place. The range of forms that asexual reproduction can take is enormous, as is the proportion or relative size of the part of the 'parent' which will form the 'new' organism. In the unicellular animal *Amoeba,* for example, the *whole organism* splits into two equal halves, each of which is a miniature adult; a *cutting* representing perhaps ten per cent of the shoot system of a flowering shrub may develop into a complete plant; whereas in certain insects, new adults may arise asexually from a tiny *unfertilised egg.*

The two reproductive processes are summarised in the following chart.

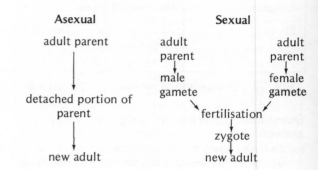

Asexual

adult parent
↓
detached portion of parent
↓
new adult

Sexual

adult parent → male gamete ↘
adult parent → female gamete ↙
fertilisation
↓
zygote
↓
new adult

8.10 Sexual reproduction in the Class Mammalia

In mammals and other higher animals, *sexual reproduction* is normally the *only* means of producing offspring. Asexual reproduction is virtually unknown under natural conditions, but *parthenogenesis* can be induced artificially in certain amphibians by, for example, pricking an unfertilised egg with a needle. The question of whether such 'virgin birth' is possible in humans has never been satisfactorily resolved.

Two important features of mammalian reproduction stem from the fact that mammals are essentially terrestrial in habit, for even seals and whales arose from land-living forms. These two features are *internal fertilisation* and *viviparity* (the birth of live young).

Most aquatic animals, from sea-snails, crustaceans and marine worms to amphibians and fish, fertilise their eggs *externally.* That is to say, the female deposits the eggs outside the body and the male releases millions of sperms into the water nearby. The sperms swim to the eggs and fuse with them. In terrestrial animals, such a simple system is not possible, since the sperms, in order to swim, must be kept in a liquid environment.

Therefore, most land-based animals have evolved specialised mechanisms for getting the sperms as near as possible to the ova before release. In earthworms, two animals are enclosed together in an elastic sheath called the *clitellum;* in some insects an envelope of sperms, the *spermatophore,* is deposited in the female reproductive tract; and in land snails and mammals the sperms are introduced into the female orifice by a special organ, the *penis.* Once inside the female tract, the sperms can swim along the mucous lining and fertilise the eggs.

Eggs which have been fertilised externally develop, of course, in a watery medium which is relatively free from fluctuations in temperature. If, however, a land-dwelling animal were simply to release the eggs which had been fertilised internally, there would be severe problems of dehydration and of over- and under-heating, since land provides a far less constant environment than water. Thus land animals that are *oviparous* (egg-laying) usually protect their eggs from the external environment by covering them with some form of *waterproof membrane* or *shell* before laying. We can see here how external fertilisation and oviparity would be unlikely to occur together in a terrestrial environment, since a shell sufficiently waterproof to prevent the exit of water molecules would be unlikely to permit the entry of sperms.

The vast majority of mammals do not, however, lay eggs. They carry the idea of protection from the environment one step further and, instead of secreting waterproof membranes around the fertilised eggs, they keep them within the maternal body. In this way they are also able to supply the embryo with all the nutrients required for development. Thus, the eggs of mammals are much smaller than the eggs of birds or even of amphibians, which need to contain very large supplies of food materials, in the form of *yolk,* to last them until hatching.

Sexual cycles

In most higher animals, the endocrine system plays a large part in reproduction. The secretion of certain hormones by the pituitary gland determines the time at which eggs can be fertilised successfully. In most mammals, the female will only allow *insemination* (the introduction of sperms into the female reproductive tract) via the penis of the male, at a time when fertilisation is likely to result. At this time, under the control of hormones, the wall of the womb or *uterus* is at its thickest, and eggs are released from the ovaries (a process called *ovulation*). Female mammals in this condition are said to be 'on heat' or 'in season', and they may produce a secretion to arouse and excite the male. If fertilisation does not take place at this time, the wall of the uterus gradually becomes thinner again, the extra material being reabsorbed into the body – usually without bleeding. The process is repeated at regular intervals which vary tremendously, depending upon the species: in mice it is five days; in dogs and cats it is about six months. The period of fertility is known as *oestrus,* and the regular cycle of events is referred to as the *oestrous cycle.*

In humans, the female sexual cycle differs from the above pattern in a number of important respects. The female is receptive to the male *almost equally at all times,* although *ovulation* occurs, from one or other

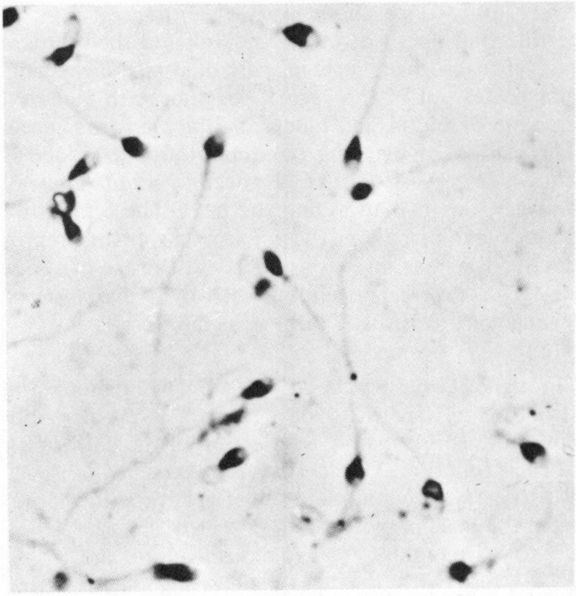

fig 8.1 A photograph of some highly magnified sperm cells (x 1 000)

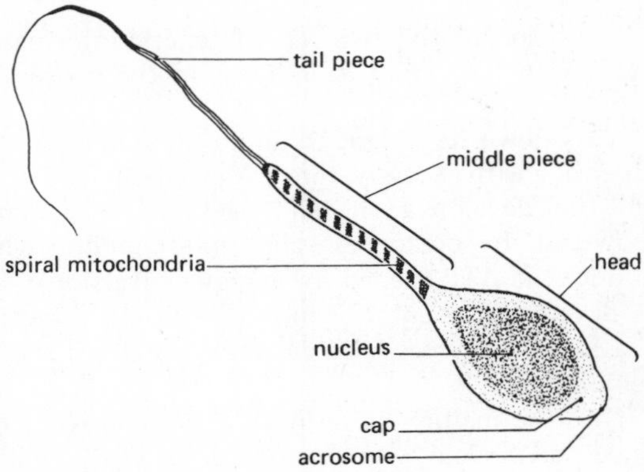

fig 8.2 A drawing of a sperm cell (x 5 000)

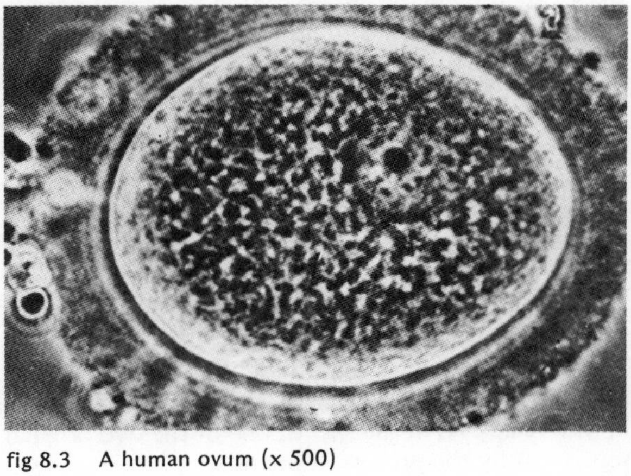

fig 8.3 A human ovum (x 500)

ovary, only once about every *twenty-eight days.* If fertilisation does not occur, the lining of the uterus is not all reabsorbed. Instead, part of it is sloughed off and passes out of the vagina, together with a certain amount of blood and mucus, as the *menstrual flow.* This period of bleeding lasts about five days and is known as *menstruation.* It occurs about midway between one ovulation and the next. These monthly periods take place over the whole of a woman's reproductive life, starting with puberty at the age of about eleven or twelve and ending with the *menopause* or 'change of life', usually during the late forties.

The gametes

Throughout the whole of the animal kingdom, the structure of spermatozoa and ova is remarkably constant. This is mainly a reflection of the uniformity of their function. The sperm is required:

1 to carry one complete set of genetic information from the male parent;
2 to swim rapidly in an aqueous medium and
3 to fuse with the egg, penetrating any covering membranes that the latter may possess.

To these ends, each sperm (fig 8.2), during the course of its development from an unspecialised body cell:

1 loses one of the two sets of genetic information (see section 9.6) from its nucleus in the process of *meiosis*;
2 cuts down its weight to an absolute minimum by dispensing with excess water and cytoplasm, and at the same time develops a long tail (flagellum) for swimming ('powered' by organelles called mitochondria which produce the ATP needed for energy conversion);
3 develops at its anterior end an *acrosome,* consisting of a sac of enzymes used for breaking down the egg membranes.

Thus the mature sperm has a head (consisting of a condensed nucleus with an apical acrosome), a middle piece containing mitochondria and a long, slender tail.

The ova, almost invariably spherical or ovoid in shape, are usually the largest cells in an animal's body. This is because they contain reserve foodstuffs in the form of yolk. Even in mammals, where the developing embryo obtains most of its nutritional requirements from the *maternal body,* the egg cell contains some yolk and may be over 100 μm in diameter (fig 8.3).

The ova of most animals are surrounded by one or more *egg membranes.* They usually serve for protection but, in the case of mammals they may also effect the absorption of nutrients from the immediate environment. Mammalian ova are surrounded by a layer of *follicle cells,* arising from the ovary.

Finally, and most important, the egg cell carries one set of *genetic information* (from the female parent) in its nucleus. As in the sperm, one of the two original sets is lost by the process of meiosis.

8.11 The male mammal

Demonstration

Dissection of the male rat

Procedure

1 A freshly-killed male rat is laid on its back on a dissecting board. Four awls are used to pin its limbs to the board.
2 The loose ventral skin of the abdomen is lifted with forceps away from the body, and a median, ventral incision is made, with scissors or a scalpel, from the rib cage to the pit of the abdomen.
3 Second and third cuts are made along the legs, and the skin is pinned back.
4 Now the muscular abdominal wall is cut and pinned back in the same way. The coils of the intestine are thus exposed.
5 The intestine is unravelled and the rectum is severed about 6 cm from the anus. The whole of the intestine is then pinned back, well away from the abdominal cavity.
6 The urinogenital system of the male rat is now clearly visible. The skin (*scrotum*) surrounding the testes is carefully removed.

Questions

1 Two tubular structures are visible leaving each testis. What is the function of the more anterior tube (the spermatic cord)?
2 What is the function of the tube leaving the posterior part of the testis?
3 The urethra is the duct inside the penis. How many structures can you see opening into it?

Fig 8.4 opposite shows a dissection of the reproductive system of a male rat, such as has been demonstrated to you. The *spermatozoa* are produced in the paired ovoid testes, each of which consists of many coiled tubules, with sperms developing in their walls. The sperms pass from the testis into the *epididymis,* which lies against the testis and also consists of a series of coiled and looped tubes. The epididymis may store the sperms temporarily and the walls of its tubules produce a secretion. The *vas deferens* then conducts the sperms to the base of the *penis,* where they receive the secretions of the *prostate* and *Cowper's glands* and the *seminal vesicles.* These secretions, together with those of the epididymis and of the testis, and of course the sperms, make up the *seminal fluid,* or *semen,* which is ejaculated through the *urethra* during copulation. The urethra, which lies inside the penis, also carries the *urine* from the bladder to the exterior and is consequently called a *urinogenital* structure.

The reproductive organs of the human male are shown in fig 8.5 on the following page. As you can see, the main difference between them and those of the rat are differences in detail of the glands that produce the seminal fluid. Note that the testes are held in the

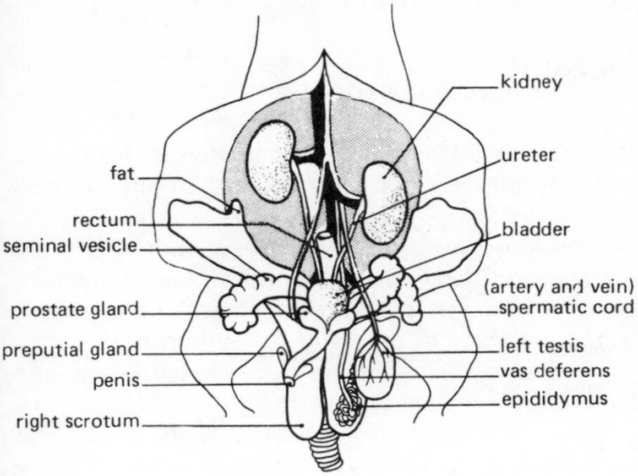

fig 8.4 The reproductive system of a male rat

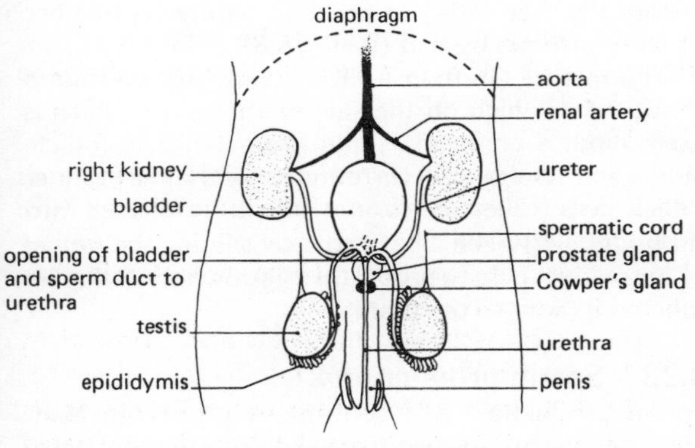

male reproductive organs (position in body)

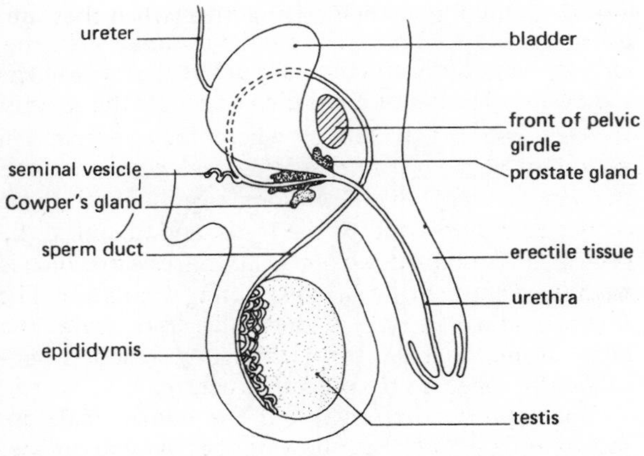

male reproductive organs (vertical section)

After D. G. Mackean

fig 8.5 The reproductive system of a male human

scrotum outside the main abdominal cavity. This enables them to be kept at the optimum temperature for sperm production, which is slightly lower than normal body temperature.

8.12 The female mammal

Demonstration

Dissection of the female rat

Procedure

The same procedure is followed as for the dissection of the male rat in section 8.11 above. On removal of the alimentary canal, the female urinogenital system is clearly exposed.

Questions

1 In what position do the ureters, bearing urine from the kidneys, enter the bladder?

2 Are the two kidneys situated at the same level?

3 What is the function of the urethra?

4 What other differences can you observe between this dissection and the dissection of the male rat?

The stem of a large Y-shaped structure runs below (dorsal to) the urethra, compared with which it is about twice as wide and 50% longer. This stem which opens to the exterior just posterior to the urethral opening, is the rat's *vagina*, a muscular structure which receives the penis during copulation.

The arms of the Y extend to just below the kidney and may be covered with fat, which should be removed. These 'arms' form the twin *uteri* (wombs) of the rat. It is quite possible that the dissected rat may have been pregnant. If this is the case, compare the uteri with those of a non-pregnant rat and note the huge increase in size. Note the positions of the developing embryos and count them.

The 'knobs' seen on the ends of the uteri are coiled *oviducts* or *fallopian tubes* which lie partly around and

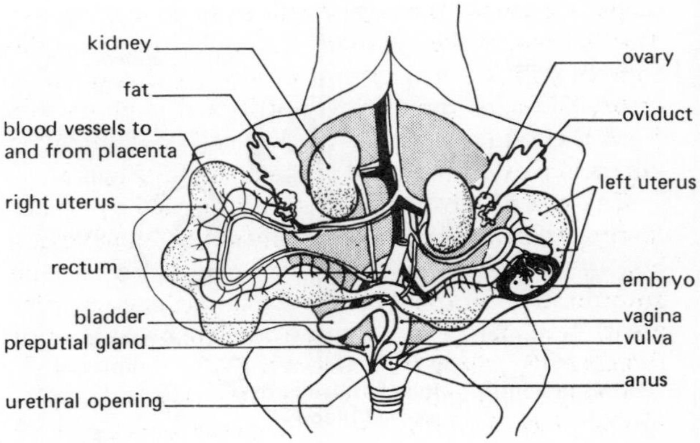

fig 8.6 The reproductive system of a female rat

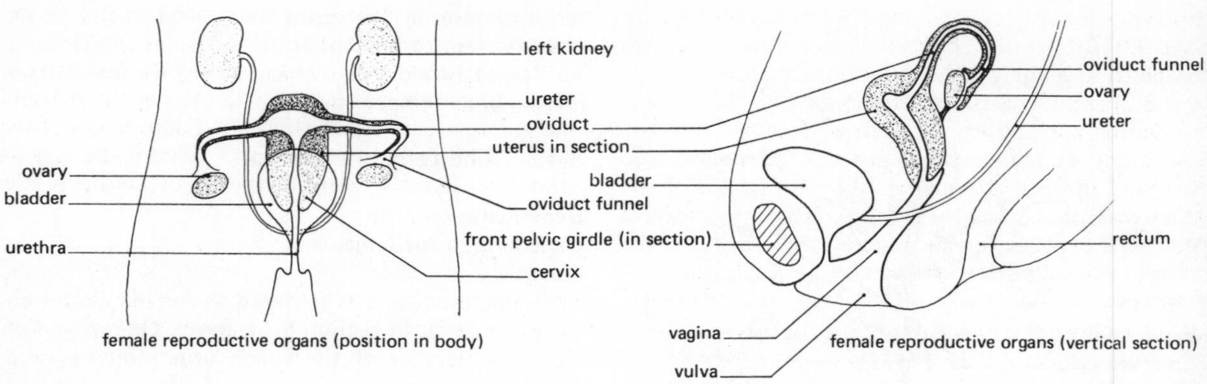

After D. G. Mackean

fig 8.7 The reproductive system of a female human

alongside the small *ovaries.* The fallopian tubes conduct the released eggs from the ovaries to the uteri.

Using a needle and forceps, try to unravel the fallopian tubes as far as possible. It is not an easy task, since the tubes are fairly delicate and are held in position by connective tissue (see fig 8.6).

The reproductive system of the human female (fig 8.7) differs from that of the rat in that there is only one uterus and the fallopian tubes are not coiled.

8.20 Sexual reproduction in humans

8.21 Ovulation

When a baby girl is born, each ovary already contains tens of thousands of potential egg cells (*oogonia*). Of these, only a small fraction (about 450) will be released from her ovaries during her reproductive life. As an oogonium begins to grow to form a mature egg, it is known as an *oocyte.* The smaller cells around it form a follicle, which eventually develops a large fluid-filled cavity. At this stage, the whole structure is known as a *Graafian follicle.* Its cells help to provide the growing oocyte with nourishment. At the end of the period of growth, the nucleus of the oocyte starts to divide meiotically (see 9.04) in order to reduce its number of chromosomes by half (see fig 8.8).

The mature Graffian follicle stands out as a bump about 0.5 cm high on the side of the ovary, which is itself about 4 cm in maximum diameter. The follicle bursts and its egg, still surrounded by a layer of inner follicle cells (called the *corona radiata*) is released into the *body cavity.* The cilia on the cells lining the *funnel* of the oviduct help to propel the egg down the fallopian tube on its way to the uterus.

8.22 Sperm production

Sperm production is a continuous process in Man. Spermatozoa are formed from certain cells (*spermatogonia*) which are found in the lining of the seminiferous tubules of the testis (see fig 8.11). Unlike oogonia, spermatogonia are constantly being produced by *mitotic* cell division (see 9.01). The cells thus formed grow to about double their original size (when they are

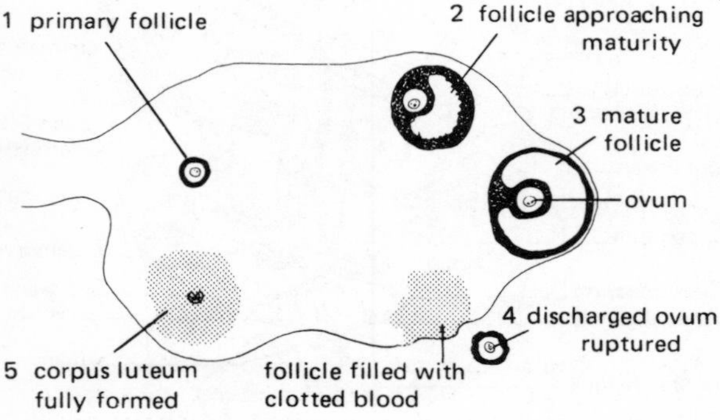

fig 8.8 Section through an ovary

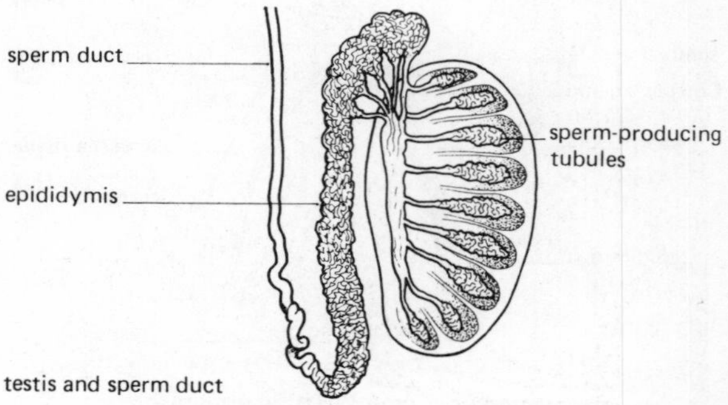

fig 8.9 Vertical section of a testis

known as *spermatocytes*) and then divide meiotically (see 9.04) to form *spermatids.* The spermatids change shape to develop into the typical spermatozoon form (fig 8.2) and then pass into the epididymis for storage.

During copulation, the sperms, together with the secretions of the prostate and seminal vesicles, are expelled through the urethra by contractions of the duct walls and associated muscles. Sperms which have not been expelled in this way are reabsorbed into the cells of the epididymis wall.

8.23 Sexual intercourse and fertilisation

Coition, copulation, intercourse, and *mating* are all alternative terms used to describe the act in which the penis of the male is inserted into the vagina of the female and *spermatozoa* are ejaculated into the female reproductive tract. In order for the penis, usually a soft and flaccid organ, to penetrate the vagina, it must become *hard and erect.* In most mammals, including Man, this is accomplished by increasing the amount of blood in the soft tissues around the urethra, thus producing turgidity. The stimulus for erection is largely psychological and depends upon arousal and excitement of the male by the female.

Penetration is also made easier by the secretion of lubricating mucus by cells lining the vaginal wall and the external genital area (the *vulva*). Movement of the penis backwards and forwards inside the vagina stimulates nerve endings in the tip of the penis (*the*

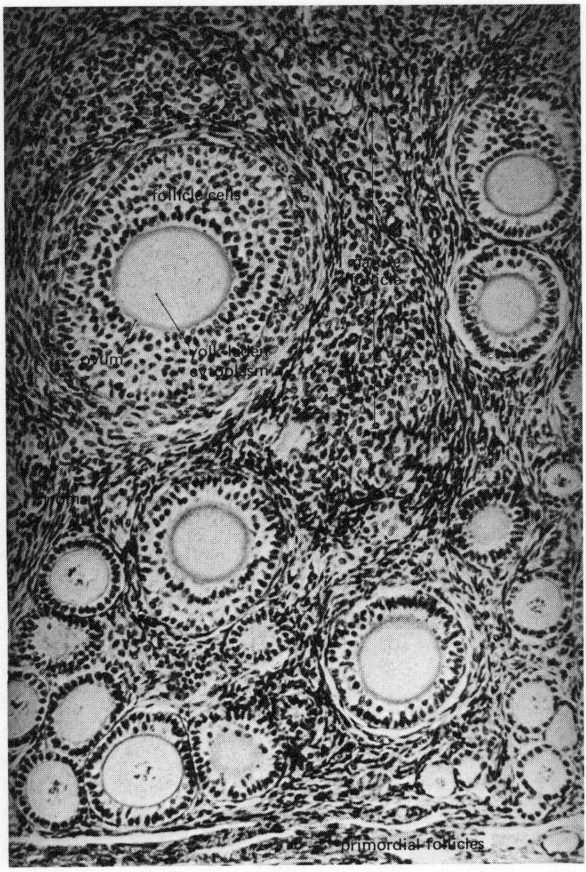

fig 8.10 Photograph of section through an ovary

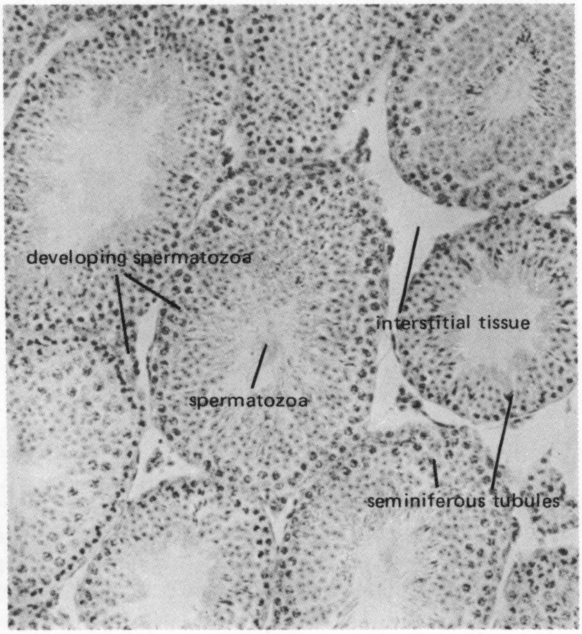

fig 8.11 Photograph of section through seminiferous tubules

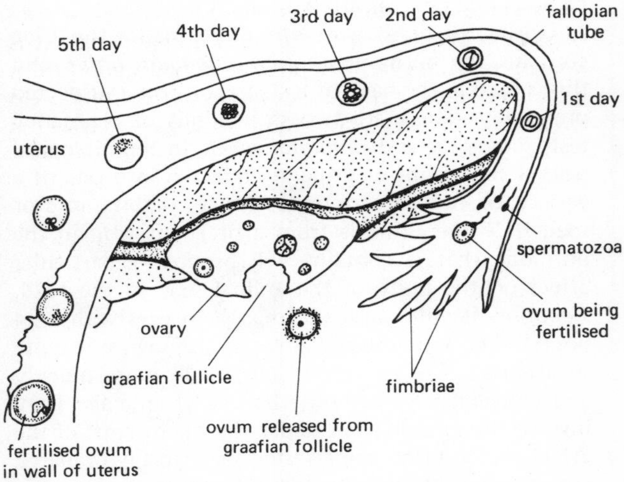

fig 8.12 Ovulation, fertilisation and implantation. Note that the lining of the womb is thickened and well supplied with blood. Had fertilisation *not* occurred the lining would have broken down and the blood flowed away during menstruation

glans), which set in motion a reflex action resulting in the ejaculation of semen.

On ejaculation, about 1.5 cm^3 (in humans) of seminal fluid containing about 100 million spermatozoa, is deposited in the vagina, near the *cervix* or neck of the uterus. The sperms *swim* up through the uterus and into the fallopian tubes. If intercourse takes place soon after ovulation, the sperms encounter a mature egg in the fallopian tube and fertilisation takes place there.

The first step in fertilisation is the breaking down of the corona radiata by an enzyme released by the acrosome of the sperm. After these cells have been dispersed, a single sperm penetrates the egg membrane and the sperm head enters the egg cytoplasm, leaving the sperm tail outside. This penetration triggers off a response in the egg, which prevents other sperms from entering. Later the egg and sperm nuclei *fuse*.

8.24 Pregnancy and birth

After fertilisation, the egg, now a *zygote*, undergoes divisions into two cells, four cells, eight cells, etc., as it passes further down the fallopian tube on the way to the uterus. It then becomes *embedded* in the uterine wall which is thick and well supplied with blood vessels.

The period between fertilisation and birth is known as *pregnancy* and lasts for about nine months. The first sign of pregnancy is usually the missing of a menstrual period, although this can also be caused by a variety of physical and emotional upsets. Later signs include 'morning sickness', which affects about two-thirds of pregnant women, enlargement of the breasts and (in light-skinned people) a darkening of the pigmented area (areola) around the nipple.

During the first few months of pregnancy the urine contains the hormone *chorionic gonadotrophin* and this is used as the basis of tests to confirm a suspected pregnancy. There are two main kinds of pregnancy test, *biological* and *immunological*. In the biological tests a few drops of urine are injected into one of a variety of female vertebrates (mice, rabbits, rats or toads). If the urine is from a pregnant woman the hormone that it contains will produce a particular effect on the body of the animal (e.g. in the toad, *Xenopus*, it will cause the release of eggs within 24 hours). The immunological tests, although not quite so accurate, can be carried out much more quickly and conveniently, without the use of animals. They involve the examination under the microscope of the effect of the urine being tested on specially prepared red blood cells, which clump together if chorionic gonadotrophin is present. Results can be obtained within two hours.

As the zygote develops into an embryo, the blood, blood vessels and heart are among the first features to

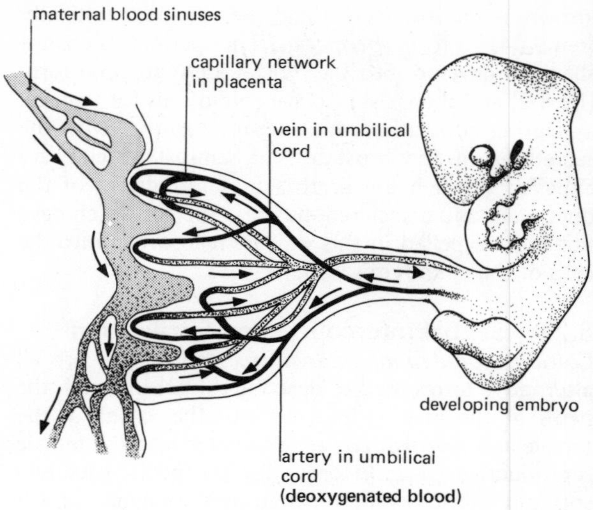

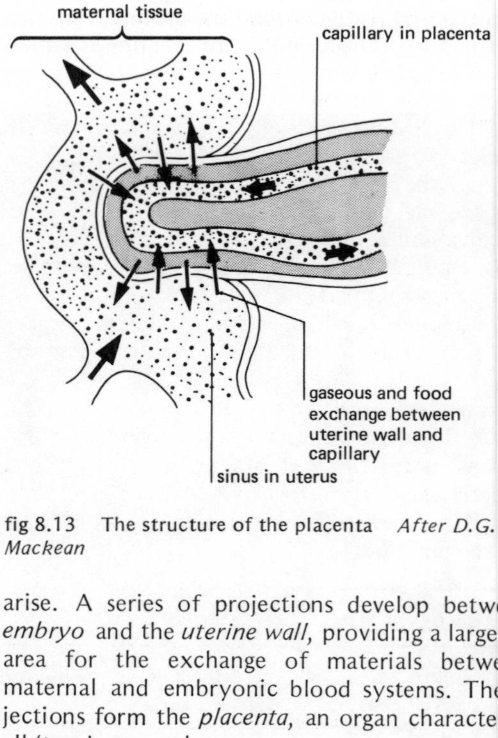

fig 8.13 The structure of the placenta *After D.G. Mackean*

arise. A series of projections develop between the *embryo* and the *uterine wall*, providing a large surface area for the exchange of materials between the maternal and embryonic blood systems. These projections form the *placenta*, an organ characteristic of all 'true' mammals.

In later development, the placenta becomes much more well-defined. It consists partly of *maternal* tissue and partly of *embryonic* tissue and is connected to the body of the embryo by the *umbilical cord*. This structure contains a main vein, which carries oxygenated blood and dissolved nutrients to the embryo, and an artery which takes deoxygenated blood and waste materials to the placenta (fig 8.13).

The embryo is contained in a fluid-filled sac, the

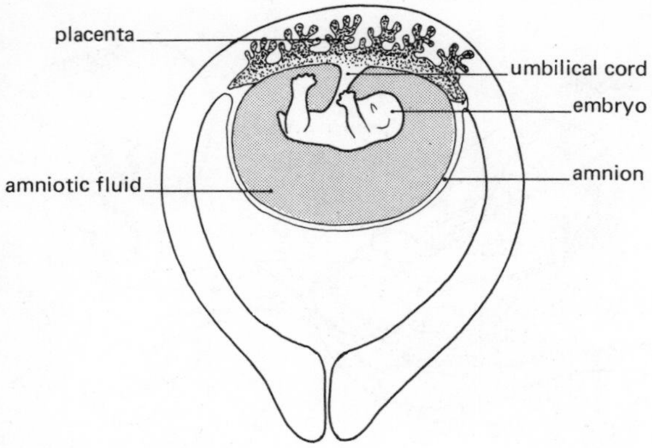

fig 8.14 A human foetus at 10 weeks

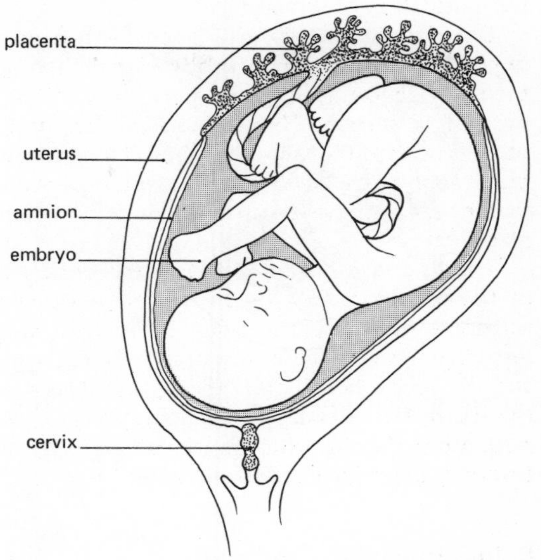

fig 8.15 A human foetus just before birth

amnion, which allows it freedom of movement during growth and protects it from mechanical damage.

At the end of nine months pregnancy, the embryo, now called a *foetus,* comes to lie with its head downwards, just above the opening (cervix) of the uterus (see fig 8.15). With a series of powerful contractions, the muscles of the uterine wall force the foetus, head first, through the dilated cervix and out of the mother's body via the vagina. Shortly after birth, the placenta is ejected in a similar manner. In pigs and certain other mammals, the foetal part of the placenta pulls neatly out of the maternal part, like a hand from a glove. In humans, however, much of the maternal part is also detached, and this results in a certain amount of bleeding.

The umbilical cord is *cut* and *tied* soon after *parturition* (birth), and the part remaining on the baby withers and drops off within a few days, leaving a small, permanent scar, the *navel.* In many other mammals, the mother bites through the cord, and may eat the placenta.

During pregnancy, most healthy women do not need to make great changes to their way of life, although they should pay particular attention to diet and rest. Since the growing foetus needs calcium for the development of bones and iron for the manufacture of red blood cells, the diet should be rich in these elements and also in vitamins (see Chapter 2). Alcohol and tobacco should be avoided and exercise should be taken in moderation. In later months, the growth of the foetus causes crowding of the organs in the abdomen, with resulting pressure on the blood vessels, nerves and bladder. This may cause varicose veins, cramps and the need to urinate frequently, all of which symptoms, although inconvenient, are not serious and are likely to disappear after the baby is born.

Time after fertilisation	Length	Stage of development
7–10 days	140–160 μm (diameter)	Hollow ball of cells, thickened in one area and implanted in the uterine wall.
3 weeks	1.5 mm	Head region obvious. Spinal cord and heart starting to develop.
6 weeks	10 mm	Brain growing rapidly. Eyes and ears developing. Arm and leg buds forming.
12 weeks	9 cm	The embryo (now called a foetus) has almost the external appearance of a miniature baby.
9 months	50 cm	Birth (parturition) occurs (see fig 8.16).

Table 8.1 Stages in the development of a human baby

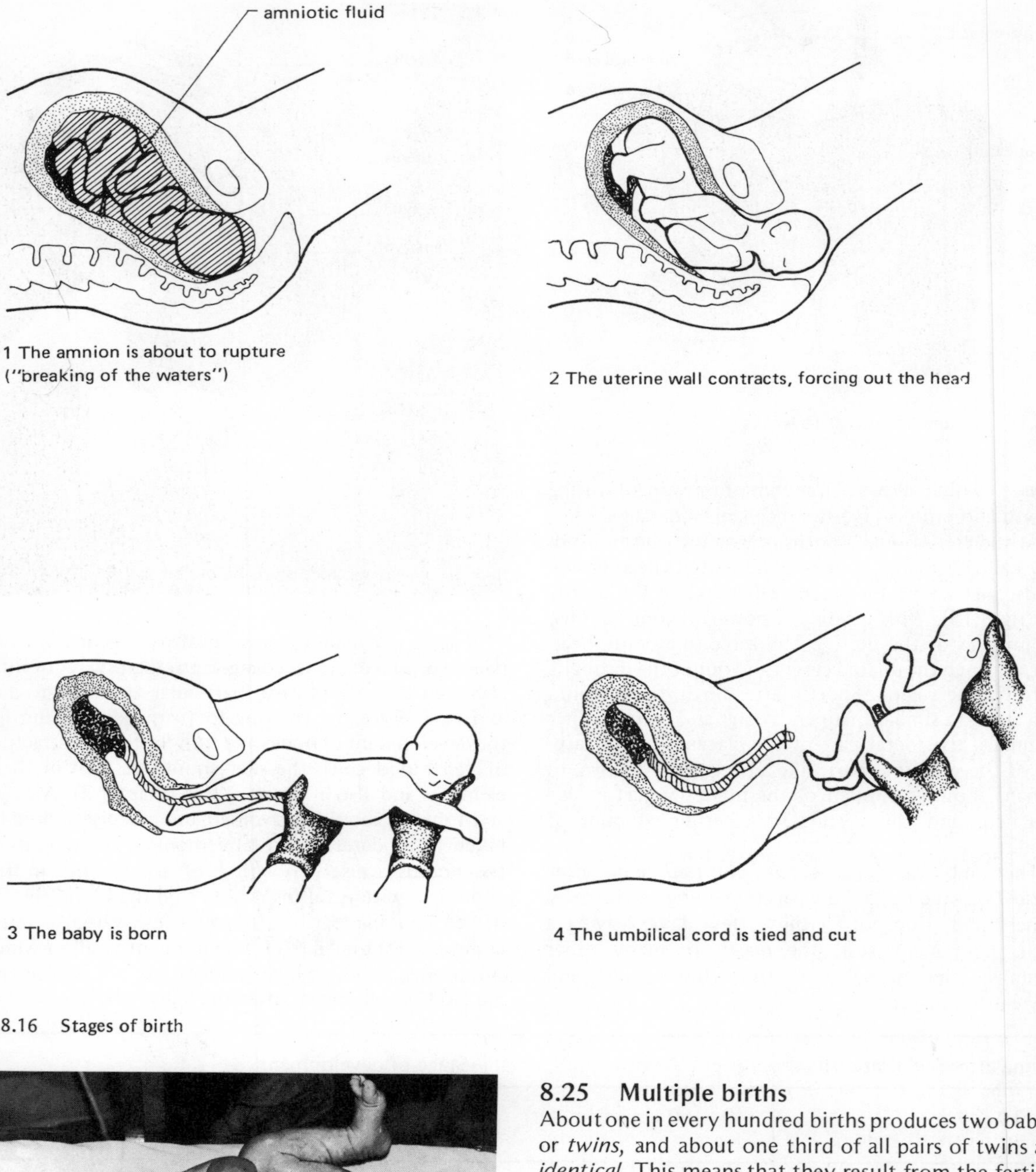
amniotic fluid
1 The amnion is about to rupture ("breaking of the waters")
2 The uterine wall contracts, forcing out the head
3 The baby is born
4 The umbilical cord is tied and cut

fig 8.16 Stages of birth

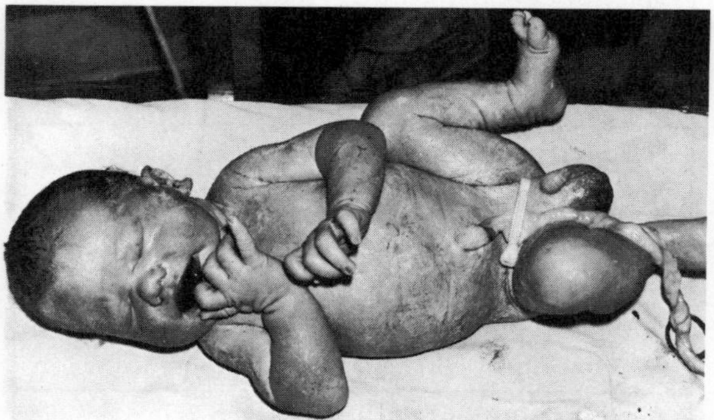

fig 8.17 A newborn baby

8.25 Multiple births

About one in every hundred births produces two babies or *twins*, and about one third of all pairs of twins are *identical*. This means that they result from the fertilisation of a single egg by a single sperm: after the first division, the two cells split to form two separate but very similar individuals. Of course, identical twins are always of the same sex. The remaining two thirds of twins are non-identical or *fraternal*. That is to say they result from the fertilisation of two eggs by two sperms. Fraternal twins are no more alike than other brothers and sisters. Triplets (three babies in one birth), quadruplets (four) and quintuplets (five) have always been rarities in the past. However, recently, 'fertility

fig 8.18 Identical twins

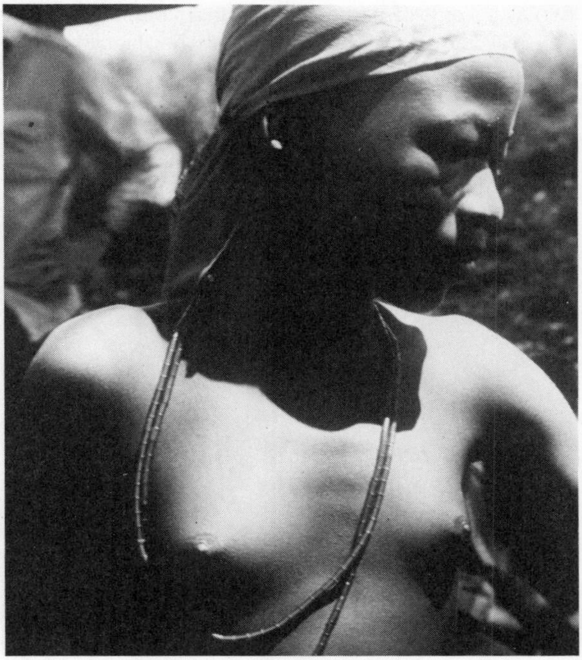

fig 8.19
(a) Non-lactating breast
For lactating (milk-producing) breast see fig 8.21

a)

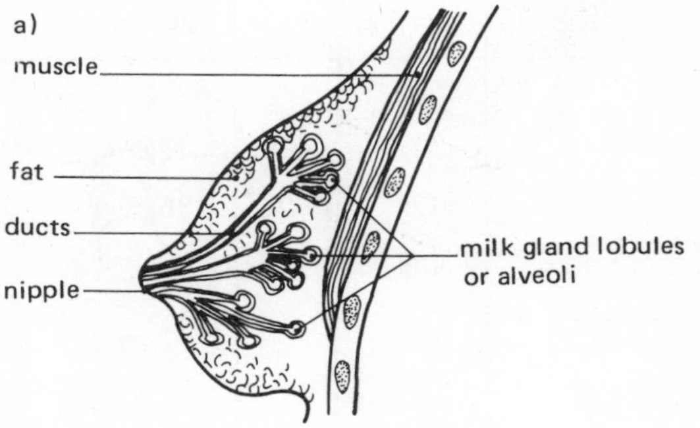

b)

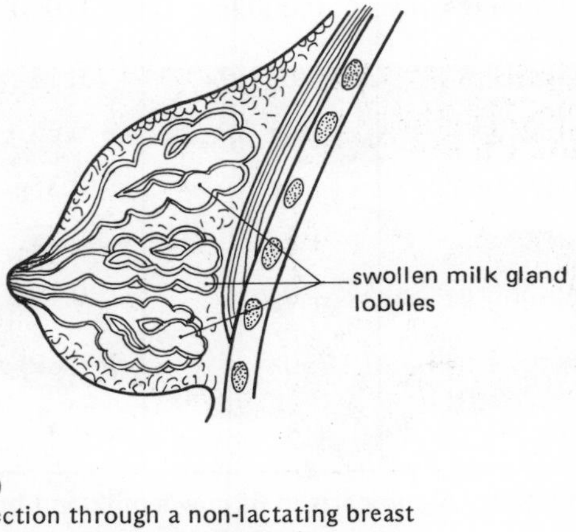

fig 8.20
(a) Section through a non-lactating breast
(b) Section through a lactating breast

drugs' have been developed to help women who are unable to have children because of a failure to ovulate. These drugs stimulate ovulation, but the doses in which they should be given are sometimes difficult to judge accurately, with the result that they occasionally cause several eggs to be released at once. This has given rise to births of six, seven and even eight babies at one time. Unfortunately, the babies are usually born *prematurely* (after less than nine months of pregnancy), are very tiny and die within a few days.

8.26 Infant feeding

Under 'natural' conditions a human baby feeds during the first months of its life on the secretions of its mother's mammary glands or *breasts.* A single breast consists of about fifteen branching, blind-ending, tubular milk glands, each opening onto the nipple via a duct. Between the branching tubules lie large amounts of fatty tissue, together with muscle fibres and connective tissue.

During the first two or three days after birth, the mother's breasts secrete *colostrum,* a yellowish fluid which contains the antibodies that protect the baby from many diseases during its early life. Colostrum also contains a different balance of proteins (less casein and more albumen) than that of the true milk secreted later.

Although great advances have been made in recent years in the production of 'bottle-feeds' based on cow's milk, there is no doubt that both mother and baby benefit from breast-feeding in several ways. First, there is the special relationship that builds up between the two as a result of perhaps three hours per day of close physical contact. Then there is the fact that the sucking action of the baby stimulates contraction of the uterine wall muscles and thus helps to bring the mother's body back to normal after the birth. Finally, there are the important nutritional advantages of

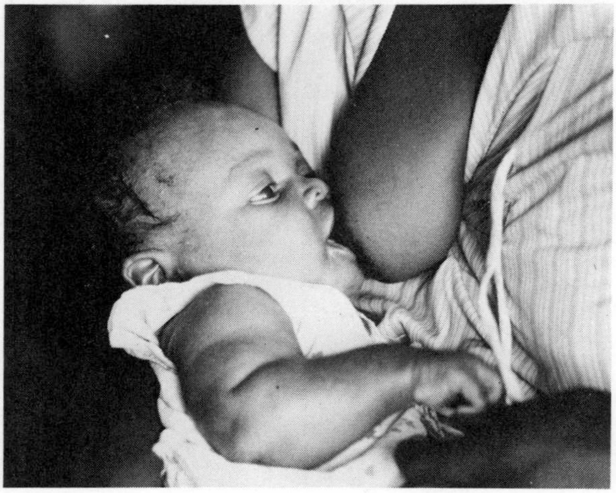

fig 8.21 Breast feeding

	Cow's milk (per kg)	Human milk (per kg)
Water	860g	876g
Protein	34g	12g
Fat	42g	36g
Carbohydrate	48g	70g
Vitamin C	15mg	52mg
Vitamin A	0.4 mg	4.5mg
Vitamin D	0.0002mg	0.0003mg
Sodium	0.6g	0.15g
Potassium	1.4g	0.6g
Magnesium	1.3g	0.35g
Calcium	1.2g	0.3g
Phosphorus	1.0g	0.15g
Energy	2 730kJ	2 730kJ
pH	6.8	7.3

Table 8.2 A comparison of cow's milk and human milk

human milk over cow's milk (see Table 8.2). The antibodies contained in colostrum have already been mentioned, but in addition to losing this protection some bottle-fed babies may become allergic to cow proteins, others develop convulsions because of the unsuitable balance of minerals in cow's milk. Also, the high protein content of cow's milk absorbs acid from the baby's stomach. This acid is a natural protection against certain harmful bacteria, which may cause intestinal infections.

On the other hand, it can be argued that the parents' social life is hindered by breast-feeding (babies cannot be left for more than a few hours) and the father is able to take a greater part in the baby's early development if he can help with bottle-feeding.

8.30 Growth – what is it?

At first glance, the answer to this question may seem obvious. 'Growth is increase in size', would be the first reaction of many people.

Consider the following examples of increase in size, and decide for yourself whether or not they agree with a biologist's idea of growth.

1 A crystal of salt increases in volume and weight by addition to its external surface.
2 A camel increases in volume and weight when it takes a long drink after travelling for several days without access to water (fig 8.22).
3 A wilted plant increases in weight and height immediately after a rainstorm.
4 A butterfly is much longer and wider than the pupa from which it has emerged.
5 A puffer fish increases in volume by blowing up its body with gas when it is disturbed.
6 After weighing yourself, you drink a litre of water: on reweighing yourself, you find that you have gained 1 kg.
7 A 36-year-old man, after following a certain diet for one year, has gained 3 kg in weight.
8 A 12-year-old boy, after following the same diet for one year, has also gained 3 kg in weight.
9 A tree in a well-watered garden increases in height by 10 cm and in trunk circumference by 1 cm in one year.

To help you with this decision, here is a definition of growth given by a leading expert on animal development.

'Growth is the increase in size of an organism or of its parts, due to synthesis of protoplasm or of extracellular substances. Protoplasm in this definition includes both the cytoplasm and the nucleus of cells. Extracellular substances are the substances produced by cells and forming a constituent part of the tissues of the organism, such as the fibres of the connective tissue, the matrix of bone and cartilage, and so on, as

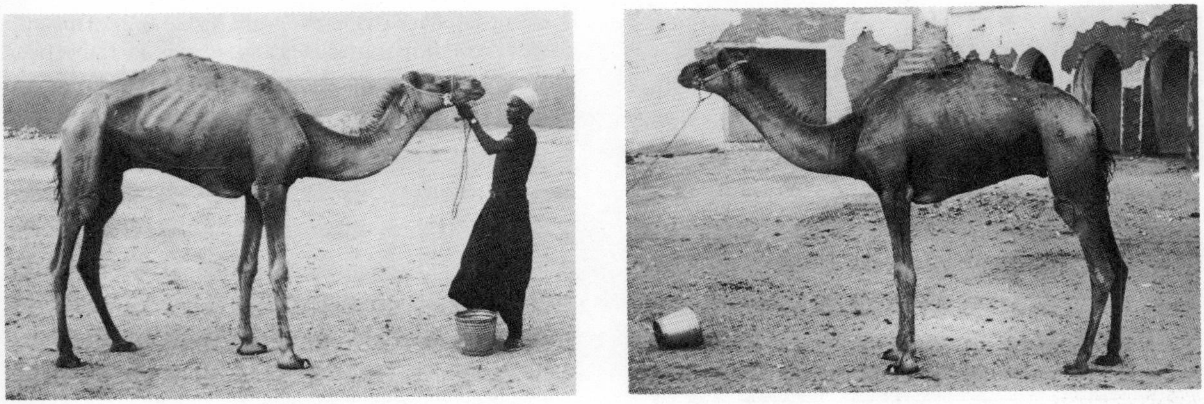

fig 8.22 Photographs of a camel (a) dehydrated (b) one hour after drinking

opposed to substances produced by the cells and subsequently removed from the organisms, such as the secretions of digestive and skin glands, or substances stored as food, such as fat droplets in cells of the adipose tissue. Absorption of water or the taking of food into the alimentary canal before the food is digested and incorporated into the tissues of the animal, although it may increase the weight of the animal, does not constitute growth.'

According to this definition, only examples 8 and 9 qualify as true growth. However, this is only one definition, and it is probable that many people would disagree with it. Some biologists would argue that an increase in complexity alone constitutes growth, others would feel that elongation of plant cells by increasing the size of the vacuole is a form of growth. There is no universal agreement on the precise meaning of this very important term.

Many small living organisms are studied in terms of the population as a whole, and the changes in number as it increases in size can be thought of as growth. Bacteria, unicellular algae and fungi can all be studied in this way, but a convenient organism is *yeast.* A single yeast cell divides into two cells at regular intervals. The yeast culture will continue to divide and double its numbers as long as none of the cells die. If the growth of this population is analysed it starts slowly, increases rapidly and then slows down again, finally levelling out. These numbers plotted against time give a characteristic S-shaped curve said to have a *sigmoid* form (see fig 8.23).

Hence it can be determined whether numbers are increasing, remaining stationary, or decreasing. The population increases in bulk, and we can apply the term 'growth', and knowing the organism and the environment in which it lives, we may be able to explain the growth changes.

Question

1 With your previous knowledge of yeast cells (see Chapter 4, section 4.40) give reasons why the sigmoid growth curve (fig 8.23) slows its rate of growth and finally flattens out at point X.

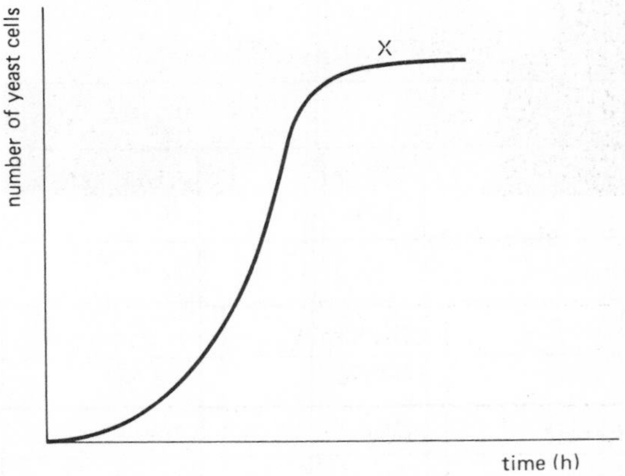

fig 8.23 A sigmoid growth curve of a population of yeast cells

8.31 Measuring growth

Experiment

To measure the growth rate of a laboratory mouse.

Procedure

1 Take a litter of new-born mice and weigh each one separately on the day of birth. Weighing should be carried out on a pan balance, with the mouse enclosed in a weighed, covered jar.

2 Repeat the weighings every two days, until there is no further increase in weight. Always carry out the weighings at the same time of the day and before a feed. After the young mice have been weaned, place them at weighing times in a jar with a weighed amount of food. This will discourage them from moving around and causing the balance to waver.

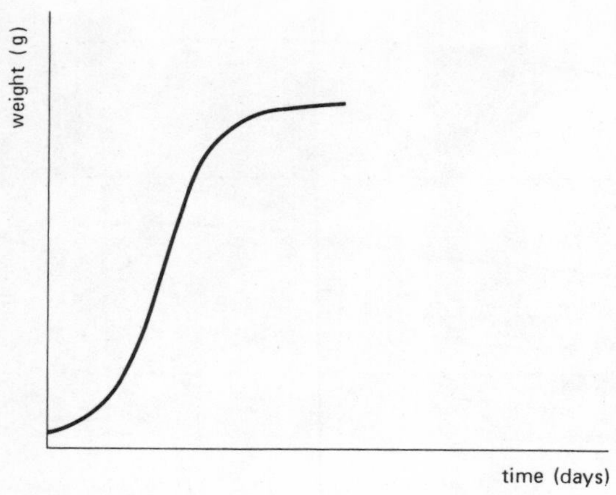

fig 8.24 Idealised sigmoid growth curve of a mouse

3 Plot your results in the form of a graph, with age in days on the horizontal axis and weight on the vertical axis.

Question

1 What is the shape of the graph that you obtain?

Human growth

Investigation

To study the rate of (i) weight increase and (ii) height increase in a population of (a) boys and (b) girls.

Procedure

1 Study carefully the chart which supplies figures for weights and heights of boys and girls from birth to the age of eighteen.

2 Use this information to plot graphs demonstrating weight increase with age in (a) boys and (b) girls.

Boys			Girls		
Age	Weight (kg)	Total height (cm)	Age	Weight (kg)	Total height (cm)
Birth	3.40	50.6	Birth	3.36	50.2
1 year	10.07	75.2	1 year	9.75	74.2
2 years	12.56	87.5	2 years	12.29	86.6
3 years	14.61	96.2	3 years	14.42	95.7
4 years	16.51	103.4	4 years	16.42	103.2
5 years	18.89	110.0	5 years	18.58	109.4
6 years	21.90	117.5	6 years	21.09	115.9
7 years	24.54	124.1	7 years	23.68	122.3
8 years	27.26	130.0	8 years	26.35	128.0
9 years	29.94	135.5	9 years	28.94	132.9
10 years	32.61	140.3	10 years	31.89	138.6
11 years	35.20	144.2	11 years	35.74	144.7
12 years	38.28	149.6	12 years	39.74	151.9
13 years	42.18	155.0	13 years	44.95	157.1
14 years	48.81	162.7	14 years	49.17	159.6
15 years	54.48	167.8	15 years	51.48	161.1
16 years	58.30	171.6	16 years	53.07	162.2
17 years	61.78	173.7	17 years	54.02	162.5
18 years	63.05	174.5	18 years	54.39	162.5

Table 8.4 Weight and height in boys and girls

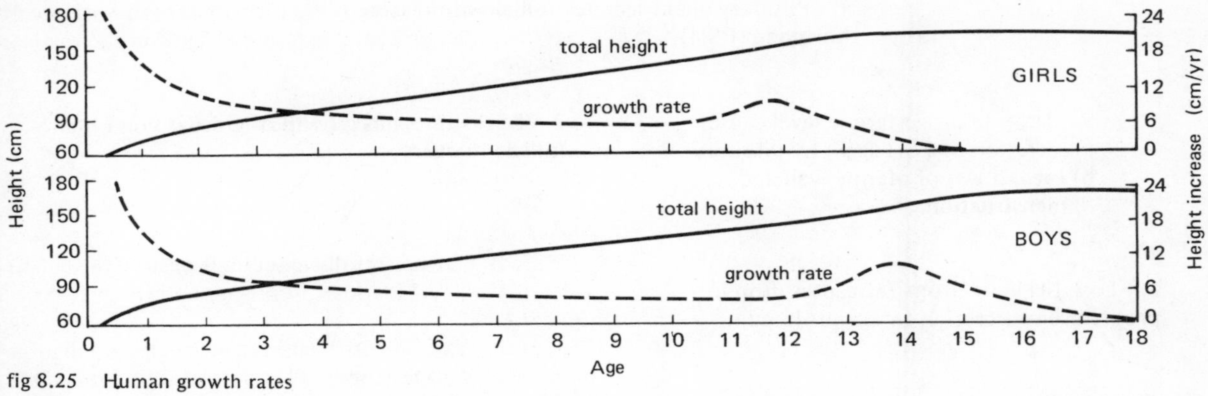

fig 8.25 Human growth rates

3 Similarly plot graphs to show height increase with age in (a) boys and (b) girls.

Questions

2 What are the shapes of the two curves obtained in procedure 2?

3 Which of these curves forms the larger sigma? How do you account for this?

4 What differences are evident in the graphs you have drawn of increasing height in boys and girls?

8.40 Puberty in humans

At the age of about 14 in boys and 12 in girls, there is a rapid increase in the growth rate of the body. At the same time, certain other changes take place: boys' *voices* become deeper and a certain amount of coarse *facial hair* may grow. Girls' *breasts* develop and their *hip girdles* enlarge; in both sexes there is a growth of hair around the genital area and under the armpits.

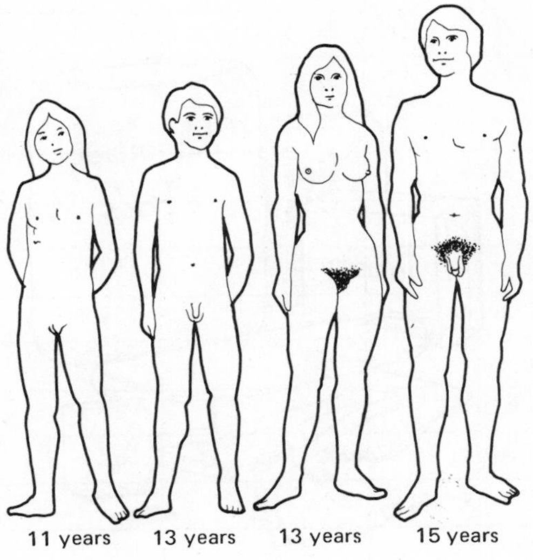

fig 8.26 Changes in body form at puberty

These physical changes are called *secondary sexual characteristics,* and the period of their development is known as puberty or *adolescence.* The development of secondary sexual characteristics is associated with the onset of ovulation and menstruation in girls and of sperm production in boys. At this stage, they are physiologically and anatomically capable of parenthood, although they are usually not ready for it on economic, psychological and emotional grounds.

8.50 Hormones in reproduction

Sperm production and ovulation are triggered by the action on the testes and ovaries respectively of the *gonadotrophic hormones* produced by the *pituitary gland* at the base of the brain. Another effect of these hormones is that they cause the gonads themselves to begin acting as endocrine organs by producing *'sex hormones'.* The testes secrete the male sex hormone (testosterone), which stimulates sperm production and male secondary sexual characteristics, while the ovaries secrete the female sex hormones, of which oestrogen produces female secondary sexual characteristics and regulates the menstrual cycle and ovulation. Progesterone, the 'pregnancy hormone' is also produced by the ovary.

If pregnancy occurs, the placenta itself becomes an important endocrine organ, interacting with the ovaries and pituitary to bring about the major changes that occur in the mother's body during gestation and birth.

At ovulation, an egg is released from a Graafian follicle in one or other of the ovaries (see section 8.31). This happens about every 28 days, and afterwards the empty follicle (now known as the *corpus luteum*) begins to secrete the hormone *progesterone.* The effect of progesterone is to cause the lining of the uterus to become swollen with blood, ready for the implantation of a fertilised egg. If fertilisation does not occur the thickened lining breaks down and the blood that it contains flows out through the vagina over a period of about five days. This is known as the *menstrual period.*

1 Pituitary gland secretes **follicle-stimulating hormone (FSH)**

2 FSH stimulates growth of **Graafian follicle** in ovary.

3 Graafian follicle secretes **oestrogen.**

4 Oestrogen
a) stimulates **repair of uterine wall** after menstruation,
b) inhibits secretion of FSH by pituitary (**negative feedback**) and so prevents the growth of more follicles,
c) stimulates secretion of **luteinising hormone (LH)** by pituitary.

5 LH a) stimulates Graafian follicle to release egg (**ovulation**),
b) stimulates change of empty follicle into **corpus luteum.**

6 Corpus luteum secretes **progesterone.**

7 Progesterone
a) inhibits secretion of LH by pituitary (**negative feedback**),
b) stimulates **thickening of uterine wall** in preparation for implantation of embryo,*
c) inhibits secretion of FSH by pituitary (**carrying on from 4b)**).

8 Low LH level (from 7a) causes **drop in progesterone** secretion by corpus luteum.

9 Drop in progesterone level causes
a) secretion of FSH again by **pituitary**.
b) breakdown of uterine wall and **menstruation.**

* If **fertilisation** occurs, the corpus luteum continues to secrete progesterone (thus preventing further ovulation and maintaining the thick, blood-filled uterine wall) for about **three months,** after which this function is taken over by the **placenta.**

fig 8.27 Hormones in the female reproductive (menstrual) cycle

Immediately before a period, some girls experience a 'bloated' feeling and become slightly irritable and easily tired — a condition known as 'premenstrual tension'. During menstruation the blood is absorbed by cotton *sanitary towels* or pads, which should be changed often. Most school and public toilets have special bins where they can be left for burning later in an incinerator. Fully-grown girls and women may prefer to use *tampons,* which are cylinders of absorbent material that fit inside the vagina and can be flushed down a lavatory.

In pregnancy, of course, there is no menstruation since the blood-filled uterine lining remains to form part of the placenta. Monthly periods stop altogether at some time between the ages of 40 and 55. This stage is known as the *menopause* or 'change of life' and when it ends a woman is no longer able to bear children.

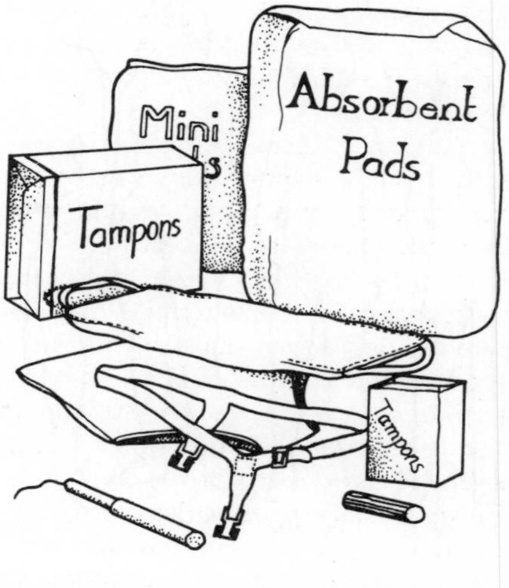

fig 8.28 Different types of sanitary protection

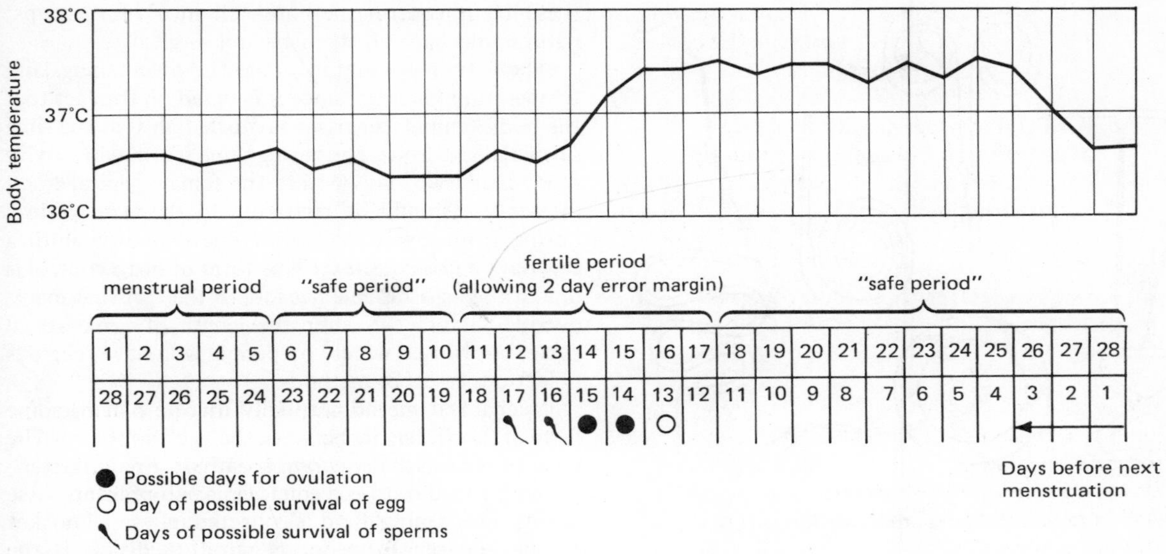

fig 8.29 The rhythm method of family planning

8.60 Birth control

For a normal, healthy woman it is theoretically possible to produce one baby every year for the whole of her reproductive life – a total of more than thirty children! In practice, because of the general 'wear and tear' on the body caused by years of continuous child-bearing, most women could manage a little less than half that figure. Even so, in today's world, families of ten or more children are becoming less and less common. The basic reason for this is quite simple, few families can afford to buy food and clothes for several children, while governments are concerned that national resources of food and space will not be able to meet demand if populations continue to increase. Indeed certain countries, such as India and Singapore, have well-organised publicity campaigns directed at limiting family size; some offer inducements ranging from transistor radios to special medical and educational facilities.

In certain cases, there are also social and medical reasons for wishing to limit family size. In areas where the level of education is high, people tend to move away from their home towns and villages after leaving school, in order to find suitable employment. This means that they are not able to rely on the support of grandparents, uncles and aunts in bringing up large families of their own. Some couples do not wish to risk having children because of the poor state of the wife's health or because of the possibility of passing on a serious genetic defect (see Chapter 9).

There are several ways in which the size of a family can be carefully planned, without the need to avoid sexual intercourse and thus risk damaging the loving relationship between husband and wife. Essentially

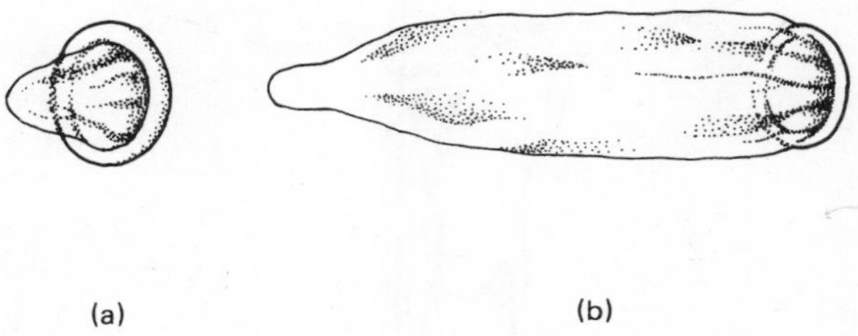

fig 8.30 A condom (a) as supplied (b) unrolled

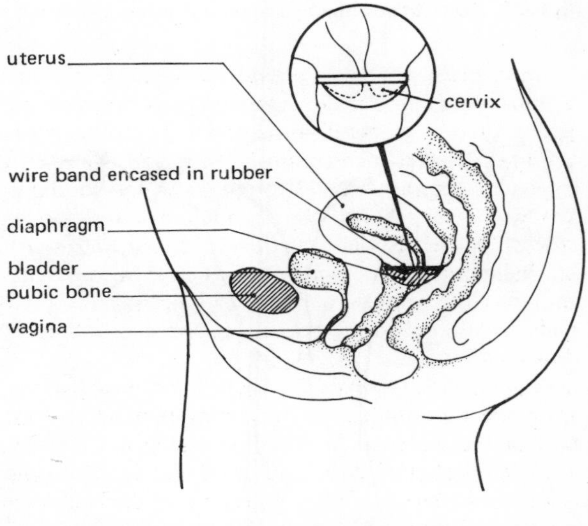

fig 8.31 Contraceptive diaphragm in position

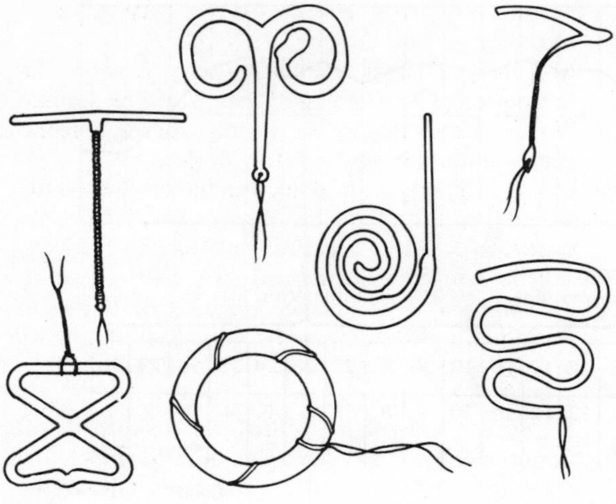

fig 8.32 A selection of intra-uterine devices (I.U.D. s)

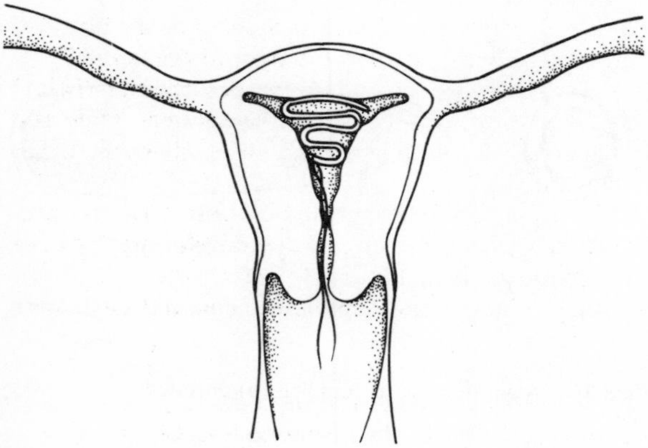

fig 8.33 I.U.D. in position

these birth control methods fall into four groups: natural, mechanical, chemical and surgical.

The natural or rhythm method involves taking daily temperature readings, since it is based on the fact that the body temperature rises as much as 1°C at the time of ovulation. Since neither egg nor sperm will survive more than two days within the female genital tract, intercourse should be restricted to the 'safe period' lasting from a few days after one ovulation until a few days before the next. This form of birth control is one of the least reliable because of the fact that many outside factors can alter the length of a menstrual cycle, but is used widely by people who have religious objections to other methods.

Mechanical methods usually involve placing some form of rubber barrier between the egg and sperm. The most popular is the *condom,* sheath or 'French letter', which is rolled onto the erect penis before intercourse begins. The seminal fluid is collected in a small pocket at the end of the condom. Another device is the *diaphragm* or 'Dutch cap', a rubber dome designed to fit over the top of the vagina so that the spermatozoa cannot reach the narrow opening of the cervix.

A mechanical method that operates on a different principle is the *intra-uterine device* (I.U.D.) which consists of a coil or loop of plastic placed inside the uterus to prevent the egg becoming implanted.

There are two main chemical methods of birth control. One involves the introduction of spermicidal creams or jellies into the vagina to kill the spermatozoa and these are often used together with a diaphragm or condom. The other and more modern method is the 'birth control pill' or oral contraceptive. In fact the idea of eating something to prevent pregnancy is a very old one. Two thousand years ago the early practitioners of Chinese folk medicine recommended swallowing 28 tadpoles. Scientifically this was not very sound,

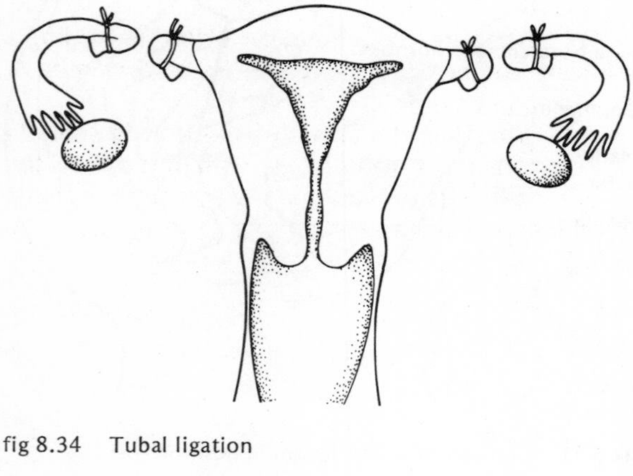

fig 8.34 Tubal ligation

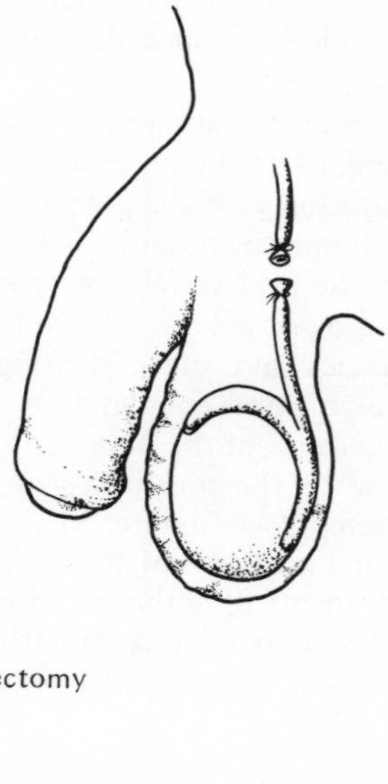

fig 8.35 Vasectomy

except in that it may have made people feel too ill for intercourse. The modern contraceptive pill contains a mixture of oestrogen and progesterone-like hormones which act on the feedback mechanisms of the endocrine system (see Chapter 7) to prevent ovulation. The pills are taken daily from the 5th to the 25th day of the menstrual cycle and cause the uterus lining to fill with blood in the same way as does progesterone from the corpus luteum. After this time a normal menstrual period occurs even though an egg has not been released.

Surgical methods of birth control (sterilisation) involve cutting or sealing the tubes through which the sex cells pass. In men the operation (*vasectomy*) takes only a few minutes and does not require a stay in hospital. A local anaesthetic is given and the surgeon cuts and ties the sperm ducts. In women, a piece of the fallopian tube (oviduct) is removed in an operation called *tubal ligation,* which involves a general anaesthetic and a few days in hospital. Recently, simplified methods have been developed of 'heat-sealing' or *cauterising* the oviducts by passing an electric current through a flexible instrument inserted via the vagina. The disadvantage of sterilisation is that it is almost always irreversible, and therefore not suitable for delaying the starting of a family, or for young people.

Finally, a more controversial method of limiting family size is by abortion. This is the ending of a pregnancy at a stage before the foetus is able to survive outside the mother's body (generally before three months have passed). Many people have moral or religious objections to this practice but in some countries, such as Japan, the current annual number of abortions is actually greater than the number of births.

Method	Pregnancies per 100 woman-years
No contraception practiced	45
Rhythm (safe period)	25
Spermicidal cream only	20
Withdrawal of penis before ejaculation	18
Condom	14
Diaphragm with spermicide	12
I.U.D.	2
Pill	0.5
Sterilisation	0

Table 8.3 Comparative effectiveness of birth control methods

Questions requiring an extended essay-type answer

1 a) Make a large labelled drawing to show the reproductive and urinary system of the human female. Label clearly the vagina, uterus, urethra, cervix and pubis of the pelvic girdle.
b) In your diagram place the letter 'f' to show where fertilisation occurs. Explain why it can only occur on a few days of the 28-day span of the normal female sexual/menstrual cycle.

2 a) Make a large labelled diagram of the male reproductive system and by means of arrows indicate the path of a sperm from its production in the testis until it leaves the penis.
b) Describe how the foetus obtains nutritional material and oxygen during its development.

3 With the aid of labelled diagrams, show the development of the human foetus from the time of fertilisation of the ovum until just before birth.

4 a) Explain how a baby is able to eliminate the products of excretion and respiration during its growth in the womb.
b) How is the foetus protected from physical damage during its development in the womb?

5 a) What are secondary sexual characteristics? What form do they take in the human male and female? Give two examples of these in other mammals.
b) Describe the hormonal control of the production of gametes and the development of the secondary sexual characteristics.

6 a) Describe one complete menstrual cycle with reference to:
i) changes in the ovary
ii) changes in the uterus and
iii) changes in the hormone levels.
b) Briefly explain how two different methods of contraception work.

7 a) What do you understand by family planning?
b) Review the methods of contraception and discuss how effective the various methods are in practice.

8 a) What are the advantages of breast-feeding a baby compared with feeding fresh cow's milk or dried cow's milk?
b) What important health-giving practices should a pregnant woman follow with respect to her physical activity, the food she eats and the things she should avoid?

9 Heredity

9.00 Cell division

All living organisms are made up of units called *cells*, and of their products. In spite of the enormous variety in size and structure of living organisms, their cells are relatively uniform. The cells of an elephant do not differ very much from the corresponding cells in a mouse; there are simply more of them.

As an organism grows, its cells increase in number in much the same way as *Amoeba* divides to reproduce itself. The process in which a cell divides to form two identical daughter cells is known as *mitosis*. It differs slightly between animals and plants, mainly because of the structural differences between the cells of the two kingdoms.

Fig 9.1 and fig 9.2 show the appearance of some typical animal and plant cells, as seen under a light microscope.

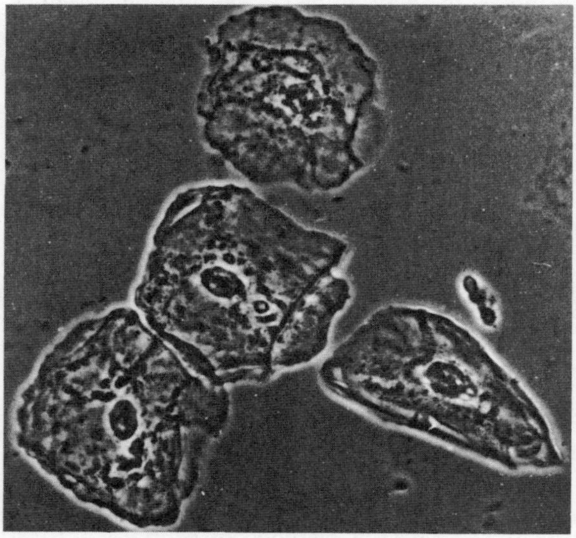

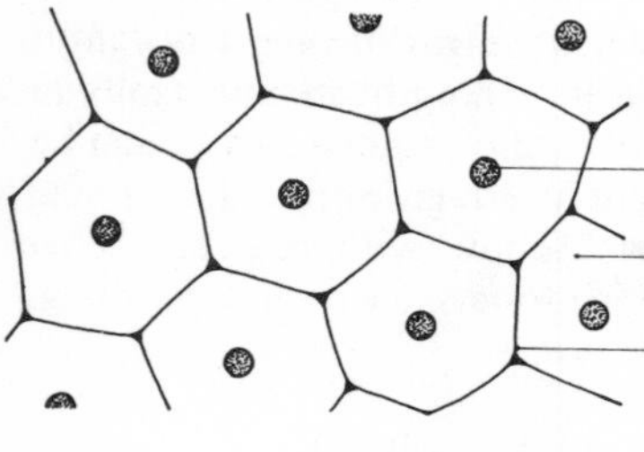

fig 9.1
(a) A photograph of a group of animal cells from the lining of the mouth
(b) A diagram of a group of cheek cells

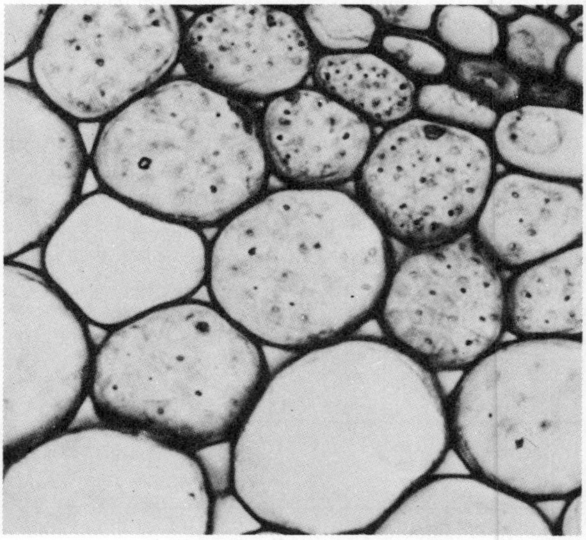

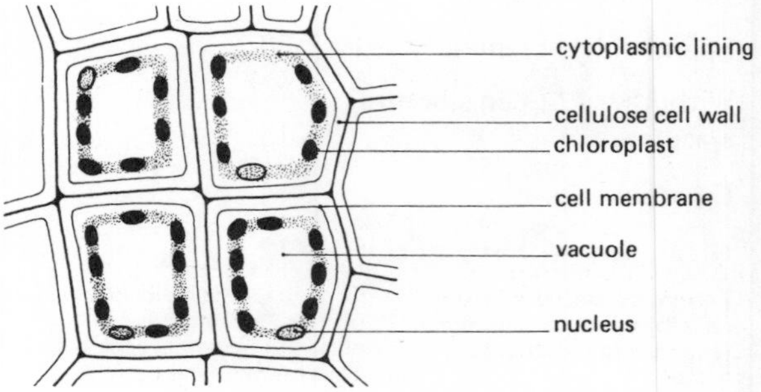

fig 9.2
(a) A photograph of a group of plant cells
(b) A diagram of a group of plant cells

Questions

1 What are the main similarities between the two cells?

2 In what respects does the plant cell differ from the animal cell?

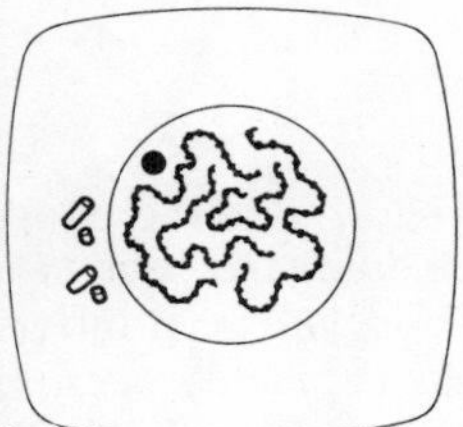

1

EARLY PROPHASE
During the interphase the cell carries out everyday activities. In prophase, the chromosomes begin to shorten and the nucleolus begins to shrink

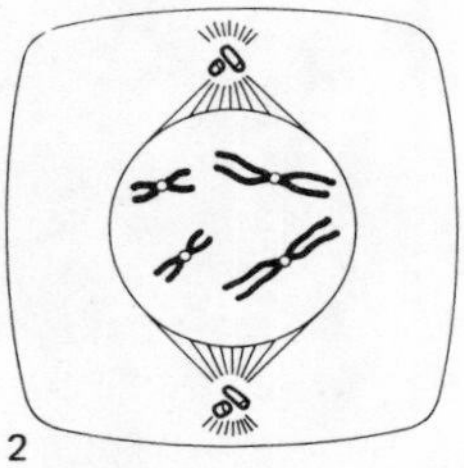

2

LATE PROPHASE
Chromosomes shorten further and duplicate themselves, each one consists of a pair of chromatids joined at the centromere. Centrioles move apart, nucleolus disappears

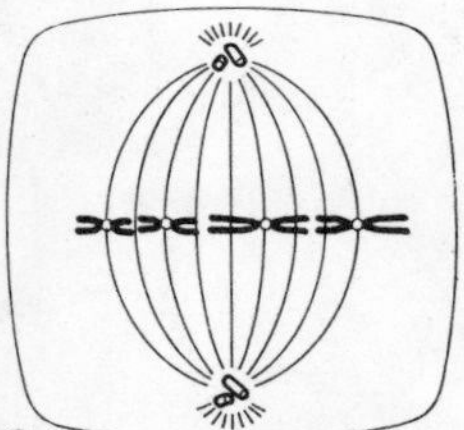

3

METAPHASE
Nuclear membrane breaks down, spindle forms, chromatid pairs line up on equator of cell. (Single chromosome per spindle.

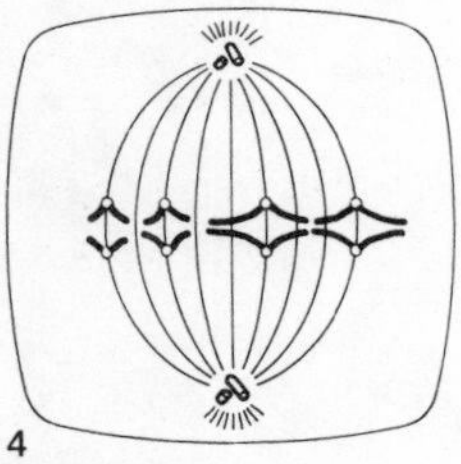

4

ANAPHASE
Chromatid pairs separate and move to opposite ends of cell.

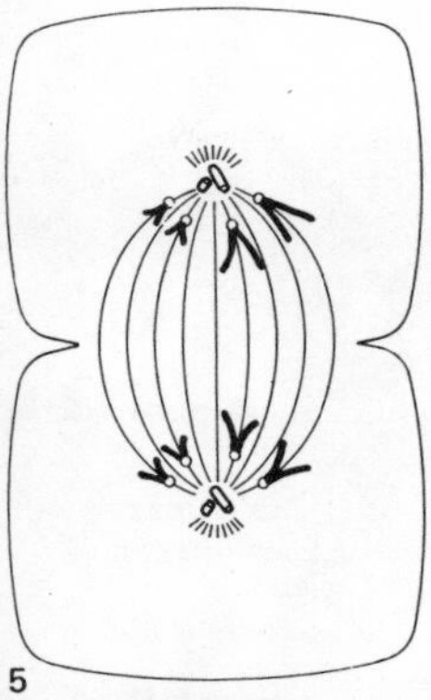

5

EARLY TELOPHASE
Chromatids reach destination and reform into chromosomes. Cell begins to constrict at equator.

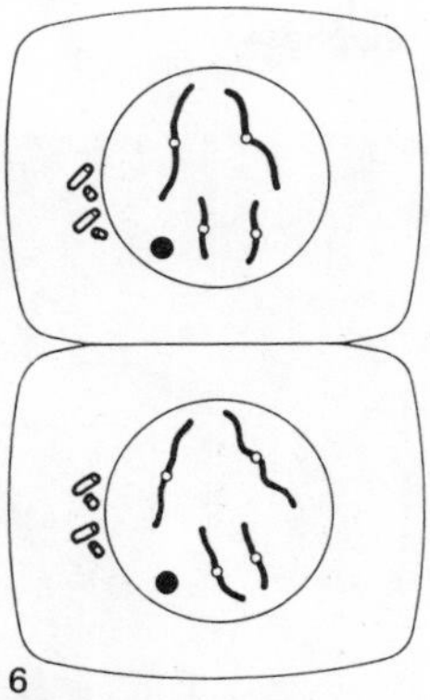

6

LATE TELOPHASE
Constriction completed and nuclear membrane and nucleolus reform in each cell. The chromosomes become invisible threads and each cell enters interphase

fig 9.3 The main stages of mitosis in a generalised animal cell

9.01 Mitosis in animal cells

In a growing animal, cells are continually dividing: one cell divides to form two, these two divide to form four, then eight, sixteen, thirty-two, and so on in a continuous progression. In between divisions, the daughter cells grow to the size of the parent cells and prepare for the next division in certain other ways. This stage is known as *interphase.* Mitosis consists of four other more or less distinct stages: *prophase, metaphase, anaphase* and *telophase.*

At the beginning of prophase, an examination of the nucleus under a light microscope reveals the presence of a number of fine thread-like structures, the *chromosomes.* Later these are seen to be double-stranded. Each pair of strands (*chromatids*) is connected at one point, the *centromere.* The chromosomes are very important structures, since they contain the coded information that will determine the physical and chemical features of the daughter cells.

Gradually, the chromosomes become shorter, thicker and, as a result, more easily seen. At the same time, two star-shaped *asters* of transparently fine fibres appear on opposite sides of the cell. Towards the end of prophase, a *spindle* of fine fibres appears between the asters, and the membrane surrounding the nucleus breaks up and disappears. The *nucleolus,* seen as a dense spot within the nucleus, also disintegrates.

During metaphase, the chromosomes become arranged on the middle or *'equator'* of the spindle, and each centromere splits into two. The daughter centromeres then begin to move towards opposite ends of the spindle, each one pulling a single chromatid with it.

In anaphase, the chromatids separate completely and travel along the spindle towards the asters.

Finally, during telophase, the cell begins to constrict (narrow) at the equator, and the chromatids become surrounded by two new nuclear membranes.

Two new nucleoli are also formed and when constriction is complete, the two daughter cells are separated completely by a membrane between them. Each new cell passes into interphase again.

Note that *one half* of each chromosome has passed into each daughter cell. By the time these daughter cells are ready for division, the chromosomal material will have doubled and the chromosomes will again be two-stranded.

9.02 Mitosis in plant cells

Experiment

Preparation of an onion root tip squash

Procedure

1 Place two drops of molar hydrochloric acid on a microscope slide and add a drop of *acetic-orcein* stain.

2 Cut about 2 mm from the tip of an onion root and place it in the stain/acid mixture.

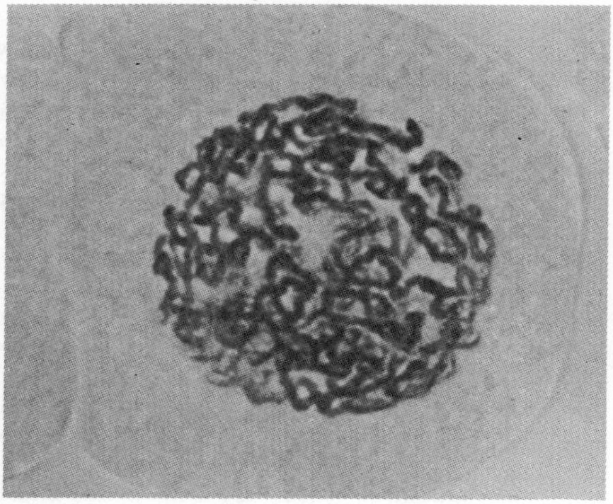

(a) Prophase

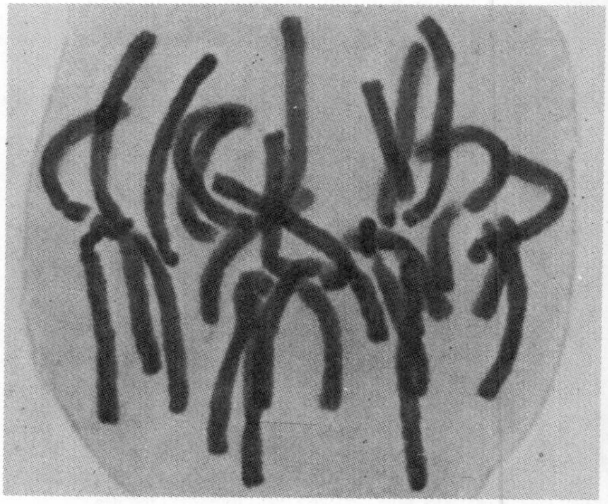

(b) Early metaphase

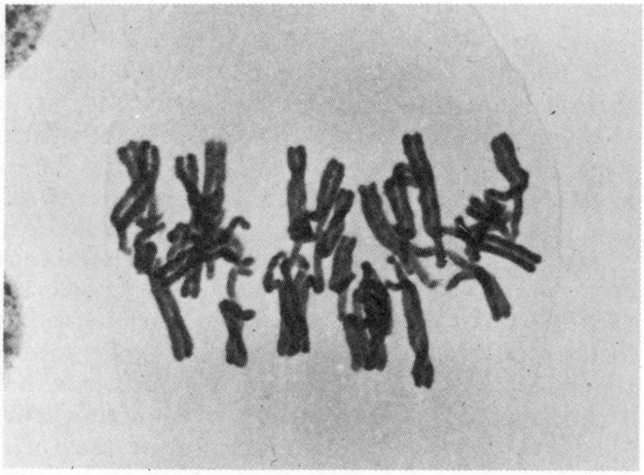

(c) Late metaphase

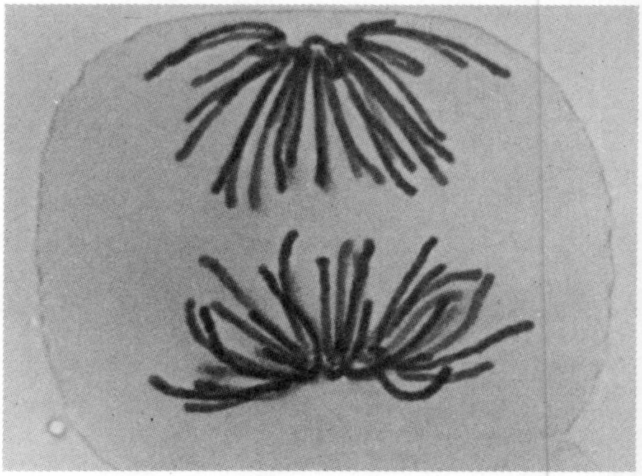

(d) Anaphase

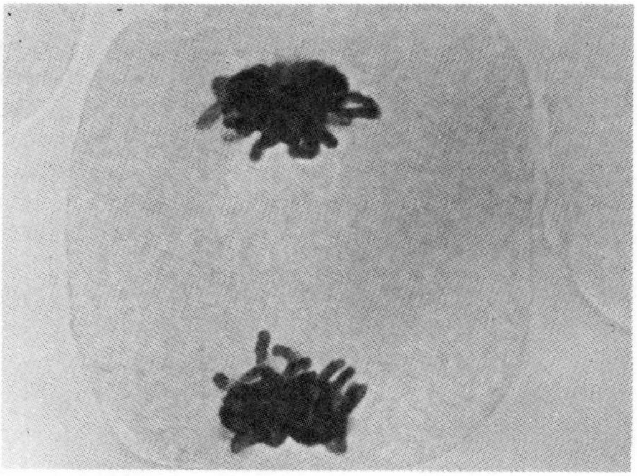

(e) Telophase

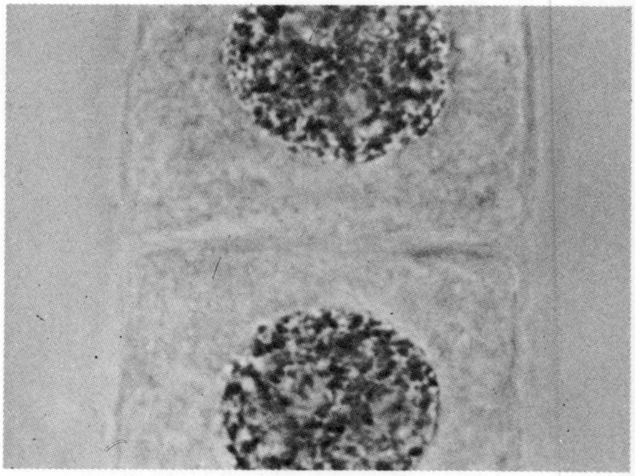

(f) Completed division

fig 9.4 The main stages of mitosis

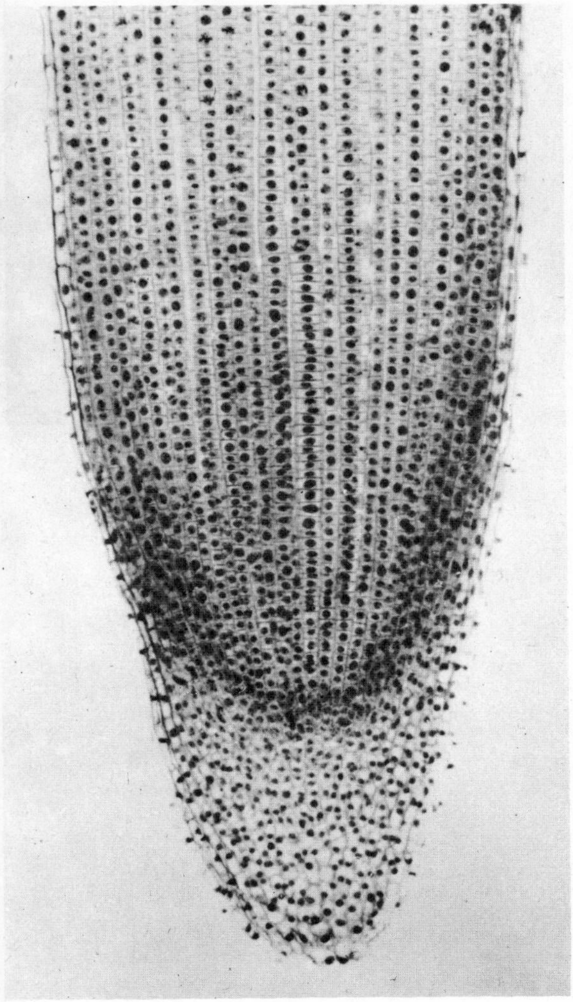

fig 9.5 The cells at the tip of an onion root

3 Warm the slide gently over a Bunsen burner for about 3 minutes using a low flame. Do not boil.
4 Draw off the stain/acid mixture carefully, using a piece of blotting paper, at the edge of the liquid.
5 Add a further drop of stain to the root tip and cover with a coverslip.
6 Place a piece of filter paper over the coverslip and apply firm but gentle pressure with your thumb in order to obtain a thin squash.
7 Remove the filter paper and examine the squashed root tip under the microscope.

Questions

1 Study fig 9.5 and your own preparation carefully. In what respects do these two examples of plant cell mitosis differ from the process in animals as described in section 9.01?
2 Why are root tips particularly suitable for the demonstration of mitosis?

9.03 Fertilisation

The number of chromosomes in the nucleus of a *somatic* (body) cell varies considerably throughout the animal and plant kingdoms, but within any one species this number is constant. For example, in humans each somatic cell contains 46 chromosomes.

If we refer back to section 8.23, we can see that sexual reproduction involves the fusion of a male gamete or sex cell with a female gamete in the process of fertilisation. This would lead us to assume that the cells of an organism resulting from this fusion have *double* the number of chromosomes contained by the cells of the parents. We know that this cannot be so, since it has already been pointed out that the number of chromosomes per cell is constant for all individuals in a species.

The apparent contradiction between the above two statements could be explained if the number of chromosomes were to be *halved* during the formation of the gametes. This, in fact, is exactly what does happen in the process known as *meiosis,* which takes place in the reproductive organs of plants and animals during gamete formation.

9.04 Meiosis

When we first considered chromosomes in section 9.01, it was mentioned that they contain the coded information that determines the structure and function of the cells, and of the organisms that they make up. In fact, each body cell possesses two sets of information, one from the male parent and one from the female parent, contained in two similar sets of chromosomes. For every chromosome originating from the mother, there is one of similar length and appearance (containing equivalent information) from the father. These are termed *homologous* chromosomes, and each human body cell has 23 pairs.

The process of meiosis consists of two divisions, during which the pairs of homologous chromosomes are separated. In many ways, these divisions are similar to mitotic divisions, and the same terms are used to describe the stages.

The more important differences between the two processes are to be found in the very first stage: the prophase of the first meiotic division (prophase I). At the beginning of this stage in animals, the chromosomes shorten and thicken, and there is a breakdown of the nucleolus and nuclear membrane, just as in mitosis. Then the pairs of homologous chromosomes come to lie together, each pair forming a *bivalent.* Later, it can be seen that each chromosome consists of two chromatids.

During metaphase I, the bivalents arrange themselves on the equator of the spindle.

At anaphase I, the bivalents split, one homologous chromosome moving to each pole or end of the cell.

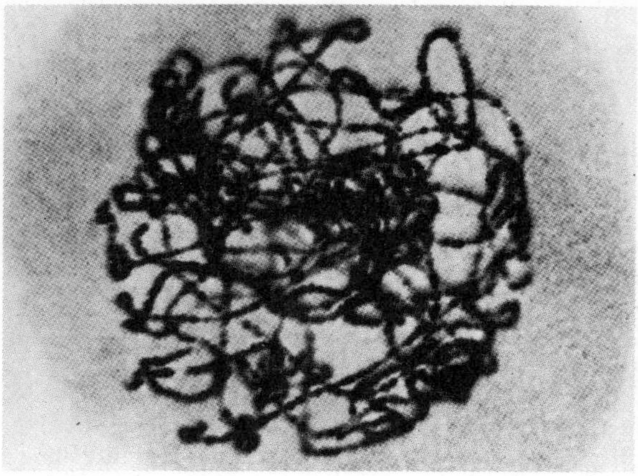

(a) Pachytene

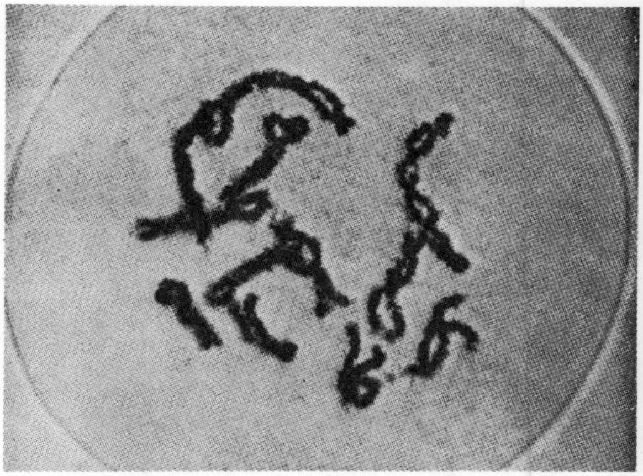

(b) Diplotene

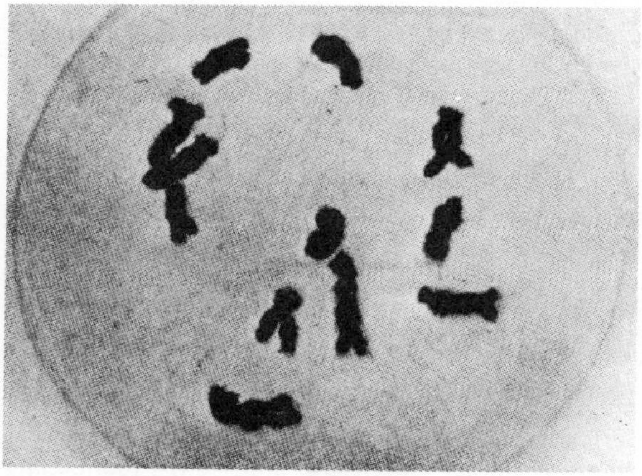

(c) Diakinesis

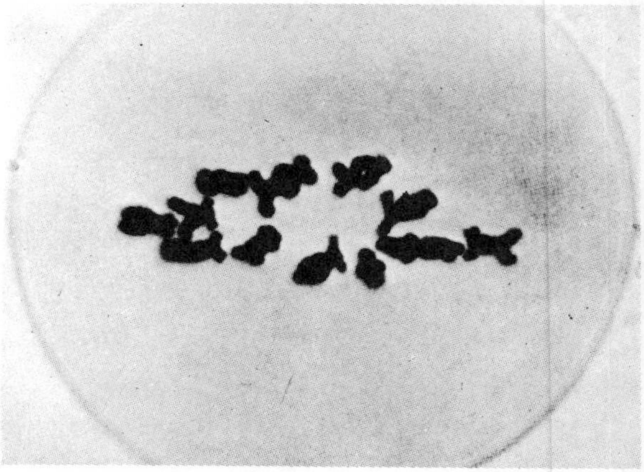

(d) First metaphase

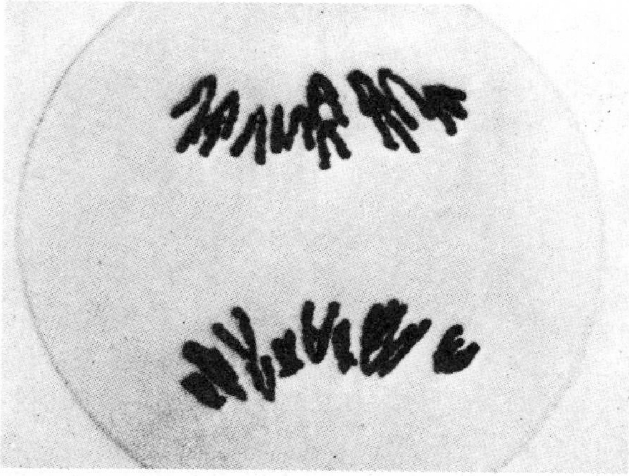

(e) First anaphase

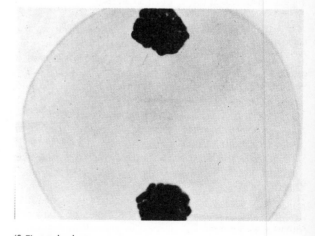

(f) First telophase

fig 9.6 The main stages of meiosis

Note that the chromosomes originating from the male parent do not all move to the same end, so that each daughter cell will probably contain some chromosomes originating from each parent.

Telophase I follows the same pattern as mitotic telophase, and the second meiotic division is very similar to mitosis, where the chromosomes split and the daughter chromatids enter separate cells.

The entire process is summarised in fig 9.7.

Questions

1 How many gametes are produced from one body cell during meiosis?

2 How does the number of chromosomes in a gamete compare with the number in a body cell of the same organism?

3 What would be the effect if, during anaphase I, all the chromosomes originating from the male parent moved to one pole, and all those originating from the female parent moved to the other?

4 What is the purpose of exchange of chromatid segments?

9.10 Variation and heredity

Consider the four people shown in fig 9.8. They come from different parts of the world and represent the four major races of mankind. In spite of the differences in appearance, they all belong to the same species, *Homo sapiens.* This means that it is possible for them to interbreed and to produce children who are themselves fertile. If an African man were to marry a European woman, their children would have some negroid characters, some caucasoid characters and some characters intermediate between those of the two races.

A more striking example of variation can be seen in the case of the dog. For centuries, dogs have been used by Man on every continent for a variety of purposes; for hunting, for guarding property, for herding livestock and even for drawing small carts. As a result, an enormous number of different shapes, sizes and colours of dog have developed throughout the world. Nevertheless, all domestic dogs belong to the same species, *Canis familiaris,* and thus are capable of interbreeding.

Fig 9.9 shows the results of crossing two pedigree animals of different breeds. The offspring, which are shown when fully grown, are all similar, having some characters in common with their dam (mother) and some in common with their sire (father). Crossing of two animals from this generation results in a much wider variety of offspring. You probably have inherited some characteristics from your mother, and some from your father. However, some of your features may appear unlike either parent. These may have been present in your grandparents or in earlier generations.

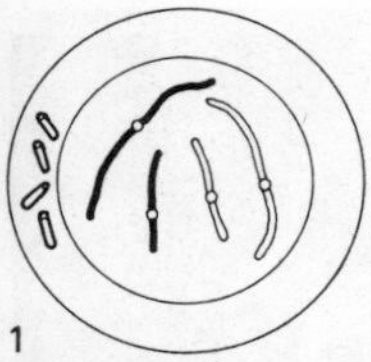

1

EARLY PROPHASE 1
Homologous chromosomes appear in nucleus.

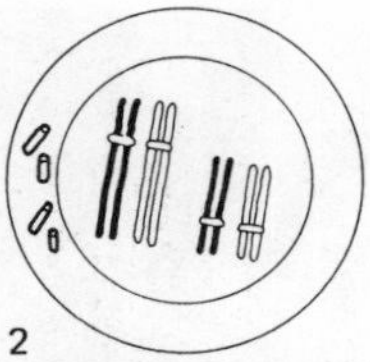

2

MIDDLE PROPHASE 1
Homologous chromosomes pair up, then split into chromatids.

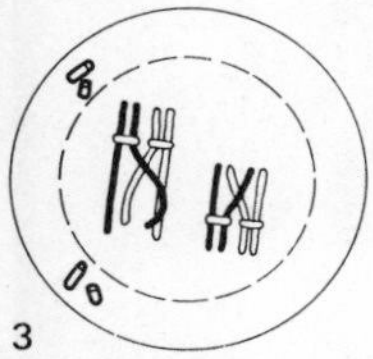

3

LATE PROPHASE 1
Chromatids of homologous chromosomes cross over each other.

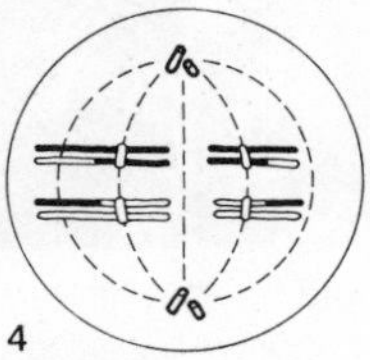

4

METAPHASE 1
Homologous chromosomes arrange themselves on equator of spindle. Segments of crossed chromatids have exchanged by breakage and subsequent rejoining.

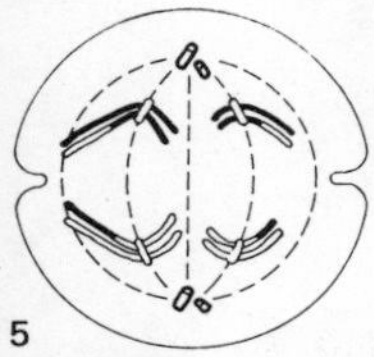

5

ANAPHASE 1
Homologous chromosomes separate.

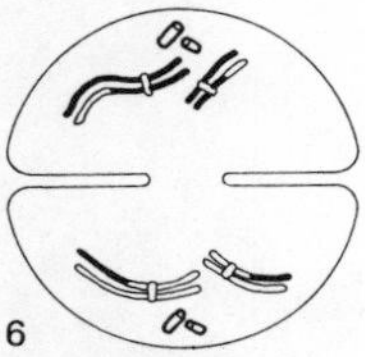

6

TELOPHASE 1
Two new cells form, each with half the number of chromosomes of the original cell. Nuclear membrane may not reform, as metaphase 2 may follow immediately.

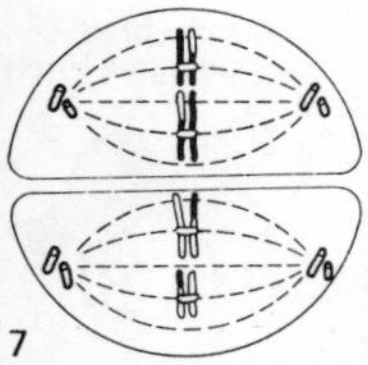

7

METAPHASE 2
A new spindle is formed in each new cell and two chromosomes line up on each equator

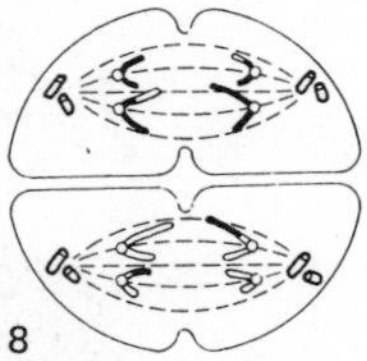

8

ANAPHASE 2
Chromatids of the two chromosomes in each new cell separate. (As in mitosis)

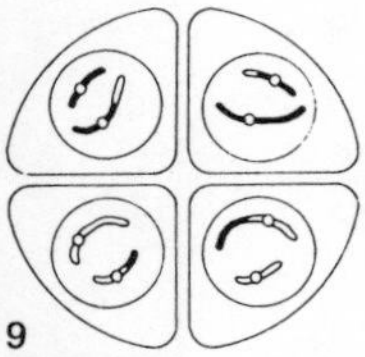

9

TELOPHASE 2
Four new cells form, each with half the number of chromosomes compared to the original cell. The composition of the chromosomes is altered.

fig 9.7 The main stages of meiosis, involving two pairs of homologous chromosomes

(a) Negroid

(b) Mongoloid

(c) Caucasoid

(d) Australoid

fig 9.8 The four races of Man

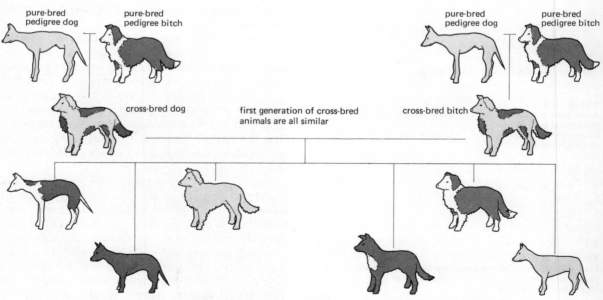

fig 9.9 Variation produced by cross-breeding pedigree dogs over two generations

(a) Arch (b) Loop (c) Whorl (d) Double whorl

fig 9.10 The main groups of fingerprints

9.11 Variation in humans

Investigation
Variation in height
Procedure
1 Measure your own height (to the nearest centimetre) against a scale marked on a wall.
2 Measure the heights of all members of your class in the same way. If possible, obtain the heights of as many as a hundred students of about the same age. You will obtain more accurate results with a large sample.
3 Plot your results on graph paper, with height shown on the horizontal axis, and numbers of students in each height group on the vertical axis. You will find the detailed method in the Introductory Chapter (section I 4.10).
Question
1 What is the shape of the curve that you obtain?

Experiment
To investigate variation in fingerprints
Procedure
1 Wash and dry your hands thoroughly.
2 Place the tip of your right index finger lightly but firmly on an office ink pad.
3 Place the inside of your fingertip on a piece of white paper, and roll it carefully to the outside before removing it. Take care not to smudge the print.
4 Note the pattern of loops, arches and whorls (see fig 9.10).
5 Take a large sheet of paper and rule it into 2 cm squares. On each square obtain a fingerprint from the right index finger of a different member of your class.
6 Examine each print with a hand lens and classify it according to the patterns shown in fig 9.10.
Questions
2 Are any of the fingerprints identical?
3 What are the proportions of each type found in your sample?

Investigation
The inheritance of tongue rolling
Procedure
1 Put out your tongue and try to roll it as shown in fig 9.11.
2 Observe other members of your class, noting how many students can roll their tongues and how many cannot.
Questions
4 Do any members of the class show a partial ability to roll their tongues?
5 What explanation can you offer for the results of your investigation?
6 This is one of the few examples of an easily identifiable human characteristic inherited *via* a single gene. (Height and fingerprints are controlled by the action of several genes.) What other factors make humans unsuitable material for studies in genetics? (See section 9.12, question 2.)

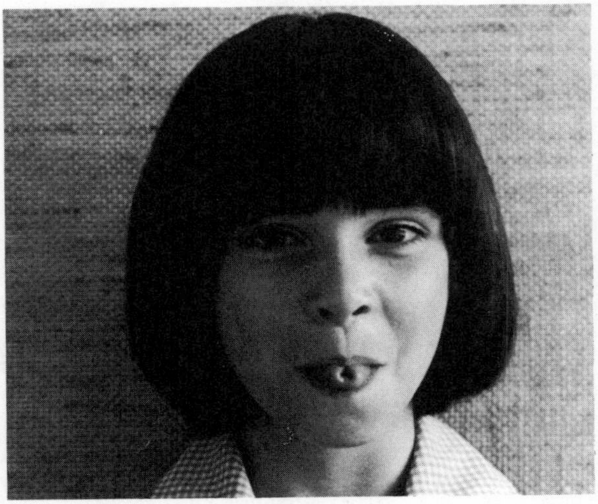

fig 9.11 Tongue rolling

9.12

Having considered several examples of variation in living organisms, we can now ask ourselves the question) 'What is it that determines whether, in a particular feature, an organism resembles its male parent, its female parent or neither parent?' In order to find an answer, we should examine as simple an example as possible.

The fruit fly, *Drosophila melanogaster,* usually has wings about twice as long as its abdomen. However, certain individuals have vestigial (very short) wings. A *long-winged* male fly was crossed with a *vestigial-winged* female fly: eighty-seven eggs were laid, all of which developed into flies with long wings. The original pair of flies are referred to as the *parental* (P) generation and the 87 progeny as the *first filial* (F_1) generation. When two members of this long-winged F_1 generation were mated, the resulting *second filial* (F_2) generation consisted of 61 long-winged flies and 20 vestigial-winged flies.

Questions

1 What was the approximate percentage of vestigial-winged flies in the F_2 generation?

2 *Drosophila* is a very suitable animal for use in experiments on heredity. Can you suggest any reasons for this?

9.13

In order to explain the results obtained in 9.12, we must refer back to the chromosomes which we encountered in our study of cell division (see sections 9.01 and 9.04). Every organism arising from sexual reproduction receives from each parent a set of coded 'instructions' for development. These 'instructions' are called *genes* and are found on the chromosomes.

Thus, each fertilised fly egg will have *two* genes affecting wing size: one at a certain position on a chromosome received from its female parent, and one at the corresponding position on the homologous chromosome received, via the sperm, from its male parent. If both genes are 'instructions' for producing long wings, the wings will be long; if both genes are 'instructions' for producing vestigial wings, the wings will be vestigial. In these two cases the flies are said to be *homozygous* for the long-winged and vestigial-winged conditions respectively.

However, if one gene of each type is present, (i.e. if the organism is *heterozygous* for wing length), the resulting wings will not be halfway between the long and the vestigial state: they will be *long.* This is because the gene for long wings (referred to for convenience as L) is able to exert its influence in the presence of the gene for vestigial wings (referred to as l), and prevents l from having any effect. The gene for long wings is thus said to be *dominant* to the gene for vestigial wings: conversely l is *recessive* to L. This is a conveniently simple example of *discontinuous variation,* involving a single pair of genes. In most cases, a given character is affected by a large number of genes acting together, the variation produced is *continuous* and the pattern of inheritance is extremely complex. Nevertheless, such simple examples are valuable in helping us to understand the principles on which heredity operates.

9.14

We can now explain the crosses mentioned in section 9.12 in terms of the genes involved. The long-winged parent can be assumed to have had two L genes. (Another way of expressing the same idea is to say that it had a *genotype* of LL.) The vestigial-winged female parent has a genotype of ll. The F_1 generation all had genotypes of Ll, since they received one gene of each type in the gametes from their parents. However, their *phenotypes* (the physical expression of the genotypes) were long-winged. The conventional way of expressing this cross diagrammatically is in one of the following forms:

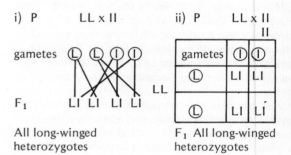

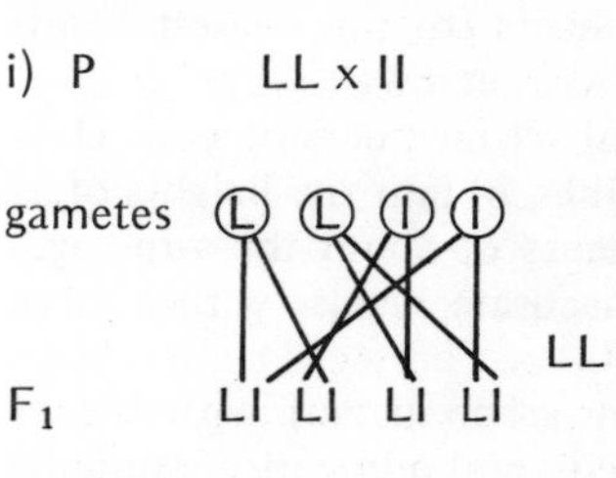

All long-winged heterozygotes

ii) P LL x ll

gametes	(l)	(l)
(L)	Ll	Ll
(L)	Ll	Ll

F_1 All long-winged heterozygotes

Remember that, during meiosis, the homologous chromosomes separate. Thus the F_1 generation produces gametes of two different types in respect of wing length, those containing the chromosome that bears the L gene and those containing the chromosome that bears the l gene. When the two organisms are crossed, any sperm can fertilise any egg: thus any gene can be combined with any other gene. The diagram can now be completed as follows, to show the F_1 cross.

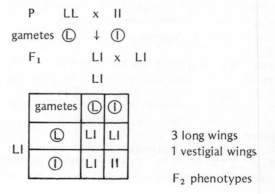

P LL x ll

gametes (L) ↓ (l)

F_1 Ll x Ll

gametes	(L)	(l)
(L)	Ll	Ll
(l)	Ll	ll

3 long wings
1 vestigial wings

F_2 phenotypes

We can see that, for every vestigial-winged fly produced in the F_2 generation, three long-winged flies are produced. This 1:3 ratio corresponds closely with the ratio of 20:61 obtained in section 9.12.

9.15

The above experiment helps to emphasise a very important aspect of all statistical work in genetics: it is essential to deal in sufficiently large numbers.

Drosophila is such a useful organism for genetics experiments partly because each female fly lays so many eggs. The fact that the ratio of 61:20 is not exactly the same as 3:1 is not important because we are dealing with such large numbers. However, if the extra one long-winged fly had occured in a brood of five, the ratio would have changed drastically to 4:1, and the figures would have been impossible to interpret.

Let us consider a more familiar example. The number of boy babies born is statistically equal to the number of girl babies. Thus, in a village where there are 160 children, the expected number of boys is 80 and of girls is also 80. It would be quite unremarkable for a village of this size to have 75 boys and 85 girls, or vice versa, and the ratio would still be almost 1:1. However, if we were to study individual families, it would be quite common to find mothers who had given birth to two, three or four children, all of whom were boys, and others who had only girl babies. If we were to consider only one of these families, then we would obtain a completely misleading picture of the ratio of the sexes.

A further example of the importance of using large numbers in studies on inheritance can be seen in the work of Gregor Mendel. Mendel was an Austrian monk, regarded by many as the father of genetics, who carried out most of his work more than 100 years ago, using pea plants. He investigated the type of cross seen in section 9.12 except that he studied several pairs of contrasting characters. The results of these crosses are shown in the table below. (Since only one pair of characters is being considered, the crosses are described as *monohybrid*).

Question

1 Draw a graph, showing the difference between the ratio obtained and the ideal ratio (3:1), plotted against the number of plants in the F_2 generation. What deduction can you make from this graph?

Investigation

To study monohybrid crosses, using models

Procedure

1 Into each of two large jars or boxes, place one hundred red beads and one hundred white beads. The jars represent two individuals of the F_1 generation in the experiment, described in section 9.12. The red beads represent the dominant gene for long wings (L) and the white beads represent the recessive gene for vestigial wings (l).

2 Shake each jar thoroughly to mix up the beads.

3 Put one hand into each jar and select a bead at random (do not look at the jars while you are making your selection). Put the two beads together on the bench: they represent a zygote. Note whether the combination is red and red, white and white or red and white.

4 Repeat this process one hundred times.

Questions

2 What wing phenotypes will be produced by the following bead combinations: red and red, red and white, white and white?

3 What is the ratio of long-winged to vestigial-winged flies produced from the hundred zygotes?

9.16

Refer back to the *Drosophila* cross covered in section

Experiment	Dominant character	Recessive character	No. of F_2	Ratio of F_2 dominants to F_2 recessives
1	Round seeds	Wrinkled seeds	7 324	2.96:1
2	Yellow seeds	Green seeds	8 023	3.01:1
3	Red flowers	White flowers	929	3.15:1
4	Inflated pods	Constricted pods	1 181	2.95:1
5	Green pods	Yellow pods	580	2.82:1
6	Axial flowers	Terminal flowers	858	3.14:1
7	Long stem	Short stem	1 064	2.84:1

Table 9.1 The F_2 results of Mendel's monohybrid crosses

9.12. One of the male flies from the F_1 generation was mated with its vestigial-winged parent.

Questions

1 What are the genotypes of the two parents in this cross?

2 What gametes are produced by the male?

3 What gametes are produced by the female?

4 How could the cross be expressed diagrammatically? State the phenotypes of the offspring.

9.17

Let us consider another type of cross, this time involving a species of pea plant which is normally self-pollinating and thus pure-breeding. The anthers were removed from a flower of a tall-stemmed pea plant, and its stigma was dusted with pollen taken from a short-stemmed plant of the same species. The resulting seeds produced F_1 generation plants which were all tall-stemmed.

Several flowers from this F_1 generation had their anthers removed and their stigmas dusted with pollen from the original tall-stemmed parent plant. The resulting seeds produced plants that were all tall-stemmed.

Questions

1 Construct a diagram to illustrate this cross, using suitable symbols for the genes involved.

2 What proportion of the plants produced from this cross are likely to be homozygous (pure-breeding) for the gene for tallness?

3 Refer back to section 9.12. Can you think of any ways in which pea plants are preferable to *Drosophila* for use in genetics experiments?

4 Another organism that is particularly useful in genetics experiments is the maize plant (*Zea mays*). Each ear of maize contains a large number of seeds or kernels, and each seed is the result of a single fertilisation. The seeds possess a number of characters which show discontinuous variation, including colour and shape. For example, the gene for *colour* is dominant to the gene for *lack of colour*, and the gene for *smoothness* is dominant to the gene for *shrunkenness*. Suppose a plant grown from a heterozygous smooth seed were pollinated by a plant grown from a shrunken seed. What proportions of seed phenotypes would you expect to be produced in the resulting F_1 generation?

5 What particular advantage does the use of maize seed characters in genetics experiments have over, for example, the use of stem length and flower colour in peas?

9.20 Incomplete dominance

In a certain plant species, a red-flowered individual when crossed with a white flowered individual, will produce pink-flowered offspring. If self-pollinated, the pink-flowered plants will produce an F_2 generation with flowers in the ratio of 1 red: 2 pink: 1 white. This is an example of incomplete dominance or blending inheritance, where the heterozygote produces a condition intermediate between those produced by the two homozygotes. Neither gene or allele is completely dominant over the other, and so the above crosses can be explained diagrammatically in the following manner.

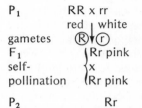

P_1 RR x rr
red ↓ white
gametes (R) (r)
F_1 self-pollination: Rr pink x Rr pink

P_2 Rr

	gametes	(R)	(r)
Rr	(R)	RR	Rr
	(r)	Rr	rr

F_2 phenotypes:
1RR red, 2Rr pink, 1rr white

Although the cells of a single individual cannot possess more than two genes of any one kind, this does not mean that a greater variety of genes is not possible. For example, the inheritance of blood groups in Man is governed by three alleles: G^A, G^B and G^O (see section 3.31). Both G^A and G^B are dominant to G^O, but G^A is not dominant to G^B nor G^B to G^A: in other words, these two genes are *co-dominant*.

Therefore, an individual will have blood of group A if his genotype is $G^A G^A$ or $G^A G^O$; he will have blood of group B if his genotype is $G^B G^B$ or $G^B G^O$; but he will have blood of group O only if his genotype is $G^O G^O$. However, since G^A and G^B are co-dominant, a genotype of $G^A G^B$ will produce a fourth phenotype, blood group AB.

Question

1 Construct a diagram to show the possible phenotypes arising in the children of a marriage between a woman with blood group A and a man with blood group B, where both these parents are heterozygous.

9.30 Sex determination

'Will our new baby be a boy or a girl?' This is the question that all prospective parents ask. We all know that the chances of any one baby belonging to a particular sex are about 1:1 or 50%, but how can this be explained in terms of chromosomes?

To answer this, we must look at the 46 chromosomes that occur in each human cell. In women and girls, they consist of 23 pairs of similar chromosomes: in men and boys, there are 22 pairs of similar chromosomes, but one member of the 23rd pair is much smaller than the other. This smaller one is called the Y chromosome, the larger one is called the X chromosome. Together they form the pair of sex chromosomes. There is no Y chromosome in human females; the pair of sex chromosomes consists of two Xs.

Thus, each one of 50% of the sperms produced by a man (as a result of separation of the homologous chromosomes during meiosis) will contain an X chromosome and each one of the other 50% will contain a Y chromosome. However, every ovum produced by a woman will contain an X chromosome. Therefore, the probability of an offspring having XX sex chromosomes is equal to the probability of its having XY sex chromosomes.

This can be represented diagrammatically as follows:

	Man x Woman
Sex chromosomes	XY XX

Man

	gametes	Ⓧ	Ⓨ
Woman	Ⓧ	XX	XY
	Ⓧ	XX	XY

2 girls 2 boys

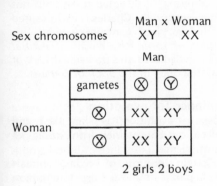

9.40 Sex linkage

When the genes controlling a certain character are located on the sex chromosomes, the character is inherited along with the sex. Such characters are described as *sex-linked.* Since the sex chromosomes in the male are the only pair of dissimilar chromosomes, a gene carried on the X chromosome need not have a corresponding gene (or allele) on the Y chromosome (in fact, the human Y chromosome is thought to contain very few genes, if any at all).

Now a boy child receives his X chromosome from his mother and his Y chromosome from his father. Therefore, he cannot inherit his father's sex-linked characters.

However, a man can pass on sex-linked characters, via his daughters, to his grandchildren.

9.41 Haemophilia

One of the best-known examples of sex-linked inheritance is shown by the disease *haemophilia* (bleeder's disease), in which the blood does not have the ability to clot. The disease is caused by a recessive gene carried on the X chromosome.

Female haemophiliacs are rare, since they can only be produced when both parents carry the gene. However, a woman with the recessive gene on one of her X chromosomes will act as a 'carrier' of the disease.

Queen Victoria was such a carrier. Only one of her four sons (Prince Leopold) was a haemophiliac, but she passed on the disease to many of the European

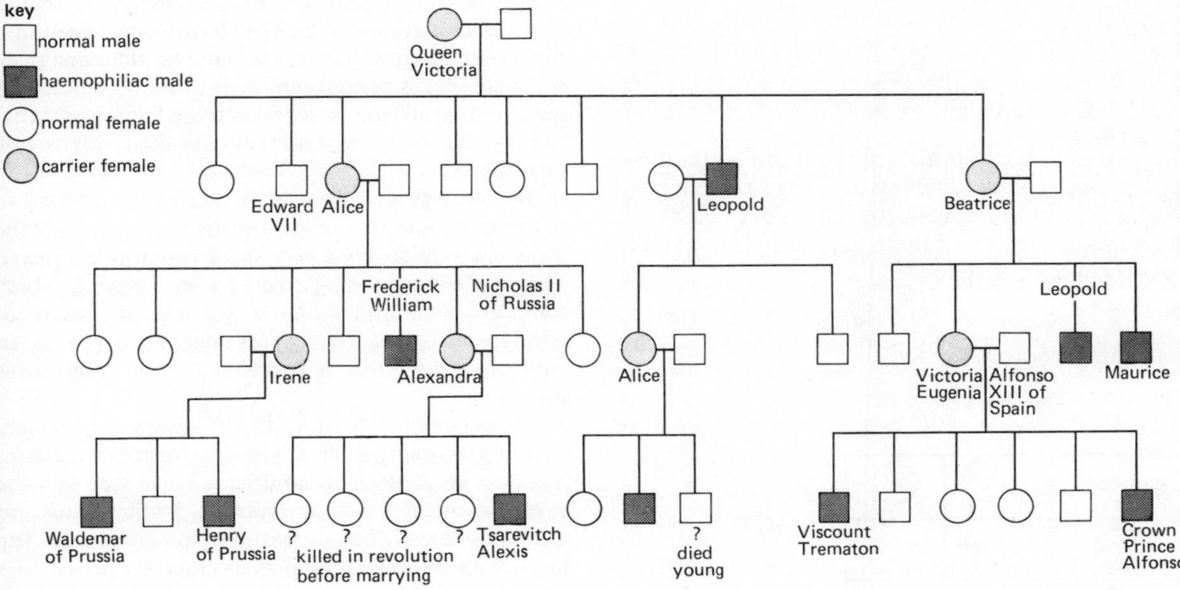

fig 9.12 Queen Victoria's family tree, showing inheritance of the recessive sex-linked gene for haemophilia

Royal Houses through her two carrier daughters, the princesses Alice and Beatrice. Fig 9.12 shows the progress of the disease through four generations of royal families. Note that all affected invididuals are males. This indicates that the sex chromosomes are involved. A haemophiliac male of the second generation produces a son who is unaffected by the disease since a boy receives the Y chromosome from his father and the X chromosome from his mother. This confirms that the gene for haemophilia is likely to be carried on the X chromosome.

9.42 Colour blindness

Abnormality of colour vision is another variation that has been shown to be due to a single gene located on a sex chromosome. There are different types of colour blindness but the commonest involves an inability to distinguish between red and green. Red-green colour blindness can present a number of problems in every-day life. For example sufferers are excluded from work on ships or planes involving recognition of port(red) and starboard(green) navigational light signals. Studies in Norway and Sweden have shown that red-blind people make up 1% of the population and green-blind 5%. The condition is more common among caucasoids than among other races. The selective advantage of this condition could have been important amongst our stone age ancestors.

Investigations of families shows that all four red-green conditions are sex-linked, since they are determined by genes on the X chromosome. Girls and women are rarely affected and thus, as with haemophilia, the allele for red-green colour blindness is recessive. The disorder occurs in the male when his single X chromosome bears this allele, and in women only when it is carried on both X chromosomes. Remember that the smaller Y chromosome bears few, if any, genes.

Let the allele for normal sight be R and let the allele for red-green colour blindness be r. Then a colour blind male is X^rY and a normal sighted female can be X^RX^R (homozygous) or X^RX^r (heterozygous). Thus a family history could be as follows:

P_1 X^RX^r x X^RY

both parents normal sighted

mother a 'carrier'

gametes	(X^R)	(X^r)
(X^R)	X^RX^R	X^rX^R
(Y)	X^RY	X^rY

F_1 genotypes:	X^RX^R	X^rX^R	X^RY	X^rY
phenotypes:	daughter normal sighted	daughter 'carrier'	son normal sighted	son colour blind

Questions

1 Draw a genetic diagram to show the production of a red-green colour blind girl.

2 What proportion of her sons and daughters would be likely to show the disease?

9.50 Evolution

The fact that variation can be inherited from one generation to the next helps us to explain the process of evolution – the gradual change in the nature of living organisms over long periods of time. That evolution occurs is almost universally accepted today, but not much more than a century ago Charles Darwin caused great scientific and religious controversy when his book *On the origin of species by means of natural selection* (1859) contradicted the traditional idea of divine creation as described in the *Book of Genesis.*

9.51 The evidence for evolution

Evidence to show that evolution has taken place, and is still taking place, can be obtained from studies of several branches of science.

1 *Anatomy* Certain groups of living animals resemble each other in a way that suggests common ancestry. For example, men, sparrows, lizards and frogs all have feet which bear five toes (or remnants of them). Close examination reveals that birds' wings, sea lions' flippers and horses' hooves are based on the same plan, and that snakes possess remnants of limb girdles. This suggests that all land vertebrates arose from forms that had two pairs of pentadactyl (five-digit) limbs (see Chapter 11).

The limb-girdle remnants of snakes are referred to as *vestigial.* Among our own vestigial structures is the appendix – relatively a very small structure compared with the corresponding structure in a rabbit, which contains symbiotic bacteria for the digestion of cellulose in plant food. This may be taken as an indication that man is descended from herbivorous ancestors.

2 *Embryology* A study of the embryos of widely differing animals, such as the five vertebrate classes reveals a great deal of similarity. Early stages in the development of fishes, amphibians, reptiles, birds and mammals have many features in common. The similarities between mammalian embryos persist to a much later stage. This suggests that all mammals arose from the same stock, and that the original, ancestral

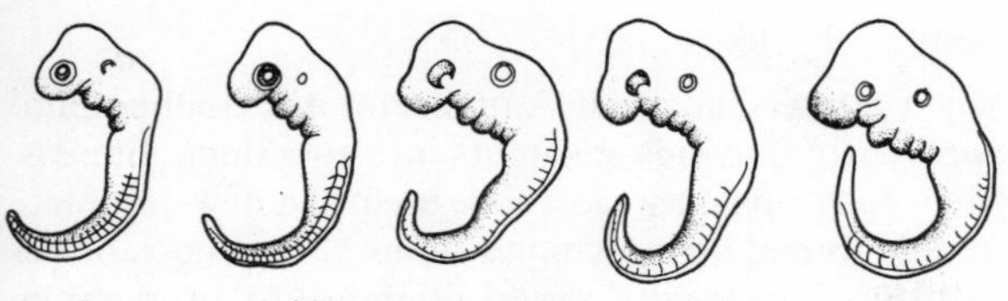

fig 9.13 Early embryos of vertebrates

mammals had ancestors in common with other vertebrate groups.

3 *History, cytology and biochemistry* Examination of the tissues and cells of related organisms under the microscope also reveals distinct similarities. At first glance, sections through, say, the blood vessels of a mouse, a pig and an elephant can be distinguished from each other only on the basis of size. A muscle cell from the foot of a snail looks very much like a smooth muscle cell from the bladder of a mammal.

At the sub-cellular level, resemblances are even more striking: chloroplasts of algae are very similar to those of higher plants, and all mitochondria are built on the same plan.

Most fundamental of all, perhaps, is the point that the molecules upon which the structure and functioning of all living organisms are based show remarkably little variation: the composition of plasma membranes in protozoans differs little from that in mammalian cells, and the biochemical pathways of aerobic respiration are similar in animals and plants.

4 *Palaeontology* Palaeontology is the study of fossils: it is probably the branch of science most closely associated in people's minds with evolution.

The fact that the flora and fauna of the earth many millions of years ago were quite different from those found today can have only two possible explanations. Either the old animals and plants died out completely and were replaced by newly-created forms, or they underwent a gradual change from generation to generation. The first alternative is plainly absurd, whereas the second is supported by the existence of many 'fossil histories', showing progressive changes in form of organisms through succeeding strata of rocks.

In this way, the evolution of Man, monkeys and apes has been traced back to a group of small tree-shrews that lived 60 million years ago. Note that monkeys and Man arose from a **common ancestor**; it is **not** true to say that Man evolved from monkeys.

5 *Geography* Lastly, it is worth noting the effect of *isolation* on flora and fauna. In Australia, the native mammals are quite different from those of other continents. They are marsupials, such as the kangaroo, wallaby, wombat and koala. One of their most striking features is that the young are born in a very immature state, and are transferred immediately to a pouch or *marsupium* on the abdomen of the mother.

This special feature of the Australian fauna can be explained by the fact that, since Australia has been separated from the other continents for many millions of years, there has been no interbreeding between its fauna and that of the rest of the world. Thus the gradual evolutionary change in the form of Australian mammals has proceeded in a unique direction.

9.52 The chemical basis of inheritance

Research by cell biologists and others has shown that genes control the manufacture of proteins, including enzymes. Thus there has developed the one gene/one enzyme hypothesis that inherited characteristics are produced and controlled by the action of enzymes on chemical processes in the living organism. Furthermore the chemical structure of chromosomes has been analysed and found to consist of protein and *deoxyribonucleic acid* (or DNA). Although the existence of nucleic acids had been known for some years it was not until 1953 that Watson and Crick in Cambridge, and Wilkins in London, suggested that DNA was 'two helical chains curled around the same axis'.

The chemical investigation of DNA showed that it consisted of a series of units called *nucleotides.* Each nucleotide is made up of phosphoric acid, the sugar deoxyribose and one only of the four bases adenine (A), thymine (T), cytosine (C) and guanine (G). Thus there are *four types* of nucleotide, differing according to the type of base that is attached to the sugar. A length of DNA molecule may have thousands of nucleotides on each of the two spiral threads and each pair of nucleotides is joined through the adjacent bases. Fig 9.13 shows a diagram of the molecular structure, looking rather like a 'twisted ladder' with the rungs formed by the junction of pairs of bases, one from each nucleotide. The diagram also shows that junction is only possible between adenine and thymine and guanine with cytosine.

If DNA is truly the physical basis of heredity it must clearly fulfill certain requirements:

i) It must be able to replicate (make a copy of itself) during mitosis and meiosis.

ii) It must be stable enough to carry the basic genetic information detailing the characteristics of the species.

iii) It must, however, be capable of some change to account for genetic variation and mutation.

iv) It must be capable of carrying, in a coded form, all the vast amount of information governing the appearance of any one organism.

Let us consider how DNA measures up to these four requirements.

i) The chain of bases on one half of the double spiral is partnered by a chain on the other half, that is A with T, T with A, C with G and G with C.

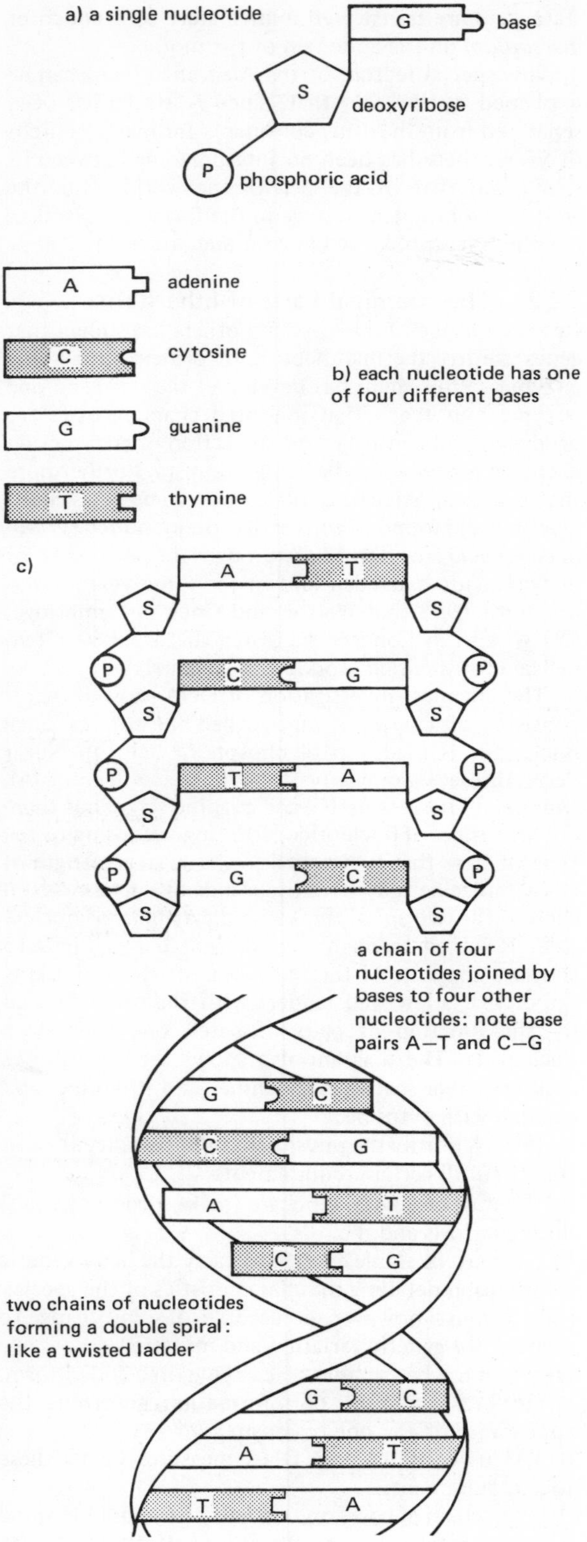
a) a single nucleotide
G
base
S
deoxyribose
P
phosphoric acid
A
adenine
C
cytosine
G
guanine
T
thymine
b) each nucleotide has one of four different bases
c)
a chain of four nucleotides joined by bases to four other nucleotides: note base pairs A–T and C–G
two chains of nucleotides forming a double spiral: like a twisted ladder

fig 9.14 The building of the DNA molecule

Thus the sequence TAGCATGCAAG———
must be partnered ATCGTACGTTC———

When a molecule of DNA replicates, the double spiral apparently unwinds and splits into two single threads. Then each single thread builds up into a new complete double thread by attaching a series of new nucleotides to itself. The original double spiral of DNA might be seen as 'unzipping' itself like a zipfastener and to each side adding a new half, forming two complete new zips (see fig 9.14).

ii) DNA is a relatively stable chemical compound and therefore will continue unchanged to pass on the information coded in its many nucleotides for generation after generation.

iii) Nevertheless it is capable of small changes due to omission of nucleotides, addition of new nucleotides, insertion of new nucleotides in the chain, breaking and rejoining of the chain and so on. Any of these small alterations could result in changes to the structure or the physiology of the organism, known as mutations.

These alterations can be caused in a variety of ways, by internal changes or by external factors such as chemicals or radiation. Mutations can cause the wrong amino acid to be added to a chain in a protein structure, with effects that can range from the death of the organism (a *lethal gene*) to an improvement in the structure or function of the organism such that it becomes more able to succeed in the struggle for existence, in other words *better adapted* to its environment.

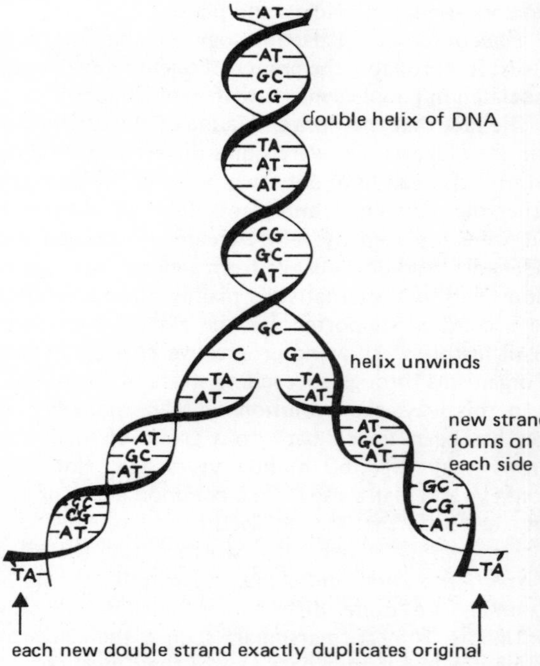
double helix of DNA
helix unwinds
new strand forms on each side
each new double strand exactly duplicates original

fig 9.15 Duplication of a double helical strand of DNA

iv) Living organisms contain an enormous variety of proteins but the construction of these proteins requires only *20 different forms* of amino acid. Very large numbers of amino acids are necessary for the construction of even the simplest protein and it is necessary for these to be arranged in a particular sequence. Therefore the DNA must act as a code designating which amino acids should be placed in which sequence to form a particular protein. If each base stood for an amino acid this would be insufficient to code for the 20 vital amino acids. If two bases represented an amino acid there would still not be enough combinations, for from four bases only 16 different pairs can be formed. Clearly then, a range of twenty amino acids requires a code with three bases per amino acid (see fig 9.15), a so-called triplet code.

Although the DNA carries the coded information for amino acid sequence, and thus protein structure, the actual formation of proteins is not within the nucleus but in the minute *ribosomes* in the cytoplasm. The coded information must therefore be carried from the nucleus to the ribosomes and this is brought about by another type of nucleic acid called *messenger RNA* (ribonucleic acid). Thus the process can be summarised:

NUCLEUS:
chromosome with
triplet code in DNA.
A long length of
DNA = one gene ⟶ messenger RNA passes to cytoplasm ⟶ RIBOSOME: amino acids assembled to form protein = enzyme
⟶ enzyme acts as organic catalyst in one stage of a chemical reaction ⟶ chemical reactions form the characteristic of a living organism

9.53 Mutation

In 9.41 (fig 9.12), we saw that Queen Victoria was a carrier of haemophilia. Since there is no record to show that any of her ancestors was haemophiliac, it seems likely that the gene causing haemophilia arose by a change in a gene responsible for blood clotting, before she was conceived. This gene change could have occurred in the X chromosome of the egg or of the sperm that fertilised it. Such changes are known as

A
G
T
C

a) CODE
One base represents one amino acid

AA	AG	AC	AT
GA	GG	GC	GT
CA	CG	CC	CT
TA	TG	TC	TT

b) CODE
Two bases represent one amino acid

There are 20 essential amino acids for protein manufacture: NEITHER of these two codes (a) or (b) could supply the necessary information.

AAA	AAG	AAC	AAT
AGA	AGG	AGC	AGT
ACA	ACG	ACC	ACT
ATA	ATG	ATC	ATT
GAA	GAG	GAC	GAT
GGA	GGG	GGC	GGT
GCA	GCG	GCC	GCT
GTA	GTG	GTC	GTT
CAA	CAG	CAC	CAT
CGA	CGG	CGC	CGT
CCA	CCG	CCC	CCT
CTA	CTG	CTC	CTT
TAA	TAG	TAC	TAT
TGA	TGG	TGC	TGT
TCA	TCG	TCC	TCT
TTA	TTG	TTC	TTT

c) CODE
Three bases represent one amino acid

ONLY THIS CODE (c) could represent the 20 amino acids, AND there are spare triplets to code for other essential items of information, e.g. "start", "stop".

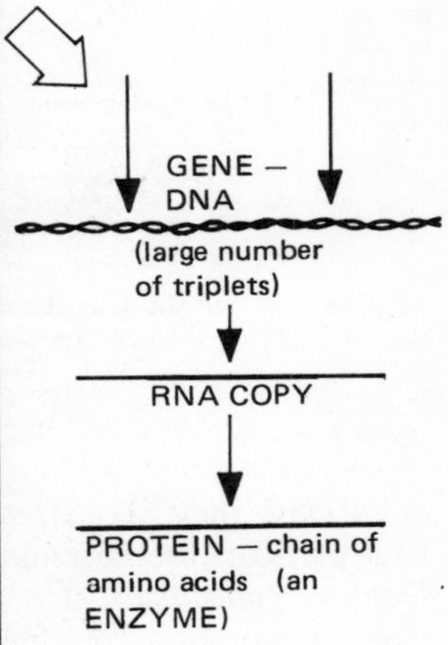

fig 9.16 The triplet code of DNA

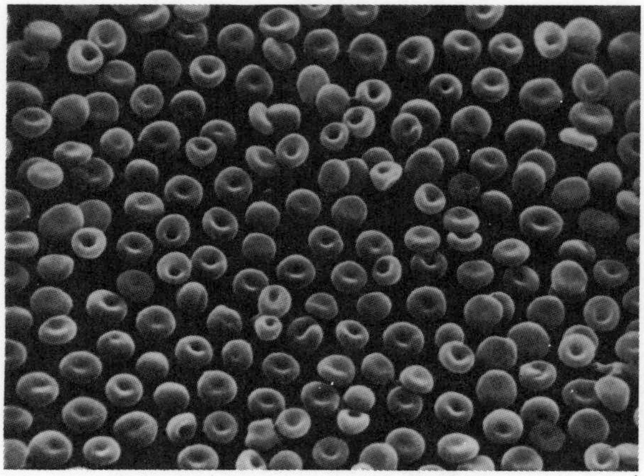

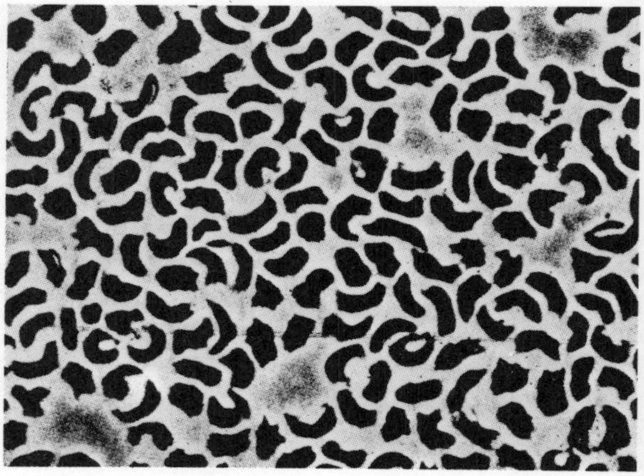

fig 9.17
(a) Normal red blood cells
(b) Sickled red blood cells

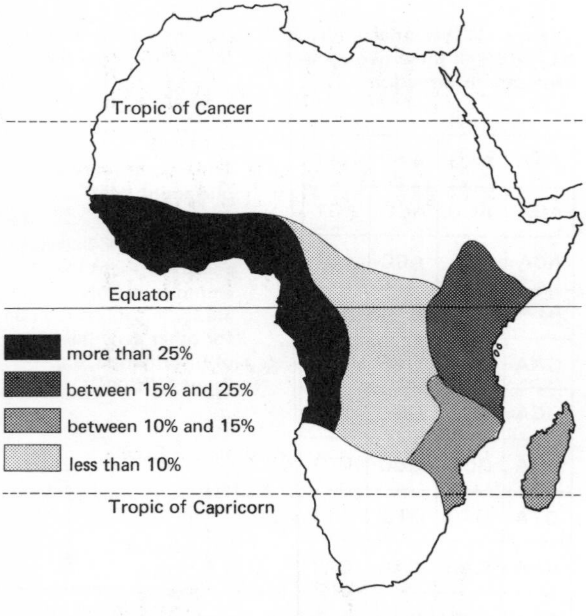

fig 9.18 A map showing the distribution frequency of people carrying the gene for sickle-cell haemoglobin

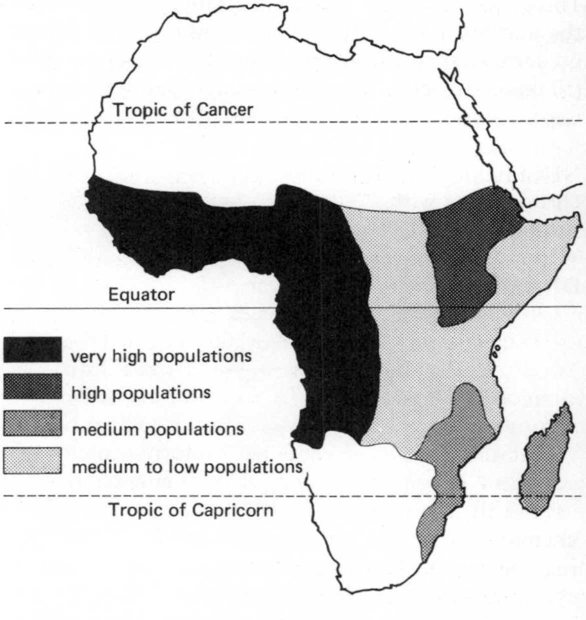

fig 9.19 A map showing the distribution frequency of *Anopheles* mosquitoes which act as vectors for the organisms causing malaria

mutations: they may affect single genes or relatively large parts of chromosomes.

Sickle cell anaemia

Not all mutations produce disastrous results in the phenotype. In some cases, their effects are beneficial, and they result in an organism being able to compete successfully with other organisms for food and for the production of more offspring. Such a mutation is said to be *selectively advantageous* and it will thus be possessed by more and more individuals in successive generations. In this way a species evolves: mutations are the raw material of evolution.

An interesting and important example of the selective advantage of a mutation can be seen in the condition known as the 'sickle cell trait'. This is a very common blood condition in many parts of the world, and may occur in up to 30% of some populations in West and Central Africa (see fig 9.17).

It arises in the following way. The blood pigment haemoglobin is essentially a protein, containing iron. One particular form of haemoglobin (haemoglobin S) is produced by a mutation of the gene responsible for the production of normal haemoglobin. This mutation causes the substitution of one amino acid (*valine*) for another (*glutamic acid*) during the formation of the haemoglobin protein, and as a result the red blood cells containing haemoglobin S collapse into the shape of a sickle at low oxygen concentrations.

People who are homozygous for this mutated S gene suffer from an often fatal disease known as *sickle cell anaemia,* where all the haemoglobin is of the S type and all the red blood cells are liable to collapse. People who are homozygous for the non-mutated gene do of course have normal haemoglobin and normal red cells. However, heterozygotes (people who possess both mutated and non-mutated genes) have some normal haemoglobin and some haemoglobin S. Thus some of their red cells are liable to collapse at low oxygen concentrations, and they tend to suffer from a milder, non-fatal form of anaemia: they exhibit the 'sickle cell trait'. This is clearly a case of the incomplete dominance discussed in section 9.20.

Study fig 9.18 carefully, and compare it with fig 9.19.

Questions

1 What serious disease is carried by anopheline mosquitoes?

2 In view of the fact that sickle cell anaemia is such a serious disease, and that even the milder sickle cell trait is clearly disadvantageous, why is the S gene so common in certain areas? (A clue to the answer may be found in the maps in figs 9.18 and 9.19).

Mutations occur infrequently, otherwise one of the important requirements for DNA (see section 9.52 number ii), the need for stability, would not be met. Nevertheless we can see by requirement number (iii) that there must be mutations to ensure adaptation to new environmental conditions. Thus all living organisms, including Man, mutate at a certain rate due to natural causes. It is true also that mutations can be increased by certain agencies. These include:

i) radiation: X-rays, gamma rays and cosmic rays
ii) chemicals such as mustard gas and nitrous acid.

Problem

The antibiotic drug Penicillin was found to be effective against many types of bacterium, particularly those causing wound infection. One such bacterium, *Staphylococcus aureus,* can infect patients after surgery and at first was quickly destroyed by Penicillin. Examine the following figures for infection of operation wounds in hospital patients.

	number of patients with post-operative infection	number NOT improving after penicillin
1946 9 month period	99	14
1947 5 month period	100	38
1948 5 month period	100	59

Questions

3 Can you suggest a reason for penicillin becoming more and more ineffective?

4 How could you control this problem?

Mongolism (Down's syndrome), a condition of

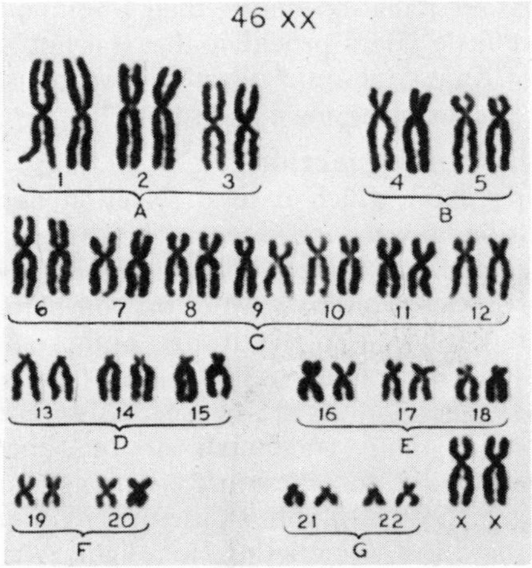

fig 9.20 Chromosomes of a normal human female

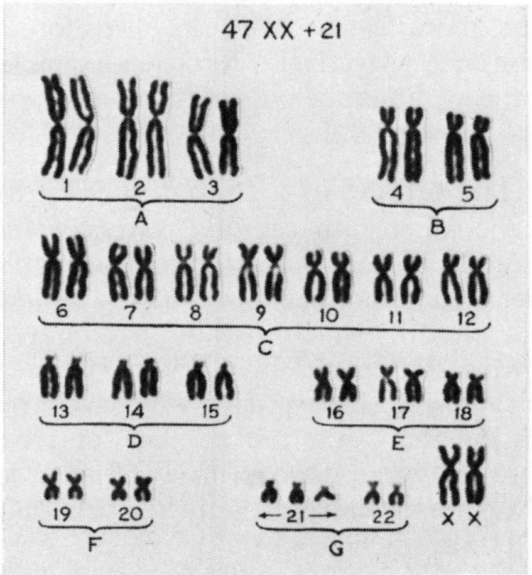

fig 9.21 Chromosomes of a mongoloid human female

mental and often physical retardation in humans, is known to result from a mutation that occurs approximately 16 times in every 10 000 births. Mongols do not usually produce children and thus normal parents apparently produce mongols in a known ratio. It has been established that this condition is a mutation present in the children of otherwise normal parents. Examine fig 9.20 and answer the following questions.

Questions

5 What is the diploid number of chromosomes in the human female?

6 How does the chromosome compliment of the mongoloid female in fig 9.21 differ from that of the normal human female in fig 9.20?

This is an example of a *chromosome mutation* which is different from the *gene mutation* which causes sickle-cell anaemia.

Humans must take care that they are not exposed to more radiation than is necessary. They cannot avoid cosmic radiation during their lifetime, but this will have little effect providing that it is not added to by use of X-rays, fall-out from atomic weapons testing, nuclear reactor breakdown and so on.

9.54 Natural selection

The basis upon which natural selection operates is competition between individuals. The stronger, faster, 'fitter' animal will be able to obtain more food at the expense of weaker animals, which may die before being able to reproduce. Also, the stronger or more attractive males will be more likely to find a mate (this is known as *'sexual selection'*). As a result, successive generations will contain more and more of the phenotypic characters of the *'fittest'* animals. Among plants, the 'fittest' are those best able to use the environment in which they live, competing for light, water and nutrients. An example of this is to be found in dense forest, where the fastest growing trees compete more successfully than the slower growing forms. They reach the light more easily and can, therefore, photosynthesise more effectively. A further example can be seen in the evolution of the giraffe from short-necked ancestors (see fig 9.22).

Ancestral giraffes probably had necks that varied in length. The variations were hereditary. (Darwin could not explain the origin of variations.)

Competition and natural selection led to survival of longer-necked offspring at the expense of shorter-necked ones.

Eventually only long-necked giraffes survived the competition.

fig 9.22 The origin of the giraffe's long neck by natural selection

9.55 The effects of a changing environment

During the nineteenth century, butterfly and moth collecting (lepidoptery) was a popular hobby. One particularly common species was *Biston betularia* (the peppered moth), which usually had greyish-white wings with scattered dark markings; from time to time, odd specimens with very dark wings were recorded. Towards the end of the century, lepidopterists noted that more and more dark forms were being collected in the North and Midlands, until the situation shown in fig 9.23 was reached.

Questions

1 Study fig 9.24. What does it tell you about the vulnerability of the moths to predatory birds?

2 How can the results shown on the map in fig 9.23 be related to the changing environment in Britain during the last century?

3 What would you expect to happen if there were to be a substantial reduction in the industrial air pollution by smoke and soot throughout the North and Midlands?

Over the centuries Man has succeeded in interfering with the process of natural selection in order to

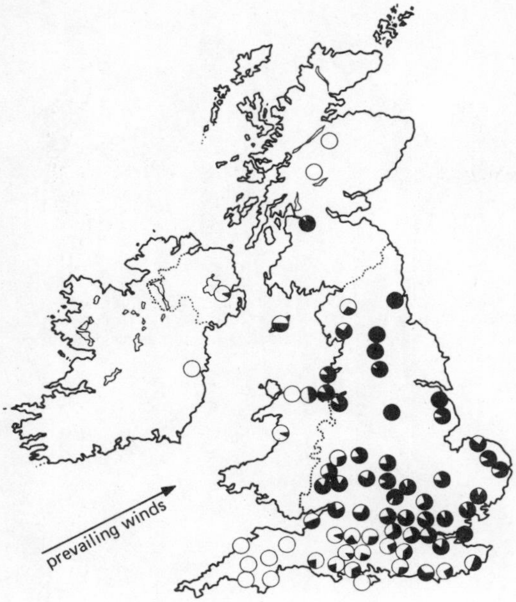

fig 9.23 The distribution of light and dark forms of the peppered moth in the British Isles. The proportion of the light form in a locality is represented by the white portion of each circle. The dark area of each circle represents the proportion of dark forms present.

fig 9.24 Light and dark forms of the peppered moth resting on tree trunks in (a) Birmingham and (b) Dorset

produce animals and plants suitable for his own agricultural purposes. He has taken animals which would not necessarily have competed successfully in the wild, but which had phenotypic characters (such as the ability to put on weight quickly or to produce large quantities of milk) which were useful to him, and has bred them over many generations. At each generation he has selected for breeding those animals which possess these qualities to the greatest degree, with the result that the domestic animals have, in many cases, become quite unlike their wild ancestors.

In the case of food plants, *storage organs* are particularly important. The potato, for example, has been produced by breeding from wild South American forms with much smaller stem tubers. Gradually, generation after generation, the plants with the largest tubers have been selected for breeding until the modern potato has been evolved, with tubers which are much larger than would be needed to supply the wild plant with enough food reserves to last over winter.

Other phenotypic characters which have been selected for artificially in the production of strains suitable for modern agricultural purposes are:

i) *early maturity* in both animals and plants (in the case of cereals, for example, this can result in the harvesting of two or more crops per growing season);

ii) *resistance to disease,* as in the case of wheat where strains resistant to the fungi causing rust diseases have been developed;

iii) *increased length* of productive season;

iv) *adaptation* to local or unfavourable conditions;

v) greater efficiency of *conversion* of plant to animal tissues in animals bred for meat;

vi) *higher yields* in terms of the production of milk, eggs, wool, fruits, etc;

vii) *ease of harvesting,* as in the case of thornless blackberries;

vii) improved eating quality, as in the cases of coreless carrots and seedless oranges;

ix) hardier constitution, such as sweetcorn that can be grown in Britain.

In many cases it has been found profitable to improve the stock by crossing locally bred animals and plants with strains adapted to simllar environments in different parts of the world.

9.60 Glossary of term used in genetics

Allele (allelomorph): one of a pair (or more) of alternative forms of a gene.

Bivalent: a pair of homologous chromosomes, situated together during meiosis.

fig 9.25 A Hereford bull, bred for beef production

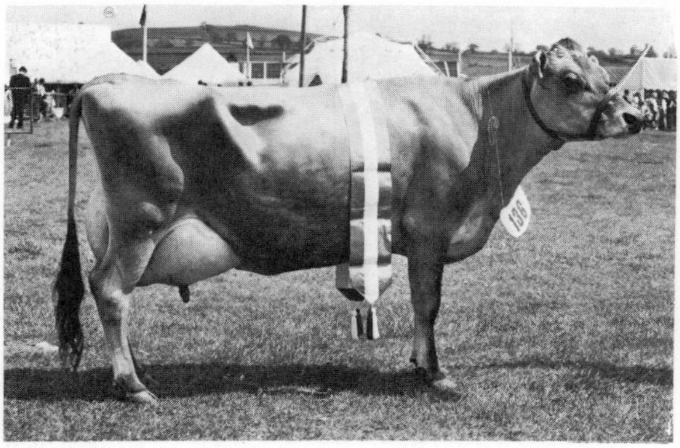

fig 9.26 A Guernsey cow, bred for milk production

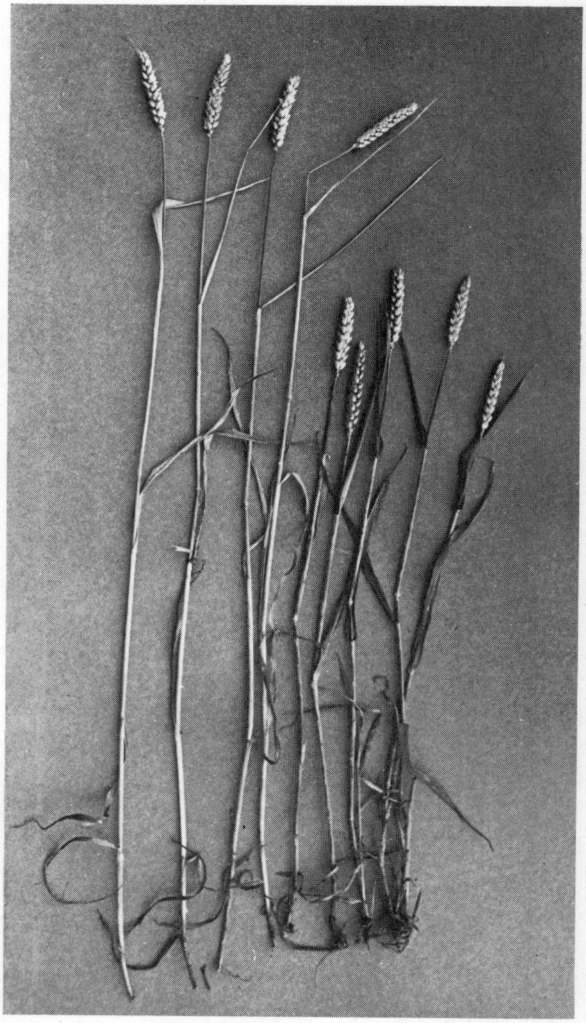

fig 9.28 Winter wheat variety with long stalk together with a short-stalked variety

fig 9.27 A British Freisian bull and cow, an all-round breed with high beef and milk yields

Centromere: a part of the chromosome which has no genes and by which the chromosome becomes attached to the spindle.
Chromatid: one half of the chromosome, when the chromosome splits longitudinally during cell division.
Chromosome: a thread-like structure, bearing genes and located in the nucleus.
Diploid: describes a cell that has two sets of homologous chromosomes.

fig 9.29 Wild pig

Dominant: describes an allele whose effect is seen in the phenotype of a heterozygote, in spite of the presence of an alternative allele.
Equator: the plane through the middle of the cell, at right angles to the main axis of the spindle.
Gene: a unit of hereditary material, located on the chromosome.
Genotype: a description of an organism in terms of certain of its genes.
Haploid: describes a cell with a single set of chromosomes.
Heterozygote: an organism which has two different alleles of the same gene.
Homozygote: an organism which has two identical alleles of the same gene.
Phenotype: a description of an organism in terms of what can be seen or measured.
Recessive: describes an allele whose effect is not seen in the heterozygote, because of the presence of a dominant allele of the same gene.
Spindle: an arrangement of cytoplasmic fibres, between the poles of a cell, along which the chromatids (or chromosomes) move during mitosis (or meiosis).

fig 9.30 English Large White pig, bred for meat production (to same scale as fig 9.29)

Questions requiring an extended essay-type anwer

1 a) Name an organ in which *meiosis* occurs in i) a flowering plant, and ii) Man.
b) Name an organ in which *mitosis* occurs in i) a flowering plant, and ii) Man.
c) How would you prepare plant cells to show mitotic figures? Draw diagrams to show the appearance of the chromosomes seen in your preparation at the four main stages of mitosis (assume that in your plant cells $2n = 8$).

2 A breed of dog has long hair dominant to short hair. A long-haired bitch was first mated with a short-haired dog and produced three long-haired and three short-haired puppies. Her second mating with a long-haired dog produced a litter with all of the puppies long-haired. Use the symbol (L) to represent the allele for long hair and (l) to represnt that for short hair.
i) What was the genetical constitution of the long-haired bitch?
ii) How could you determine which of the long-haired puppies of the second mating were homozygous?

3 a) A specimen of *Drosophila* has red eyes and when crossed with a purple-eyed mutant, all of the offspring were red-eyed. The offspring were mated among themselves and the following proportions of flies were produced: 224 red-eyed and 76 purple-eyed. Define the symbols that you will use to represent the alleles of these flies, and then by means of diagrams explain fully the genetics of these two crosses.
b) Draw diagrams to show the genetic details of a cross between red-eyed males of the parental generation and red-eyed females of the F_2 generation.

4 a) What do you understand by the term *incomplete dominance*?
b) What are the possible blood groups produced by a group B mother and a group A father? Give genetic diagrams to account for your answer.
c) Explain why a child of blood group O could not be produced from a mother of blood group AB and a father of blood group A.

5 a) How is sex determined in human reproduction?
b) A man suffering from haemophilia marries a woman whose blood has normal clotting properties.
i) Give a genetic diagram to show the different children that they could produce, indicating the sex of each child and whether or not it is a haemophiliac.
ii) If their daughter married a non-haemophiliac man what would be the possible blood-clotting characteristics of their children?

6 a) What is a mutation and what is its evolutionary significance?
b) Describe
i) a human condition caused by a gene mutation
ii) one caused by a chromosome mutation.

7 Write an essay on the chemical basis of inheritance.

10 Health and disease

10.00 Introduction

Most people have a fair understanding of what is meant by the term *health* or more precisely *good health.* In general terms, it is a condition where the body is functioning normally, with all of its chemical reactions proceeding in the right direction and at the correct speed. (See also section 10.92.)

Disease, however, is rather more difficult to define. A person with leprosy is diseased, certainly, but what about someone with a broken leg? A dictionary definition of disease is 'a disordered state of an organ or organism'. A broken leg is definitely in a 'disordered state', although it does not conform to the conventional idea of disease any more than does alcoholism or a nervous breakdown.

It is convenient, therefore, to classify diseases into the following six main groups.

1 Diseases caused by *other living organisms,* such as bacteria, viruses and various kinds of worm. These organisms live *parasitically* in or on the human body, interfering with its normal working.

2 Diseases caused by the *ageing and degeneration* of the body tissues. Chapter 3 deals in some detail with degenerative diseases of the circulatory system. Ageing of the tissues in the joints often leads to arthritic conditions, weakening of the eye muscles causes long-sightedness in many older people and at least some types of cancer may result from an age-related change in the division mechanisms of certain cells.

3 *'Human-induced' diseases* are diseases and disorders that people bring upon themselves, either individually or collectively. This group is one of growing importance in modern societies and includes industrial and domestic accidents, pollution-related disorders, alcoholism and drug abuse.

4 *Deficiency diseases,* which are still a major problem in developing countries, are covered in Chapter 2.

5 *Genetic and congenital disorders* (inherited diseases and other defects present at birth) are raising problems for the medical and social services. This is because, with the advances in medical science that have taken place over the past decades, many children who would previously have died in infancy are surviving into adulthood and having children of their own. In the cases of people with severe mental or physical handicaps, priority must be given to providing the means for them to lead happy and fulfilled lives. For patients with inherited diseases, it is necessary to provide adequate genetic counselling to minimise the risk of their passing on the conditions to their children. Inherited diseases are referred to in Chapter 9.

6 *Mental illness* covers a variety of important disorders, many of which also belong to groups 2, 3, 5 and even 4. In certain countries the majority of hospital beds are currently occupied by patients suffering from some form of mental illness.

10.10 Organisms that cause disease

10.11 The discovery of micro-organisms

Every secondary school pupil is familiar with the idea that some diseases can be caused by 'germs', tiny organisms that are too small to be seen with the naked eye. The existence of these micro-organisms was first discovered in the late seventeenth century by the inventor of the microscope, a Dutchman called Anton von Leeuwenhoek.

The magnification of his instrument was no more than 200 times, but he was able to see tiny organisms collected from pools of water. At this time the appearance of these organisms was attributed to *spontaneous generation,* the continual creation of new life from non-living things. Later Spallanzani, an Italian, showed that decay was due to micro-organisms and that substances usually teeming with micro-organisms could be prevented from developing them by immersion in boiling water. *Louis Pasteur* (1822–1895) continued this work in France. He first developed the *germ theory of disease* which postulates that all disease must be caused by micro-organisms. In section 10.12 we shall see the simple experiment performed by Pasteur, which showed decay could only be caused by living material coming from the air. He also developed a method of culturing micro-organisms by putting them into a sterile soup or on a sterile jelly; the organisms being transferred by a sterile needle from decaying matter.

Robert Koch (1842–1910), a German, designed many of the techniques for handling bacteria still used today. He investigated the disease *Anthrax,* cultured

(a) Leeuwenhoek (b) Spallanzani (c) Pasteur (d) Koch (e) Lister

fig 10.1 Pioneers of medical science

the bacteria and proved convincingly that this disease of cattle was caused by specific micro-organisms. He reasoned that the bacterial spores from the dead animal were released into the soil, and concluded that all dead cattle must be burned to destroy the spores.

Joseph Lister, (1827–1912), a famous British surgeon, was the first man to try to remove dangerous bacteria from the operating theatre by using a disinfectant, *carbolic acid.* His efforts greatly improved the chances of patients surviving an operation and not dying from bacterial infection of the wound. Antiseptic surgery involved the disinfection of all surgical instruments, the surgeons' hands, and the atmosphere.

10.12 Sterile techniques and the investigation of bacterial growth

Micro-organisms of different kinds are found everywhere. In every gram of soil there are about 100 million bacteria, while the collections of dead cells on the scalp (scurf) contain about 500 million bacteria. The air, bench tops, clothes, skin, finger nail scrapings and gut, as well as many other places, all shelter bacteria.

Fortunately most of these are harmless to human beings, i.e. they are not *pathogenic.* Nevertheless, when experimenting with micro-organisms, we must always take extreme care, for cultures of harmless types can often contain bacteria which cause disease.

When culturing bacteria, even if they are thought to be harmless, always follow the rules below.

i) Wash hands before touching a sterile Petri dish.
ii) Open the Petri dish as little as possible, and replace the lid quickly.
iii) Never cough or sneeze near the dish.
iv) Never touch the infected jelly with your fingers.
v) When cultures are no longer required they should be flooded with strong disinfectant.
vi) After cleaning out the nutrient from Petri dishes, they should be washed and disinfected, and then, if they are glass, heat sterilised.

vii) Wash your hands thoroughly after all operations. Use plenty of soap.

viii) Never put hands near the mouth during experimental work (on no account consume food in the laboratory).

Experiment

Where do micro-organisms come from? Pasteur's experiment.

Procedure

1 Set up four test-tubes as shown in fig 10.2. Tube 2 is not heated, whereas tubes 1, 3 and 4 are heated above the boiling point of water by placing them in a pressure cooker for 15 minutes.

2 Place the tubes in a rack and examine daily.

3 Make a table of results, and record the appearance of the tubes as the days go by.

Questions

1 Which tubes have changed in appearance?

2 What change is noticed in each nutrient broth?

3 What do you think causes this appearance? Smell the tubes. Is there any relation between appearance and smell? (Do not open tubes 3 and 4).

4 State the factors which have prevented any change taking place in the other tubes.

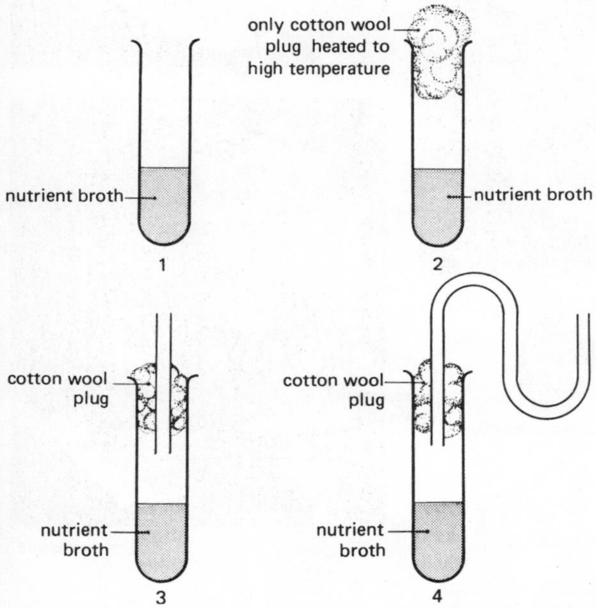

fig 10.2 Pasteur's experiment to investigate the origin of micro-organisms

Experiment

Are there micro-organisms around us in the air and on the body?

Procedure

1 Use sterile glass Petri dishes or plastic Petri dishes which have been pre-sterilised. Pour into five dishes nutrient broth (made from agar jelly – a seaweed extract) containing beef stock. Allow the broth to coagulate at room temperature.

2 Number the dishes 1 to 5. By means of a chinagraph pencil (or felt tip pen), draw lines across the bases of three dishes so that each has two halves A and B.

3 Prepare the Petri dishes as follows:

Dish 1 part A Press dirty fingers on the jelly.
part B Wash the fingers well, using soap, dry and press them on the jelly.

Dish 2 part A Place nail scrapings onto the jelly.
part B Place scrapings from between the teeth onto the jelly.

Dish 3 part A Place some hairs on the jelly.
part B Leave this half of the dish clear.

Dish 4 Open, and cough or sneeze over the jelly.

Dish 5 Open the dish in the laboratory for 30 minutes.

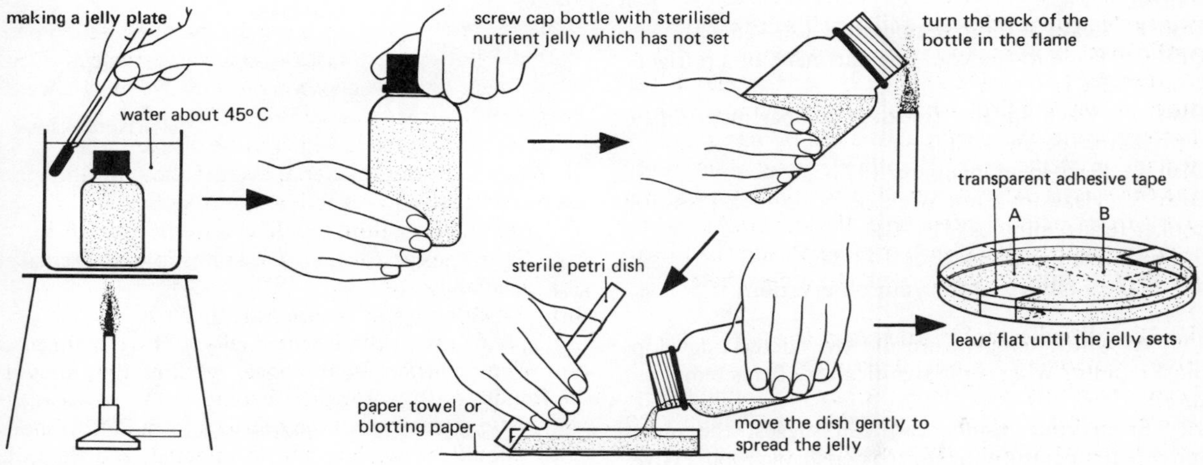

fig 10.3 The preparation of culture plates

4 Now fix the lids tightly to the bases of the Petri dishes with clear adhesive tape.
5 Place the dishes upside down in an oven at 37°C.
6 Examine the plates in your next lesson (or after 2–3 days.) **Do not open** the plates. You may have cultured human pathogens on them.
7 Record your results in the form of a table.

Questions

5 Why are the dishes marked on the under surface of the base, and placed upside down in the incubator?
6 Why are the lids tightly fixed with adhesive tape?
7 What appears in the dishes?
8 What is the function of dish 3 part B? Did anything appear in this section? Do you think it worked well?
9 Compare 1A with 1B. What is the significance of your observations?
10 If you were a health officer in a large food factory, what methods would you use to prevent workers from contaminating food?

The micro-organisms grow well in the incubator at 37°C and our experiments have shown that they are present in a number of places in the laboratory and on the body. They can thus be spread easily by the wind and by contact with animals such as insects and Man. We can ask ourselves whether we can do anything about their presence and prevent their multiplication. We know that temperature is a factor in their development, since they grow best at a temperature above that of the laboratory.

Experiment

Does temperature have any effect on the growth of bacterial cultures?

Procedure

1 Prepare five Petri dishes with nutrient broth and when cool open them in the laboratory for 30 minutes.
2 Place the lids on the dishes and then treat them as follows:

Dish 1 Leave at room temperature (20–25°C).
Dish 2 Place in the cool compartment of a refrigerator (3–5°C).
Dish 3 Place in the freezing compartment of a refrigerator (below 0°C).
Dish 4 Place in an incubator (37°C).
Dish 5 Heat in a hot oven or place in a pressure cooker for 15 minutes (121°C+).

3 Examine the plates after 2–3 days.
4 Make a table to show your observations.

Questions

11 At which temperature did the bacteria develop most rapidly? What is the significance of this temperature?
12 From your results, can you suggest the most suitable temperature for storing perishable foods? Give your reasons.

10.13 Bacteria

Bacteria are single cells but they do not have a nucleus. The nuclear material (DNA) is spread throughout the cell. The shape of bacteria is their most recognisable character and is used for identification.

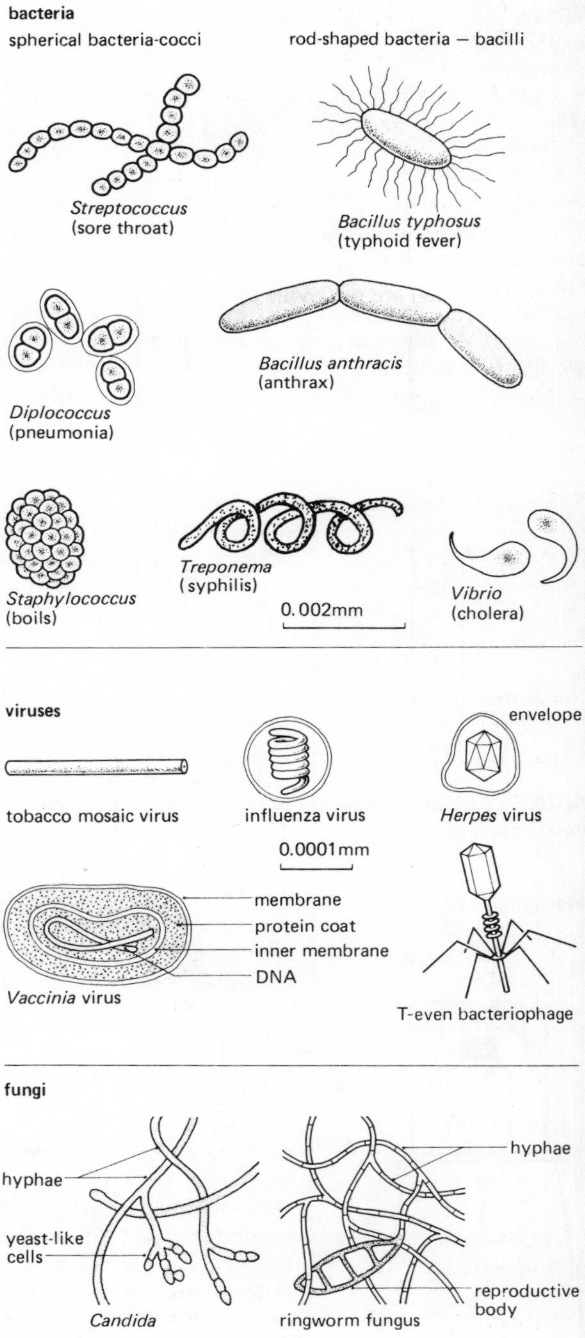

fig 10.4 Bacteria, viruses and fungi

i) *Cocci* (singular – coccus) are spherical with a diameter of about 1 μm.
a) *Streptococci* stick together to form a chain.
b) *Staphylococci* stick together in irregular bunches.
c) *Diplococci* stick together in pairs.
ii) *Bacilli* (singular – bacillus) are rod-shaped. Some are motile with one or more flagella.
iii) *Vibrios* are comma-shaped.
iv) *Spirilla* (singular – spirillum) are rod-like but twisted into a spiral cork-screw shape.
v) *Spirochaetes* are much finer, flexible and also twisted into a spiral. They are motile.

Bacteria differ in their requirements for oxygen during respiratory processes.
i) Aerobic bacteria require oxygen for respiration.
ii) Anaerobic bacteria do not require oxygen and produce their energy by anaerobic respiration (putrefaction).
iii) Facultative anaerobic bacteria, which constitute the majority of bacteria, can live under aerobic or anaerobic conditions.

Bacteria reproduce asexually by simple division into two cells. Resistant *spores* can be produced which often remain dormant for long periods. Bacteria produce enzymes which digest the surrounding organic material, and the simpler products resulting are absorbed into their cells. Thus their mode of life is *saprophytic.* The by-products of the metabolism of parasitic bacteria are often poisonous. These toxins may be the cause of disease in the host organism.

Although a few bacteria are pathogenic, there are also some which are positively useful to Man. Products of milk such as certain cheeses and yoghurt are made with the help of micro-organisms. The *Acetobacter* group of bacteria are used to produce vinegar from wine. Bacteria are most important as agents of decay. The change of dead organisms to harmless materials which become part of the air and soil has great significance for the further growth of new plants and animals. The work of bacteria is thus part of the great cycle of nature. The decay of organic matter by bacteria is a disadvantage as far as our food is concerned but we must not forget the importance of this work in nature.

Table 10.1 Diseases caused by bacteria

Disease, bacterium and method of spread	Symptoms and course of disease	Treatment and control
Tuberculosis (TB) *Mycobacterium tuberculosis* (bacillus) Airborne, occasionally via infected cow's milk	Disease may show itself many years after first infection. Tubercle bacilli may infect many organs, but pulmonary (lung) TB is the most common. General weight loss and cough, sometimes with sputum containing blood; slight afternoon fever.	Treatment: antibiotic drugs, e.g. streptomycin Control: 1 detection by mass radiography 2 vaccination with BCG (Bacille Calmette-Guerin) 3 eradication of cattle TB 4 pasteurisation of milk
Leprosy *Mycobacterium leprae* (bacillus) Airborne, occasionally by direct contact over long periods	Incubation period of many months or even years. Two forms: 1 nodular, with thickened areas of skin; 2 anaesthetic, in which the peripheral nerves are destroyed.	Treatment: drugs, e.g. streptomycin, dapone, sulphone Control: segregation and isolation. Now less common than in previous centuries
Diphtheria *Corynebacterium diptheriae* (bacillus) Airborne	Incubation period 2–4 days. Bacteria grow on mucous membranes of respiratory tract, releasing powerful toxin. Slight fever and sore throat are followed by severe damage to heart, nerve cells and adrenal glands.	Treatment: 1 injection of antitoxin 2 antibiotics, e.g. penicillin and erythromycin Control: mass immunisation with diphtheria toxoid

Disease, bacterium and method of spread	Symptoms and course of disease	Treatment and control
Typhoid and Paratyphoid *Salmonella typhi* and *S. paratyphi* (bacilli) Waterborne and foodborne	Incubation period 6–7 days. Mild fever and slight abdominal pains with constipation, followed by step-like increase in fever level, increasing pain and diarrhoea. Ulceration and rupture of the intestine may occur.	Treatment: antibiotics, e.g. chloramphenicol and ampicillin (semi-synthetic penicillin) Control: 1 purification of water supplies 2 safe disposal of sewage 3 pasteurisation of milk 4 vaccination with killed bacteria
Cholera *Vibrio cholerae* Waterborne	Bacteria release powerful toxin, causing inflammation of the gut and severe diarrhoea ('rice water'). The resulting loss of water and mineral salts may lead to death.	Treatment: 1 injection of saline solution to replace water and salts 2 drugs, e.g. tetracycline, chloramphenicol and sulfadiazine Control: 1 purification of water and treatment of sewage 2 vaccination (3–12 months protection only)
Whooping cough *Bordetella pertussis* (bacillus) Airborne	Occurs mainly in young children. Incubation period two weeks. Severe coughing bouts, with cough followed by 'whoop' sound as air is inspired through narrowed air passage.	Treatment: antibiotics Control: vaccination with killed bacteria
Bubonic plague *Pasteurella pestis* (bacillus) Spread by rat flea bite	High fever. Swelling of lymph nodes in groin and armpits to form 'buboes'. Capillary haemorrhages give dark appearance to skin, hence the name 'black death'.	Treatment: antibiotics such as streptomycin Control: eradication of rats, and storage of grain etc. in rat-proof containers

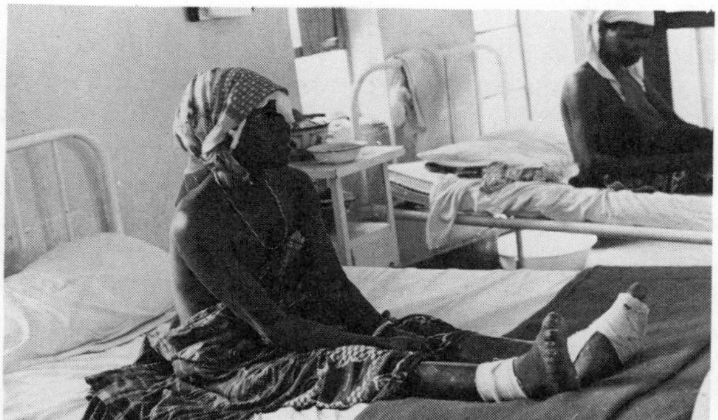

fig 10.5 Leprosy victim

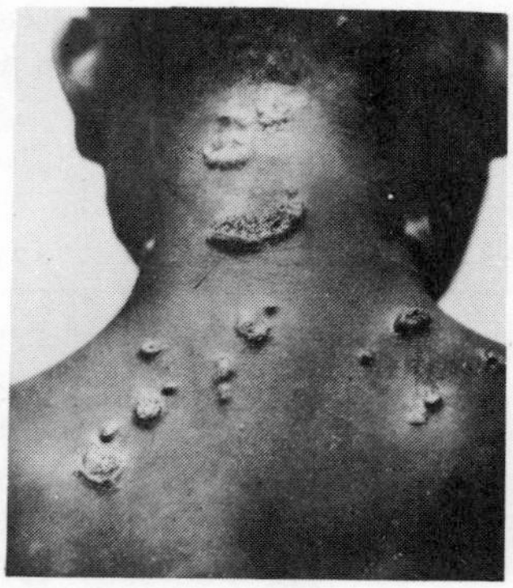

fig 10.6 Yaws victim

Disease, bacterium and method of spread	Symptoms and course of disease	Treatment and control
Yaws *Treponema pertenue* (spirochaete) Spread by direct contact with sores	Occurs mainly in children under 15. Open sores (lesions) spread over body surface. Bones also affected in severe cases.	Treatment: antibiotics such as penicillin Control: improved standards of hygiene and prevention of overcrowding
Tetanus (lock jaw) *Clostridium tetani* (bacillus) Bacteria present in soil, dust or animal faeces contaminate skin wounds	Toxins produced by bacteria cause muscular spasms (strong contractions), first in the region of the mouth and neck, then throughout the body. Eventually, convulsions may be so severe and frequent that patient dies of exhaustion or lack of oxygen. Commonest in war (infected wounds) and newborn babies (infected stump of umbilical cord).	Treatment: 1 injection of anti-toxin from horses 2 injection of muscle-relaxing drugs 3 antibiotics such as penicillin Control: Immunisation with tetanus toxoid (toxin which has been treated so that it is no longer harmful)
(Lobar) Pneumonia *Diplococcus pneumoniae* (coccus) Airborne	Short incubation period, followed by coughing, fever and chest pains. Fluid builds up in lungs. Bacteria often present in throat, but become active and invade lungs when patient is weakened by other illness. Several types of pneumonia exist, including some caused by viruses.	Treatment: antibiotics (e.g. penicillin) and sulphonamides Control: good ventilation and prevention of overcrowding
Meningitis (spotted fever) *Neisseria meningitidis* (coccus) Airborne	Bacteria multiply in mucous membranes of nose and throat, producing a sore throat, then enter bloodstream and infect brain coverings (meninges). High fever, vomiting and severe headaches, with general body rash.	Treatment: antibiotics (e.g. penicillin) and sulphonamides Control: prevention of overcrowding, especially of sleeping accommodation

10.14 Viruses

Viruses are too small to be seen by a light microscope but there is a wide range of size from 200–300 nm to 30 nm. Each virus particle is composed of nuclear material (DNA or RNA) enclosed in a coat of protein. The shape is varied (see fig 10.4).

They are all parasites and cannot be cultured outside living cells. They multiply very rapidly and can be transferred through the air or by contact between organisms.

Certain types of virus are known as *bacteriophages* since they cause fatal infections of bacteria. The phages are one of the suitable materials for the study of viruses, and over the last twenty years detailed studies have been made of their behaviour.

Viruses produce disease by attacking particular groups of cells in the animal or plant body. They multiply within these cells and destroy their structure as well as inhibiting their activity. Rabies virus attacks nerve cells in the brain while poliomyelitis virus enters nerve cells of skeletal muscle. Yellow fever virus enters the cells of the mammalian liver while leaf mosaic virus enters and damages the outer cells of the leaf.

Table 10.2 Diseases caused by viruses

Disease and method of spread	Symptoms and course of disease	Treatment and control
Measles Airborne	Occurs mainly in young children. Incubation period of two weeks, followed by sore throat, runny nose, watery eyes, cough and fever. Small, white spots (Koplik's spots) appear inside mouth on wall of cheek. Two days later, reddish rash appears at hairline, on neck and behind ears, spreading over rest of body. If no complications, patient recovers completely one week later, but virus can damage heart muscles, kidneys or brain, and secondary infections by bacteria may cause pneumonia etc.	Treatment: injection of gamma globulin a few days after exposure Control: 1 live, attenuated vaccine 2 isolation of patient and avoidance of overcrowding in schools
Rubella (German measles) Airborne	Occurs mainly in older children and young adults. Incubation period of 2½ weeks, followed by slight fever and body rash which disappears after 3 days. Complications rare except in women during first four months of pregnancy, when there is a 20% chance of blindness, deafness or other serious defects in the baby.	Treatment: 1 mild analgesics, such as aspirin and paracetamol 2 for pregnant women, injection of gamma globulin within 8 days of exposure Control: vaccination of girls 11–14 years with live, attenuated vaccine
Mumps Airborne, or spread by contact with infected saliva	Occurs mainly in children. Incubation period of 2–3 weeks, followed by fever and swelling of the parotid (salivary) glands on one or both sides, lasting about ten days. Other organs may be affected, including testes, ovaries and pancreas. Inflammation of the testes (orchitis) resulting from mumps in a male after puberty may produce sterility.	Treatment: aspirin etc. to relieve symptoms Control: a vaccine has been developed, but as yet availability is limited
Chicken pox Airborne	Occurs mainly in children and young adults. Incubation period 2 weeks, followed by fever, headache, sore throat and rash (typically on chest, spreading over body). Blister-like spots are very itchy; if scratched, may become infected secondarily by bacteria and leave permanent scars. Virus may migrate up sensory nerve fibres and lie dormant for years, becoming active again when immunity weakens, to produce *shingles* (a painful condition affecting the skin supplied by the nerves concerned).	Treatment: no drugs effective against virus. Antibiotics used to treat bacterial secondary infections

Common cold Airborne	Nasal and bronchial irritation, resulting in sneezing and coughing.	Treatment: aspirin etc. Control: colds can be caused by many different viruses so vaccine production is impracticable, and people who have recovered from a cold caused by one virus are not immune to colds caused by others
Influenza (sweating sickness) Airborne	Incubation period 1–3 days. Sudden fever with headache, sore throat and muscular aching. Recovery within one week, but after-effects such as tiredness and depression (especially in older people) may last well over a month. Secondary infection of the lung tissue by bacteria, leading to pneumonia, may occur in some cases.	Treatment: 1 analgesics 2 antibiotics to prevent secondary infections Control: vaccination with inactivated virus may give immunity for 1–2 years
Rabies Contracted from bite of rabid mammal (cat, dog, bat, etc.)	Incubation period varies between one week and several months. Slight fever, sore throat and headache, followed within a few days by convulsions and inability to swallow. Accumulated saliva causes 'foaming at the mouth'. Usually fatal.	Treatment: 1 wash and disinfect wound 2 inject horse serum containing antibodies near wound 3 inject vaccine Control: 1 strict import and quarantine controls on animals entering rabies-free areas 2 vaccination of domestic dogs
Poliomyelitis (infantile paralysis) Airborne, foodborne or waterborne, also contagious	Incubation period 7–21 days, followed by fever, headache and feeling of stiffness in neck and other muscles. Virus destroys nerve cells that supply muscles, causing paralysis and muscle-wasting. If breathing muscles are paralysed an 'iron lung' may be needed. Many people have a very mild form of the disease which they do not notice but which makes them immune. Most cases of paralysis occur in children 4–12 years, but adults may also be affected.	Treatment: no known drug or chemical treatment Control: vaccination with formaldehyde-treated virus (Salk vaccine) or mutated virus (Sabin vaccine) taken orally
Smallpox Airborne, also contagious	Incubation period about 12 days, followed by high fever and generalised aching. Rash seen on face two days later and spreads over body. Secondary infection of spots by bacteria causes permanent scarring.	Treatment: antibiotics to control secondary infection Control: vaccination – WHO campaign begun in 1968 has almost completely eliminated the disease within ten years
Infective hepatitis Waterborne or foodborne	Incubation period 15–40 days, followed by loss of appetite, vomiting and jaundice (yellow colour of eye white, due to bile pigments, showing liver disorder. In pale-skinned people the skin may turn yellow).	Treatment: injection of human gamma globulin soon after exposure Control: boiling of drinking water and attention to hygiene. Injection of gamma globulin usually too expensive for prophylactic (preventative) use

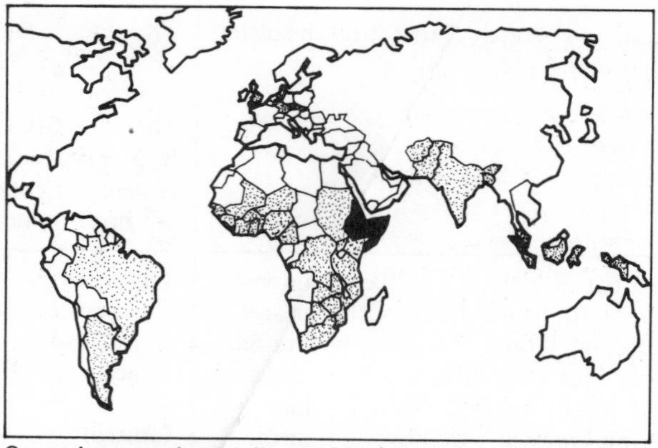

Countries reporting smallpox 1967 (shaded) and 1976 (black), showing how the disease has been pushed to the point of no return since 1967. It is hoped that the disease has now been eradicated. However, nine countries still keep stores of the virus for research purposes and in 1978 two cases were reported in Birmingham (U.K.) following an accidental release from one of the laboratories concerned.

fig 10.7 The fight against smallpox

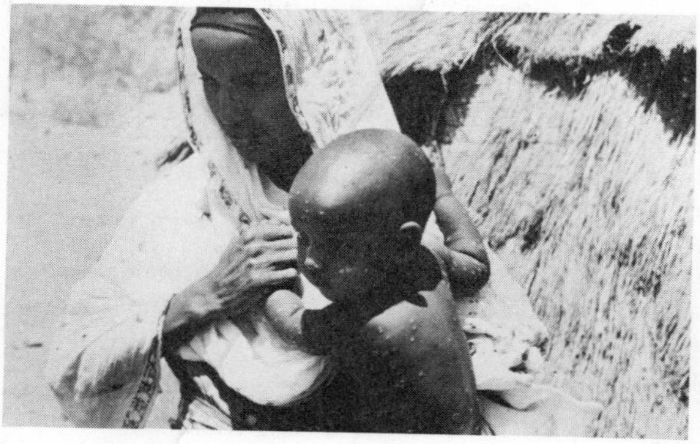

fig 10.8 Smallpox victim

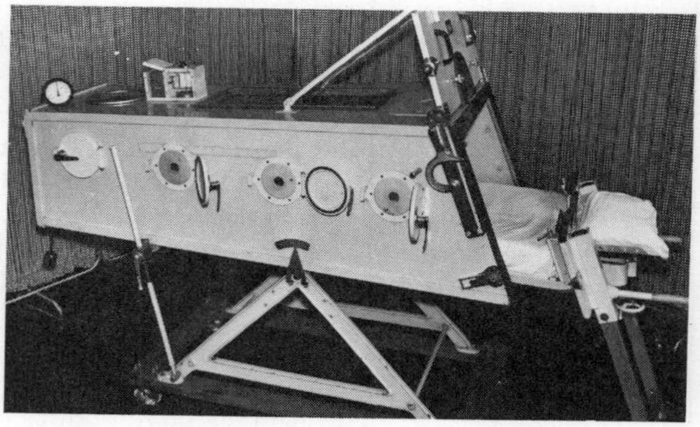

fig 10.10 Iron lung

fig 10.9 Muscle-wasting due to poliomyelitis

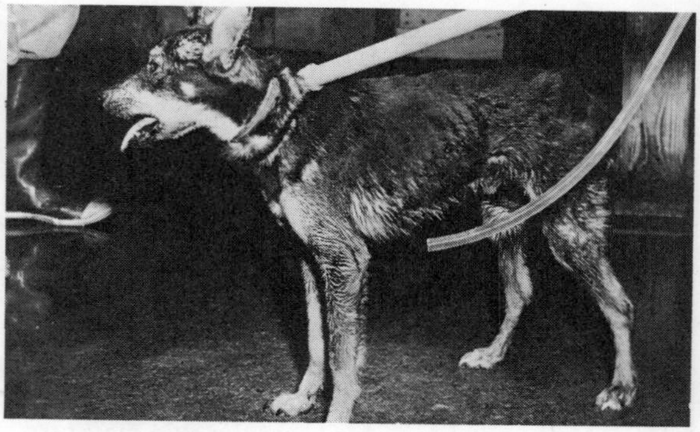

fig 10.11 A rabid dog

10.14 Rickettsiae

These are micro-organisms similar in size and structure to the smallest of the coccus or bacillus types of bacterium. However, like viruses, they cannot survive outside living cells, and are thus obligate parasites.

The most important disease caused by rickettsiae is epidemic typhus. This is spread by the body louse (*Pediculus humanus*) and its symptoms include a rash, severe headaches, high fever and mental depression. A less serious form of the disease, where the symptoms are similar but milder, is known as endemic typhus and is carried by the rat flea (*Xenopsilla cheopsis*).

Rickettsial diseases are treated with drugs such as tetracyclines and chloramphenicol, and may be prevented by vaccination with rickettsiae that have been grown on chick embryos before being killed with formalin.

A group of organisms very similar to the rickettsiae are the chlamydiaceae. *Chlamydia trachomatis* causes the eye disease trachoma, which affects some 400 million people in the world today. It is particularly common in hot, dusty climates and involves severe inflammation of the conjunctiva, leading to permanent scarring and shrinking. The cornea may also be affected and untreated victims often become blind. Tetracycline ointment should be applied to the eyes and sulphonamide drugs given orally. Vitamin A in the diet tends to improve natural resistance to the disease. *Chlamydia* species are also responsible for certain genito-urinary infections (see section 10.24).

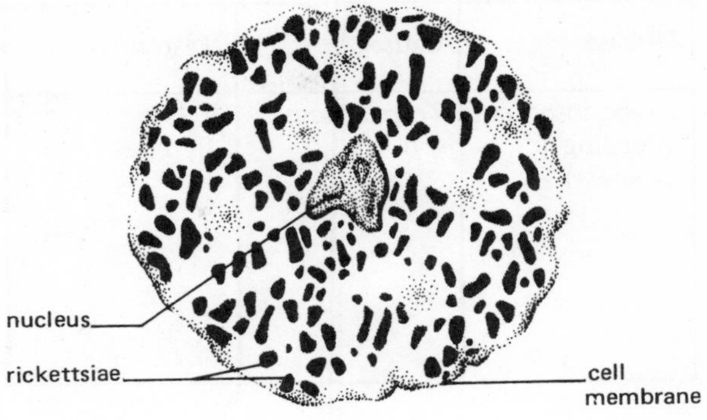

fig 10.12 Rickettsiae

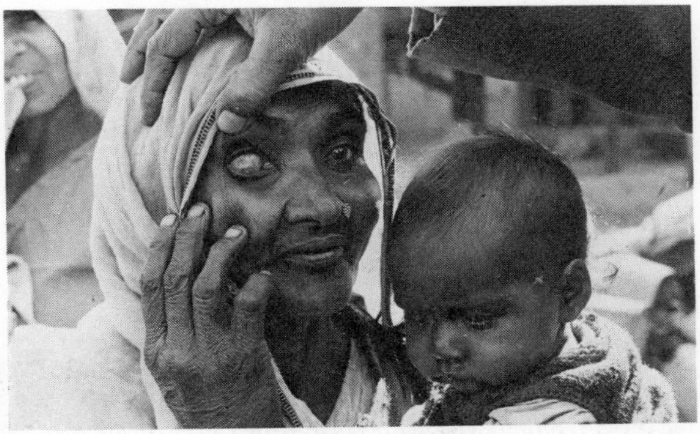

fig 10.13 Trachoma victim

Table 10.3 Protozoan diseases of Man

Disease	Cause	Transmission	Symptoms, other characteristics and treatment
Malaria	*Plasmodium* spp.	*Anopheles* mosquito bite	Plasmodia injected into the blood multiply rapidly. After ten days, high fever develops which may be continuous, irregular or occur twice a day. Control: 1 Drainage of the breeding places of mosquitoes 2 Destruction of larvae with an oil spray 3 Destruction of adults with insecticide 4 Destruction of the parasites in Man by drugs e.g. chloroquine and quinine 5 Preventive drugs e.g. paludrine and daraprim
Amoebiasis (Amoebic dysentery)	*Entamoeba histolytica*	Uncooked food, unhygienic preparation of food	Causes diarrhoea with loss of blood, fever, nausea and vomiting – can lead to death. Control: 1 Hygienic food handling 2 Prevention of flies that can spread the disease 3 Drugs – e.g. emetine, antibiotics and sulphur drugs

Disease	Cause	Transmission	Symptoms, other characteristics and treatment
Trypanosomiasis (sleeping sickness)	*Trypanosoma spp.*	Tsetse fly bite	A painless lump develops at the bite, lymph glands become enlarged, fever, enlargement of spleen and liver follow. Later the parasite invades the nervous system resulting in sleepiness and muscular spasms. Control: 1 Control of flies and limitation to certain areas 2 Fly screens in human dwelling places 3 Drugs e.g. pentamidine in the early stages

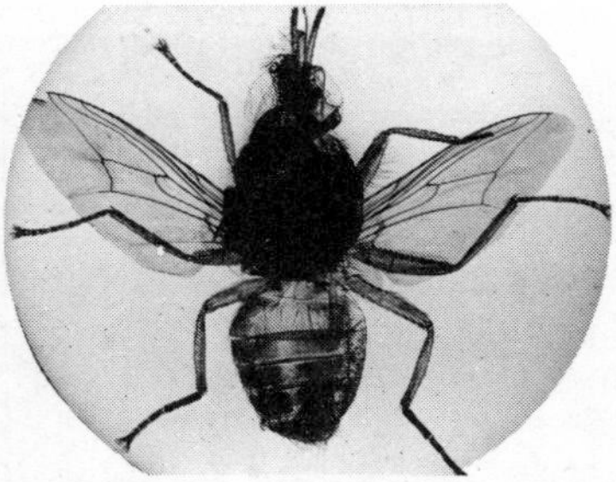

fig 10.14 Tsetse fly

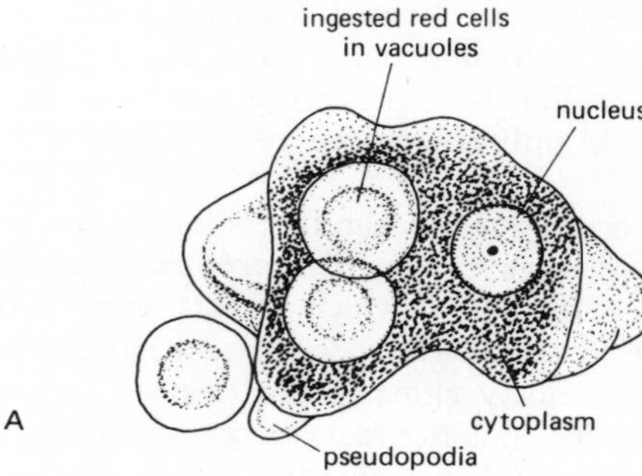

fig 10.15 *Entamoeba histolytica* — the causative agent of amoebic dysentery A

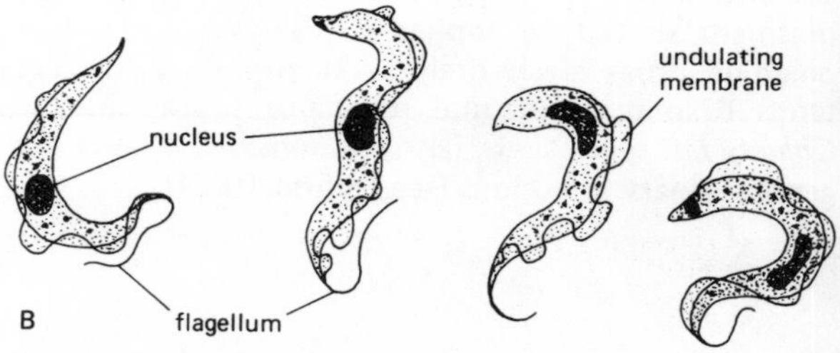

fig 10.16 *Trypanosoma* — the causative agent of sleeping sickness B

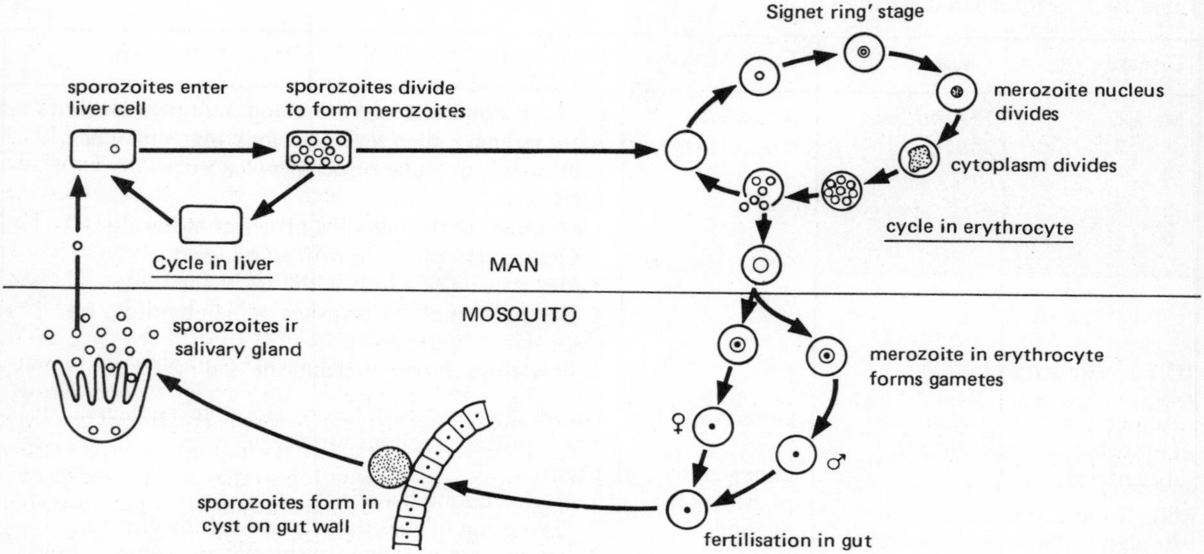

fig 10.17 Life history of the malarial parasite *Plasmodium*

Table 10.4 **Fungal diseases of Man**

Disease	Cause	Symptoms, other characteristics and treatment
Ringworm of the scalp	*Microsporium audouini*	A highly contagious disease by contact, combs, hats etc. amongst children. It begins as a small scaly spot which enlarges and older patches are covered with greyish scales. Control: 1 Exclusion of infected children from school 2 Drugs e.g. antibiotic griseofulvin taken by mouth
Ringworm of the skin	As for scalp or *M. canis*	Lesions on the skin are seen as pale, scaly discs. There is more inflammation around the edges, causing swelling and blistering. Control: 1 Drugs e.g. griseofulvin
Athlete's foot	*Tinea pedis*	Shows as sodden, peeling skin between the toes that can be subject to secondary bacterial infection. Cure rate is low. Control: 1 Exclude sufferers from swimming pools and changing rooms 2 Griseofulvin is only used in extreme cases
Candidiasis (Thrush)	*Candida albicans*	A yeast-like cell, 2–4 microns in diameter. Commonly harmless in the body but infection results from some local reduction in resistance of the tissues. This may occur in the mouth, intestine, vagina, etc. Control: 1 Establish the predisposing factor and change this to clear up the infection 2 Drug e.g. antibiotic nystatin used as a local cream

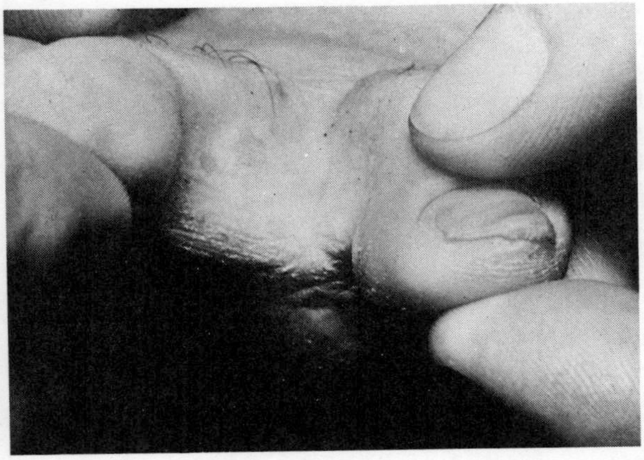

fig 10.18 Athlete's foot

10.16 Protozoa

Protozoa are small animals made up of a single piece of cytoplasm with a nucleus (acellular or unicellular). A living membrane surrounds the cytoplasm. Some protozoa are amoeboid and can change their shape while others possess a whip-like extension of the cytoplasm (flagellum) in order to help them move. They feed *holozoically*, ingesting their food and then digesting it, or *saprozoically* by digesting it externally and absorbing the products. Parasitic protozoa can have two hosts, with sexual reproduction occurring in the primary host and asexual reproduction in the secondary host.

10.17 Fungi

Many bacteria and viruses can parasitise Man and so cause illness, but comparatively few protozoa and fungi disturb Man's normal functions in this way. Fungi are either *saprophytes* or *parasites* and both types secrete enzymes to digest the external organic matter. The fungus produces fine filaments called hyphae which together form a dense mass, the mycelium. The cell wall is made of fungal cellulose and within is a lining of cytoplasm containing many nuclei (see fig 10.4). In addition, food stores such as oil droplets are present.

Fungi are responsible for many *plant diseases,* destroying large areas of crops if conditions are right for their reproduction. They reproduce very rapidly by asexual spore formation so that the disease spreads quickly. Parasitic fungi include *smuts, rusts, leaf spot* and *blights* of different plant species and cause huge losses in cereals, potatoes and many other crops.

10.18 Platyhelminthes and Nematodes

A large variety of worms infest Man. They fall into two main groups (i) *Platyhelminthes* (flatworms, tapeworms and flukes). Examples include liver fluke; pork, beef and fish tapeworm. (ii) *Nematoda* (roundworms). Examples include large round worms (*Ascaris*) and threadworms (*Enterobius*).

These worms contribute to a great deal of suffering and disease in the world. They have a complicated life cycle, often including a secondary host as well as the primary host (Man).

Table 10.5 Worms infecting Man

Disease	Transmission	Hosts	Symptoms, other characteristics and control
Tapeworm (*Taenia* spp.)	Through food – undercooked meat and fish	1 Man 2 Pig, cattle or fish	Encysted embryo in the flesh of secondary host is consumed in undercooked meat or fish. Tapeworm develops in the gut attached to the intestinal wall. Fertilised eggs are passed out with the faeces, eggs are then eaten by animals. The tapeworms cause few symptoms and relatively little damage in Man. Control: 1 Meat and fish to be well cooked at high temperatures 2 Inspection of meat at slaughterhouses 3 Proper processing and disposal of sewage 4 Drugs e.g. similar drugs to those used against malaria: mepacrine and chloroquine
Bilharzia (*Schistosoma* spp.)	Infected water	1 Man 2 Water snail	The worms live in blood vessels of intestines and bladder. Eggs penetrate the intestines and bladder and pass out with urine and faeces. Eggs hatch and the larvae penetrate water snails. In the snail the larvae multiply and change their form. When released into the water they penetrate the skin of Man and then develop into mature worms. The worms cause inflammation of intestines, blood in faeces and urine, anaemia and weakness. Control: 1 Proper processing and treatment of sewage 2 Avoidance of infected water 3 Elimination of snail 4 Preventative drugs
Ascariasis (*Ascaris lumbricoides*)	Infected food and water	1 Man (no secondary host)	The worms live in the bowel of Man and produce vast numbers of eggs that are very resistant when shed with the faeces. When eaten by Man, the eggs hatch and the larvae burrow into the lungs and from there reach the gut by way of the pharynx. The worms may obstruct the bowel and the larvae damage the lungs causing malnutrition and death. Control: 1 Proper processing and treatment of sewage 2 Hygienic food and water supply 3 Drugs; piperazine is the most effective
Threadworms or pinworms (*Enterobius vermicularis*)	Eggs swallowed	Man (no secondary host)	Very common, especially in children. Adults live in large intestine. Females migrate to anus to lay eggs, causing itching. Scratching, followed by placing of fingers in mouth, causes reinfection. Control: Washing hands after using toilet or touching anal area

GATEWAY SCHOOL
LEICESTER

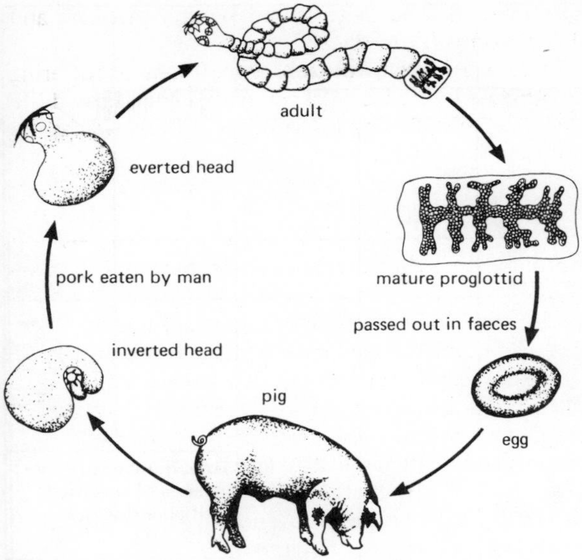

fig 10.19 The life cycle of a tapeworm

10.19 Arthropods

The Phylum Arthropoda includes insects, crustaceans, ticks and mites, many of which may carry micro-organisms that cause disease in Man. A few arthropods, however, actually cause disease themselves.

Scabies is a disease caused by the 'itch-mite' *Sarcoptes scabiei.* Skin damage is caused by the burrowing of a female carrying fertilised eggs. The burrow can be seen, often on the hands or feet, as a dirty grey line, a few millimetres in length. The colour is caused by the faeces of the mite, which also lays eggs along the burrow. The eggs hatch, and the immature stages (called nymphs) wander over the surface of the skin, feeding in hair follicles and causing small spots to appear. The original burrow usually itches, and scratching may result in secondary infection. Although infected clothing may spread the disease, it is most commonly passed between people sharing a bed.

The most effective method of prevention is careful attention to personal hygiene. Underclothing and bedding of infected persons should be boiled, and the skin should be treated with special anti-scabies lotions or ointment. These include sulphur preparations and emulsions of benzyl benzoate, which should be applied to the whole body below the neck, after careful bathing.

10.20 The transmission of disease

Diseases considered in this section are those caused directly by harmful organisms that enter the body of Man. These have been dealt with in sections 10.13 to 10.18, but we can now look at their methods of transmission in greater detail.

10.21 Airborne diseases

Some disease organisms are transmitted from person to person in tiny droplets of moisture. Coughing, sneezing, talking and ordinary breathing project moisture particles into the air. Larger droplets can settle onto food and on other objects in the home or the school. Smaller droplets evaporate quickly leaving bacteria or virus particles suspended in the air so that they can be inhaled. Droplet-borne infection is spread rapidly under conditions of high humidity and over-crowding, such as are encountered in schools, buses, trains and public meetings. Thus head colds and influenza tend to spread rapidly and produce epidemics.

10.22 Waterborne diseases

Drinking water is a source of many diseases, such as dysentery, cholera, typhoid and paratyphoid, which affect the alimentary canal. The spores and active organisms are liberated with faeces and can be spread to water supplies by insanitary conditions. Large numbers of people can become infected quickly in this way, particularly when floods, typhoons and earthquakes have seriously damaged water supplies and sewage disposal systems. Under normal conditions in home and school it is essential that hands should be washed after defaecating or urinating, for infection can be transferred on unclean hands used to prepare food or handle eating utensils.

Rivers can spread disease very quickly, so that populations remote from a source of infection can develop cholera and typhoid.

10.23 Food borne diseases

Many organisms transmitted by water can also be carried by some foods. Unwashed hands, exposed septic sores, contaminated water and flies can also spread infection to food during its preparation. Bacterial, viral and worm infections can all come from contaminated food.

10.24 Contagious diseases

These are diseases spread by direct contact between people or by objects handled by people. Fungal infections, ringworm and athlete's foot, can be transferred from skin to skin, by infected towels, floor surfaces, clothing etc. Smallpox is another disease that can be caught by contact.

10.25 Sexually transmitted diseases (S.T.D.)

These are a particular group of contagious diseases (also known as *venereal diseases*) which are spread by sexual contact. The most widely known of the sexually transmitted diseases are syphilis and gonorrhoea.

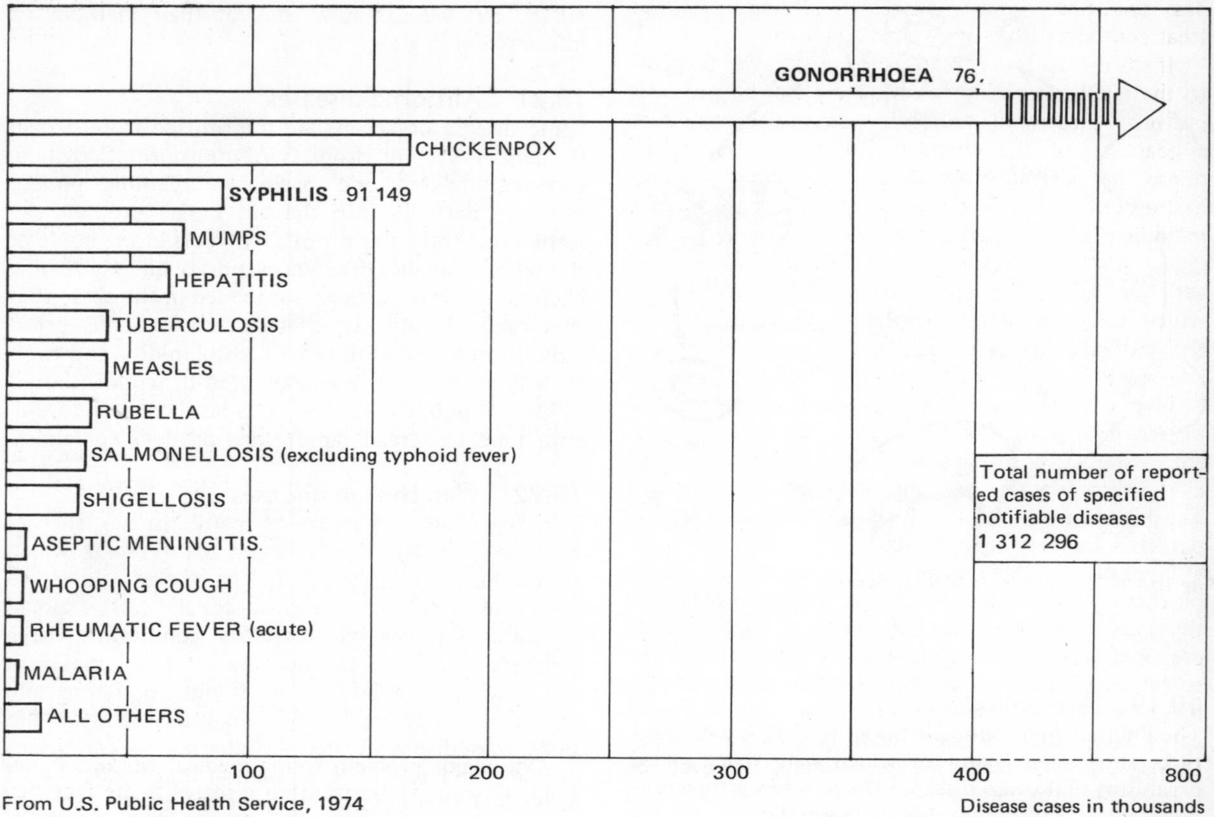

From U.S. Public Health Service, 1974

fig 10.20 The place of STD among the major communicable diseases

The organism that causes syphilis is a spirochaete bacterium called *Treponema pallidum.* It cannot live for very long outside the human body and, contrary to certain popular beliefs, the disease cannot usually be caught from such objects as toilet seats, infected clothing or drinking cups. Infection almost always occurs during sexual intercourse, but the spirochaetes can pass from the blood of a pregnant woman to that of her unborn child, across the placenta. In such cases, the baby may die before birth, or it may be born with *congenital syphilis,* resulting in blindness, deafness, heart disorders, mental or other serious handicaps. Syphilis can also be transmitted by transfusion of infected blood, and thus no syphilis patient is allowed to become a blood donor.

The course of the disease follows a series of three phases.

The first or *primary* phase is marked by the appearance of one or more painless sores or *chancres,* usually between ten and thirty days after infection. These sores may occur at any point on the surface of the body, but most often they appear on the genitalia. In women, chancres sometimes arise inside the vagina or on the cervix, and this primary phase of syphilis may

fig 10.21 A warning about venereal disease on a poster

pass unnoticed since chancres usually heal without treatment after little more than a month.

If the disease is not treated, however, it will progress to the highly infectious *secondary* phase at any time between three and fourteen weeks after the first appearance of the chancre. During this secondary phase the patient often shows signs and symptoms easily confused with those of several other diseases, including a slight fever, sore throat, swollen lymph glands and a general body rash. These symptoms are very variable and generally last about two years, after which time they may completely disappear, leaving the patient with the false impression that he has recovered from the disease itself.

Over the next several years, however, severe internal damage may take place in many of the body's organs and tissues. Eventually this damage becomes obvious as the third, *tertiary*, non-infectious phase of syphilis. The patient may suffer heart failure, blindness, insanity, paralysis and eventually death.

Syphilis is easily diagnosed by means of special biochemical tests on the blood. In some countries these tests are carried out as a matter of routine on all pregnant women, blood donors and people intending to get married. Once syphilis is confirmed, treatment by injections of penicillin produces a cure in most cases, particularly if the disease is at an early stage. Damage due to tertiary syphilis may, however, be completely irreversible.

In spite of the ease with which diagnosis and treatment are possible, syphilis is still a major problem throughout the world. Even in countries with well-established health services, such as the United Kingdom, there has been very little decrease in the numbers of syphilis cases over the past twenty years.

Even more alarming are the statistics for gonorrhoea, of which there are an estimated seventy million cases throughout the world each year. This disease, which has less drastic effects upon the individual than does syphilis, is caused by the bacterium *Neisseria gonorrhoeae*, a coccus type. Like *Treponema pallidum*, this organism cannot survive outside the human body and so infection almost always occurs during intercourse. Occasionally, a baby may catch the disease from its mother during birth, as it passes through the infected vagina.

The symptoms of gonorrhoea, which usually develop within a week of infection, are a 'burning' sensation on urinating, followed by a yellowish discharge. The symptoms are similar for both men and women, but in women they often go unnoticed, partly because any discharge may be masked by the normal vaginal secretions. The long-term consequences of the disease, however, are quite as serious for women as for men. The infection may spread from the urethra and vagina to the uterus and fallopian tubes, resulting in peritonitis and eventual sterility. In men, the most common complication is a narrowing of the urethra, leading to difficulty in urination.

Gonorrhoea is between ten and fifty times more common than syphilis. This means that the amounts of time and money spent by medical services on diagnosis and treatment are very great indeed. It has been estimated that, in the United States of America alone, the cost of treating complications of gonorrhoea in women is currently about fifty million dollars per year.

Diagnosis of gonorrhoea is much more difficult than diagnosis of syphilis. There is no simple blood test for the disease, and gonorrhoea can only be confirmed by identification of the bacteria from the discharges. Thus mass screening is not possible. Treatment is usually by means of antibiotics such as penicillin and by sulphonamide drugs. In spite of the effectiveness of this treatment, the incidence of gonorrhoea is increasing rapidly throughout the world. In the U.S.A., for example, it is at least five times more common than chicken pox, the second most common communicable disease.

Question

1 What explanations can you suggest to account for the great increase in the number of gonorrhoea cases reported over the past few years?

One major problem facing medical workers in the field of sexually transmitted diseases is the fact that very little natural immunity results from attacks of either syphilis or gonorrhoea. Thus many patients are constantly being re-infected. Not surprisingly, therefore, there has been very little success in the production of vaccines for S.T.D. An anti-syphilis vaccine has been developed which is effective to some degree in rabbits, but this has to be injected regularly over several months. It also produces undesirable side effects and thus is not yet suitable for use on humans.

A third sexually transmitted disease which still poses a serious health problem in many tropical countries is *chancroid* or 'soft sore'. This is caused by a bacillus, which produces painful ulcers on the external genital organs, with associated swellings and possibly abscesses in the groin. It is treated by the use of sulphonamides and tetracycline tablets.

Certain other conditions are also regarded as being spread primarily by sexual contact. Most common of these are forms of urethritis (inflammation of the urethra) caused by organisms other than gonococci. Many are caused by the rickettsia – like *Chlamydia*, others by the protozoan *Trichomonas*, and yet others are grouped together under the heading 'non-specific urethritis' or N.S.U. Finally, also included under the general heading of S.T.D. are such disorders as genital warts, pubic lice or 'crabs', and *Herpes genitalis* (a condition related to the cold sore).

10.26 Insect borne diseases

Insects can transmit disease organisms in two ways: (i) on the outsides of their bodies, and (ii) on the insides of their bodies.

i) By virtue of its feeding habits, the **housefly** (*Musca* spp.) is the most important vector of intestinal diseases for it often feeds on animal dung and human faeces. Micro-organisms present on this material will cling to its legs and body and, as the insect walks across food, they drop off. Furthermore, the fly feeds by passing out digestive juices from the gut. If its previous meal had been faeces, any bacteria present are vomited out onto the food. Any human eating the food can pick up the bacteria which have been left by the fly. The blowfly (*Calliphora* spp.) is similar to, but larger than, the housefly. It feeds on decaying meat or dead animals and so can easily be a vector of disease organims. The Cockroach (*Periplaneta* or *Blatta* spp.) is also a vector of disease, mostly intestinal, although it does not feed on human faeces.

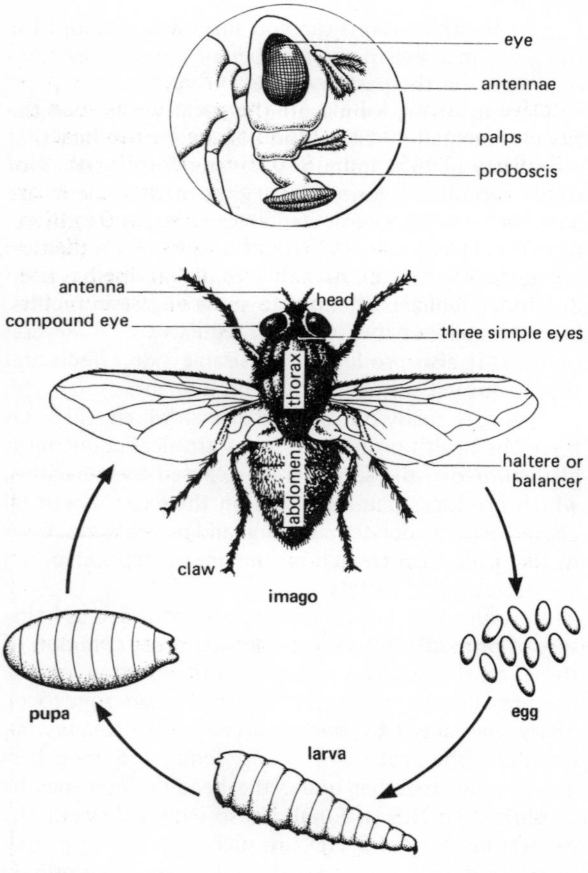

fig 10.22 The life cycle of the housefly

Investigation

To examine the life history of the blowfly (*Calliphora* spp.)

Procedure

1 Place a piece of meat or a dead animal in the shade and out of reach of scavengers (e.g. dogs – cover meat or animal with a piece of wire netting).

2 Examine the meat at regular intervals and take samples of eggs, larvae, pupae and adult insects that appear.

3 Place the specimens into specimen boxes or petri dishes and observe. Make drawings of the larvae and pupae.

Questions

1 How long do the eggs take to change into larvae?

2 How long do the larvae take to change into pupae?

3 What difference is there between the activity of the larvae and pupae?

4 What happens when the adult fly hatches from the pupa?

ii) Of the diseases carried by insects, malaria is probably the most important. About 500 million people are at risk, some 200 million have severe attacks and there is an annual death rate of 2 million people. The disease is caused by the protozoan *Plasmodium* transmitted by the *Anopheles* mosquito. It was in 1895 that a British doctor, **Manson**, showed that mosquitoes transmitted *filariasis*, and this led **Ross**, an Indian Army doctor, to search for a similar vector for malaria. In 1897 he found a malarial parasite in the stomach of a mosquito, and established the fact that the parasite goes through a developmental stage in this vector. This was the sexual phase of *Plasmodium;* the asexual phase takes place in the liver and blood cells of Man. Prevention of malaria involves attacking both the insect vector and the plasmodial parasite.

1 Control of the insect vector

a) The adult mosquito can be killed by insecticides, for example, *gamma B.H.C.* or *Dieldrin.* These are sprayed on the outer walls of buildings, under roof overhangs and inside huts and rooms. The insecticide remains effective for weeks or even months, killing any insect that lands on the sprayed surface.

b) The larvae are destroyed by spraying *oil* on the surface of stagnant water. The mosquitoes lay their eggs in lakes, ponds, swamps and in any man-made object such as tin cans, pots, barrels, drains and gutters that retains water. The latter should be emptied regularly or sprayed with oil. This substance reduces the surface tension of the water and the larvae sink below the surface where they are unable to obtain oxygen.

c) The adult mosquito can be prevented from reaching Man. The latter can protect himself at night

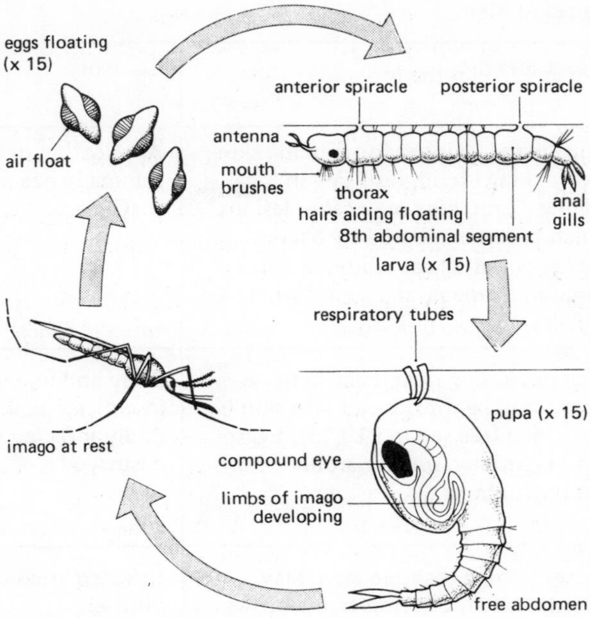

fig 10.23 The life cycle of the *Anopheles* mosquito

with mosquito nets over beds, or by fly gauze, carefully fixed over doors and windows.

2 Control of the Plasmodium parasite

a) The use of drugs to cure malarial patients by killing the parasites in the liver and the blood cells, has long been used. The patients are then no longer a source of further infection. *Quinine* was used first but later *Chloroquine* and *Mepacrine* have been developed to kill the asexual stages in the body of Man. *Paludrine* is more effective still and is taken daily, whereas the new drug *Daraprim* is only taken once per week.

b) These latter drugs are also effective as a preventative measure, killing off the parasites as soon as they are injected into the blood of the human host.

Between 1945 and 1970 three-quarters of the population of the world living in malarious areas were freed from the disease, but even so, 500 million are still liable to contract malaria. Constant vigilance is required, even in areas made free of the disease, and all efforts should be made to prevent it where the mosquito and its parasite are still endemic.

Cause	Disease	Course of the disease
Plasmodium vivax	Benign tertian malaria	Seldom fatal with attacks of fever every 48 hours
Plasmodium falciparum	Malignant tertian malaria	Severe disease and more often fatal with attacks of fever every 48 hours
Plasmodium ovale	Tertian malaria	Relatively mild attacks of fever every 48 hours
Plasmodium malariae	Quartian malaria	Mild disease compared with the others and with attacks every 72 hours

Table 10.6 Types of malaria

Table 10.7 Arthropod parasites of Man

Parasite and method of spread	Effect on host	Control
Head louse (*Pediculus humanus* var. *capitis*) Spread by direct contact.	Mouthparts adapted for piercing skin and sucking blood, resulting in local itching. Scratching may cause lesions which become infected by bacteria. Lice eggs ('nits') are white, about 1 mm in diameter and found stuck to hairs.	Application of an insecticide like gamma benzene hexachloride (gamma BHC) to hair after washing.
Body louse (*Pediculus humanus* var. *corporis*) Spread by direct contact, or by infected clothes	Typhus-causing rickettsiae in louse faeces may be introduced into skin by scratching (see section 10.15). Eggs laid in clothing, usually around seams and buttonholes.	Body and underclothing dusted with insecticide such as gamma BHC powder. Badly infected clothing may be destroyed.
Crab louse (pubic louse) (*Pthirus pubis*) Spread by direct contact (often during sexual intercourse, or, from lavatory seats.	Causes itching in pubic area. May migrate to hair of armpits, eyebrows or beard (but not to head hair). Strong attachment to hairs.	Infected areas dusted with gamma BHC powder.
Bed bug (*Cimex lectularius*) Spread by infested clothes, suitcases, secondhand furniture, or by the bugs crawling through crevices in walls between bedrooms.	Pierces skin and sucks blood. Lesions itch and may become secondarily infected through scratching. Feeding occurs at night, after which bugs return to hiding places in mattresses, wall cracks and furniture crevices. Bugs may survive several months on a single blood meal.	Possible hiding places sprayed with insecticide in solution. In serious cases, fumigation with very poisonous hydrogen cyanide gas, repeated after one or two weeks to kill nymphs hatched from resistant eggs after first fumigation.
Human flea (*Pulex irritans*) Spread by jumping from host to host.	Pierces skin and sucks blood. Local irritation and possible secondary infection of puncture sites.	Treatment of bites with antiseptic solution. Clothing, furniture, bedding and houses of infested persons sprayed with insecticide. Application of insect repellant (e.g. dimethyl phthallate) to clothing of persons exposed to possible infestation.
Mites and ticks (Acarina). Spread by direct contact, by other animals acting as vectors or by attacks from mobile stages in the life history.	The itch-mite (*Sarcoptes scabiei*) causes scabies (see section 10.19). Some acarines are important as vectors of diseases such as typhus, relapsing fever and Rocky Mountain spotted fever. Others, such as the harvest mite, simply cause local irritation. Still other tiny mites live harmlessly on the surface of the skin, unnoticed by the host and feeding on cell debris.	Wearing of protective clothing. Use of repellants such as dimethyl phthallate and diethyltoluamide. Application of calamine to bites.

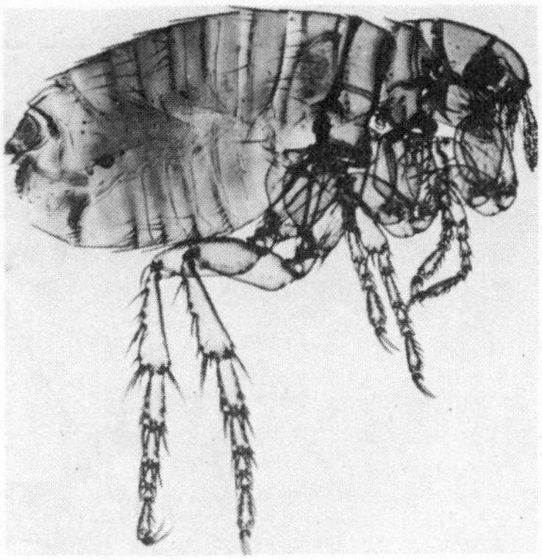

fig 10.24 The human flea

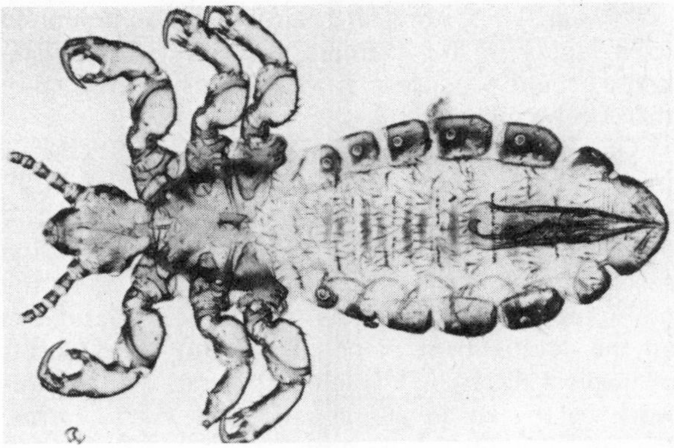

fig 10.25 The head louse

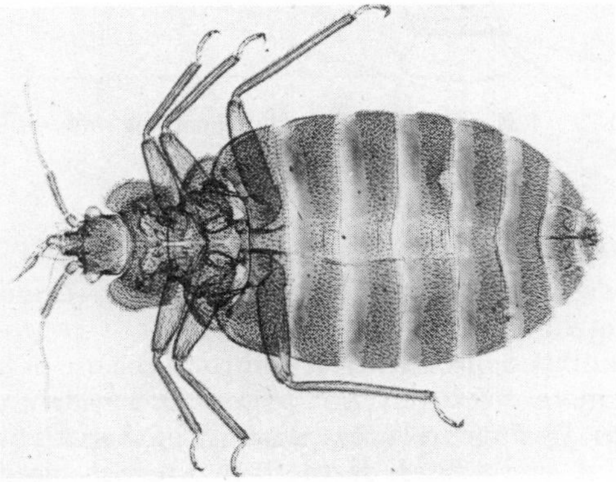

fig 10.26 The bed bug

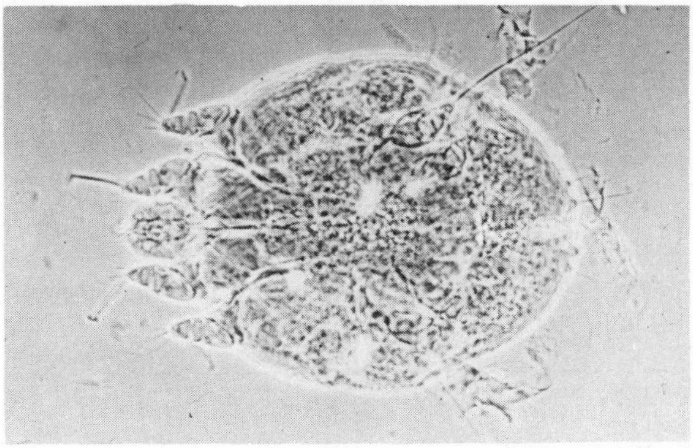

fig 10.27 The itch mite

fig 10.28 The spraying of exposed water surfaces against mosquito larvae

10.27 The rat

Of the many different types of rat that exist throughout the world, there are two that are of particular interest to us because of their effect on human health. They are the so-called black rat (*Rattus rattus*) and the brown or Norwegian rat (*Rattus norvegicus*). Both species have an almost world-wide distribution because of their association with Man. Their names are rather misleading in fact, since there is a considerable overlap between their ranges of colour. However, the brown rat is rather heavier than the black rat and has smaller ears. It is the more common rat in northern Europe and Asia, whereas the black rat is dominant in the tropics and, because it is carried on ships, in seaports throughout the world.

Rats are omnivorous and cause great damage to food supplies by eating grain and other stored foods, by contaminating these foods with their urine and faeces and even by attacking poultry. They also have an extremely high reproductive rate, a typical female starting to breed at the age of four to six months and producing four or five litters of six to ten offspring every year. Thus, a breeding pair, if given an unlimited food supply in, for example, a dockside food warehouse, could produce a thousand rats in little more than twelve months.

On the credit side, this high level of fecundity means that they are very useful for work in genetics. Also, the resemblance between their omnivorous diet and our own is in part a reflection of the fact that the general physiology of the rat is very similar to the physiology of Man. Thus rats are used for experiments on the development of new drugs and other aspects of medical research. The laboratory rats used for this work belong to an albino variety of *Rattus. rattus*, specially bred for this purpose.

The main importance of wild rats, however, concerns their role in the spread of disease. We have already mentioned bubonic plague and endemic typhus, which are carried by the rat flea. The rickettsiae that cause endemic typhus are also found in rat lice, which spread the disease within the rat population but do not bite humans. Another form of typhus, scrub typhus, is spread by certain rat mites.

Two important bacterial diseases are spread directly by rats: salmonellosis and Weil's disease. The former is perhaps the most common type of food poisoning and is caused by certain species of the bacterial genus *Salmonella*. The bacteria are often present in the alimentary canals of rats and the disease may arise in people who have eaten food contaminated by rat faeces.

Weil's disease is one of a group of diseases that come under the general heading of leptospirosis. They are caused by spirochaete bacteria of the genus *Leptospira*, which live in the kidneys of various mammals. Humans contract the disease usually from food or water that has been contaminated by rat urine. It is an occupational disease of sewage workers, and may cause serious liver and kidney damage. Another type of leptospirosis is transmitted to pet owners via dog urine.

The control of rats is a major problem in both rural and urban areas. The main reason for the success of rats as a group is their ability to adapt to changes in their environment. The black rat is descended from tree-dwelling ancestors and is a good climber. It can move from house to house along telephone wires and can board a ship by climbing steel cables. It is a patient gnawer and no wooden door or wall can be permanently rat-proof. Thus food stores should be inspected periodically and any damage repaired upon discovery. Many devices aimed at preventing rats from climbing into houses, stores or ships are based on the principle of an inverted funnel. Thus the rat is presented with a smooth surface at an angle of greater than 90°, which it finds impossible to scale. Such devices are found on ship's cables, on telephone wires and on the supporting pillars of buildings raised high off the ground. The outside steps of these buildings are often removable to prevent entry of rats.

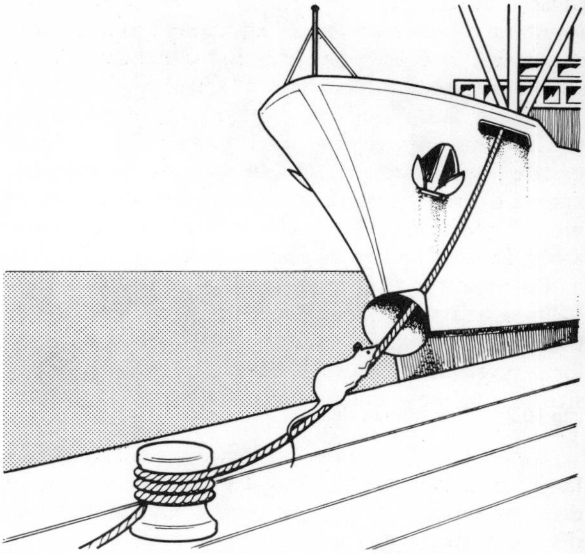

fig 10.29 Rat guard, to prevent rats boarding ships via cables

Another approach to rat control is the use of traps and, more effectively, rat poison. One widely used poison, Warfarin, reduces the ability of the rat's blood to clot and causes death from internal bleeding. Unfortunately, in recent years, certain strains of rat, adaptable as ever, have developed a resistance to Warfarin.

10.30 Prevention of disease

Our bodies are surrounded by micro-organisms and many can live on the surface of the skin, in our buccal cavity and in the alimentary canal. The remainder of our body, muscles, nervous system, glands and skeleton should normally be free of micro-organisms unless disease is present. How can we take action to prevent, or make it more difficult, for pathogenic micro-organisms to invade our organs?

Personal hygiene is very important. Washing with soap and water removes sweat, body oils and dirt which provide a fine breeding ground for bacteria. Clothes must also be kept clean by frequent washing with soap or detergents, both of which are lethal to bacteria.

The house and its surroundings can contribute towards disease of its occupants if attention is not paid to cleanliness. Food preparation areas must be kept clean and free from stale pieces of food. Household waste must be cleared away and *not* thrown on the ground in or around the house. Flies can breed in organic waste, and carry bacteria onto food so that *intestinal diseases* are spread e.g. cholera, typhoid and dysentery. If possible, waste food and other matter should be placed in a bin with a tight-fitting lid.

Human faeces and urine are an important link in the life histories of many different parasites. Inadequate disposal of sewage is responsible for the rapid spread of many intestinal diseases and a number of worm infections. See section 11.30 for domestic sanitation and sewage disposal.

Community hygiene plays an important part. Villages and towns have tended to grow slowly over the years in a rather haphazard fashion, but today local governments exercise control over the type and site of any new building. (See sections 11.11 and 11.12).

Water supply is a most important provision for any large or small community. The contamination of drinking water is one of the most frequent causes of disease. Water supply is covered in section 11.20.

Over 40% of all deaths in developing countries occur amongst children under five years of age. These are from a combination of malnutrition, parasitic infection, diarrhoeal disease and other infections such as pneumonia. The WHO diarrhoeal advisory team found that in many regions, only 20% of children were free from helminth (worm) infections. The very young in the developing countries carry such a high incidence of infection that many experience only a few days free of illness in the course of a year. Many diseases now confined to sub-tropical and tropical areas were once widespread in temperate regions. Europe has known malaria as well as hookworm, and leprosy used to be prevalent in Iceland and Scandinavia. The first stages in the eradication of these diseases in the developed world began with the gradual changes in human behaviour that led to the interruption of infection transmission. This began the process which resulted in their eventual disappearance.

10.31 Food preservation

All decay is due to the activity of micro-organisms, particularly bacteria and fungi. Food consumed by Man is susceptible to decay and , furthermore, his diseases are often transmitted by way of food. The proper handling, preparation, storage and cooking of food are all important in preventing wastage, and transmission of disease organisms. In addition to diseases such as cholera, typhoid and dysentery, other mild and sometimes severe illnesses come under the general title of *food poisoning.*

Precautions which serve to reduce the incidence of food poisoning

1 Storage of food in vermin-proof rooms avoids contamination by faeces of rats and mice.

2 Slaughter and evisceration of animals away from carcase meat or cooked food.

3 Regular disinfection of equipment in slaughter houses and butchers' shops.

4 Exclusion of any human 'carriers' of disease from slaughter houses and food shops.

5 Refrigeration of meat and fish, then thorough thawing before cooking.

6 Thorough roasting or pressure cooking to achieve high temperatures; many spores of bacteria are resistant to boiling for several hours.

7 Avoiding the reheating of food that has been standing in warm conditions.

8 Reducing the handling of food to a minimum and excluding food handlers with septic sores on any part of the body.

9 Education of food handlers in personal hygiene.

Food processing in relation to food hygiene

1 *Killing micro-organisms* This is possible by thermal processes such as boiling, roasting and pasteurisation. Cans of food are heated and then sealed. The food will remain unharmed unless the tin of food is pierced. Milk contains a wide variety of micro-organisms which cause souring within a short time unless the milk is pasteurised (kept at 72°C for 15 seconds then cooled rapidly to 10°C) and refrigerated.

2 *Prevention of growth of micro-organisms*

a) *Freezing* A great variety of food is marketed in frozen form and large quantities are shipped across the world in this state. Spore-bearing organisms are particularly resistant to freezing. Micro-organisms are not killed by freezing; they simply stop growing. Frozen food removed from the freezer will begin to thaw and bacterial growth commences immediately. Thus the food must be eaten as soon as possible.

b) *Dehydration* Dehydrated foods are useful since they are light in weight and keep for a long time. Water, which is essential for bacterial growth, has been removed. The foods can reabsorb water from the atmosphere so that they must be packed in air-tight containers (e.g. plastic bags).

3 *Inhibitors of the growth of micro-organisms*

a) *Acids* Lactic acid and acetic acid (vinegar) are used to preserve food. They inhibit growth of organisms which cannot stand an acid environment.

b) *Salt* Meat and fish can be kept for long periods without decay in salt since this stops bacterial growth and enzyme action in the tissues.

c) *Smoke* Meat products are cured by smoking, which dries the surface and also coats the meat with substances retarding the growth of micro-organisms.

d) *Sugar* This acts as a preservative when present in high concentrations. Water is not available for the

Causative agent	Source	Symptoms	Prevention of transmission
Salmonella spp. causing Salmonellosis	Many animals carry the disease organisms e.g. pigs, calves and poultry. Main source for Man is meat of these animals.	Within 12–14 hours, fever followed by vomiting and diarrhoea. Firm diagnosis needs laboratory tests on faeces. Rarely fatal.	Flies can transmit the bacteria which are excreted in the faeces, therefore environmental control is important. This is difficult to achieve on farms. Fish and shellfish can transmit *Salmonella* especially if in contact with sewage. Refrigeration, complete thawing and thorough cooking will contain and eventually kill bacteria.
Clostridium welchii causing clostridial food poisoning	Widely distributed in nature e.g. soil, sewage and water. Thus many possibilities of food contamination. Spores can survive several hours of boiling water. Many outbreaks traced to meat.	12–24 hours incubation; followed by fever and vomiting, abdominal pain and diarrhoea. Infection only lasts about 24 hours; rarely fatal.	Meat should be thoroughly roasted in small quantities, not more than 2–3 kg. Meat should be eaten after cooking, any remaining refrigerated within 1½ hours.
Clostridium botulinum causing Botulism	Anaerobic bacteria living where air is excluded, as in canned, potted or pickled food. The food generally has been treated but spore forms survive.	Rare disease, but a high mortality rate – 50% in reported cases. Takes 24 hours to develop with vomiting, muscle paralysis and constipation.	Adequate heating of food will destroy the spores. Growth of the organism can be inhibited by complete drying, refrigeration, thorough salting or reduction of pH i.e. acid conditions.

Table 10.7 Food poisoning

micro-organisms and so growth is inhibited. Both sugar and salt exert an osmotic action on the organisms. Honey and jam are examples which do not spoil readily.

4 *Radiation* The use of ionising radiation has increased since it has been shown to prevent spoilage and destroy disease-producing organisms. Small doses of radiation can also destroy animal parasites present in food. The *'shelf life'* of raw or cooked foods can be prolonged without changing the taste or appearance.

10.40 The treatment of disease

10.41 How do parasitic organisms cause disease?

This question is relatively easy to answer with regard to some parasites. Tapeworms, for example, deprive their hosts of food and thus produce some of the symptoms of malnutrition. Hookworms suck blood, causing anaemia as well as mechanical damage to the intestinal wall. *Plasmodium* damages red blood corpuscles. Fungi break down skin cells. (They are much more important pathogens of plants, where they use up food reserves and block conducting tissues). Viruses invade specific body cells, where they multiply and cause cell damage.

The most common causative agents of human disease, though, are bacteria. However, only a tiny fraction of all bacteria are pathogenic. Many are free-living in the air, water and soil, where they play important roles in the carbon and nitrogen cycles. Others live harmlessly on the surface of the human body and in the alimentary canal (human faeces are composed very largely of bacterial cells). However, certain of these bacteria can cause disease if introduced in sufficiently large numbers into other tissues of the body. For example, the bacterium that causes tetanus lives harmlessly in the intestine, but produces disease when it enters the bloodstream as a result of a cut being contaminated by soil containing faeces. Also, many of the unpleasant symptoms of a cold are caused, not by the cold virus itself, but by the bacteria that normally inhabit the nose and throat. These enter the mucous membranes after the cells of the protective surface layer have been damaged by viruses.

Other bacteria, such as the spirochaete that causes syphilis, are strictly pathogenic. They do not live harmlessly in or on the body, but must be transferred directly from one disease sufferer to another person.

The way in which bacteria actually cause disease is rather complex. It involves the production of *toxins* (poisonous organic chemicals) which affect the host's body in a variety of ways. Some toxins, such as those produced by the plague bacillus, damage mitochondria and thus affect cellular respiration. Other toxin effects include the breaking down of different kinds of blood cell, formation of plasma clots and digestion of the jelly-like substance that holds connective tissue cells together.

10.42 Destruction of micro-organisms

The discovery of micro-organisms by Pasteur, and their identification as causing disease, enabled an attack to be made upon them during surgical operations. Joseph Lister, in 1860, developed techniques to kill micro-organisms in wounds and on instruments by using carbolic acid. His work resulted in more successful operations, but the surgeons who followed, reduced their dependence on these antiseptics and concentrated on eliminating the organisms before the operation began. These *aseptic* practices involved sterilising instruments, gowns, gloves, masks and every other item in contact with a patient.

Sterilisation means the destruction or removal of *all* micro-organisms. *Disinfectants* and *antiseptics* are chemical substances that destroy micro-organisms, but generally the manner of their use means that they do not destroy all. Most of the harmful organisms are eliminated by these substances. Disinfectants are chemicals used on non-living surfaces such as crockery, cooking utensils, cutlery and drains, whereas antiseptics are used on a living surface such as skin.

Sterilisation is best carried out by *thermal treatment* using an *autoclave* or pressure cooker. These enable temperatures of 115°C to 135°C to be reached which will destroy bacteria and most spores. Radiation sterilisation is now used for medical and surgical instruments. Gamma radiation from radio-isotopes is a powerful penetrating form produced by an isotope of Cobalt (*Cobalt 60*).

Disinfectants are commonest in the liquid form, and their usefulness depends on the resistance of the bacteria and on other factors such as temperature and concentration of the chemical. *Hypochlorites* are readily available as calcium hypochlorite or sodium hypochlorite and their action depends on the formation of hypochlorous acid which liberates oxygen in a highly active stage. They were first used to reduce the incidence of childbirth fever, but are now commonly used in the home, laundries, dairies and the food industry in order to clean equipment. *Iodine* (dissolved with potassium iodide in 90% alcohol) is used as an antiseptic for superficial cuts and scratches to the skin. Phenol is rarely used as an antiseptic these days, but a chlorinated phenol (*Hexachlorophene*) is widely used as a skin antiseptic. *Chloroxylenol* (present in 'Dettol') is an antiseptic used to prevent sepsis during childbirth.

10.43 Natural defences of the body

The body is protected externally by the *skin* and *mucous membranes* that act as a barrier against disease organisms. The skin has a dead, horny layer difficult to penetrate, but it also possesses its own population of harmless bacteria together with certain chemical secretions. The respiratory tracts are continually exposed to atmospheric pollution, certain bacteria and viruses. These tubes are covered internally by a layer of mucus which is driven upwards by minute protoplasmic processes called *cilia.* Most of the airborne particles are trapped in the nose or in the mucus of the tracts, and are destroyed by anti-bacterial enzymes present. The mouth, buccal cavity, alimentary canal, vagina and the surfaces of the eye have their own defence mechanisms. The *saliva* has a cleansing action, but, if any bacteria are swallowed, then the *acid* of the stomach will kill them. The acid conditions of the vagina inhibit the growth of pathogenic bacteria. The *tears* secreted across the surface of the eye also destroy bacteria.

Internally, living tissue reacts to injury in a complex way, although this often begins as a simple local reaction to damage. This is called *inflammation* and is recognisable when our skin is burned, cut or infected with pathogenic bacteria. The skin becomes reddened, warm, painful and often swollen. This is due to the dilation of the small blood vessels increasing the supply of blood. As a result, the blood plasma oozes from the capillary walls to accumulate in the tissues. In addition, white blood cells squeeze through the capillary walls and are present in the tissue fluid.

Thus the tissue receives any anti-bacterial substances that are in the plasma while the white blood cells are able to engulf bacteria and cell debris. The *polymorph leucocytes* are the first to begin the scavenging process and are followed by *monocytes.* At the same time, the escaping plasma contains *fibrinogen* which forms blood clots helping to unite the damaged tissues and form a barrier to the movement of bacteria. If the polymorphs can overcome the local invasion of pathogens then the tissue gradually returns to normal. Tissue fluid is absorbed back into the blood stream, fibrin clots break down and the blood flow decreases.

The dead tissues are softened by protein-digesting enzymes from the polymorphs and the resulting fluid, called *pus*, accumulates beneath the skin. This is released by bursting through the surface of the skin accompanied by dramatic easing of pain in the tissues, as in the case of a boil or abscess.

10.44 Immunity and vaccination

The mechanisms described above may be regarded as the body's first line of defence against injury and/or invasion by pathogenic micro-organisms. What happens, however, if disease-causing bacteria or viruses succeed in penetrating this defence and entering the bloodstream in fairly large numbers? The answer depends very largely on the individual concerned. In some cases, after an *incubation period* during which the micro-organisms multiply rapidly, the patient begins to show the signs and symptoms of the disease. In other cases, he or she will show no ill effects: such people are said to be *immune* to the disease.

The way in which this immunity works is as follows. Certain large 'foreign' organic molecules (mainly proteins but also some carbohydrates), when introduced into the human body, cause the white blood cells to produce substances called *antibodies*. The type of white blood cell that produces antibodies is the *lymphocyte*, and the function of antibodies is to act specifically against one or other of the foreign organic molecules (known as *antigens*). Viruses and bacteria are souces of antigens, as are organ transplants and skin grafts from other people.

Antibodies act in a variety of ways: some are antitoxins, which 'neutralise' the toxins produced by bacteria; others cause the chemical breakdown of the invading micro-organisms; still others cause bacteria to clump together, thus preventing them from reproducing and enabling other white blood cells (*phagocytes*) to ingest them. It is the possession of antibodies of the appropriate kind that produces immunity to a particular disease.

When a person catches a disease there follows a 'battle' between the invading pathogens and the defence mechanisms of the body. If the pathogens win, the patient dies: if the defences win, the patient recovers and the antibodies produced during the course of the disease remain in the bloodstream, giving immunity from immediate further attacks. This immunity may last for a few days only or for a lifetime, depending upon how long the lymphocytes retain the ability to produce the antibodies.

Some fortunate people inherit a *natural immunity* to certain diseases such as typhoid. They possess the appropriate antibodies, or the ability to produce them readily, without ever having had the illness.

A third type of immunity is acquired artificially by *vaccination*. This involves injecting into the bloodstream a *vaccine* consisting of antigens that will stimulate the production of antibodies without causing the disease itself. Several types of vaccine are in use, including mild strains of pathogenic micro-organisms, pathogens killed with formalin, pathogens that have been *attenuated* or weakened (by culturing them outside the human body for example) and cell-free extracts of bacterial toxins that have been treated chemically to make them harmless (when they are referred to as *toxoids*).

All the above methods produce immunity described as *active*, because they result in the production of antibodies by the patient himself. *Passive immunity*, on the other hand, is the term used to describe the situation where a patient receives antibodies produced by another person or animal. In these cases, small quantities of serum from a person who has recently recovered from a disease (or from an animal, such as a horse, which has been given the appropriate antigens) is injected directly into the bloodstream. Immunity obtained in this way is relatively short-lived, as the antibodies are gradually excreted. A similar kind of passive immunity is received by babies from their mothers, as certain antibodies can pass across the placenta and others are contained in the first secretions of the mammary glands.

Questions

1 One of the major problems with heart transplant surgery is the fact that the donated heart is often 'rejected' by the patient. Why should this be so?

2 Natural immunity to a disease may be very desirable from the point of view of the individual, but may be dangerous to the community as a whole. Can you explain this?

3 On which occasions is passive immunisation likely to be a useful procedure?

	Active (involves production of antibodies by patient's lymphocytes)	*Passive* (involves use of antibodies produced by another animal)
Natural	(i) inherited (ii) acquired on recovery from disease	antibodies received via placenta or mammary glands
Artificial	from vaccination	from injection of serum containing antibodies

Table 10.8 Punnett square, showing types of immunity

Disease	Type of vaccine	Age for immunisation	Age for booster doses
Diphtheria	toxoid	given in	
Tetanus	toxoid	3 doses	5 years } 17 years
Whooping cough	killed bacteria	at 4, 6 and	
Poliomyelitis	attenuated virus (given by mouth)	12 months	17 years
Measles	attenuated virus	18 months	none – lifelong immunity
German measles (Rubella)	attenuated virus	13 years (girls only)	none – lifelong immunity
Tuberculosis	attenuated bacteria	14 years (after testing to show no natural immunity)	none – lifelong immunity

Table 10.9 Typical immunisation schedule for children

The history of immunisation

Perhaps the most important single achievement of the World Health Organisation has been the virtual elimination of smallpox from the face of the earth by a carefully planned programme of vaccination. This can be looked upon as the successful climax to several hundred years of medical research and development, since the very first recorded attempts to produce immunity artificially were directed against the smallpox epidemics that occurred in ancient India and China.

The technique used was called *variolation* (variola is another name for smallpox) and consisted of introducing matter obtained from smallpox sores into scratches made on the skin of a healthy person. The idea was that this would produce a mild form of the disease and thus protect the patient from further, more serious, attacks. Variolation was first brought to England in the early eighteenth century and it succeeded in reducing the death rate from smallpox quite considerably, although it was by no means safe and produced fatal attacks in several people.

At about the same time, it was commonly believed by country people that milkmaids were in some way protected against the worst effects of smallpox. Some recognised that this was more than just another 'old wives' tale' and might be related to the fact that many people who worked with cows caught a fairly mild disease known as *cowpox*.

At the end of that century, a country doctor, Edward Jenner of Gloucestershire, took this idea an important stage further by combining it with the technique of variolation. He took some matter from a milkmaid's cowpox sore and introduced it into a scratch made on the arm of a young boy. The scratch became inflamed and a sore developed on the arm but this soon healed. Six weeks later, matter from a smallpox sore was scratched into the boy's skin, but

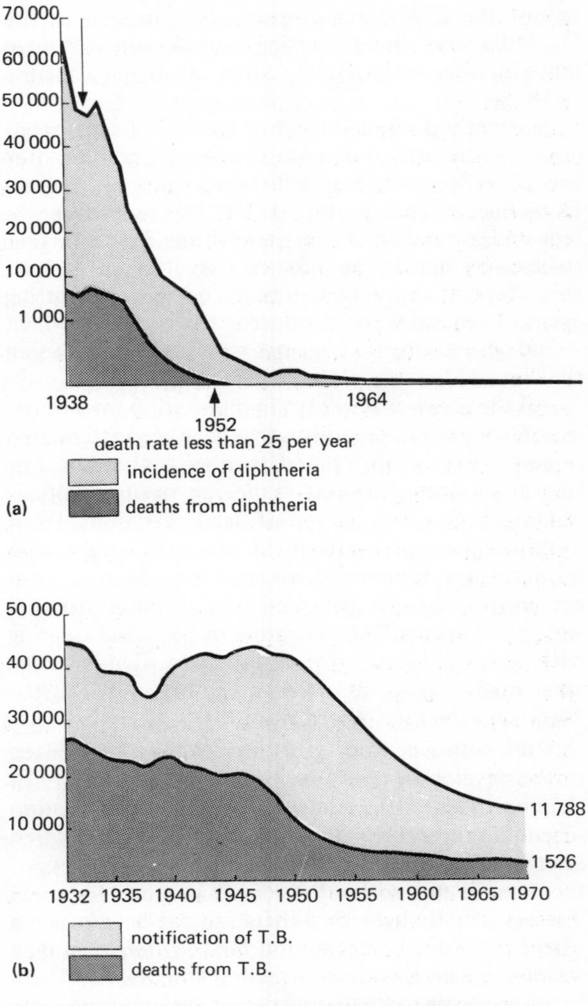

fig 10.30 Graphs showing decrease in (a) diphtheria (b) tuberculosis

he remained perfectly well. This method of producing immunity was called *vaccination,* after the Latin word for cow (vacca).

The next major step in the development of immunisation techniques was made about seventy years later by Louis Pasteur (see sections 10.00 and 10.12), who put forward the germ theory of disease. He discovered that more or less complete immunity to certain diseases was obtainable by injecting micro-organisms whose *virulence* (ability to produce disease symptoms) had been considerably reduced.

This reduction in virulence was achieved in several ways. Fowl cholera bacilli became attenuated (less virulent) after having been cultured in the laboratory for long periods. Anthrax bacilli were unaffected by such treatment, but lost their disease-producing effects if kept for several days at a temperature 5°C above that of the normal mammalian body. Most important of all, Pasteur succeeded in producing vaccine effective against the killer virus disease *rabies,* in spite of the fact that the disease-causing micro-organisms were quite invisible to him under the microscopes available in 1885.

His method was to inject brain tissue from a rabid dog into the central nervous system of a rabbit. After a while, the rabbit was killed and some of its brain tissue injected into another rabbit. This procedure was repeated a number of times and virulence was further reduced by drying the infected tissue for up to fourteen days at room temperature, using a desiccating agent. Eventually, he produced a living virus which would give protection against rabies without causing the disease.

All the above examples of immunisation involve the use of living micro-organisms that have been attenuated in some way or are related to a virulent strain. For certain diseases, however, a degree of immunity is obtainable by the injection of dead pathogens. Work in this field was pioneered by Sir Almroth Wright, who produced a vaccine effective against typhoid fever at the end of the nineteenth century. He used bacteria which had been killed by heating to 53°C and treating with a weak solution of disinfectant. This vaccine saved many thousands of lives in the Boer War (1899–1902) and the First World War (1914–18).

During this century great strides have been made in the development of vaccines for a wide variety of diseases. Table 10.9 shows a typical immunisation schedule currently in use for protection of children against serious diseases. Needless to say, progress of this kind is not without its drawbacks and for some diseases it is thought that the stage has been reached where the risks of vaccination *for the individual* may outweigh the risks of contracting the disease. There have been some particularly tragic cases in recent years involving brain damage to children as a result of vaccination against whooping cough. Nevertheless, anyone who has witnessed the effects of this most distressing and dangerous disease on a young child would surely think twice before advising parents against vaccination, especially in view of the recent increases in the numbers of whooping cough cases reported.

10.45 Treatment of diseases caused by micro-organisms

So far we have been concerned mainly with the *prevention* of disease by immunisation. Now we shall look at what can be done once a person has actually caught a disease.

For bacterial diseases, the problem can be tackled in three ways: by neutralising the bacterial toxins, by preventing the bacteria from multiplying and by killing the bacteria. We have already seen that cases of diphtheria can be treated by inoculation with horse serum containing specific anti-toxins.

The second and third types of treatment involve the use of drugs, or *chemotherapy.* Drugs obtained from naturally-occurring substances have been part of folk-medicine for thousands of years, but modern chemotherapy began in the mid-1930s with the discovery of a group of synthetic compounds called *sulphonamides.* These were found to be effective against a range of bacterial diseases, including pneumonia, meningitis and gonorrhoea. The sulphonamides acted by preventing bacteria from growing and multiplying and thus were known as *bacteriostatic* drugs. After the Second World War, the use of sulphonamides declined, partly because of an increase in the number of sulphonamide-resistant strains of bacteria and partly because of the availability of *bactericidal* drugs, which acted more quickly by killing the bacteria. More recently sulphonamides have been used in conjunction with trimethoprim, another synthetic, bacteriostatic drug, since the combination of the two actually produces a bactericidal effect.

Antibiotics are a major group of drugs which are obtained from extracts of certain fungi and bacteria. Many of them are bactericidal in action. The special properties of these drugs were discovered in 1928 by Sir Alexander Fleming, who found that an extract (called Penicillin) of the mould fungus *Penicillium* attacked the bacterium *Staphylococcus,* but antibiotics did not become generally available until the 1940s. Penicillin has the two major advantages of being (i) harmless to humans even in large doses and (ii) effective against a wide range of diseases, from boils and sore throats to syphilis and gangrene.

Other antibiotics, such as *tetracyclines* and *chloramphenicol,* may be used against certain rickettsiae and even some of the larger viruses, while *griseofulvin* is an example of an antibiotic that acts against fungi. During recent years it has become possible to prepare a wide range of antibiotics synthetically in the laboratory.

Question 1

A patient suffering from a bacterial intestinal infection was given an antibiotic medicine and told to take 5 cm^3, three times per day, for two weeks. As he disliked taking drugs, he thought that he would be cautious and took only two doses per day. Instead of recovering, he became even more unwell. Can you explain this?

Diseases caused by viruses present particularly difficult problems in respect of chemotherapy. It is true that certain of the larger viruses are destroyed by antibiotics but most are so intimately connected with the human host cells in which they multiply that a drug affecting the virus would also damage the host cell. Thus one of the most successful methods of treating viral disease is by the use of gamma-globulin extracted from human blood. Gamma-globulin is a concentrated solution of antibodies and so will attack the disease-causing viruses and give short-term immunity, provided that the donor of the gamma-globulin has recovered from the disease concerned. This technique is used extensively in the treatment of pregnant women who have been exposed to German measles (rubella), because of the high risk of damage to the child. Unfortunately, human gamma-globulin is very expensive to produce.

Question 2

Why does gamma-globulin give only short-term immunity?

Finally, certain viral diseases may be treated with a substance called *interferon,* a protein which was discovered in 1957 and which prevents the multiplication of viruses. Unfortunately, interferon is species-specific, that is to say it is effective only in the species in which it is produced. Production of human interferon is extremely costly and so research is being directed towards producing substances which will stimulate human bodies to produce more of their own interferon.

10.50 Cancer

In Chapter 9, we looked at cell division. This is a process that occurs more or less continuously in most of the body tissues in order to replace cells as they become damaged or worn out. The rate at which cells divide depends upon the rate of breakdown of old cells. Too fast a rate of cell division is prevented by a phenomenon known as *contact inhibition,* which is really a system built into the body to prevent overcrowding – cells in too-close contact with other cells do not divide. Occasionally, this system breaks down and a group of cells continues to divide in spite of close contact with surrounding cells. This leads to a swelling or *tumour.*

There are many different kinds of tumour, but for our purposes we can divide them into two groups, *benign* (harmless) and *malignant* (harmful). Benign tumours are little more than an inconvenience. At a certain point they stop growing and they will not recur if removed surgically. Usually, such an operation is necessary only if the tumour is unsightly or if it causes discomfort by pressing on other organs.

Malignant tumours are referred to in everyday language as *cancers.* They are capable of invading surrounding tissues, and cancer cells may spread by means of the blood and lymphatic systems to other parts of the body, where secondary tumours may become established.

Cancer, along with cardiovascular disease and accidents, is one of the major causes of death in the developed world. Generally, it is a disease of old age, although young and middle-aged people tend to be affected by particular forms such as leukemia and breast cancer.

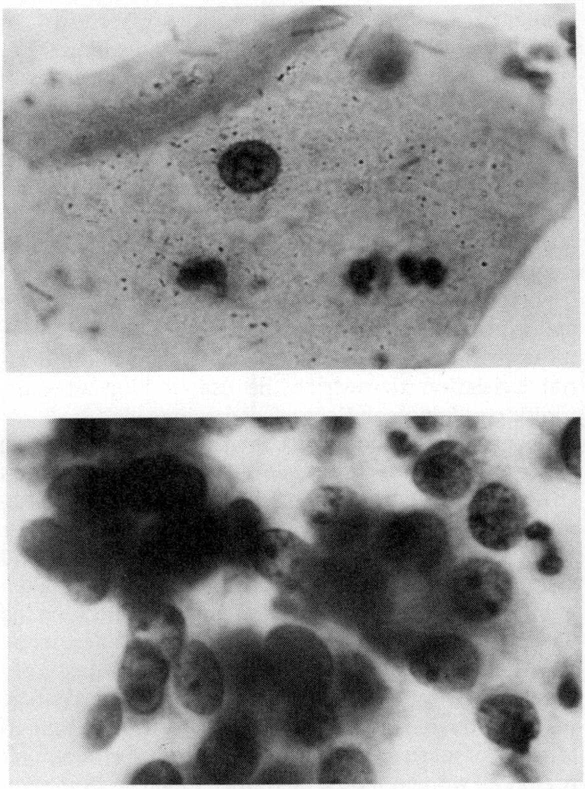

fig 10.31
(a) Cervical smear showing healthy cells
(b) Cervical smear with cells showing a cancerous condition

Càuse	Men (%)	Women (%)
Lung cancer	9	2.3
Other cancer	13	17
Heart attack	30	23
Stroke	10	17
Respiratory disease	14	12
Childbirth/pregnancy	–	0.03
Road accident	1.7	0.8
Other accident	1.7	2.0
Suicide	0.8	0.6
War/murder	0.1	0.06

Table 10.10 Causes of death in the United Kingdom

In relatively rare cases, such as some forms of intestinal cancer or cancer of the breast, there is a tendency for predisposition to the disease to be inherited. Other cancers are caused by a wide variety of agents, known as *carcinogens.* There is now no doubt that cigarette smoke contains carcinogens and that certain types of asbestos fibre can have a carcinogenic effect. Ionising radiations (including X-rays) can also cause cancer, which is why hospitals are careful to limit their use of radiography for any one patient, particularily in the case of pregnant women, where the foetus may be harmed.

Efforts to control cancer are concentrated on three areas: prevention, screening and treatment. Preventive measures include anti-smoking campaigns and industrial legislation to restrict the use of blue asbestos, ionising radiation and chemical carcinogens.

Considerable importance is attached to diagnosing cancer at an early stage, since this greatly increases the chances of curing the disease. Mass screening techniques have been particularly successful against cervical cancer, where women are encouraged to attend every three years a clinic at which a smear is taken painlessly from the epithelial tissue at the neck of the womb. The cells are examined under the microscope in order to check for a 'pre-cancerous' state. Early detection of cancer in this way gives a high percentage of cures by means of a straightforward operation to remove or destroy the affected tissue. For a similar reason, all women should be trained in self-examination of the breasts. Most lumps detected in this way are in fact benign, but in all cases a *biopsy* should be performed. This involves a minor operation to remove a piece of tissue and inspect it microscopically for signs of cancer.

Treatment of cancer may be by *surgery* to remove the tumour, by *radiotherapy* to destroy the cancerous cells, by *chemotherapy* with drugs, or by combinations of all three.

10.60 Drug abuse

In Chapter 7 we touched briefly on the subject of mood-influencing drugs. The abuse of these substances is an increasing problem in almost all modern societies. They can be classified conveniently under the following five headings:

a) Domestic, e.g. caffeine, nicotine and alcohol
b) Hallucinogenic, e.g. cannabis and LSD
c) Barbiturates, e.g. seconal, nembutal and phenobarbitone
d) Stimulants, e.g. amphetamines and cocaine
e) Opiates, e.g. morphine and heroin

10.61 Addiction

The pattern of addiction is surprisingly uniform for almost the whole range of addictive drugs. It might be useful, therefore, to consider the way in which the all-too-familiar addiction to nicotine develops.

The first attempts to inhale cigarette smoke may produce a *pleasant feeling* of light-headedness and relaxation (once the urge to cough has been controlled). This leads to *repetition* and the smoking of several cigarettes per day. Soon a *tolerance* to nicotine develops and more cigarettes are needed in order to obtain the same effect. The next stage is *withdrawal,* when the smoker becomes irritable and nervous when not smoking and needs a cigarette in order to achieve a feeling of normality. Heavy smokers do not enjoy cigarettes after the first two or three puffs: smoking is simply a habit that keeps them from feeling worse than everyone else.

The same is true of almost all drug addiction. A crucial stage is reached when the drug is taken not to give pleasure but to give relief from the symptoms of withdrawal.

10.62 Domestic drugs

Caffeine is the drug found in coffee and to a lesser extent in tea. It tends to give a slight 'lift' by stimulating mental activity, possibly by acting on the synapses in the brain. Addiction to caffeine is rare, but excessive coffee drinking can cause heart and kidney damage.

Addiction to nicotine (the drug in tobacco) is of course very common. The effect of nicotine is to stimulate the sympathetic nervous system. In past years cigarette smoking has been socially acceptable on the grounds that it does not cause intoxication and is harmful only to the user. More recently however serious diseases such as lung cancer and cardiovascular disorders have been linked to the coal tar and other compounds found in tobacco (see Chapter 3). This has led to educational and publicity campaigns directed against the habit.

One of the major social problems in almost all developed countries is alcoholism – the addiction to alcohol, in particular to ethyl alcohol or ethanol. This is formed by the fermentation of sugars by yeast, a unicellular fungus. It is drunk in the form of beers, wines and spirits. Alcohol acts on the brain to depress the functions of the cerebral cortex, leading to clumsiness, giddiness and slurring of speech – the well-known symptoms of drunkenness or alcoholic intoxication. Heavy drinking may also cause severe liver damage.

Alcoholics usually start drinking heavily because of a severe personality problem since the alcohol produces a feeling of well-being or *euphoria* which makes the problem seem unimportant for a while. Alcoholism cannot be cured since a single drink, even after a long period of abstainance, can lead to a long bout of excessive drinking. 'Alcoholics Anonymous' is a world-wide charitable organisation devoted to encouraging people to overcome alcoholism by self-help. It is generally agreed that true reform is possible only by solving the problem that caused the addiction in the first place. If this should be, for example, the death of a wife or the loss of a job, chances are improved by remarriage and new employment, respectively. Reform is much less likely if the original problem is of a permanent nature.

10.63 Hallucinogenic drugs

Cannabis (also known as marihuana, hashish, pot, grass, etc.) is obtained from the yellow, sticky resin covering the female flowers and topmost leaves of the hemp plant *Cannabis sativa.* Its use is widespread in Asia, Africa, North America and the Caribbean area. It is usually smoked in hand-made cigarettes, but it may be eaten or drunk. In most countries it is illegal.

The action of cannabis resin on the brain is not fully understood, although it certainly produces both euphoria and hallucinations ('waking dreams', often with distorted images and vivid colours). Reports of damage due to cannabis smoking are controversial, and include references to brain damage, chromosome damage, personality disorders and male impotence. The drug is probably not addictive, but illicit suppliers may encourage transfer to heroin, which they also supply.

LSD (D-lysergic acid diethylamide) is a very powerful synthetic drug. A tiny drop taken on a sugar lump is sufficient to produce the most vivid hallucinations and a number of suicides have resulted from the exaggerated emotions that it triggers off. LSD has been used under very carefully controlled conditions by psychoanalysts, but there is a fear that it may cause permanent personality changes.

Another hallucinogenic drug is mescaline, which is obtained from certain cactus plants. It has been used for centuries in the religious ceremonies of North and Central American Indians.

10.64 Barbiturates

This is a group of *depressant* drugs, including phenobarbitone, seconal, nembutal, amytal and luminal. They are used extensively as sleeping tablets and for the treatment of epilepsy. Their intended effect is to produce drowsiness, but if they are taken under conditions which normally induce wakefulness, for example at a party, they cause rapid intoxication. Addiction to barbiturates, an increasing problem among elderly people, is very dangerous, since withdrawal may cause convulsions and even death.

10.65 Stimulants

The stimulant drug cocaine is obtained from the leaves of the coca tree (*Erythroxylum coca*). It is not usually considered to be addictive, but overdoses can cause death from respiratory failure. For centuries Peruvian Indians have chewed coca leaves to ease the fatigue brought on by long and heavy work, while those who abuse the drug may sniff it as a powder to induce a short-lived 'high' feeling.

Amphetamines such as benzedrine are used to counteract hunger in people who are trying to lose weight, and also to prevent fatigue in soldiers on active service. Its effect is to increase the rates of breathing and heartbeat. They are particularly widely used in developed countries.

10.66 Opiates

Opium is a product of the poppy *Papaver somniferum.* It has been used for centuries in Asia, both medically and socially. It contains the pain-killers morphine and codeine. Morphine has been used for over 150 years to ease pain and to suppress coughing and diarrhoea.

The drug heroin is in turn derived from morphine and presents the greatest single addiction problem in the developed world. The recovery rate from heroin addiction is very low and the death rate high, not only from organic damage, but from overdose, hepatitis caused by dirty hypodermic needles, venereal disease and tetanus. Obsession with obtaining and using the drug leads to inadequate diet and lack of hygiene. The sheer expense of obtaining illegal heroin supplies often leads to crime and violent deaths among addicts are common.

10.70 Accidents

Accidents are the most common cause of death in people between the ages of one and forty-five. Every year in the United Kingdom there are more than 20 000 fatal accidents, and these account for about 5% of all deaths. In developing countries the accident statistics are somewhat less striking, partly because of the higher infant mortality due to malnutrition and tropical diseases, and partly because there are not so many mechanical, chemical and electrical hazards as in developed countries. Nevertheless, the situation is rapidly worsening in Africa and Asia, where rapid change from a rural to an urban society (or from an agricultural to an industrial one) tends to increase the problem.

In the United Kingdom, about one third of all accidental deaths occur in the home and another third on the roads.

The main causes of accidental death in the home are falls, burns, scalds, poisons, suffocation, electrocution and hypothermia (low body temperature).

Questions

1 Which types of domestic accident are likely to affect old people particularly, and what can be done to prevent them?

2 Which types of domestic accident are likely to affect children particularly, and what can be done to prevent them?

Since 1976, all places of work in the United Kingdom have been subject to the Health and Safety at Work Act. This places responsibility on the employer to make special provision for the safety of employees, and for their training in matters related to safety. Each school may have a member of the teaching or ancillary staff appointed as a 'safety representative' and there may also be School Safety Committees.

fig 10.32 All medicines should be kept out of reach of small children

Table 10.11 shows the places in a school where accidents are likely to occur.

Place of occurrence	Number of pupils per million involved in accidents
Indoor	
Classroom	328
Laboratory	35
Handicraft room	61
Domestic science room	14
Hall (excluding P.E. sessions)	88
Gym (or hall used for P.E.)	916
Corridor	122
Staircase	92
Cloakroom	111
Toilets	41
Other indoor	62
Outdoor	
Playground	2 098
Playing field	1 009
Swimming bath	16
School garden	39
Other school grounds	86
School steps	82
School journey	14
Elsewhere	101

Table 10.11 Accidents in schools

fig 10.33 Many serious accidents result from falls

Most schools are also concerned with the education of children in two other vital areas: road safety and swimming. The Royal Society for the Prevention of Accidents produces a termly publication called Safety Education which gives information on a wide variety of subjects, including the Green Cross Code, the handling of pets, cycling proficiency and lifesaving.

10.80 Mental health

In many developed countries, more than 50% of all hospital beds are occupied by people suffering from some kind of psychiatric disorder. In developing countries, the problem may be equally severe, particularly where there has been a rapid and large-scale movement of population from rural areas to the cities.

Classification of mental disorders is extremely difficult, and no scheme is without its exceptions, but for our purposes they can be considered conveniently under three main headings:

a) Congenital handicaps, e.g. mongolism, brain damage during birth
b) Senility – the feeble-mindedness of old age
c) 'Control disorders' – neuroses and psychoses

Congenital mental handicap may be due to chromosomal abnormalities (as with Down's syndrome), to inherited disorders (such as phenylketonuria), to hormonal imbalance (as in cretinism) or to brain damage during pregnancy or birth. Common causes of brain damage are lack of oxygen (anoxia) because of interruption to the foetal blood supply, rubella (German measles) during the first weeks of pregnancy, and the use of forceps during difficult deliveries.

As medical science advances, an increasing number of people survive into old age. In some cases the brain begins to deteriorate before the other organs, and elderly people who are 'sound in wind and limb' may become feeble-minded and incapable of looking after themselves, often because of damage to the brain's blood supply.

The third type of mental disorder includes neuroses and psychoses. Typically these may affect people of normal intelligence who are unable to exercise sufficient control over quite usual mental states. For example, everyone feels depressed from time to time, and most of us experience fantasies or daydreams, where we imagine what it is like to be exceptionally rich, famous or talented. However, we are usually able to control these thoughts: we try to cheer ourselves up if we feel depressed, and we snap out of our daydreams and get back to the everyday tasks of homework, cooking or earning a living. Unfortunately, some people are unable to exercise this control and are termed *neurotic* if their problem is relatively mild or *psychotic* if it is more severe, the difference between the two states being one of degree.

Of course, this classification is a great over-simplification and many cases overlap two or all three of the categories. For example, brain damage either at birth or in later life may cause severe psychoses, and psychotics may also become senile.

Common neuroses include depression, over-anxiety, claustrophobia (fear of enclosed spaces), agoraphobia (fear of open spaces) and compulsive cleanliness. Some neuroses lead to physical illnesses (termed *psychosomatic*) such as migraine, asthma and peptic ulcers. The mechanisms by which psychosomatic illnesses arise are not always understood, but one fairly common example arises from *hyperventilation* (over-breathing). A person suffering from anxiety may, without realising it, begin to breathe very deeply and quickly, often emitting loud sighs. This has the effect of increasing the oxygen concentration of the blood and lowering the carbon dioxide concentration. This in turn produces involuntary muscular spasms, causing severe pains in the chest and limbs. Spasms of this kind in the walls of the blood vessels in the brain may be one cause of migraine headaches.

Psychoses usually represent more serious lack of control and are correspondingly more difficult to treat. They include *schizophrenia, manic-depressive illnesses* and *psychopathic* personality. In schizophrenia, which accounts for about half of all cases of severe mental illness, the patient tends to lose contact with reality and suffer strong hallucinations. Manic-depressives swing in mood between extremes of depression and elation. The behaviour of psychopaths on the other hand is governed almost entirely by impulse. This leads to a great deal of criminality and anti-social behaviour without any trace of remorse.

Mental health is very difficult to define, since individual, quite normal personalities differ greatly from each other. However, it is generally accepted that a mentally healthy person should have sufficient intelligence and control over his emotions to care adequately for himself and his family, and to make a positive contribution to society.

The importance of a happy and stable family to good mental health cannot be overemphasised. Many of life's difficulties are made bearable by the support of relatives and close friends. Conversely, loneliness is one of the chief causes of mental illness. When a family breaks up, in addition to loneliness there is often an increase in financial problems. These in turn lead to malnutrition and inadequate housing with overcrowding, noise and lack of hygiene combining to make life very difficult.

People faced with these problems often suffer from depression and anxiety neuroses, which they may try to ease by resorting to alcohol or drugs (see section 10.60). Others may turn to crime or prostitution in an attempt to escape from their environment.

fig 10.34 Many handicapped people can do useful work in 'sheltered employment'

In past years it was the practice to isolate mentally ill and handicapped people in institutions. This tended to lessen the chances of recovery, since patients became 'institutionalised' and less able to cope with the outside world. More recently the emphasis has changed to community-based care, with treatment at day centres for less severe cases and support for the families of patients. Above all there has been increasing recognition of the fact that all of us may suffer from mental illness at some time in our lives, and that mental handicap is deserving of just as much understanding and help as is physical handicap.

10.90 Administration and coordination of health services

In developed countries as many as 2% of the population may be involved in medical care. This figure includes doctors, dentists, nurses, midwives, ambulance drivers, psychologists, technicians and health inspectors, as well as a wide variety of ancillary staff such as clerks, typists, cooks and cleaners.

In the United Kingdom alone, spending on health for the year 1977 was almost seven thousand million pounds. It is not surprising, therefore, that the administration of a country's health service can be extremely complex.

10.91 Community health care

The systems of health care within the community vary considerably throughout the world, the differences being based mainly on the different priorities of the areas concerned.

To illustrate this point it is useful to look at the figures for 'cause of death' over the last hundred years in the United Kingdom. In 1870, 32% of all deaths occurred from infectious disease, and 6% from cancer, heart attack or stroke (the major diseases of old age). In 1970, the figures were 0.5% from infectious disease and 57% from cancer, heart attack or stroke. In 1870, 40% of all United Kingdom deaths occurred in children under the age of five, whereas in 1970 the figure was 3%. No such major changes have taken place in developing countries, where about 50% of all deaths occur in children under five. The main causes of death in these children are infectious disease and malnutrition.

In some respects, the differences in health care *within* the developing world are almost as striking as those illustrated above, since the biggest problems are to be found in the rural areas where 75% of the population lives. In many third world cities, the ratio of doctors to patients is as high as in Western Europe, where primary health care is based on a system of family doctors. In country areas, however, the situation is quite different. Doctors are few and far between, and many people in the past have received too little medical attention or have received it too late.

Fortunately, the emphasis has now shifted from curing a few people who are very ill to preventing as many people as possible from becoming ill in the first place. The following priorities have been identified:

1 provision of safe drinking water
2 provision of hygienic waste disposal
3 provision of adequate nutrition
4 provision of some measure of primary health care for all.

Obviously, in most countries it is impossible to provide sufficient expensively-trained doctors to give primary health care to everyone. In any case it is recognised that many family doctors in the developed world waste a great deal of their time in dealing with trivial complaints and, worse still, in carrying out large amounts of paper work. Therefore several countries have decided to place their primary health care in the hands of *medical auxilliaries.* These workers, who have a much shorter training period than do doctors, may also be known as community health workers, medical assistants, public health deputies or assistant medical officers. They may travel from village to village on bicycles, carrying their medical supplies with them, or they may use cars or public transport. Their duties are as follows:

1 to diagnose common diseases
2 to refer patients to a medical centre or hospital where necessary
3 to treat simple ailments
4 to carry out vaccinations
5 to educate patients and their families with regard to correct nutrition, hygiene and other matters relating to good health.

A second member of the community health team is the *midwife,* who is responsible for the delivery of

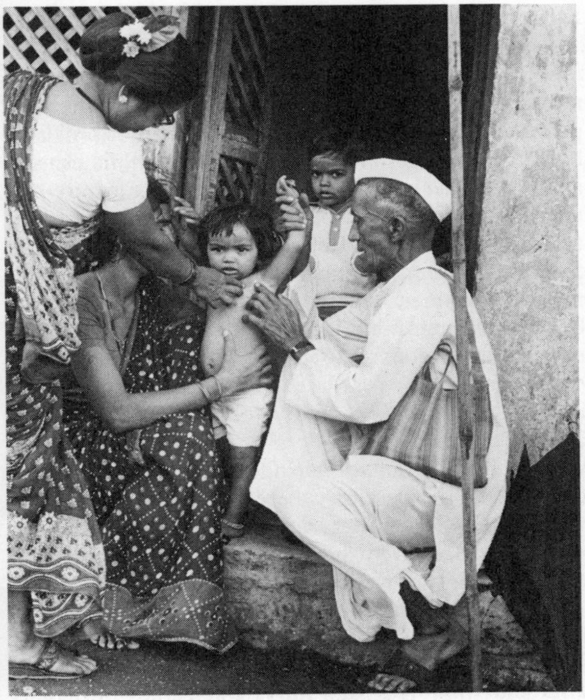

fig 10.35 A community health worker on his rounds in India

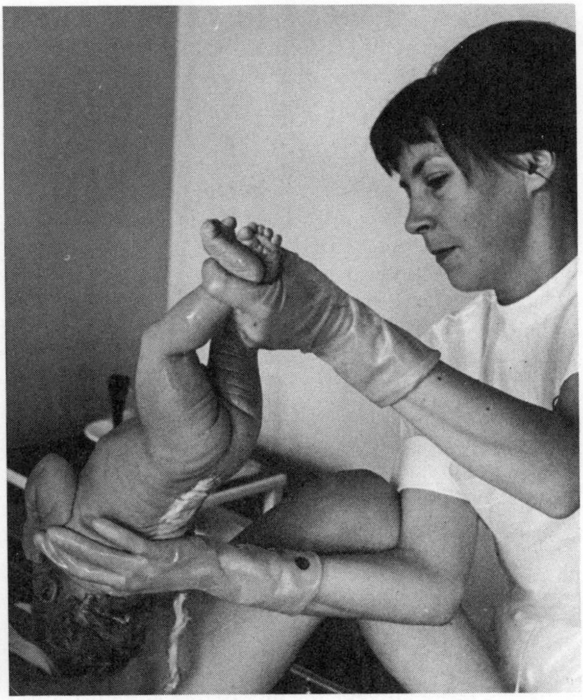

fig 10.36 The midwife

babies and the health of all pregnant women in her area. In some countries she may also care for the health of young children under the age of five and give advice to mothers on such topics as infant feeding and family planning. The midwife may visit all her patients in their homes or she may work from antenatal and child welfare clinics.

The person to whom both medical auxilliaries and midwives are usually answerable is the *Medical Officer for Health* (M.O.H.), a doctor who is also an administrator. He keeps a record of major health problems and reports *notifiable diseases* to his Ministry of Health. These are usually serious, infectious diseases such as smallpox, tuberculosis, typhoid and cholera: the list varies from time to time and from country to country.

The M.O.H. has responsibility for all the officially controlled medical care within his area, including school medical services, clinics and health centres. In urban areas, there is a greater centralisation of services and many clinics etc. are associated with large hospitals, such as the one illustrated in fig 10.38.

10.92 The National Health Service

The U.K. National Health Service was founded in 1946 (although it actually came into operation two years later) on the principle that the best medical care should be available to everyone at little or no cost. Since that time the infant mortality rate has halved and there

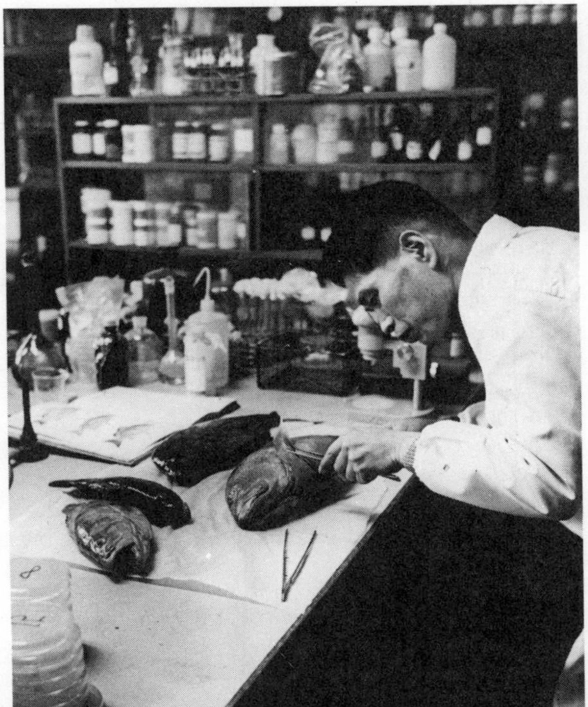

fig 10.37 A food inspector

have been dramatic falls in the numbers of deaths from tuberculosis, poliomyelitis and diphtheria. Of course, many of these improvements have been the direct result of advances in medical science, but few would deny the important of the contribution made by the National Health Service.

In 1974, the service was completely reorganised. Politically, it is the responsibility in England of the Secretary of State for Social Services (the arrangement is rather different in Wales, Scotland and Northern Ireland), the Department of Health and Social Security being responsible for strategic planning.

Under the D.H.S.S. come the Regional Health Authorities. There are fourteen of these throughout the country, each one associated with a medical school and carrying responsibility for research, capital building expenditure and major services such as blood transfusion and ambulances.

Operational control is in the hands of the ninety Area Health Authorities. These are statutory agents of Central Government and cooperate closely with the local authorities responsible for social work, environmental health and education. Most Area Health Authorities have at least two Districts, the District being the basic unit for the day-to-day administration of services. Each District has a District General Hospital and serves between 200 000 and 500 000 people.

Approximately 88% of the cost of the N.H.S. is met from the general budget of the country, while the remainder is made up by the N.H.S. contributions paid with National Insurance Contributions and by certain charges. These are mainly nominal charges made for medicine, for dental treatment (but *not* inspection), for dentures, for spectacles and for certain other articles. Exemption from such charges is given to children under 16, men over 65, women over 60, pregnant women, nursing mothers, Armed Service Disablement pensioners, people suffering from certain medical conditions and families with incomes below a minimum level.

Most of the 26 000 family doctors, 14 000 dentists, 7 000 opticians and 11 000 pharmacists working within the N.H.S. act as independent agents under contract to the Service, whereas district nurses, midwives and health visitors are employees of the Health Authority. On average there is one general medical practictioner to every 2 500 patients.

There are 2 700 hospitals in the N.H.S., catering for up to half a million in-patients at any one time. The 39 000 hospital doctors include 15 000 consultants and there is a total nursing and midwifery staff of almost 425 000.

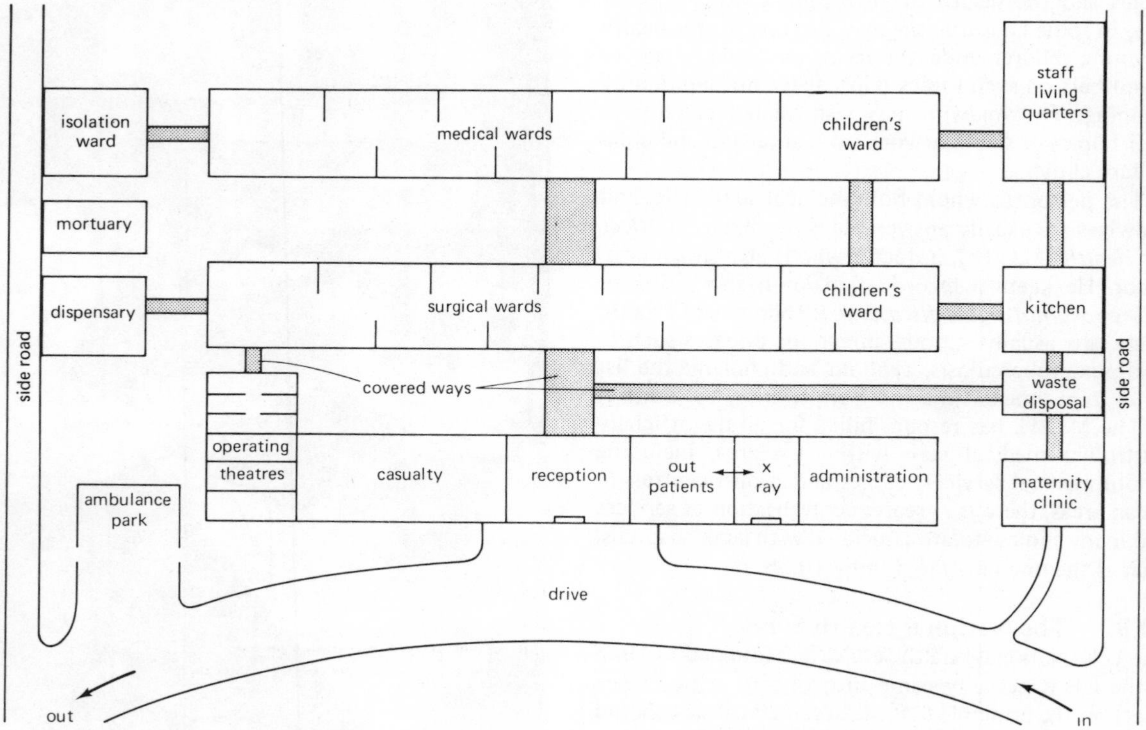

fig 10.38 Plan of a large, modern general hospital

10.93 International health control

During recent years, travel between countries has become much easier for ordinary people, and this has led to an increase in the risk of spreading infectious diseases. To keep this risk to a minimum, international cooperation is necessary. When there is an outbreak of a serious disease in a country, surrounding countries are warned and informed of the ports and other places which may be centres of infection. Anyone travelling from such a centre may be placed in *quarantine.* Originally, quarantine meant isolation for forty days, but now the period is long enough to cover the incubation of the disease and it may involve simply reporting at intervals to the Public Health Authorities. Diseases requiring quarantine are cholera, yellow fever, plague, smallpox and relapsing fever.

Another example of international cooperation in disease control is the issue of certificates of immunisation (see fig 10.39), which guarantee that a person has been vaccinated against certain diseases. Until, recently, a certificate of vaccination against smallpox (valid for three years) was the most commonly required, but fortunately this is no longer necessary. Cholera certificates (three months) and yellow fever certificates (ten years) are still needed for travel between certain countries.

10.94 Disease outbreaks

The following terms may be used to describe the extent of an outbreak of an infectious disease.

Sporadic: a few cases occur here and there in a community.

Epidemic: large numbers of cases occur in one locality. For example, there may be an epidemic of measles in a school, an epidemic of cholera in a town or even a epidemic of influenza in a whole country.

Pandemic: very large areas are affected, several countries or even more than one continent. Influenza pandemics occur from time to time.

Endemic: an endemic disease is one that is always present in a community to some extent. Bilharzia is endemic to most of Africa and malaria is endemic to many tropical countries.

10.95 The World Health Organisation

This branch of the United Nations Organisation was established in 1948 for the purpose of furthering international cooperation in 'promoting the highest possible level of health for all peoples'.

The W.H.O. defines health as 'a state of complete physical and mental well-being and not merely the absence of disease or infirmity'.

The headquarters of W.H.O. are in Geneva, Switzerland, and there are six regional offices, each responsible for an area with a similar range of health problems. These offices are located in Alexandria (Egypt), Brazzaville (Republic of the Congo), Copenhagen (Denmark), New Delhi (India), Manila (Philippines) and Washington (U.S.A.).

The work of the W.H.O. falls into three main categories.

1 *Coordination*

This involves acting as a clearing-house for health

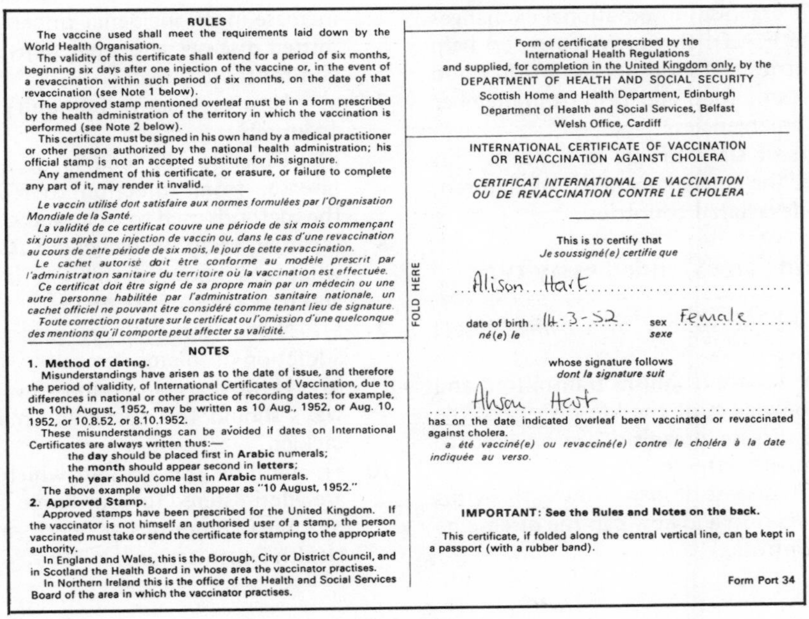

RULES

The vaccine used shall meet the requirements laid down by the World Health Organisation.

The validity of this certificate shall extend for a period of six months, beginning six days after one injection of the vaccine or, in the event of a revaccination within such period of six months, on the date of that revaccination (see Note 1 below).

The approved stamp mentioned overleaf must be in a form prescribed by the health administration of the territory in which the vaccination is performed (see Note 2 below).

This certificate must be signed in his own hand by a medical practitioner or other person authorized by the national health administration; his official stamp is not an accepted substitute for his signature.

Any amendment of this certificate, or erasure, or failure to complete any part of it, may render it invalid.

Le vaccin utilisé doit satisfaire aux normes formulées par l'Organisation Mondiale de la Santé.

La validité de ce certificat couvre une période de six mois commençant six jours après une injection de vaccin ou, dans le cas d'une revaccination au cours de cette période de six mois, le jour de cette revaccination.

Le cachet autorisé doit être conforme au modèle prescrit par l'administration sanitaire du territoire où la vaccination est effectuée.

Ce certificat doit être signé de sa propre main par un médecin ou une autre personne habilitée par l'administration sanitaire nationale, un cachet officiel ne pouvant être considéré comme tenant lieu de signature.

Toute correction ou rature sur le certificat ou l'omission d'une quelconque des mentions qu'il comporte peut affecter sa validité.

NOTES

1. Method of dating.

Misunderstandings have arisen as to the date of issue, and therefore the period of validity, of International Certificates of Vaccination, due to differences in national or other practice of recording dates: for example, the 10th August, 1952, may be written as 10 Aug., 1952, or Aug. 10, 1952, or 10.8.52, or 8.10.1952.

These misunderstandings can be avoided if dates on International Certificates are always written thus:—

the **day** should be placed first in **Arabic** numerals;
the **month** should appear second in **letters**;
the **year** should come last in **Arabic** numerals.

The above example would then appear as "10 August, 1952."

2. Approved Stamp.

Approved stamps have been prescribed for the United Kingdom. If the vaccinator is not himself an authorised user of a stamp, the person vaccinated must take or send the certificate for stamping to the appropriate authority.

In England and Wales, this is the Borough, City or District Council, and in Scotland the Health Board in whose area the vaccinator practises.

In Northern Ireland this is the office of the Health and Social Services Board of the area in which the vaccinator practises.

FOLD HERE

Form of certificate prescribed by the International Health Regulations and supplied, for completion in the United Kingdom only, by the
DEPARTMENT OF HEALTH AND SOCIAL SECURITY
Scottish Home and Health Department, Edinburgh
Department of Health and Social Services, Belfast
Welsh Office, Cardiff

INTERNATIONAL CERTIFICATE OF VACCINATION OR REVACCINATION AGAINST CHOLERA

CERTIFICAT INTERNATIONAL DE VACCINATION OU DE REVACCINATION CONTRE LE CHOLERA

This is to certify that
Je soussigné(e) certifie que

Alison Hart

date of birth 14-3-52 sex Female
né(e) le *sexe*

whose signature follows
dont la signature suit

Alison Hart

has on the date indicated overleaf been vaccinated or revaccinated against cholera.

a été vacciné(e) ou revacciné(e) contre le choléra à la date indiquée au verso.

IMPORTANT: See the Rules and Notes on the back.

This certificate, if folded along the central vertical line, can be kept in a passport (with a rubber band).

Form Port 34

fig 10.39 International Certificate of Vaccination against Cholera information and research services. Quarantine and vaccination measures are standardised internationally in order to make travel between countries easier. W.H.O. also organises the pooling and distribution of information on such topics as nutrition, vaccines, drug addiction control and radiation hazards.

2 *Disease Control*
W.H.O. sponsors the control of epidemic and endemic diseases by promoting vaccination programmes, early diagnosis, pure water supplies and health education. It has achieved notable successes against smallpox, tuberculosis, malaria and certain sexually-transmitted diseases. Progress is also being made in the fight against trachoma, cholera, yaws and yellow fever.

3 *Health administration*
The organisation makes major contributions towards improving the public health administrations of its member nations by giving technical advice and aid in planning. It also conducts field surveys, helps to set up local health centres and contributes to the education of doctors and nurses, both by direct teaching and by providing travelling fellowships.

Two other branches of the United Nations organisation are concerned particularly with health, the Food and Agriculture Organisation (F.A.O.) and the United Nations International Children's Emergency Fund (UNICEF).

The F.A.O., whose headquarters is in Rome, gives international support to national programmes for the increase of efficiency in agriculture, forestry and fisheries. It carries out a continuous review of world food problems, and organises international exchanges of information. The F.A.O. also provides expert help for developing countries and sponsors the World Food Programme. Reserves of food and cash from member nations are used in emergencies.

The headquarters of UNICEF are in London. This fund is used to meet the emergency needs of children, particularly in war-devastated countries.

Questions requiring an extended essay-type answer

1 a) Describe the life history of a *named* insect vector of disease.
b) Name the disease organism transmitted and show how a knowledge of the life history of the insect and the disease organism can be used to implement control methods.

2 a) *Name* a virus disease of man. How is this virus transmitted to its host and how can the disease be prevented or controlled?
b) How would you show that bacteria are present on your skin? (Consider that no microscope is available.)
c) *Name* a disease caused by a protozoan organism transmitted by an invertebrate vector. Describe the life history of the vector and show at which vulnerable stages in its life history it could be controlled.

3 a) Food poisoning can sometimes result from unhygienic handling of food. Give six rules that could be observed in kitchens and shops, and discuss the importance of personal hygiene in all stages of the preparation and handling of food.
b) Describe three methods of food preservation to prevent the growth of micro-organisms. Why are these methods successful?

4 a) Why does damp food (e.g. bread) go mouldy and then become almost liquid?
b) Describe how you would test experimentally your observations in part a).
c) Describe briefly three different methods that you could use to prevent food from going mouldy and decomposing.

5 a) Why is it important to keep an area where food is prepared free from house-flies (*Musca domestica*)?
b) Give a detailed description of an experiment to confirm your statements in part a), and explain the results that you obtain.
c) What methods are available for controlling these pests?

6 a) List three sexually transmitted diseases.
b) What are the main reasons for the recent increase in the incidence of certain sexually transmitted diseases and what steps are being taken to control these diseases?

7 a) How may *active* immunity from a disease be acquired?
b) Describe the course of a named bacterial disease, from infection to recovery, indicating the methods used to treat and control the disease.

8 Write an essay on *malignant tumours,* explaining what you know of the way in which they arise and how they are treated.

9 a) Name three drugs responsible for major addiction problems.
b) How does addiction develop and in what ways should this social and medical problem be tackled?

10 a) Outline the way in which medical services are administered in your home country.
b) What functions are performed by the World Health Organisation?

11 Man and his environment

11.00 Ecosystems

The study of ecology is a relatively new science and today it is becoming increasingly mathematical. Mathematical models can be set up to represent the balance between the different factors that determine the survival or death of organisms in any one biological community. For example, the energy requirements and the energy flow through communities can be calculated (see fig 11.6).

Although it might appear that any one natural community is a stable system, it should be realised that communities are continually changing. Over a period of many thousands of years a forest, for example, may have developed from an area of bare rock and the survival of that forest community is dependent upon a balance of constantly changing dynamic forces. Charles Darwin understood this when he wrote the following in his book *The Origin of Species.*

> I find from experiments that humble-bees are almost indispensable to the fertilisation of the heartsease (Viola tricolor), for other bees do not visit this flower. I have also found that the visits of bees are necessary for the fertilisation of some kinds of clover; for instance, 20 heads of Dutch clover (Trifolium repens) yielded 2,290 seeds, but 20 other heads protected from bees produced not one. Again, 100 heads of red clover (T. pratense) produced 2,700 seeds, but the same number of protected heads produced not a single seed. Humble-bees alone visit red clover, as other bees cannot reach the nectar. It has been suggested that moths may fertilise the clovers; but I doubt whether they could do so in the case of the red clover, from their weight not being sufficient to depress the wing petals. Hence we may infer as highly probable that, if the whole genus of humble-bees became extinct or very rare in England, the heartsease and red clover would become very rare, or wholly disappear. The number of humble-bees in any district depends in a great measure upon the number of field-mice, which destroy their combs and nests; and Col. Newman, who has long attended to the habits of humble-bees, believes that "more than two-thirds of them are thus destroyed all over England." Now the number of mice is largely dependent, as every one knows, on the number of cats; and Col. Newman says, "Near villages and small towns I have found the nests of humble-bees more numerous than elsewhere, which I attribute to the number of cats that destroy the mice." Hence it is quite credible that the presence of a feline animal in large numbers in a district might determine, through the intervention first of mice and then of bees, the frequency of certain flowers in that district!

Thus no organism, plant or animal, lives in isolation from either its living or its non-living environment. Any member of a species is always in competition with other members of the species, and the species itself is in competition with other species for survival in its own small habitat.

We therefore come to this concept of the balance of nature and the idea that the apparent equilibrium is in fact a dynamic and not a static one. The organisms are balanced in that, in the *food chains* at any one level, there are limited numbers of individuals feeding on members of the link (stage) below them. Not only are there a number of food chains in each community but these chains are interwoven to form a *food web.* A food web is thus a series of inter-connected food chains with a number of organisms at each feeding level. The interruption or the elimination of one part of a food chain or web can cause considerable disruption in the remainder of the biological system (see fig 11.3 and 11.4).

In a natural community, the vegetation is the first *producer* of living material, since the incorporation of radiant energy from the sun is possible only by green plants. Feeding on these producers are *herbivores,* which are usually less numerous than plants. Feeding on the herbivores are *carnivores* and these are fewer still. Feeding in turn on the carnivores there may be tertiary consumers or 'top' carnivores. Hence we have a *pyramid of numbers* with plants at the base and the top carnivores at the tip. We see below that only a small proportion of the energy (in food terms) taken up by each level of organisms is transferred to the next

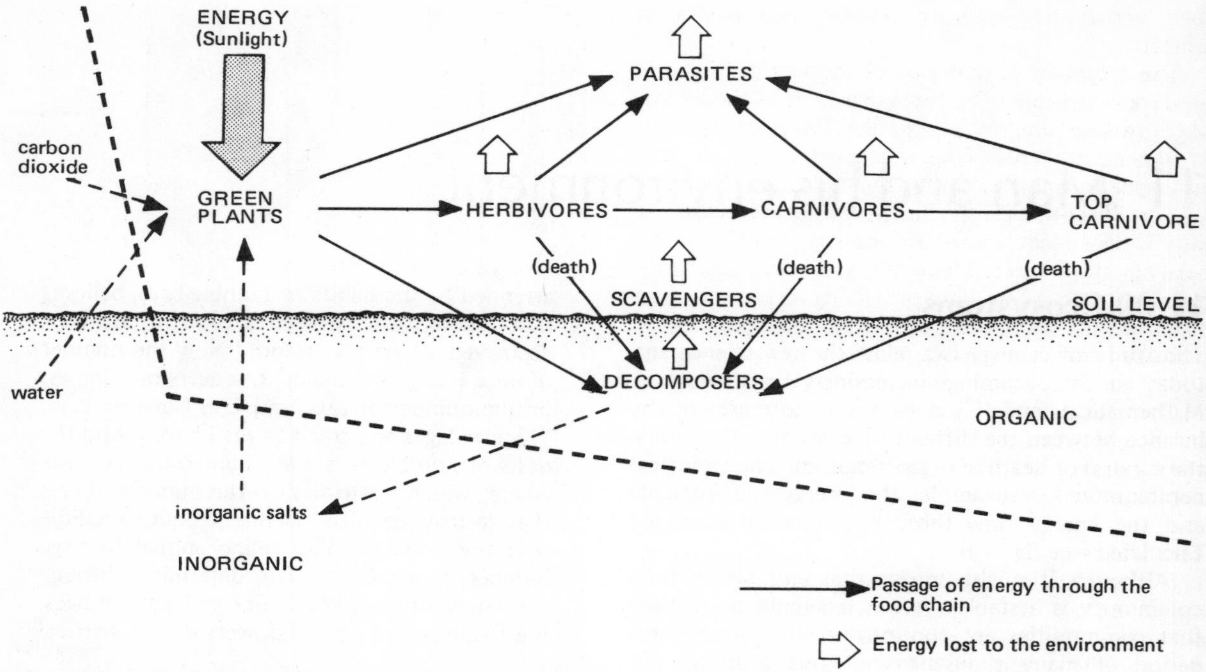

fig 11.1 The cycle of nature relating organic and inorganic parts of the ecosystem

step of the pyramid (see section 11.01). Most of the energy (about 80%) at each transfer is lost as heat, and thus only 20% of the food material becomes incorporated into the chemical structure of the tissues of the consumer. Therefore a better picture than a pyramid of numbers is provided by a *pyramid of biomass* which indicates the total weight of living matter that can exist at any level.

Ecological units that are recognisable on the surface of the earth are known as ecosystems, and in them a whole series of life processes are operating. We can see large ecosystems such as tropical rain forests, temperate grass land, temperate forests, moorland and swamp. Within any one country smaller ecosystems can be recognised as functioning within the broad ecosystem for that part of the world.

11.01 The Biosphere

Fig 11.1 shows the relationship between the inorganic and organic parts of the ecosystem. Note that the energy input is to the green plants which convert only about 0.2% of incident sunlight to usable chemical energy, and there is a further gradual loss of this energy throughout the food chains. By the time the top predator is reached only 2 to 3% of the original energy intake of the plants has been passed on and eventually even this is lost as the animal dies and is decomposed to form its constituent organic matter. *Parasites* and *decomposers* also play their parts in extracting food and energy and are themselves finally broken down after their death, so that the organic materials are returned to the soil.

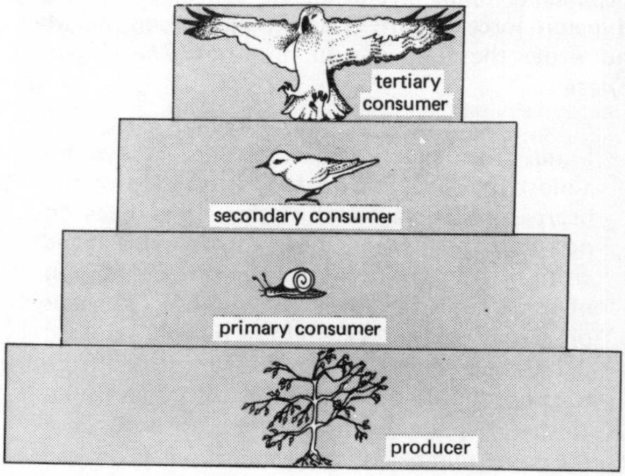

fig 11.2 The relationship between producers and consumers

As far as Man is concerned, he can obtain more energy from a given area of land by eating the plants he grows than by feeding them to cattle and sheep and thus losing energy in the transference. In other words, the shorter the food chain the more energy reaches the final consumer. This is why large numbers of people in the world must eat plant material since they cannot afford to waste the energy available from their agricultural land by feeding their plants to animals.

The *biosphere* is that part of the world, from the hard rocks beneath us to the upper limit of the atmosphere, within which life can exist. The upper portion of the earth's crust (the lithosphere), the air (the atmosphere) and the waters (the hydrosphere) can all support life within certain limits, and furthermore that life is dependent upon the exchange of materials between these three. There are, as a result, cycles of material that move between living organisms (the biotic world) and the air, earth and water (the abiotic world). The most important of these cycles are (i) *water*, which composes about 95% of the structure of all living organisms, (ii) *carbon*, on which all organic compounds are based, and (iii) *nitrogen*, which is a part of all proteins. Many other elements, e.g. *sulphur*, also circulate in nature.

The water cycle

This most important cycle, shown in fig 11.7, depends mainly on the two processes of evaporation and precipitation. The major sources of evaporation are the oceans, with a combined loss of some 900 km³ per day. The land masses also lose water in the same fashion to the value of some 150 km³ per day. Evaporation depends upon the sun's energy and thus it decreases with increasing distance from the equator.

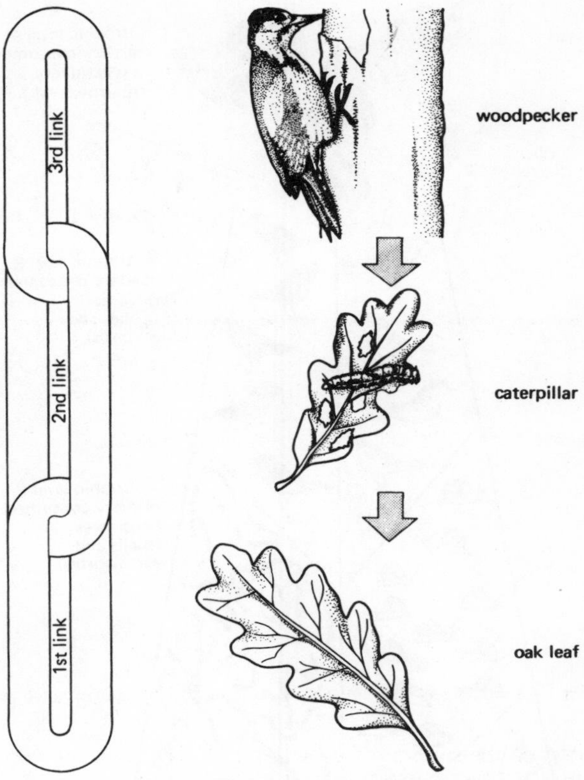

fig 11.3 A simple 3-link food chain

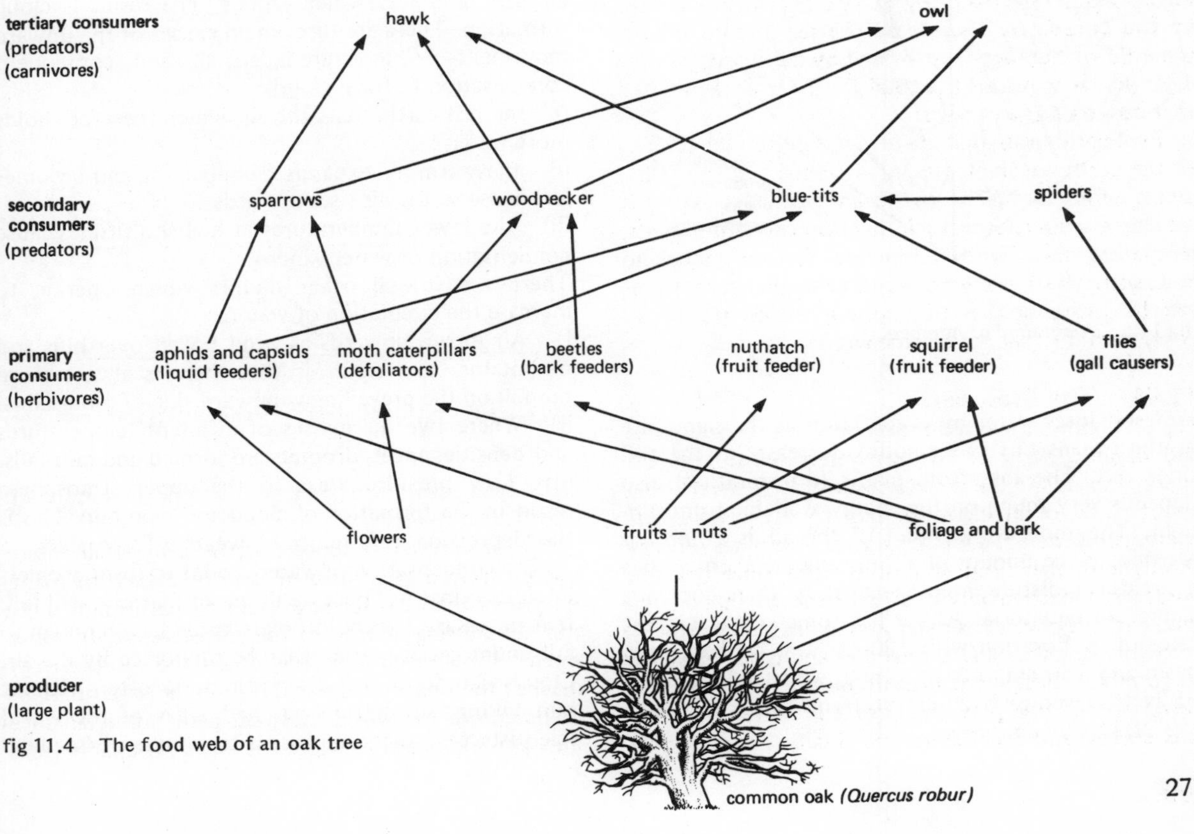

fig 11.4 The food web of an oak tree

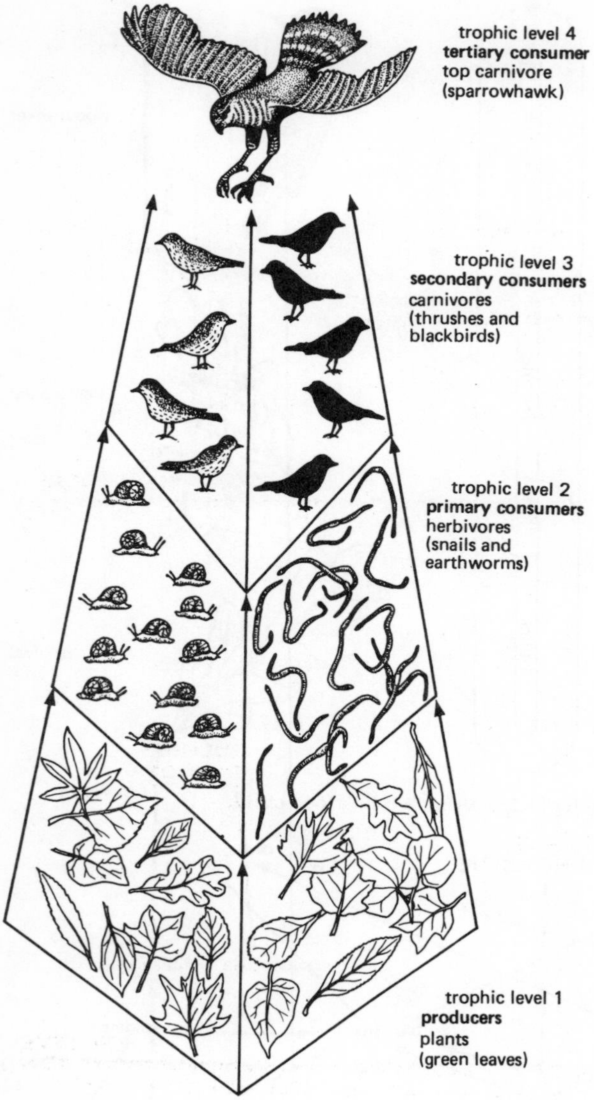

fig 11.5 A pyramid of numbers

Wet soil loses water by evaporation at the same rate as the oceans but this amount decreases as the soil dries out. The loss from plants (transpiration) also depends very much on the energy available from sunlight. The moisture content of the air is a *limiting factor* as is the amount of air movement. Warm air has a greater moisture holding capacity than cool air. Therefore equatorial Africa has some of the highest rates of evaporation with 120 cm/year in the Congo basin and 150 cm/year in Kenya, while in Finland at 65°N the average is 20 cm/year and in S.E. England (50°N) it is 50 cm/year.

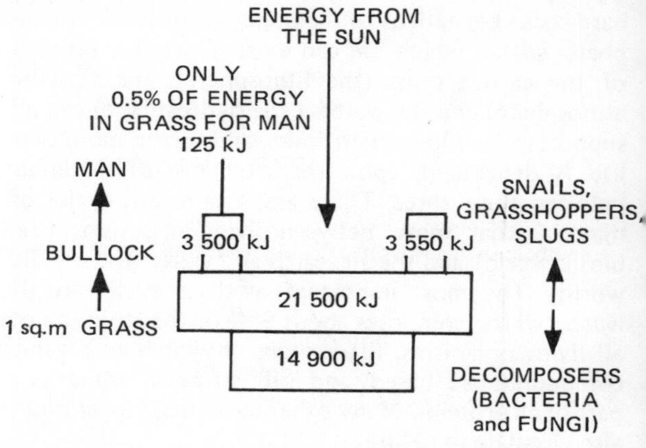

fig 11.6 Energy flow through a food chain involving Man and his food

The metabolic activities of plants and animals produce water during the process of respiration. Decomposition also adds water vapour to the air, whilst combustion of fossil fuels and wood contributes small amounts.

Water collects as small droplets in the upper atmosphere where the lower temperatures mean that the air can hold less water vapour. The result is cloud formation. There are three main causes of the upward movement of moisture-laden air and consequent condensation to form clouds:

i) the hot earth heats the air, which therefore holds more water
ii) the warm air expands from heating and becomes less dense so the air rises upwards
iii) the lower temperatures at higher altitudes cause condensation of water vapour.

There are several other factors which operate to increase the circulation of water.

i) Air moves upwards as wind travels over hills and mountains. The result is that there is always more rainfall on the prevailing windward side of mountains.
ii) Where two air masses of different temperatures and densities meet, droplets are formed and rain falls.
iii) Low pressure areas in the upper atmosphere result in the formation of cloud and then rain. This is the 'depression' referred to by weather forecasters.

The condensation of water vapour to form droplets in clouds does not necessarily mean that rain will fall. It is necessary for the droplets to be large enough to fall under gravity rather than be supported by the air. Clouds can form and disappear quite often without rain falling, so that in parched areas of the world scientists can sometimes use a technique of 'seeding'

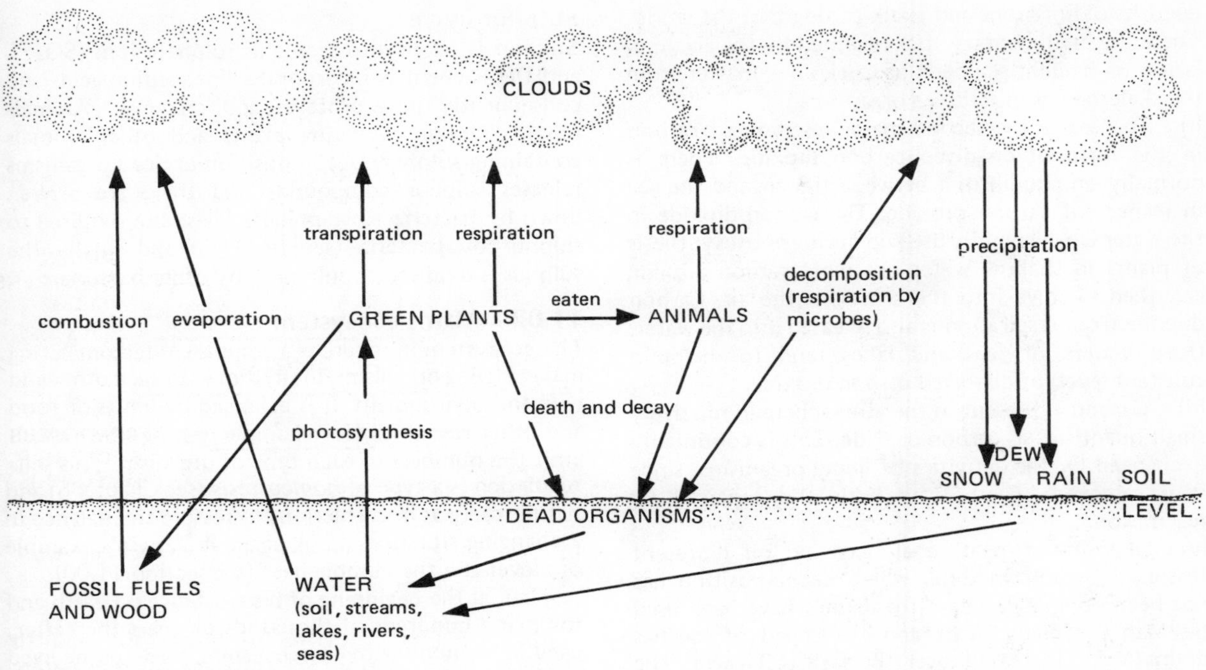

fig 11.7 The water cycle

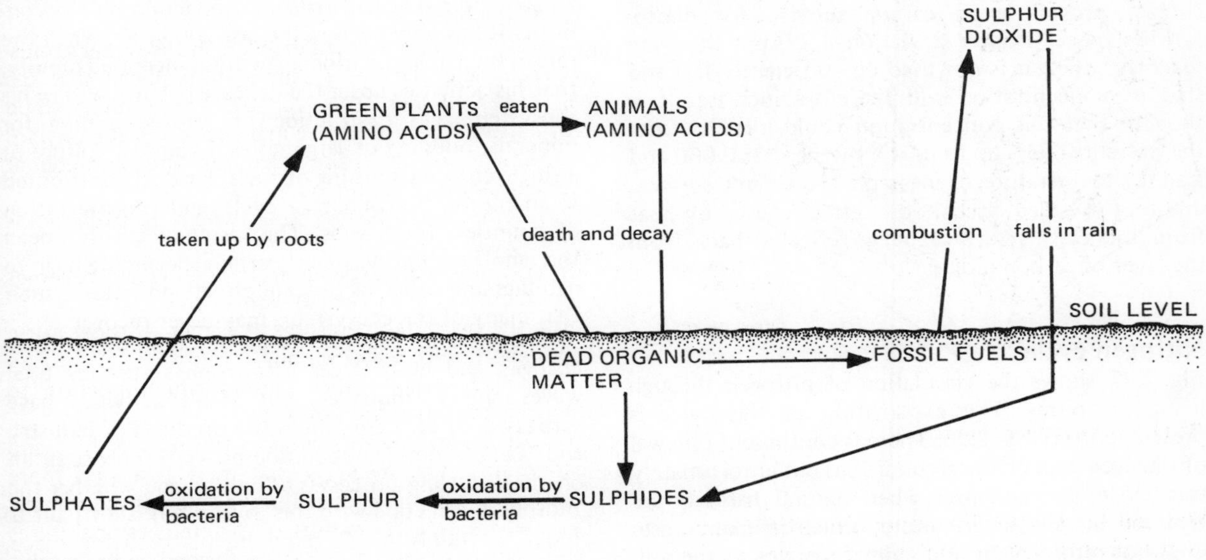

fig 11.8 The sulphur cycle

clouds with chemicals from aeroplances to enlarge drops and thus start precipitation.

Water may descend out of the air as rain, hail or snow depending on the air temperature. Cooling of moisture-laden air at the surface of the earth can also liberate water as dew or frost, provided that there is little air movement.

Carbon cycle

Fig 2.28 shows the circulation of carbon through animals and plants, but most of the carbon on the earth's surface is 'locked up' in material other than living organisms.

i) The majority of the carbon occurs in chemical combination with calcium and oxygen in the vast deposits of limestone and chalk throughout the world. These are sedimentary rocks formed by the accumulation on ancient sea beds of calcareous matter from the skeletons of minute sea organisms.

ii) The sea is the second major reservoir of carbon in the form of dissolved carbon dioxide. There is normally an equilibrium between the air and the sea in respect of carbon dioxide. The carbon dioxide in the water can be rapidly used up during photosynthesis of plants in shallow water but more carbon dioxide can then dissolve into the sea from the air. Carbon dioxide from respiration is also released into the water. Deep waters of seas and lakes tend to maintain constant levels of dissolved carbon dioxide.

iii) Carbon is present in the atmosphere in relatively small quantities as carbon dioxide. This is continually exchanged by the activities of living organisms, since it is removed by photosynthesis and replaced by respiration.

iv) Deposits of peat, coal, gas and oil represent fossilised organic material where decomposition has not been complete. These fossil fuels have been used by Man for many years and the speed of use has dramatically increased over the last 100 years. The burning of these fuels discharges large quantities of carbon dioxide into the atmosphere. This does, of course, provide extra carbon dioxide for photosynthesis (see Chapter 2 section 2.25) but there are fears that the gas is not used up sufficiently fast and that its concentration will therefore increase. It is thought that this concentration could increase from the present 0.03% up to 0.04% by the year 2 000 and lead to temperature changes on the earth's surface. due to a so-called 'greenhouse' effect caused by heat from the earth's surface being reflected back from the layer of carbon dioxide.

Nitrogen cycle

Fig 2.27 shows the circulation of nitrogen through living organisms. The explanation of this cycle is presented in section 2.26. There is continuous renewal of nitrogen under natural conditions but unfortunately this cycle can be upset when natural habitats are replaced by Man's agriculture, which in many cases does not return plant and animal remains to the soil. Man compensates for this by the industrial production of nitrogen-containing fertilisers, which currently amount to almost 30 billion kilograms per year. This is probably equal to the amount produced under natural conditions. The extra fertiliser has, of course, permitted a great increase in food supplies, but it has brought with it problems resulting from unused fertilisers running off the land into water systems and causing deoxygenation and the death of water organisms (see section 11.02 no. 3).

Sulphur cycle

Sulphur is taken into plants as sulphate ions (SO_4^{2-}) and the element is incorporated into amino acids and consequently into proteins and enzymes. Animals obtain sulphur by eating plants and other animals containing sulphur. Decomposition of dead organisms releases sulphur compounds and these are broken down by bacteria to sulphides. These are oxidised to sulphur by bacteria (see fig 11.8) and finally the sulphur is oxidised to sulphates by other bacteria.

11.02 Man's ecosystem

The ecosystem in nature is a complex interconnecting mass of living organisms, in rhythm with each other and with the environment. It is balanced by limits on food and other resources (for example nesting sites) which limit the numbers of each type of organism. This self-regulation is a type of *homeostasis* (see Chapter 6) and nature can put in corrections to restore the balance in a changing situation. Look again at Darwin's example of clover and the humble-bees (see section 11.00).

Man, at the beginning of his existence on earth and for many hundreds of thousands of years thereafter, lived in harmony with his ecosystem. Early on he lived only in small groups, hunting and gathering such food as he needed, and even when he began to develop as an agriculturist he had little real impact on his environment. It has only been in the last 200 years, when the rate of human population growth has increased rapidly, that his activities upset the delicate balance within his ecosystem. His destruction of forests, mining for minerals, building of large cities, extensive culture of a single crop and burning of fuels, have all contributed to an interruption of natural cycles and a disintegration of countless food webs. The result of this has been the wholesale extinction of certain species, damage to habitats and other large scale effects on nature, such that many of the ecosystems may never recover.

The natural ecosystems can accommodate local changes and even larger catastrophes such as tidal waves and earthquakes, but Man's activities have surpassed all of these. The waste products of industry and Man's booming population puts an immense strain on the recycling processes of nature, and whether the biosphere can cope with this in future it is difficult to judge. Certainly Man's requirement for living space, more food and industrial production means that this strain is unlikely to be reduced in the foreseeable future.

Man's interference with natural cycles can cause many problems, which often outweigh or cancel out the efforts that Man has made. These problems are often referred to as the 'biological backlash'. Here are three examples:

1 Egypt and the Aswan Dam

The first dam across the River Nile was built in 1902

(a) Freshwater

(b) Temperate forest

(c) Savanna

(d) Desert

fig 11.9 Different types of habitat

and, together with a number of barrages downstream, it was the first attempt to control the flooding every year in the lower reaches of the river. The irrigation channels that were developed to use the stored water were able to provide a steady flow throughout the year. More crops could be grown over a longer period of time. The old basin irrigation system that had operated for five thousand years was therefore abandoned and the danger from flood waters and that of crop failure were reduced. The latest dam, larger than all the rest before it, was conceived by the late President Nasser and it holds back a lake 200 miles (320 km) in length.

These benefits also brought losses. No more silt was deposited as it was held back by the dam. Eventually this could result in the dam silting up completely. Evaporation from the enormous surface area of the water caused increased salinity which decreased crop yields. There was no seasonal drying of the soil that had in the past destroyed weeds. Soil fertility became exhausted from over-cultivation and more fertilisers were required to maintain yield.

Other undesirable effects included an increase in disease. The incidence of malaria increased sharply. Guinea worm and certain bacterial infections were also favoured by the change in irrigation methods. Most dangerous of all, the irrigation water provided a habitat for the snails that harbour the dangerous parasitic worm *Schistosoma,* which causes the disease bilharzia. Details of the life history of this parasite are in Chapter

10 (see section 10.18). This has become the world's most widespread disease afflicting some 120 million people.

The spread of this disease may well cancel out the benefits of such a dam, that is its irrigation and its agricultural products. The disease causes fatigue and thus lack of output in workers, and consequently the agricultural production also decreases. Although the life cycle of the parasite is well known, eradication has proved difficult. Copper sulphate does destroy the snails, but costly attempts to do this have not succeeded in eliminating all of the snails in one area, and within a short time the population has become just as numerous.

Another big dam which has been built in Africa is the Kariba dam on the Zambesi river. Irrigation channels here are similarly colonised by the host snail of *Schistosoma*. In Zimbabwe an irrigation scheme has been practically abandoned because of the incidence of schistosomiasis.

fig 11.10 Irrigation channels in Egypt

2 Insecticides and herbicides

The position of world agriculture today is an example of extreme interference with natural ecosystems. Plants have been bred which have lost their natural resistance to attack by insects. Furthermore single crop varieties are planted over very large areas, providing ideal habitats for the insect pests that attack those particular varieties. The percentage of crop losses to insects thirty years ago was very large, estimated by a number of specialists to be up to 10 to 15% of any crop. In order to overcome this problem in the early 1950s, *insecticides* were developed which were sprayed onto crops in order to destroy the insect pests. By the late 1950s vast quantities of insecticides were being used on crops, but in spite of this the losses from insect pests did not drop dramatically. It is probably true that the development of high yield strains of crops has been responsible for increasing production rather than the use of insecticides.

DDT is the most well known, and the oldest and most widely used, insecticide. It is found everywhere, not only at the point of application but all over the earth. The seas, the land and almost every animal on this planet have been exposed to it. DDT residues have been found in the body fat of the people of India, in Americans, in Eskimos and in Antarctic penguins. Seals from the east coast of Scotland have been shown to have concentrations of DDT as high as 23 p.p.m. Because DDT breaks down so slowly it lasts a very long time in soils. For example, in the late 1940s and early 1950s marshes were sprayed with DDT to control the breeding of malarial mosquitoes. Nearly thirty years later, large amounts of DDT are still found in the upper mud layers of these marshes. As a result of its concentration at the bottom of the food chain and its uptake by plants, it moves up the food chains

fig 11.11 The Aswan Dam, a large dam in Egypt

through the primary consumers and finally to the top carnivores or tertiary consumers. This is noticeable in large predatory birds such as eagles, hawks and falcons. As a result of the concentration of DDT in their tissues the chemical has interfered with the metabolism of the birds so that their eggs are often infertile and furthermore the shells have become so thin that they are crushed by the weight of the brooding hen bird whilst incubating the eggs. This information obtained from wild birds has been confirmed by feeding DDT to domestic birds, which also produced thin shells and suffered reduced hatch rates. The insecticide Dieldrin produces similar results.

The effects of insecticides are not confined to terrestrial food chains but are noticeable in aquatic habitats where DDT is first taken up by all the tiny green plants that float in fresh water and the sea. These plants are the first link in the food chains and the insecticide moves up through the food chain to larger and larger organisms (see fig 11.13).

The complete effect of insecticides on soil is very difficult to evaluate. The living soil contains vast quantities of invertebrates such as insects, worms, nematodes, arthropods and their larvae, in every square metre. The washing of insecticides into the soil must result in the death of these organisms and therefore the fertility must become less, since all the organisms play a vital role in the soil ecosystem, as can be seen in the nitrogen and carbon cycles.

In addition to these insecticides there are *herbicides*, which are now used extensively to control the plant growth alongside roads and railway embankments and under power lines. Herbicides were used as defoliants to remove leaf cover from opposing forces during the Vietnam war, and those forests have been damaged beyond repair for many years to come. Defoliation of jungles and roadside verges inevitably leads to the local extinction of large populations of animals that are part of the ecosystem. In the soil, the herbicides probably also have an effect on the bacteria and fungi.

3 Fertilisers

A great deal of the nitrogen in the soil is contained in the humus, which, besides its chemical value (it can be broken down to provide nitrates), increases the capacity of the soil to retain water. It acts like a sponge, holding water and releasing it over a long period of time. Roots in the soil not only require water and nitrates, but also need oxygen in order to respire and obtain energy to do the work necessary for the uptake of dissolved salts in the soil water. The humus assists in all of these activities by allowing the soil to remain porous.

With large scale monoculture (one type of crop) the soil gradually loses its fertility and thus Man has had to find means of replacing the *nitrates* which are essential for a high level of production. Therefore nitrate fertilisers have been used and the effect of these on the soil is quite different from that of humus. The nitrogen in the humus is bound tightly into the soil and not much is removed by surface run-off of rain water, or by rain water moving downwards through the soil. Nitrate fertilisers, however, dissolve very easily in rain water, which carries away a great deal of the fertiliser. With the decline in amounts of humus, the soil becomes more compacted, oxygen availability decreases and this in turn interferes with the processes by which roots can take up salts from the soil. The consequence is that vast amounts of fertiliser placed on argicultural land tends to be flushed into rivers and lakes before it can be absorbed by the crop.

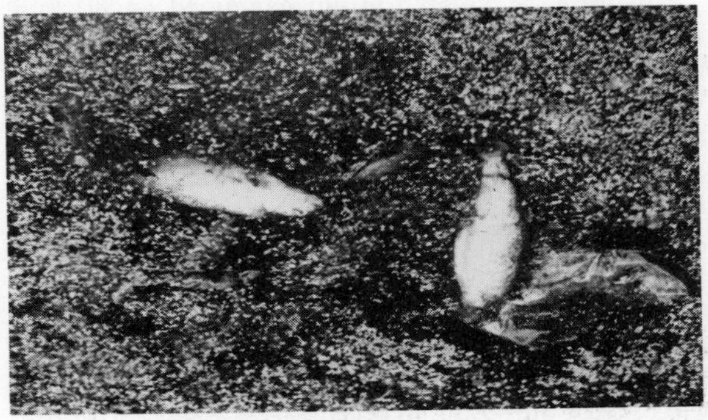

fig 11.12 Fish killed by nitrate 'bloom'

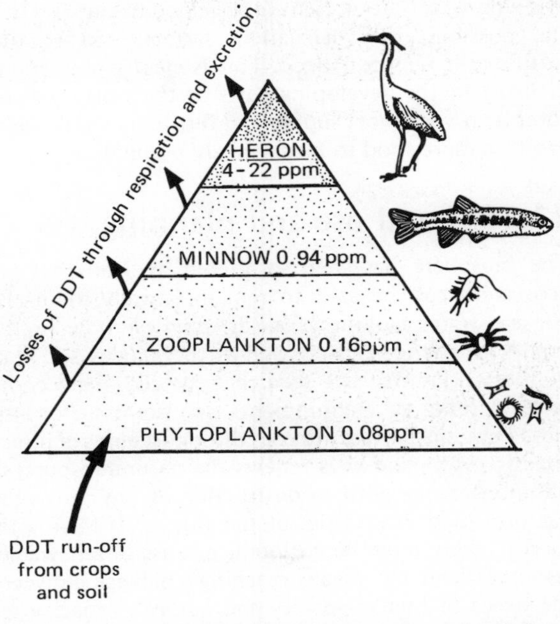

fig 11.13 The concentration of DDT up through a fresh-water food chain (units in parts per million)

The fertilisers can be taken up by the phytoplankton (very small plants) in the water, and as a result their growth increases so much that vast 'blooms' of algae appear, covering large areas of lakes and ponds. Eventually these masses of algae die and decay, resulting in an increase of bacteria, which use up the oxygen in the water and reduce the amount available for the plants and animals. This state of affairs is known as *eutrophication* and results eventually in the 'death' of the lake. Normally this process takes place over thousands of years, but through Man's intervention it is speeded up and takes only a few years. With no oxygen left the water's capacity for purifying itself has gone. Furthermore, the high concentration of nitrate in the water makes it quite unsuitable for drinking.

In 1968, a research group examined an 80 kilometre stretch of the River Ouse, which is bordered by agricultural land that had been heavily fertilised. It was found that 90% of the nitrate in the river had come from the fertilisers and this was equivalent to approximately 7 kg per hectare. Farmers had applied their fertiliser at the rate of 27 kg per hectare so that approximately *one quarter* of the fertiliser was finding its way into the river.

Although nitrates themselves are relatively harmless in drinking water they can be converted to nitrites in the body, and this damages the oxygen-carrying capacity of the blood. In certain areas water authorities have had to supply pure water in bottles for small children, since they are particularly susceptible to this contamination of water.

These problems of eutrophication and nitrate in drinking water are present in developed countries but the possibility of persuading farmers not to use fertilisers is remote indeed. This is clearly even more difficult in the developing parts of the world, where food is in very short supply and fertilisers would help produce more food to feed a needy population.

11.10 Land use and planning

The biosphere has been so altered by Man that no ecosystem can be said to remain entirely 'natural'. Air and water pollution have affected even the Arctic and Antarctic ice caps as well as the jungles and seas of the world. We have seen that the ecosphere comprises a series of environments that are more or less closed systems, where life goes through a series of interlocking cycles and thus renews the natural resources. The interference with, or destruction of, any one cycle has profound effects on all the others. If Man is to survive, there must be an intelligent use of renewable resources, and this means reaching a balance between the needs of human society and nature's capacity to satisfy them. This must involve the management of renewable resources in order to give economic and continuous production. Some of our renewable resources e.g. soil and timber, are deteriorating because of past exploitation. In addition the problems have been intensified in the developed world through industrial pollution and the use of toxic pesticides.

The management of the ecosystem will vary throughout the world in developed and developing countries and it will depend on many factors, but if it is to be successful it must be based on sound ecological principles. Therefore the planners must obtain research information about the basic ecological situation before making any decisions. Problems of land management in many tropical areas are accentuated because of the thinness of the soils and the quick breakdown of organic matter after the removal of plant cover. The subsequent exposure of this soil to hot sun and heavy rains results in serious erosion.

The following conservation measures must be undertaken to ensure the renewal of natural resources:

Soil

1 Restoration of plant cover to halt processes of erosion
2 Renewal of forest and grassland
3 Restriction of timber felling
4 Measures to counter erosion on arable land
5 Restoration of the biological processes that result in soil formation and the continuance of soil cycles.

Fresh water

1 Thorough ecological investigations preceding any new irrigation or hydro-electric schemes, to ensure that there are no dangerous consequences to Man in the future
2 Reafforestation and conservation on hills and mountains, to ensure a steady drainage of water
3 Maintenance of a continuous pollution check on waterways
4 Development of water supplies in arid regions together with careful consideration of the ecological problems.

Seas

1 International agreements to regulate fishing in national and international waters
2 Protection of marine species against collection for the tourist and commercial curio trades
3 Maintenance of a continuous check on the sea for pollution caused by land-based factories and towns, and by ships releasing oil.

11.11 Regional and city planning

Most of the people who live in developed countries live in cities; 78% in the United Kingdom, 87% in Australia, 73.5% in North America, 78% in Germany and almost 100% of the inhabitants of Hong Kong.

The drift of people to the towns has been characteristic of industrialisation in all Western nations and indeed has become a problem of the developing countries. The population sizes of cities can be enormous, for example 14.5 million in New York, 11.5 million in London and 13.5 million in Tokyo.

The principal problems of such large collections of people are:

i) transport of people, goods and materials
ii) disposal of waste products
iii) housing for the growing population
iv) energy supply
v) air, noise and water pollution.

These problems have been aggravated by the random, unplanned process by which cities have grown, especially if they have been in existence for hundreds of years. Population growth will continue to enlarge old cities and produce new ones but the increasing urban sprawl and the formation of outer city slums should not be allowed to continue. For this reason most countries now have national, regional and city planning authorities, whose job is to exercise control over the future development of land and buildings.

National and regional planning usually involves the following stages:

1 An inventory is drawn up, showing resources, projected population, topographical and climatic factors, possible hazards and so on, and these elements are surveyed relative to their quality and quantity within the region.
2 The relationship between human and natural resources within the region is analysed.
3 A plan is drawn up and proposals are made for the effective use of the resources.
4 Recommendations are made concerning the timing of the different stages of the development of the plan.

Once the plan has been completed for a country or region it must be subject to continuous review, since as development proceeds some aspects of the environment or plan may change. In democratic societies, the authorities are answerable to the public and therefore, certain decisions may have to be put to the vote. In order that the public should be kept informed, there must be an effective information and education service to explain the plan and the different stages of its growth.

Once a regional plan has been established, the planning of cities is the next important step. This again should have important environmental objectives:

1 There must be a division of the city into an orderly arrangement of residential, business, industrial and open space areas, so that it functions efficiently with a minimum of cost and effort.
2 All types of transport moving freely within the city, and to and from the rest of the region must be available.
3 There must be minimum standards imposed on building construction.
4 Different types of living accommodation must be provided in residential areas. This accommodation must be both comfortable and healthy for all sizes of family, with particular provision for the old (see section 11.13).
5 Each residential area must be provided with schools, shops, meeting halls and open spaces (parks).
6 An adequate water and electricity supply must be provided for the whole city, together with other services such as sewerage, hospital, fire and police (see sections 11.20 and 11.30).

Obviously, most old cities cannot be torn down and replanned, since the cost and destruction involved would not be tolerated by the inhabitants. Furthermore the old house and street patterns must be preserved for aesthetic and historical reasons. Thus local authority planners must generally be content with gradual change and strive towards slow development and improvement. In any case, the ideal plan is not a fixed objective but one which will itself change as development proceeds. In old cities the clearance of slums and sub-standard housing must always be a major objective.

The shortage of building space in and around large cities has encouraged planners and architects to build tower blocks containing large numbers of flats. These satisfy the demand for a greater number of individual dwellings and also for the maximum use of land. Such flats have no private gardens, although there is often a balcony on which plants can be grown in pots. Unfortunately, many of the people who have moved into these tower blocks from small houses in overcrowded slums have not come to terms with their new style of life. They miss the community spirit of their old neighbourhoods, where families have known each other for generations. The greatest problem, however, is lack of playing space for small children, where mothers can watch their activities and ensure that they do not come to any harm. These families like to be low down in the block so that they can reach the communal gardens and open spaces. There are considerable problems of emotional stress in dwellers in tower blocks, particularly young mothers. The incidence of this type of illness is about double that for people living in houses.

Project

To make a study of your local community, town or village

Procedure

1 Make a plan or obtain a plan of your town or village.
2 Mark on the plan in coloured pen or pencil
i) the housing areas

	Urban population (millions)			Annual growth rate (in percentage)	
	1970	1985	2000	1970-1985	1985-2000
World	1 352	2 169	3 329	3.3	2.9
Developed regions	717	943	1 174	1.8	1.5
Developing regions	635	1 226	2 155	4.5	3.8
Eastern Asia	266	461	722	3.7	3.0
Southern Asia	238	458	793	4.4	3.7
Latin America	158	291	495	4.1	2.6
Africa	77	160	320	5.0	4.7
Western Africa	20	45	96	5.4	5.2
Eastern Africa	9	22	50	5.7	5.7
Central Africa	6	14	33	6.0	5.8
Northern Africa	30	62	113	4.8	6.1
Southern Africa	10	17	29	3.5	3.5

Table 11.1 Urban population and its annual growth rate (%)

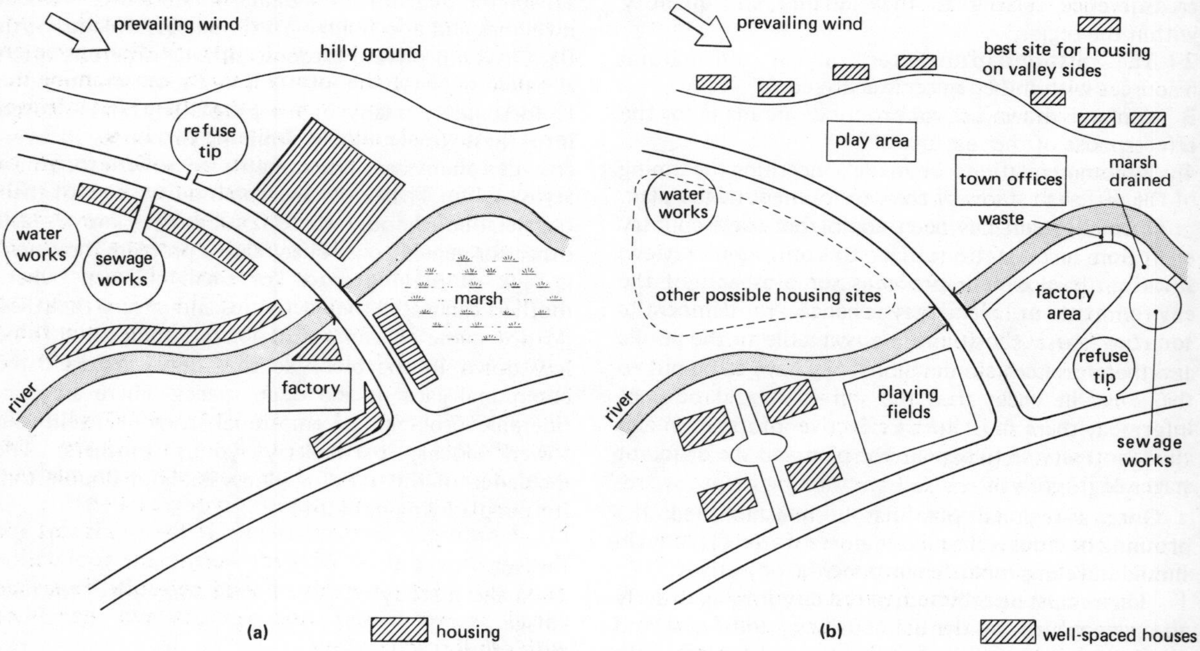

fig 11.14
(a) Plan of a town with poor siting of buildings and services
(b) Plan of the same site based on hygienic principles

ii) the community services, e.g. hospital, water works, sewage works, fire station, police station, rubbish dump, railway station, bus station, schools and so on

iii) the industrial area

iv) the commercial area; show the shopping areas for housing estates

v) the high ground, prevailing winds, rivers and streams and their direction of flow; indicate where water extraction is made

3 Try to find out something of the history of the community to show how long it has been situated there, and how the population has increased over the last hundred years.

Questions

1 Do you think that the community has developed slowly over the years or has it been planned? Do the more recent developments show evidence of planning?

2 Were the new housing estates planned so as to be provided with all the necessary services?

3 Have the community services been sited to the best possible advantage?

4 Do the local shopping areas provide all that is required by the people?

5 Have the factories all been sited in an industrial area, or are they mixed into the housing areas?

6 How would you alter the plan of the town given a free hand? What small changes could make a great deal of difference to the lives of the people?

11.12 Siting of houses

There are two chief considerations:

i) soil and underlying rock

ii) aspect.

i) **Soils**

Those that allow water to pass through freely are termed *permeable* and include those derived from sandstone and limestone. Those through which water does not pass are *impermeable* and usually include a high proportion of clay particles. Clearly it is best to build a house on permeable soil assuming that the layer of soil is thick enough. If it is thin and overlays impermeable rock, then the soil will be continually soaked with water. Furthermore a flat site will be damper than a sloping one. Such sites can be improved by laying drains in the soil and sloping them towards a cleared water course before building. Surface drains around the house should carry off rainwater from the house and its immediate surroundings.

Houses can be protected against damp rising through the porous walls by the use of a *damp-proof course* at the base of the walls. This is an impermeable layer made of tiles, polythene, slates or tarred cardboard. Ground floors of houses are likewise protected from damp by an impermeable layer underneath the concrete base.

ii) **Aspect**

Light and free circulation of air are important factors in the siting of a house and are essential to health. Therefore the house should not be closely surrounded and overshadowed by buildings or trees. Houses in the country can be protected from wind by belts of trees or by hills but they should not be so close that they keep the house in shadow. Many countries have laws governing the heights of houses and the amount of open space that surrounds them. The amount of sunlight received by a house will differ according to the region of the world in which the house is built. In the Northern Hemisphere the house should, ideally, have a clear southern aspect to obtain as much sunlight as possible. Living rooms and bedrooms should face south to receive the warmth of the mid-day sun, while the main roof area should face south-west to take the rain from the prevailing south-west wind. Kitchens and food stores used to be on the northern, shaded side for coolness, but the use of refrigerators means this is no longer necessary.

11.13 Construction of houses

The following considerations are important in housing construction.

i) **Plans** A set of professionally drawn, detailed plans are essential, whether the house is to be built privately or by local government. They mean that (a) the owner knows exactly what he is going to get in terms of size of rooms, doors, windows and so on, (b) the builder can follow the specifications given on the plan, and (c) the local authority can examine the plans in advance and so ensure that there is no infringement of town planning or building bye-laws.

ii) **Foundations** These must be solid enough to prevent the weight of the construction material distorting the final structure. The nature of the ground must be considered. The foundations must also protect against rising damp. Three types are in common use:

a) *Concrete-filled trenches* If the soil is deep the trenches are dug about 1m in depth and then filled with concrete. The load-bearing walls are built on them.

b) *Concrete raft* This is a slab of concrete, larger than the house and at least 12 cm in thickness. This type of foundation is vital on a site that is on made-up soil, such as a disused rubbish tip or gravel pit.

c) *Reinforced concrete piles* If the soil is soft and the building will be of great weight, the foundation must extend downwards to solid bed-rock. Holes are bored or excavated to the bed-rock and then filled with reinforced concrete.

iii) **Walls** The materials which are used for a house wall depend very much on the country and its local raw materials. The Canadians and Scandinavians build of wood from their extensive forests, the river dweller from mud and reed and the plain dweller from wood and grass. Whatever the material, the function of the outer walls is to keep out rain, wind, cold and the rays of the sun.

Bricks are a common type of building material. They are made of clay, which is ground and mixed, moulded, dried and then baked. In some parts of the world they are still made of mud. The size is such that they can easily be grasped in the hand, for no mechanical means has yet been found to lay bricks. They are laid in overlapping rows and bonded together with mortar (sand and cement).

Limestone, sandstone and granite have been used for centuries as building materials. The acid atmosphere of large cities tends to attack limestone so that the surface gradually crumbles. Many cathedrals built of limestone have needed restoration after hundreds of years of slow decay. Certain modern buildings are made of pre-cast *concrete slabs* that fit into each other and thus speed the time of erection.

Outer walls should always have an air space between two layers of material in order to provide heat insulation and to prevent the entry of damp through the pores of the bricks as a result of driving rain. This cavity may be filled with plastic foam for insulation. In addition a *damp-proof course* at the base of the wall prevents moisture rising up from the foundations (see fig 11.15). A *party wall* may divide two adjacent houses and, since it is weight bearing, this must be of similar construction to an outer wall. In addition it should be fire resistant and sound proof. *Partition walls* divide a house into separate rooms. They are not weight-bearing and may thus be of light construction, without an air-space.

To protect outer walls of stone or brick against damp they can be faced with a layer of cement, tiles, slates or wood cladding which may also have an ornamental function.

iv) **Roofs** The roof, like the walls of a house, must provide protection against wind, rain, snow and extremes of temperature. It is traditionally made of *non-absorbent* material in the form of *clay tiles, slates* or *thin stone slabs* which are laid on roof timbers so that they overlap. Asbestos sheeting or corrugated galvanised steel may also be used. Some other metals, particularly *copper* sheeting, are also used and these are not only relatively invulnerable to atmospheric corrosion, but are also good conductors and reflectors of heat. *Bituminous felt, glass* and *wood* are less widely used. Many houses in the British Isles are still thatched with *straw or reeds* and these are sought after and preserved in the countryside, although their inflammability is a source of danger. The same is true of grass roofs used in tropical regions.

The design of the roof varies according to the material of which it is made and the weather conditions likely to affect it. Thatched roofs have a steep *pitch* (angle of slope) so that the water will not run through the straw. Countries with a great deal of snow have house roofs with a steep pitch and large over-hanging eaves so that the snow will both slip off and fall clear of the walls. The weight of 15 cm of snow can be so great as to crack supporting beams in timber-framed houses. In tropical and temperate countries roofs generally have a pitch less than 45° and in fact many roofs are made almost flat, but with a slight tilt so

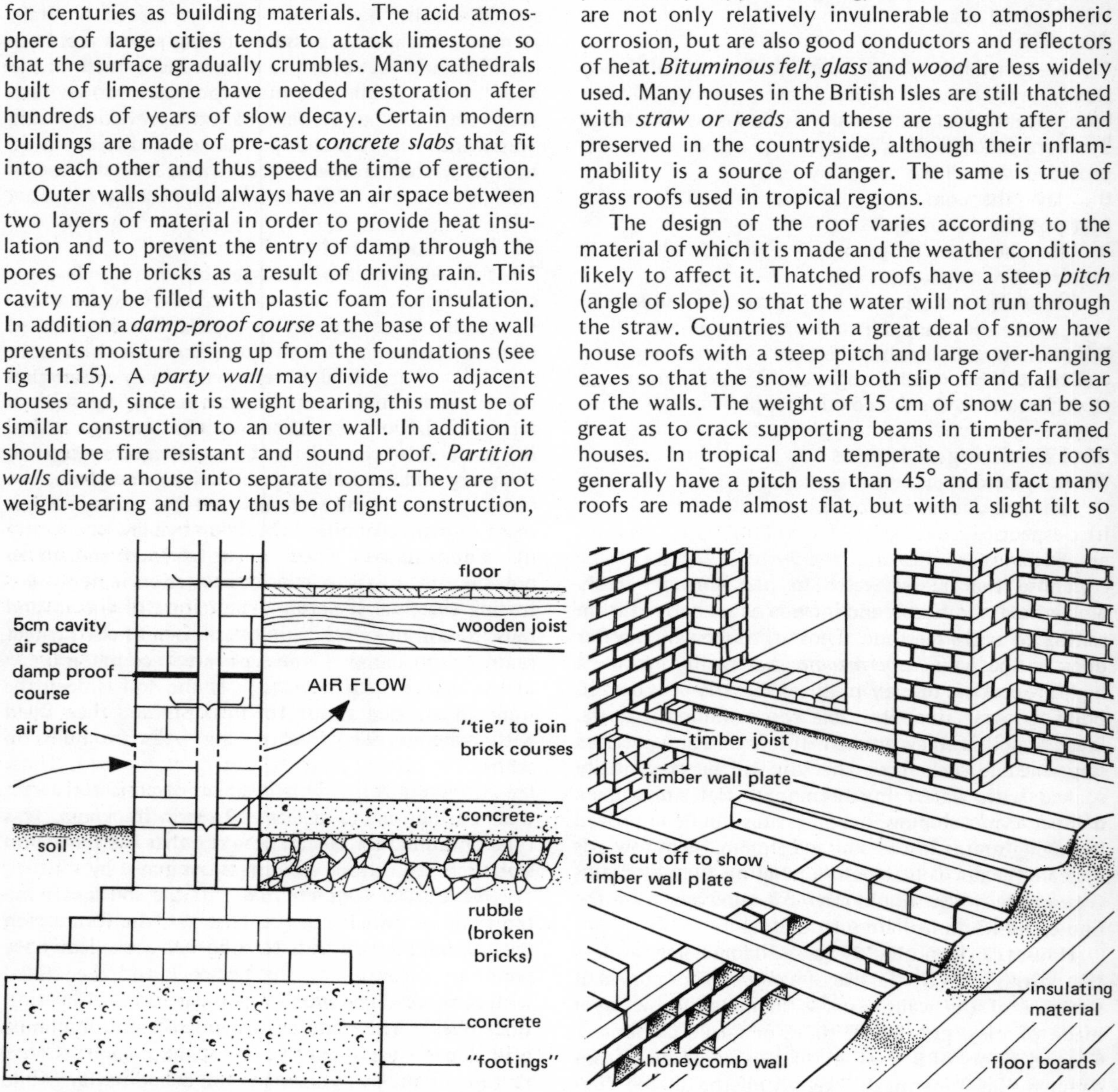

fig 11.15 An outer cavity wall. Notice the damp-proof course above the soil level but below the joists

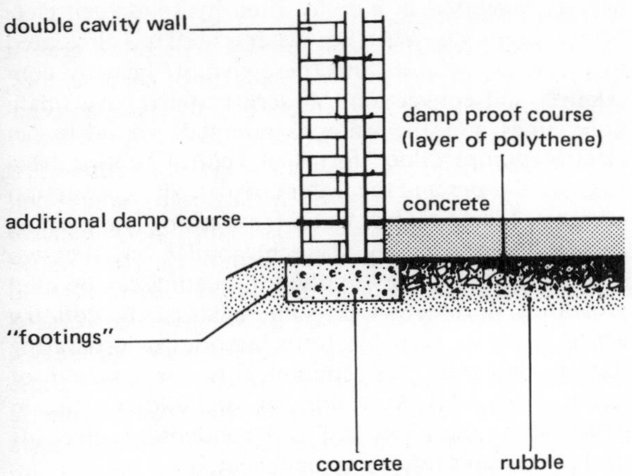

fig 11.16 An outer cavity wall without a suspended floor but with a solid concrete floor

that the rain water runs into the *gutterings.* These are a most important feature, placed around the edge of all types of roof to prevent the rain water running onto the walls and so making them damp.

All roofs in temperate and colder regions should be insulated to prevent heat loss. There is generally a loft, which is the space between the roof and the ceilings of the upper rooms. This can be used for storage purposes, and its floor should be covered with insulating materials such as glass fibre or with granulated plastic which is spread over the ceilings between the timber joists.

v) **Floors** These may be of the solid variety, where a hard-core (broken rocks or other rubble) base is covered with a damp-proof course of polythene and then with concrete. For a wooden floor, the timber *bearers* which hold the *floor boards* must have an adequately ventilated space between them and the earth in order to keep the wood dry and prevent rotting. Modern building techniques generally involve the use of solid ground floors rather than the type with an air space. The concrete can be covered with wood-blocks or tiles of synthetic material. Upper floors are generally made of wood, supported on beams (joists) of wood, steel or reinforced concrete.

vi) **Windows** These have two main functions: (i) they allow *light* to enter a room and are therefore constructed of a transparent material such as glass, (ii) they can be used for *ventilation* and therefore need to open in some way. One method is for the window to be of two equal parts that slide vertically up and down, known as a *sash window.* The modern type of metal *casement window* has panels which open outwards and they are hung in a frame like a small door. The amount of glass used in housing varies considerably. Older houses have smaller windows, whereas modern houses tend to have large areas of window space. In order to prevent heat loss, modern windows in temperate regions are often made of two layers of glass with an air space between them. This is known as *double glazing.* Every effort is now made in temperate climates to conserve energy in these times of energy shortage.

Ventilation can be adjusted by windows having different sized sections which can be opened by varying amounts. In addition, ventilators or air bricks can allow air into a room, while a room with an open fire or gas fire connected to a chimney also ensures a movement of air indirectly by drawing it through any available opening.

Central heating systems do not require a supply of air in all rooms and therefore rooms with radiators do need some system of ventilation, although there is often sufficient air movement through badly fitting doors and windows. Gas boilers or oil boilers need oxygen for combustion, so grills must be fitted in windows or walls of the rooms where they are located to allow sufficient air to enter.

11.14 Methods of heating

i) *Direct heating* This is a method where the source of heat is in the room to be heated. The commonest method is by means of an *open fire* which burns in a grate with the air for combustion drawn in underneath the grate and the smoke and fumes rising up through a chimney. The fire *radiates warmth* but a great deal of heat is lost when warmed air goes up the chimney with the smoke. The open fire is, therefore, uneconomic in the use of fuel and the room itself is heated unequally. Also the fire must be constantly made up with new wood or coal and dust and smoke are released into the atmosphere unless a smokeless fuel is used. In certain parts of industrial countries smokeless zones have been established to decrease air pollution and in these areas only smokeless fuels can be used.

In order to decrease the disadvantage of an open fire various types of *stove* have been produced in which the rate of burning can be controlled. These are also more economic in heat production in that the warmed air can circulate through the room, and furthermore they require less attention as they burn slowly. They can be kept alight overnight by shutting down the draft but they do need more efficient ventilation, since there is a greater chance of fumes entering the room if they are free-standing with a small chimney pipe.

Gas and *electric fires* are popular forms of heating, since they can be turned on and off with minimum effort, and do not involve the space and dirt of fuel storage. The gas fire with a small chimney (flue) radiates plenty of heat besides warming air which circulates through the back of the fire. There is also a flueless type, which is fixed and attached to the gas main, and a portable type which can be moved around and receives its gas from a gas cylinder. Since these have no flue, adequate ventilation is important. With electric fires oxygen is not consumed so ventilation is not important. The fire may be of the *radiator* type where the heating element is wound around a clay former. The element produces radiant heat which is reflected by a polished sheet of curved metal. The *convector* type, which again may be fixed or portable, has a heating element but in this case cold air enters the bottom of the cabinet and passes out at the top as warm air which heats the room by convection.

Oil is used in some country districts for direct heating of rooms and, provided that the appliances are kept clean, they are a satisfactory form of heating. As with electric fires these come in two forms, radiators and convectors. The oil is burned from the top of a wick on which it vapourises. The burning consumes oxygen and therefore additional room ventilation is necessary. Great care must always be taken to ensure that oil stoves are in such a position that they cannot be overturned and start a fire in the house.

ii) *Indirect heating* This is a system where the heat is generated at some distance away from the rooms that are to be warmed. The heat is transferred to the rooms by water, steam or ducted warmed air. This system is commonly known as *central heating* and the heat is generated in a boiler fired by solid fuel, electricity, oil or gas. When hot water is used it is circulated in pipes to radiators and these give off heat by conduction and convection. Modern systems have small-bore pipes and the water is pumped around by an electric pump. Older forms of central heating have large-bore pipes and the water circulates by convection currents. Domestic hot water for washing and bathing is usually provided by the same boiler but the two systems of hot water supply and heating can be used separately or together. In some areas of the country waste steam is available from factories or generating stations and thus it is economical to use a system of steam pipes to heat a number of buildings. Steam passes along the pipes and as it condenses it gives off heat. The water returns to be reheated.

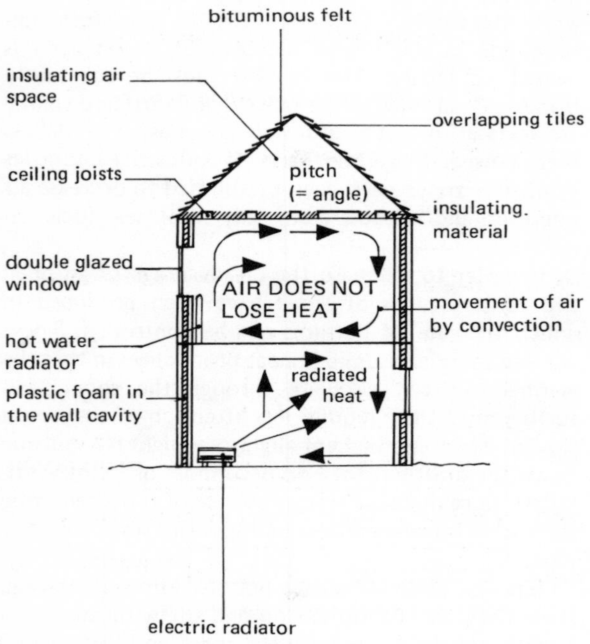

fig 11.17 Roof design, wall insulation and double glazing to conserve heat

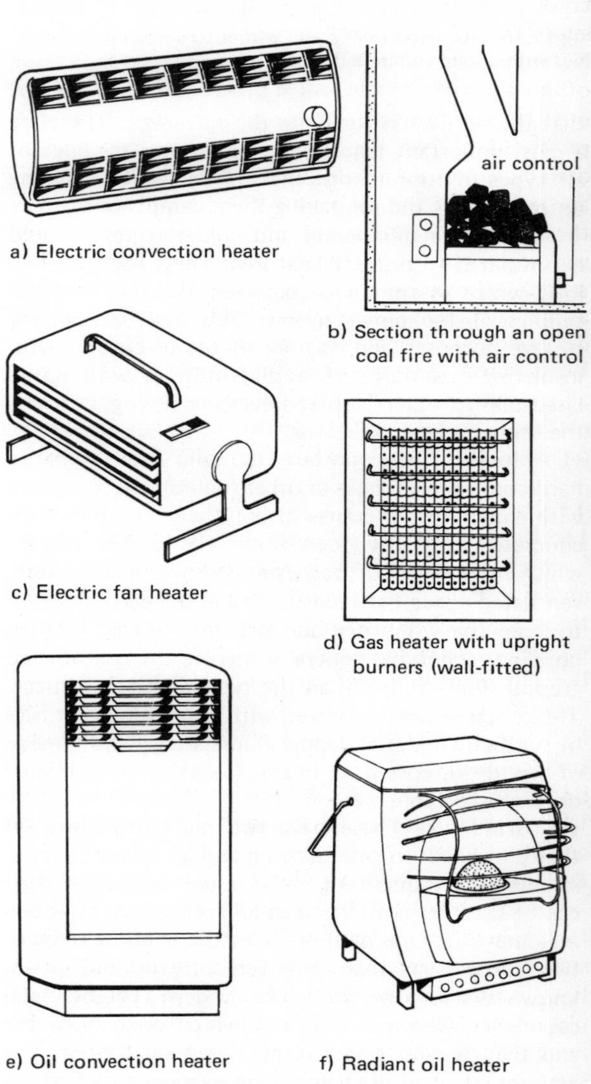

fig 11.18 Forms of heating

11.15 New sources of energy

Solar heating The world's energy originates from the sun. In the past this energy produced living material which has become fossilised, forming the coal and oil reserves that we use today. These fuels are limited and will be used up within a few hundred years, so alternative sources of energy must be found. One method is to harness the solar energy directly. Where there are long periods of sunlight in tropical or subtropical areas, the light can be trapped by 'solar panels' which cover water compartments. The heated water is stored in a hot water tank. Experimental work has shown that even in temperate latitudes sufficient sunlight is present to provide up to half of the heat required for domestic and central heating. It is clear that, with mass-production of solar panels and the consequent lowering of costs, this method of energy supply could effectively supplement traditional fuels.

Nuclear energy Nuclear fission resulting in the release of energy is already in use in power stations throughout the world to generate electricity. The heat energy produced heats water to drive conventional turbine dynamos. Industrial nations with these power stations are faced with the problem of disposal of waste fuels that are highly radioactive, and in addition there are widespread fears as to the safety of these power stations. Environmentalists in America and Europe as well as Japan, campaign actively against the further development of this source of energy.

Wind power, tidal power and wave energy These are all the subjects of considerable research as sources of electrical current but as yet there is no large scale production.

11.16 Light

Lighting is of two kinds, *natural* and *artificial.*

Windows admit sunlight into a room and also play an important part in ventilation, but unfortunately the amount of natural light available depends upon a number of variable factors:

i) The proximity and height of neighbouring buildings and trees.

ii) The presence or absence of fog, mist, cloud, smoke, haze, etc.

iii) The intensity of daylight which depends upon the season of the year (giving short or long daylength) and also the latitude.

Provided that a house has plenty of windows, and that some of these are facing south (in the Northern Hemisphere, or vice versa in the Southern Hemisphere), sunshine will enter the room and provide adequate light during the day. Sunlight is also important as its ultra-violet rays will kill bacteria, but it must be remembered that this can only happen through an open window, not through glass. Therefore it is important that windows should be opened from time to time in sunlight. In temperate climates large windows can give plenty of sunlight in the summer but in winter they allow a great deal of heat to escape unless they are double glazed.

Modern Man does not cease his activities when the sun has set, and so some degree of artificial lighting is essential. Furthermore in certain buildings a high light intensity is required at all times; for example in laboratories, factories and hospitals where close and detailed work demands clear vision. In highly industrialised societies nearly every building can be reached by an electricity grid system, so electric lighting is supplied by filament lamps or by fluorescent tubes. The intensity of illumination required will differ, but insufficient or badly arranged lighting should be avoided. It is always important that light should fall directly on the working surface or reading area and this is achieved by the light source being directed over the shoulder.

Remote rural areas in any country often find that electricity is not available. In these places lamps use some alternative fuel such as paraffin or gas. The burning of each of these fuels is used to heat an incandescent mantle. Candles are a last source of light that can be used in the event of no other source being available.

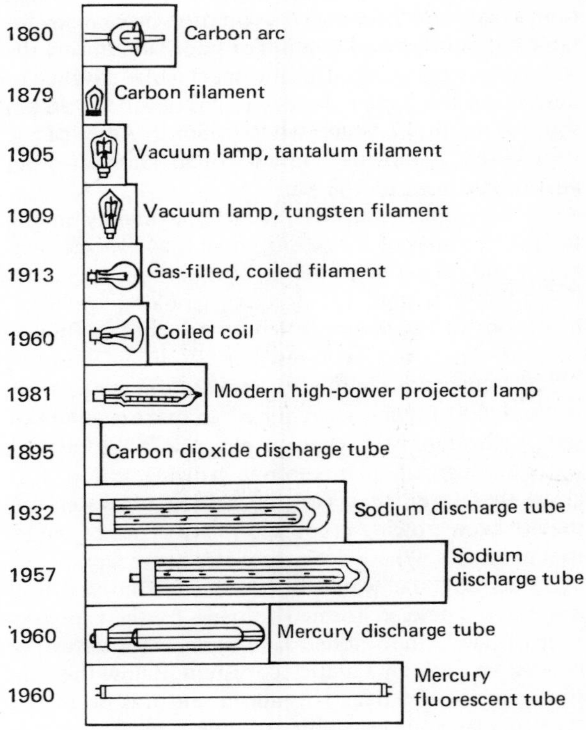

fig 11.19 The development of electric lighting. The length of each column shows the efficiency of each source for the same amount of electricity

11.20 Water supply

Some 14% of the world's population (630 million people) live in dry areas, and one-tenth of these are affected by the gradual spread of deserts. Many of these people constantly face death and starvation, as in the arid lands on the southern edge of the Sahara Desert. On the other hand, in developed countries such as the U.S.A. and Australia there are scientific and other resources to help cope with the declining productivity of land in dry areas, so their populations are protected against the problems faced by poorer societies.

Only one-third of the earth's land area is capable of growing crops, but in fact only one-third of this area is actually cultivated. With the population of the world increasing and more food required, it is estimated that Man has already lost about 2 500 000 square kilometres through land becoming deserts. The trends of world climate are difficult to predict, but meteorologists tend to deny that it is climatic trends that are responsible for the spread of deserts. The semi-arid lands have an annual rainfall of only 10–40 cm and the slow cycle of vegetation renewal, together with wind and rain, are the natural factors that have been aggravated by Man's activities. For example, by too many flocks of sheep, goats, cattle and camels overgrazing, the burning of vegetation by nomads for their fires and the cultivation of poor land during the wet years have resulted in the spectacular advance of deserts. In the Sudan the desert has advanced 96 km southwards in 17 years, while in South America the Atar desert is moving one and a half kilometres per year over a front of 160 km.

In developed countries, which lie mainly in the temperate zones of the world, there is an average rainfall of 100 cm per year. Let us consider how it is used. Fresh water is used for drinking, cooking, industry, irrigation of agricultural land and cooling power stations. The latter requires 4×10^7 litres of water, 2 500 tonnes are required to make a tonne of paper, while 4 400 tonnes are required to make a tonne of steel. A hectare of sugar cane passes 2 700 tonnes of water through the plants in one growing season. It is clear therefore, that vast quantities of water are required for growing crops and many types of industrial processes. Where does the water come from?

1 Much of the water is extracted from artificial *reservoirs.* These are formed on rivers by the construction of dams, the flooded land replacing a terrestrial ecosystem with an aquatic ecosystem. Below the dam the water in the river is reduced and this decreased flow increases the possibility of pollution because any effluent is less diluted. Reservoirs have a limited life because they silt up and gradually their water holding capacity is reduced.

fig 11.20 Famine in the Sahel

2 Many towns extract water along the length of *rivers.* Water is removed, with treated sewage being returned, so that the process of extraction and return is repeated many times. This makes it doubly important that Man does not pollute the rivers he uses, and in order to obtain more water in the future the gap between use and re-use must be shortened. Water will need to be extracted, used, purified, returned to the river and immediately used again.

3 Rain water percolates through the soil and eventually finds its level at the *water table.* When water is removed from these underground sources (by wells or pumping), the level of the water table falls. This can profoundly influence the growth of vegetation over a wide area. Plant roots find it increasingly difficult to grow down and reach the water. This water is not unlimited, and unless it is replaced as it is used, the water level sinks lower and lower. In the U.S.A., wells orginally pumping water from a depth of 36 m had to pump from 72 m ten years later.

11.21 Sources of water

Water described as 'pure' from the point of view of hygiene is different from laboratory 'pure' water which is distilled or deionised. Distilled water contains no dissolved substances, whereas 'pure' drinking water has in it small quantities of dissolved salts that are not injurious to Man. The following features are important in 'pure' drinking water:

1 colourless
2 neutral pH (7)
3 no excess salts causing hardness
4 no organic matter or pathogenic organisms
5 no suspended matter
6 no harmful metals
7 free from taste and smell, and well aerated.

Fresh water This makes up only 3% of the earth's total water resources and two-thirds of this is locked up in the Polar ice caps. The majority of water on earth is in the salt water of the oceans, but this is, however, the source of our rain water, for it evaporates, condenses, forms clouds and is finally precipitated in the form of rain, hail or snow. Much of this water eventually returns to the seas in rivers and streams. Rain water is the purest form of natural water although it contains dissolved gases and other impurities from the air, particularly over towns. Water has great solvent properties and therefore in its passage through the soil and rocks it dissolves some of the substances with which it comes into contact (see fig 11.21).

Rain water is thus very suitable for drinking, and throughout history Man has collected water for this purpose. In tropical countries rain water is collected from roofs during the rainy season, using gutters and pipes to direct water into storage tanks. The first rain water should not be used as this contains dirt and dust from the roof, so during the first rains the collecting pipe should be directed away from the tank. Tanks must always have a cover to keep out mosquitoes which would otherwise breed there. Lead tanks and pipes should never be used since they can contaminate the water.

Surface water The rain falls on the earth and some of it runs off as surface water to form lakes, ponds, streams and rivers which finally flow back into the oceans. Artificial lakes can be made by building dams to hold back river water in a natural valley. Streams and lakes in upland mountain areas have water that is relatively pure but as the water flows down into rivers and across populated areas it collects sewage and industrial waste, and thus becomes unfit for drinking. River water extracted before it reaches populated areas or cultivated land is generally suitable for drinking. Even as it becomes contaminated it is capable of some degree of self-purification by:

i) oxidation of impurities, using oxygen produced by water plants

ii) decomposition and removal of organic impurities by bacteria

iii) dilution of impurities as tributary streams flow into the main river.

Provided that the concentration of organic waste is not too great, rivers can purify themselves even after flowing through large towns. Nevertheless, 'downstream' water should never be drunk before it has been purified artificially by treatment at a waterworks or at least boiled in the kitchen.

Springs As rain water soaks into the ground it moves through the upper layers until it reaches the first impervious layer (clay or non-porous rock). The water follows the slope of this layer and emerges as a spring where the non-porous layer reaches the surface, often at the foot of a hill or the side of a valley. The water can be regarded as pure enough for drinking but its content of dissolved substances depends on the layers through which it has passed. Chalk or limestone produces 'hard' water, which will not easily lather with soap. Many towns have developed at the sites of springs which were thought to have medicinal properties because of the salts dissolved in the water.

11.22 Quality of water

The problems of water contamination are due mainly to the following:

i) *Natural impurities*, which can be both chemical and biological. Gases such as oxygen, nitrogen and carbon dioxide are picked up even by raindrops, but they are not dangerous to living organisms. Soluble salts in rock strata are picked up by percolating rain water. These include bicarbonates, sulphates and chlorides of calcium, magnesium and sodium. They are not dangerous but make water 'hard'. These salts are also deposited in water pipes, boilers and kettles causing problems of water flow and maintenance.

Natural waters containing salts are often used for irrigation and the concentrations allowable vary according to the crops supplied, the air temperature and the evaporation rates. A suitable range of salt concentration is 70 to 3 500 mg/litre but in regions of high evaporation rates the higher ranges should not be used, as evaporation causes increased concentration of salts near the top of the soil. Large areas of productive land in Africa, India and Australia have lost their fertility because of salt contamination through irrigation.

ii) *Mining* often results in the release of salt water (brine) into normally fresh water.

iii) *Sewage* is obviously a major source of contamination and many diseases can be contracted through drinking affected water (see Table 11.2). Sand filtration and chlorination contribute to the control of water-borne disease.

Organic waste in water provides food for bacteria, which multiply and in turn increase the total oxygen requirement of living organisms in the water (this is known as the Biological Oxygen Demand (B.O.D.)). The amount of dissolved oxygen is thereby decreased and this affects all organisms. Fish are the most sensitive, but in extreme cases this depletion of oxygen results in the death of all living organisms.

Normally, natural purification can cope with these problems when human population density is low. Sedimentation, sunlight (ultra-violet radiation kills bacteria) and natural consumption of pathogenic bacteria all play their part. As human population

Pathogen	Disease
Bacteria	
Bacillus typhosum *(Eberthella typhi)*	Typhoid fever
B. paratyphosum	Paratyphoid fever
B. dysenteriae	Bacillary dysentery
Vibrio cholera	Cholera
Protozoa	
Entamoeba histolytica	Amoebic dysentery
Giardia lamblia *(G. intestinalis)*	Giardiasis
Viruses	
Liver-infecting virus	Infectious hepatitis
Polio virus	Poliomyelitis
Virus (spread by mosquito needing water for part of life cycle)	Yellow fever
Platyhelminths and Nematodes	
Tapeworms	Onchocerciasis
Schistosoma	Bilharzia
Algae	
Euglena and other algae	Gastroenteritis

Table 11.2 Waterborne disease organisms

increases, human waste, and that of farm animals, overloads this natural purification process.

iv) Major pollution problems arise from the large scale *industrialisation* of many countries. Paper mills, petroleum refineries, food processing plants, etc. all produce contaminating substances. Organic waste, toxic metals, salts, acids, pesticides and herbicides are all examples of pollutants related to industrial processes. Excessive use of fertilisers and their subsequent run-off into water catchment areas presents further problems (see section 11.02).

Modern water treatment must take into account all these contaminants and the effective provision of pure water depends not only on water treatment but also on sewage treatment. The latter reduces the impact of large quantities of sewage on water supplies and ensures that water can be used over and over again. Water from the River Thames in London goes through the following sequence an average of *five* times before it reaches the sea.

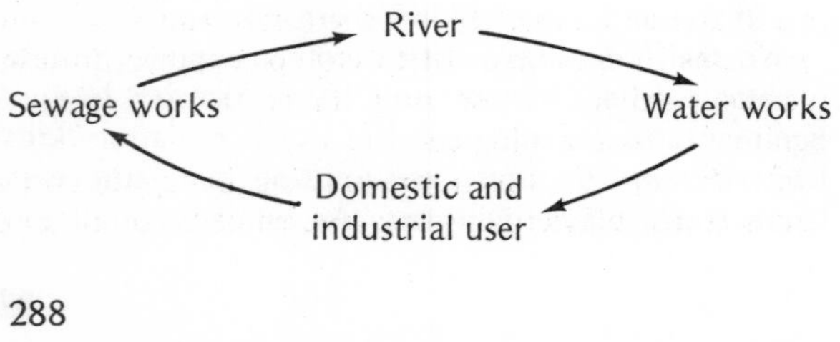

11.23 Wells

There are three types of well.

i) *Shallow wells* These are dug down to the top layer of impervious rock which retains the rain water. This type of well is liable to pollution from nearby cesspools, deep latrines or farmyards. Heavy rains may cause a sudden rise in the underground water level which may result in sewage being carried into the well. Contamination can be avoided by the following precautions:

a) siting the well away from possible sources of pollution

b) lining the well with concrete or brick to make it impervious to pollution from the sides

c) designing the well so that the lining can be easily cleaned from the top

d) covering the ground around the opening of the well with water-tight concrete or paving stones to a distance of two metres from the well

e) projecting the well lining above the soil level to a height of about 50 cm to prevent surface drainage, and protecting the opening by a tightly fitting cover

f) testing the water regularly for chemical and bacterial pollution.

ii) *Deep wells* These wells are dug through the top impervious layer to tap the water table over the next impervious layer of rock (see fig 11.22). This water is usually cleaner and better filtered, having travelled a long way through the deeper porous rocks. Deep well water is always safer for drinking than surface well water. The deep well should be protected in the same way as a shallow well (see above (a) to (f)).

iii) *Artesian wells* These wells tap water that is trapped under pressure between two impervious layers. The top of the well must be below the underground water level and for this reason the well is usually in a basin or valley (see fig 11.21). The water rises to the top of the well or even above it like a fountain.

11.24 Water purification plants

Demand for water on a large scale in towns has necessitated the storage of enormous quantities of water in reservoirs. From these a steady supply of water is obtainable throughout the year, even in periods of dry weather. The upland reservoirs and lakes have large areas of land surrounding them through which run drainage streams. These are called catchment areas and great care has to be taken in controlling human occupation and farming to ensure that there is no contamination of drinking water. Reservoirs supplying industrial communities may be many miles away and distribute their water through many kilometres of pipelines. In the United Kingdom, the Midland and northern industrial areas obtain their water from flooded valleys in North Wales and the lakes of Cumbria.

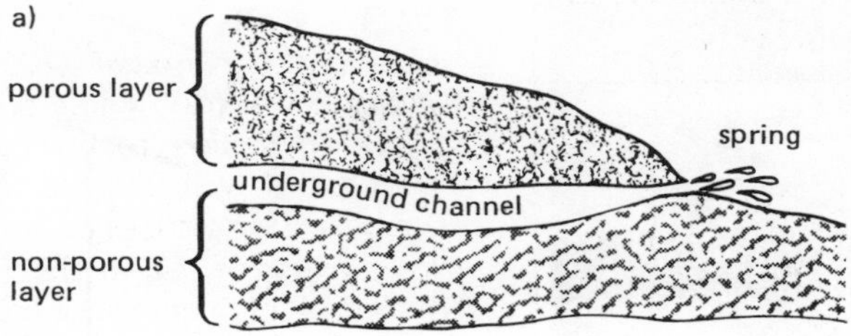

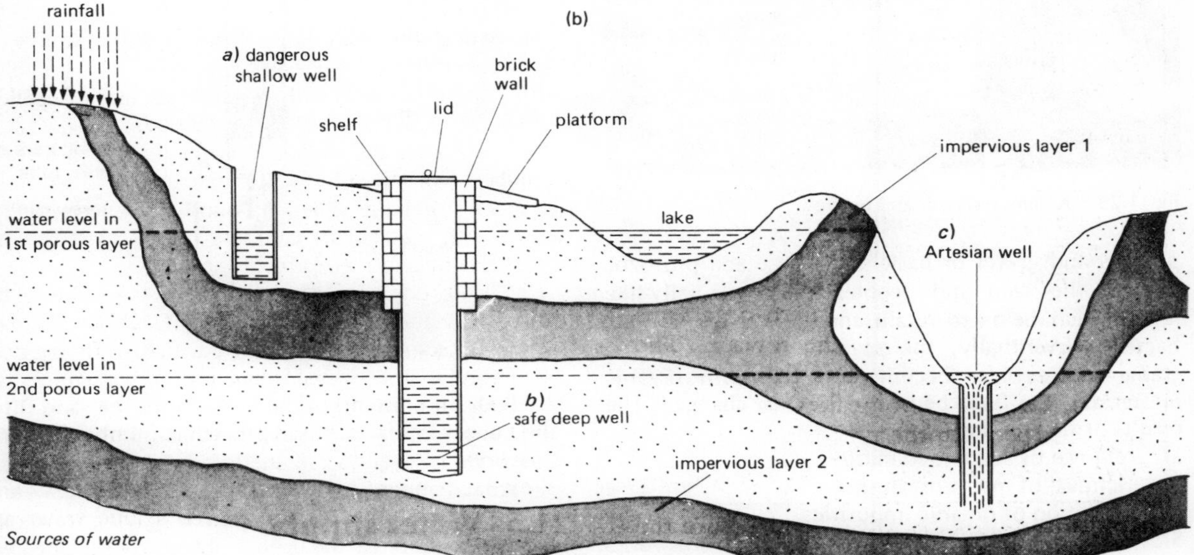

fig 11.21
(a) Rock formations for a spring source
(b) Rock formations and types of well (note precautions against contamination of the deep well)

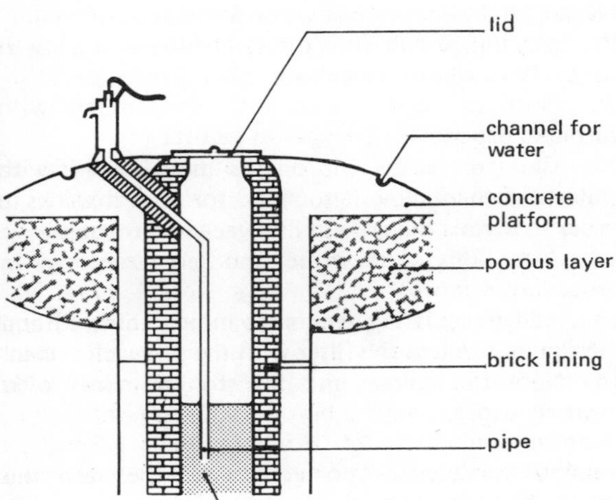

fig 11.22 A protected well with a pump for withdrawing water

Industrial communities and towns that draw their water from large rivers also need man-made reservoirs along the rivers' lengths to provide reserves when the flow of the rivers decreases.

The water obtained from reservoirs must be treated in various ways to make sure that it reaches the consumer in a pure state. It goes through the following processes:

i) *Sedimentation* In lakes and reservoirs, disease-causing bacteria in the water are killed by sunlight and, together with other suspended matter, sink to the bottom and are decomposed by saprophytic bacteria. River water is drawn off into large artificial concrete reservoirs and left for sedimentation to occur.

ii) *Coagulation* The water is then pumped into further tanks where the next stage of purification begins with the addition of solutions of aluminium sulphate and lime. These chemicals cause the fine particles of clay and silt to clump together, forming larger particles. These sink under the influence of gravity onto a settling bed.

iii) *Filtration* From the settling beds, the water runs through filter beds constructed of layers of sand,

fig 11.23 A dam, reservoir and outflow

pebbles and gravel or basic slag. Algae and protozoa live on the sand and pebbles forming a jelly-like layer which helps to retain and destroy pathogenic bacteria. Eventually, the spaces between the sand grains and pebbles becomes clogged with organic matter and the filter bed must then be changed. The results of filtration are:

a) removal of suspended matter
b) removal of micro-organisms
c) oxidation of organic impurities.

Slow sand filters are often replaced in modern works by rapid pressure filters which need to be cleaned every two or three days by back washing with water.

Investigation

To make a model of a sand filter and with it to purify a sample of contaminated water

Procedure

1 Set up the apparatus shown in fig 11.24.

2 Add some clayey soil to a beaker of tap-water and mix it thoroughly.

3 Pour this slowly into the filter and collect the effluent in a beaker. Keep the beaker and its contents for reference purposes.

Questions

1 How does the effluent compare with the original sample of water?

2 What is the difference between this filter and a working sand filter at a water treatment plant?

Procedure

4 To another large beaker half full of tap-water add some clayey soil and mix it thoroughly.

5 Add a flocculating agent (about 3g or half a teaspoonful of aluminium sulphate to 500 cm^3 of distilled water) to the water sample until the beaker is about ¾ full, then fill the beaker by adding lime water.

6 Allow the beaker to stand and then examine the contents.

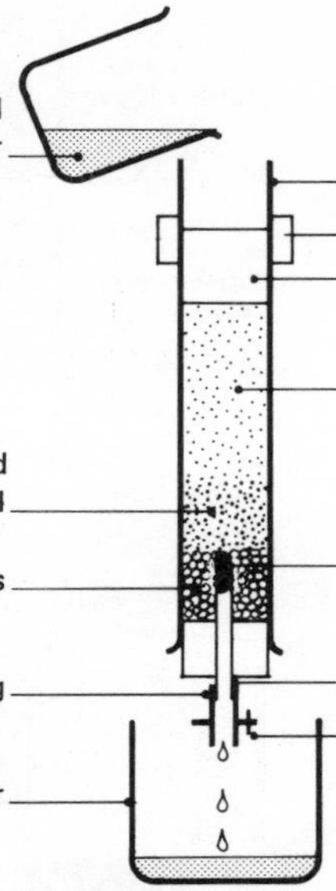

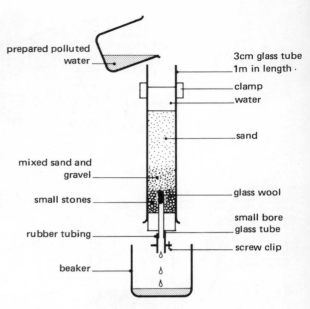

fig 11.24 A model filter bed

7 Pour the contents of the beaker into the sand filter and collect in a beaker as in procedure number 3 above.

Questions

3 What happens to the fine suspension of clay and silt when the flocculating agent and lime water are added?

4 Is the effluent of this sample any cleaner than the effluent from the first sample in procedure 1 to 3?

5 What would be the effect of the flocculating agents on any micro-organisms in this sample?

Procedure

8 In another beaker, add some clayey soil, animal faecal matter and urine to some water and stir well.

9 Pour this sample into the sand filter and allow it to run through into a beaker.

10 Note the colour and test for ammonia with turmeric paper.

11 Close the screw clip on the model. Fill it with water and then allow it to stand for several weeks in order to permit the growth of algae and protozoa.

12 After this time make another sample as in procedure 8 above.

13 Add flocculating agents (as in procedure 5) and then pour through the filter with the screw clip open.

14 Note the colour and test for ammonia with turmeric paper.

Question

6 What differences do you notice between the appearance and ammonia content of the two samples tested in sections 10 and 14 above?

iv) *Chlorination* After filtration the water is passed to the chlorinating plant. Small quantities of chlorine (0.25 parts per million of water) are passed into the

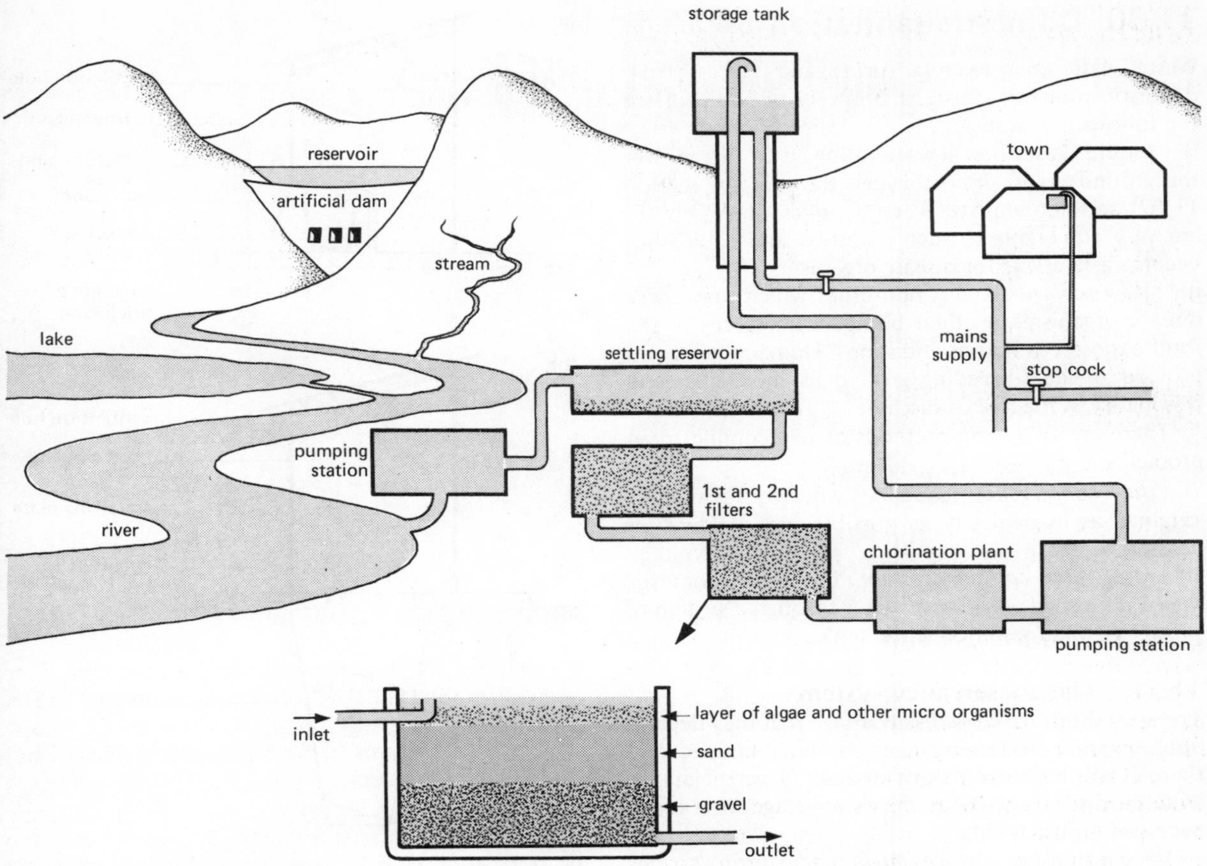

fig 11.25 The water supply of a town, with detail of a filter bed

water to kill any remaining bacteria. The chlorine combines with water to form hypochlorous acid which releases oxygen and destroys the bacteria. Water sometimes has an excess of chlorine and this gives it an unpleasant, 'medicinal' taste.

Chloride of lime (calcium hypochlorite) is sometimes used instead of chlorine (3 g of chloride of lime to 1 000 litres of water).

v) *Service reservoirs* The clear, filtered water that has been rendered germ-free is now pumped through water mains to service reservoirs where there is no further risk of contamination. The reservoirs are generally on high ground, from where the water can flow by gravity to neighbouring towns or villages.

vi) *Storage tanks in buildings* Tanks of varying capacity (according to demand, about 300 litres for a house) are placed in the roofs of buildings. Water under pressure enters the tank by a valve that closes when the tank is full. Water can be drawn by gravity from the tank for washing and cleaning purposes. The tank must be covered to prevent contamination by airborne dirt, birds or rodents, but it should not be used for drinking. Generally the cold water tap in the kitchen is direct from the water mains and this water only should be used for drinking.

In underdeveloped areas houses may not have an individual supply but water is delivered by means of a standpipe. The ground around the standpipe should be cemented or paved (as with a well) to prevent the ground becoming muddy and to allow waste water to be drained off.

Water can be contaminated at source, in transit, during storage, during filtration and during distribution, so there must be vigilance at all stages to ensure that water is kept pure for community use. If the water supply to a house is found to be impure there are two methods of purification.

i) By *boiling* for *at least five minutes.* Pathogenic bacteria and eggs of parasitic worms are destroyed and hardness in the water is also removed. Boiled water loses its oxygen and becomes flat and insipid to the taste.

ii) By the use of chemicals such as *bleaching powder* and *hypochlorite solutions.* Proprietary brands of these compounds can be purchased with the necessary instructions for use.

11.30 Domestic sanitation

Waste water and excreta (urine and faeces) from domestic premises must be properly disposed of for the following reasons:

i) Liquid waste (i.e. sewage) allowed to spread over the ground could drain through the soil (see section 11.22) and contaminate water supplies. Shallow wells are very vulnerable to such pollution and could thus become a reservoir for disease organisms.

ii) Raw sewage attracts houseflies which may carry disease organisms on their bodies from faeces to the food exposed in houses and shops. Human faeces also provide an ideal breeding ground for houseflies and this increases the risk of disease.

There are two main methods of overcoming these problems:

i) *the conservancy system,* whereby sewage is retained as hygienically as possible near to dwelling houses

ii) the *water carriage system,* by which domestic waste is flushed away with water through a system of pipes to a main sewer or septic tank.

11.31 The conservancy system

These methods have the disadvantage that they depend upon excreta remaining near dwelling houses, and there is always some risk of disease. Nevertheless, in country districts without mains sewerage, it is often necessary to use them.

Very primitive *surface latrines* and *drop latrines* are very insanitary. The excreta is exposed to flies and other insects and the liquid can seep into the soil and into rivers.

Bucket latrines are simple and can be used in a hygienic fashion. The contents should be emptied regularly into trenches about 1 m wide and 0.25 m in depth, and then covered with soil. The faeces are decomposed and enrich the soil with salts and thus a grass crop can be grown on the trenching ground. Root crops or any other crop that can be eaten should not be grown.

The earth closet works on a similar principle to the bucket latrine and trench system. The power of the top-soil to disintegrate faecal matter is used in that each time the closet is used, fertile soil is thrown on top of the waste matter (see nitrogen cycle section 11.01).

For the *deep trench* or *pit latrine* a trench is dug from 2–4 metres in depth and 1 m wide. The length of the trench depends upon the number of holes or seat openings, since each opening needs about 0.75 m. The latrine should not be constructed near dwelling houses or within 40 m of any well. The trench or pit is covered by a holed wooden or concrete platform which can be cleaned and later moved to a new site when required.

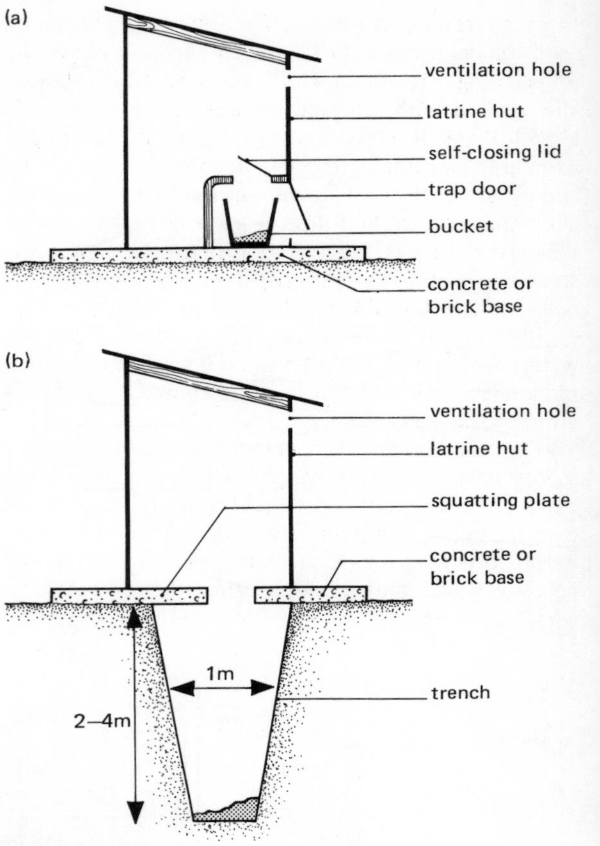

fig 11.26
(a) A bucket latrine
(b) A trench or pit latrine

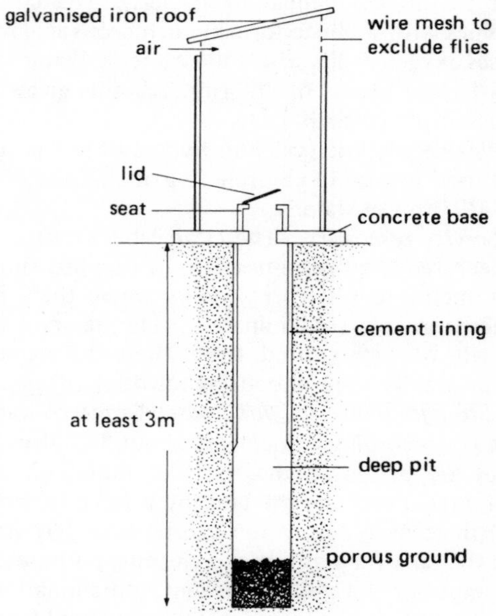

fig 11.27 A deep bore-hole latrine

In most tropical countries, the squatting position is used for defaecation but a seat can be built above the pit opening if required. The cover should be occasionally washed with chloride of lime (calcium hypochlorite). Water is added daily to liquify the faeces and to help in decomposition. No disinfectants should be added, since these will kill the useful bacteria that decompose faeces and destroy pathogenic organisms. When the pit has filled up the platform and superstructure (hut and partitions) should be removed to a new site and the old trench filled in. A clear marker should indicate the position of the old pit. If the soil water level is high the pit should be constructed on a mound about 1½ m tall in order to avoid contamination of the water supply.

The *bore-hole latrine* operates on the same principle as the deep trench, but fills more slowly. It is made by means of a special auger that produces a hole about 6 m deep and ½ m wide. The edge of the hole must be protected by a base or platform and there should be a cover, and a suitable superstructure to give privacy.

11.32 Water carriage systems

Without doubt a great reduction in intestinal diseases over the past century has been achieved by a very simple system whereby excreta can be kept away from insects and carried to a central sewage works or septic tank. This is known as the water carriage system which can generally cope with domestic waste and water, and other materials from industrial premises. Rain water is often kept separate, but in some large towns it can be carried in the same system.

At the beginning of the system is a lavatory pan or basin, which must have a smooth, readily-cleansed, non-absorbent surface. At the base of this the excreta pass through an effective trap and thence, without storage, pass via a soil pipe into a drain. There is also an apparatus for flushing the pan with water, known as a cistern. This is a reservoir of water which empties into the pan by means of a siphon. The siphon is in the form of an inverted U-tube, the longer arm of which acts as a flush pipe. Once the cistern has been discharged of water, a ball-valve tap refills the tank of

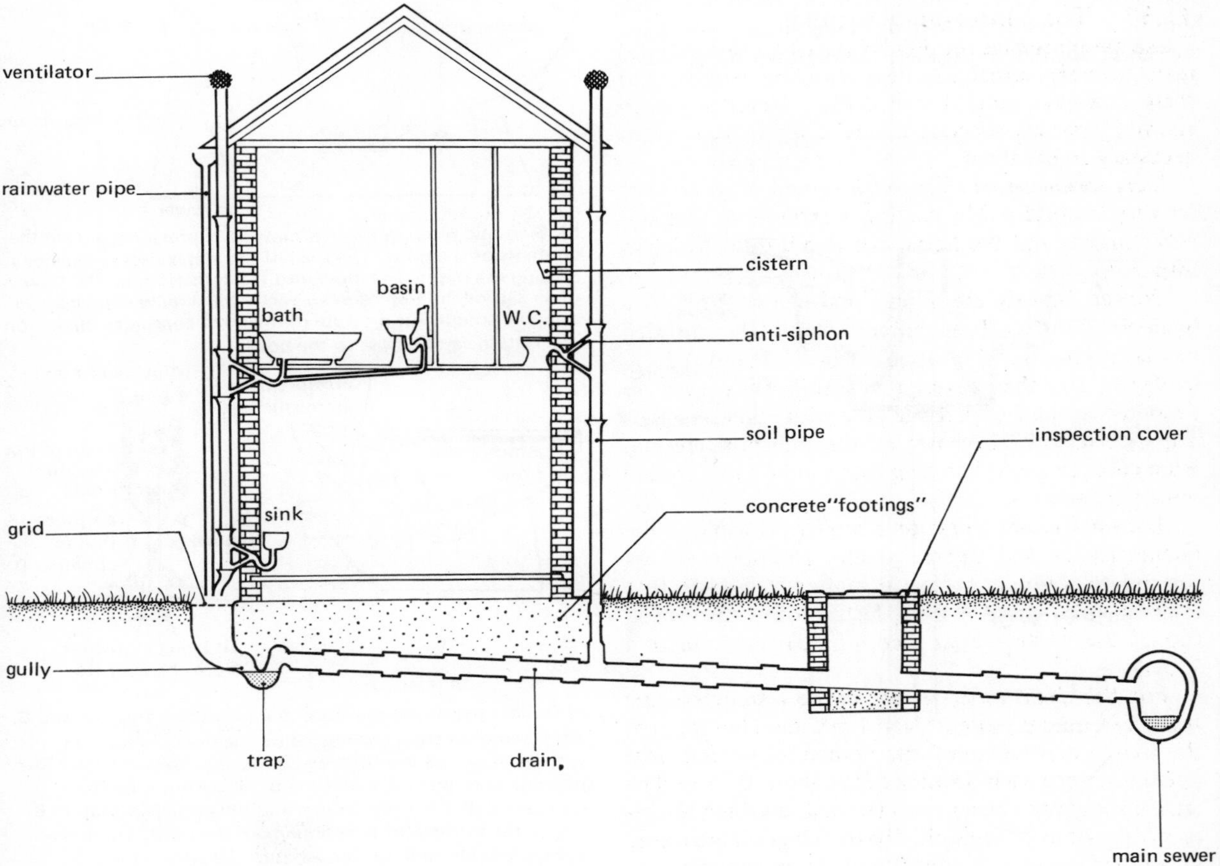

fig 11.28 A typical house drainage system

the cistern automatically. The cistern is usually made of enamelled iron or glazed clay. It delivers approximately 10 litres of water each time that it is used to flush the lavatory clean. The pan itself uses the principle of the siphon, since the disappearance of the water and excreta is due to the difference in pressure either side of the trap (see fig 11.29). The urine and faeces are carried away by means of a cast iron or plastic soil pipe connected to the drain. The lavatory is usually on the outside wall of the house and the soil pipe runs directly into the drain without connecting with any other pipe. The soil pipe also runs upwards above the roof of the house and opens to the air through a wire grid. This prevents the pipe becoming blocked, while allowing air into the system.

Generally all waste discharged from baths, wash basins, sinks and lavatory pans runs directly into a soil pipe, which then discharges into a gully trap outside the building. The pipes underground, within the building and grounds, are known as drains and these are usually made of glazed stoneware. Several house drains may lead into the main drain which passes outwards from the property into a large sewer pipe under the street. This receives drains from all the houses in a street. All drains slope slightly towards the sewer so that the sewage flows down by gravity.

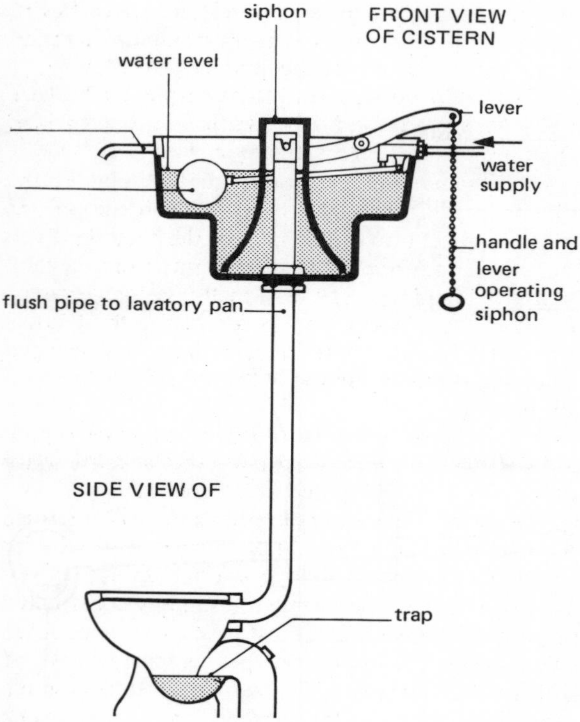

fig 11.29
(a) Lavatory pan (side view) and cistern (front view)

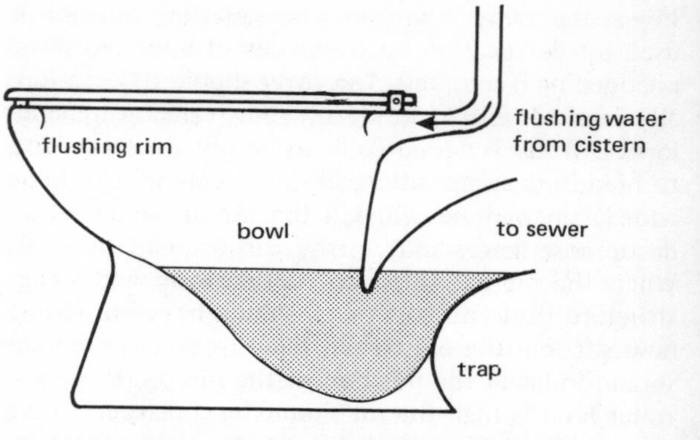

a) **Common type:** relies on the momentum of water discharged from the cistern to flush and clean the pan.

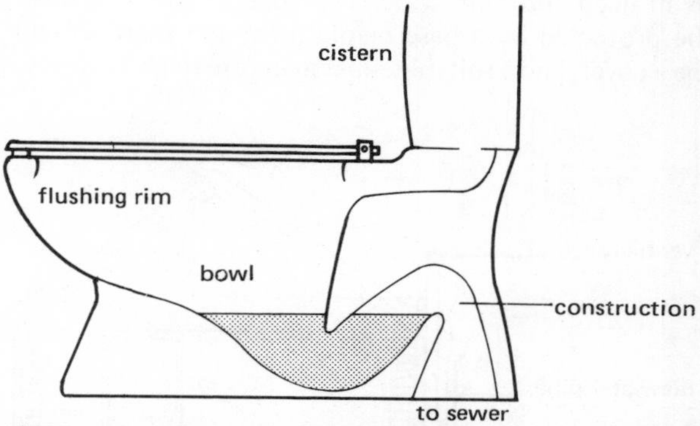

b) **Siphonic type:** relies on atmospheric pressure to flush the contents of the bowl. The level of the water rises in the bowl causing the trap to overflow into the constriction. The flow of air out of the trap causes a partial vacuum which begins a siphonic action. Air pressure in the bowl continues the action. The cistern can be close to the bowl.

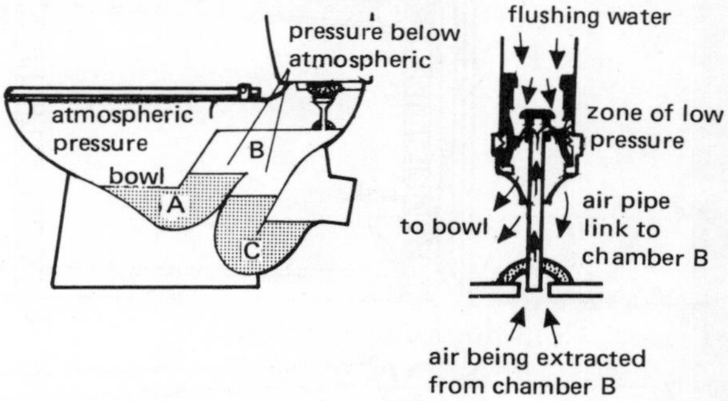

c) **Double trap type:** the latest design has two traps, A and C. Air is removed from chamber B by the device shown in section. It can be seen that water rushing down creates a low pressure zone and air is sucked up the central tube from chamber B. Water flows into A and the partial vacuum in B begins the outflow of the contents of the bowl. There is a continuous cleaning action. As in b) the cistern is close to the bowl.

(b) Advance in lavatory pan design

Bends are avoided as far as possible and where two drains meet there is an inspection chamber to allow the drains to be checked. If any blockage occurs flexible rods can be pushed through from the inspection chamber in order to free the obstruction. The chambers have brick walls, a concrete bottom and an iron lid. Sewers are usually vaulted brick tunnels with concrete bottoms and 'man-hole' inspection covers placed at intervals. There must be ventilation at each house to allow air to enter the system and to ensure that blow-outs of gully or sink traps do not occur (see fig 11.30). In the inspection chambers there is usually a trap to prevent the passage of gases backwards along the pipe from the sewer into the drain, or from the drain into the house (see fig 11.28). The drains and every sanitary fitting should have water traps which retain the water until it is flushed away by another flow of water.

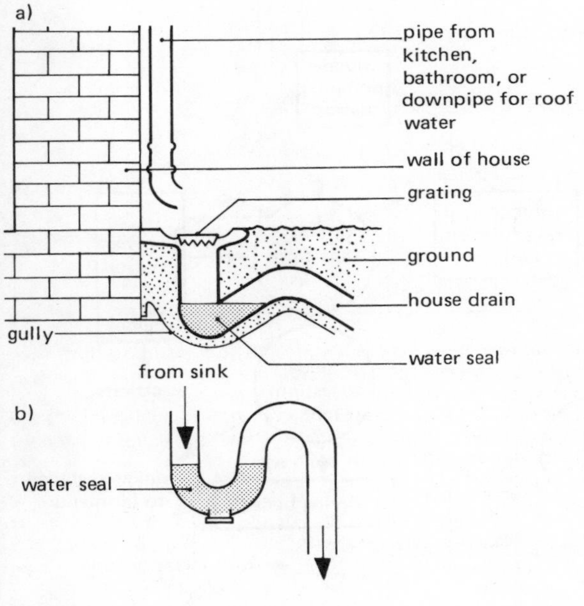

fig 11.30
(a) Gully trap
(b) Sink trap

11.33 Community sanitation

The sewers carry the raw sewage from towns to the sewage works where it is treated. In coastal areas sewage is sometimes dumped into the sea through very long pipes beyond the low tide mark, although efforts are continuously being made by local authorities to discontinue this very unhealthy practice. In the sewage works the treatment takes place in five stages (see fig 11.31).

1 The sewage is *screened* by passing through metal grids to remove all solid matter. The grids are cleared by power driven mechanical rakes which remove any object caught between the bars and the residues are burned or buried.

2 The screened sewage then passes through *detritus (sedimentation) tanks* where grit and stones settle to the bottom. Storm tanks can be included at this stage in order to cope with a sudden rush of water.

3 From the detritus tanks the sewage runs into *settlement tanks* where it is retained longer in order that lighter suspended matter can settle to the bottom. The size of this tank will depend on the amount of sewage treated and the whole process can take two to four hours. The sediment that sinks to the bottom is known as sludge and the whole process can be speeded up by the addition of chemicals such as aluminium sulphate or iron (III) chloride. The sludge is eventually removed and spread over the land to dry for use as manure. Modern treatment works have sludge dewatering units in which water is drawn off and the sludge compressed.

4 The liquid effluent from the sedimentation tanks still contains a large amount of organic matter, which is removed biologically by means of a *'trickling filter'*. The effluent exits from slowly rotating perforated arms (see fig 11.33) above the surface of a coke or clinker bed. This bed has smaller grades of clinker at the top and coarser grades at the bottom. Bacteria, algae and protoza on the very large surface area thus provided decompose the organic matter in the effluent. Furthermore the process is given maximum aeration by the spraying and trickling action of the filter.

5 The liquid now passes into a *humus tank* where the dead bacteria and algae settle and the purified liquid is drawn off into a stream or river.

In the alternative *'activated sludge'* method the effluent is run into tanks through which compressed air is forced to provide oxygen for the bacteria which cause decomposition. Then it is run into a settling tank in which the sludge settles. Finally, before entering the river, the effluent may be left for up to 60 hours in oxidation ponds. These further reduce the suspended solids and bacterial content of the effluent (see section (iii) below).

In rural areas where there is a piped water supply to houses but no local sewage works, the waste water and sewage can be disposed of in three ways.

i) *Cesspool* This is a bricklined container for sewage and foul water built below ground level. It may be for one house only or of a larger size for several houses. The walls of the cesspool must be soundly constructed to prevent leakage and it must have a roof with ventilation spaces. The inspection opening must be covered with a tight cover-plate. The cesspool must be emptied regularly by pumping the contents into a tanker truck or by the use of a modern vacuum gully-emptier mounted on a lorry.

ii) *Septic tank* This is also constructed of non-porous bricks or concrete and is generally covered with concrete slabs. The tank is air-tight so anaerobic bacteria can decompose the faeces. During decomposition material gradually settles to the bottom and forms sludge. The liquid above the sludge can pass out through an outlet pipe to a soakage pit. The pit contains fine stones over coarse stones and the aerobic bacteria present on the surface of the stones complete the final decomposition of the organic matter, leaving a non-polluting liquid. The sludge in the tank must be removed at intervals.

The septic tank is much safer than the cesspit since the decomposition of the excreta and the consequent removal of disease organisms is much more efficient.

iii) *Waste stabilisation ponds* In many tropical countries this method is used for small villages or workers' estates. Sewage from a water carriage system is passed by a drainpipe into the middle of a large pond known as an oxidation pond. This primary pond is about 1–1.5 m deep and twice as long as it is broad. Its capacity should be equal to about 30–40 days volume of sewage. The primary pond has an overflow to a secondary pond. The sewage provides salts,

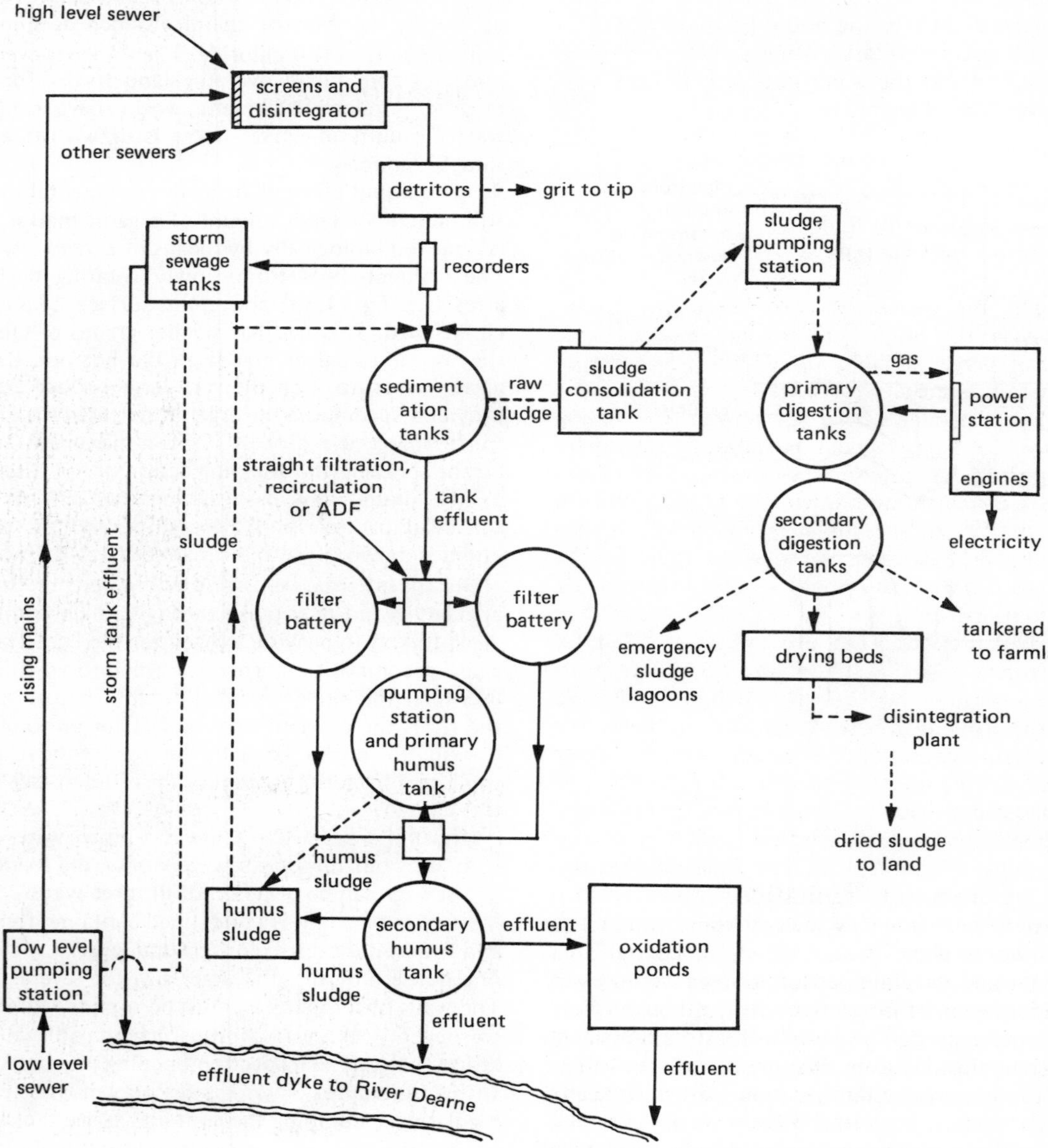

fig 11.31 Flow diagram of sewage treatment works

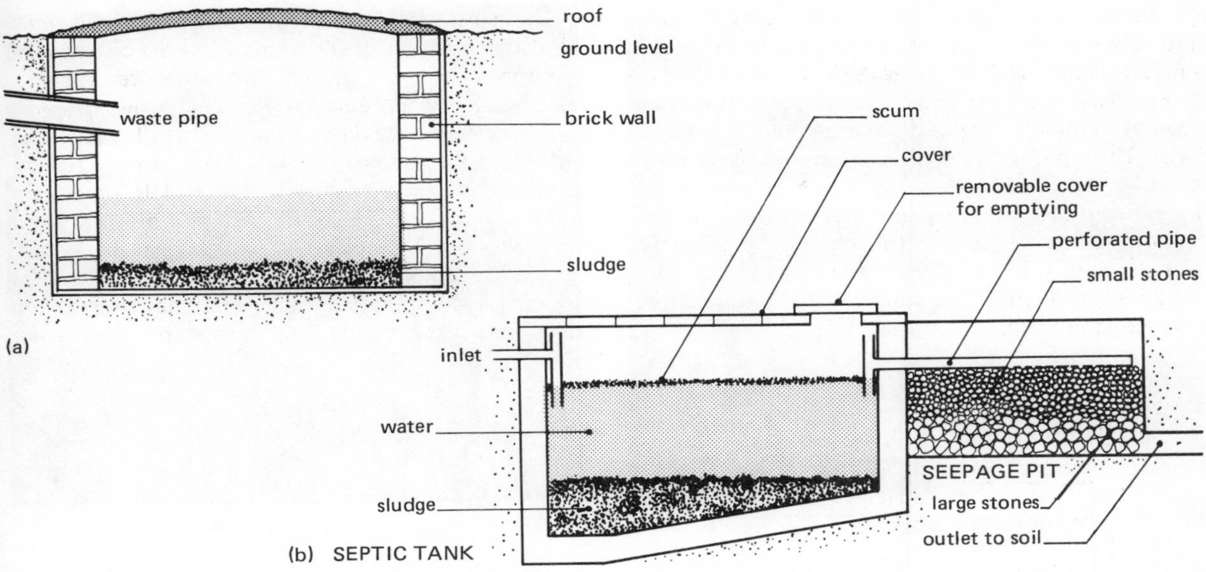

fig 11.32
(a) Cesspool structure
(b) Septic tank and seepage pit

encouraging the growth of green algae that photosynthesise in the bright sunlight and produce large amounts of oxygen. The gas is used by aerobic bacteria, which decompose the organic waste and at the same time reduce the smell of the raw sewage. The mixing of sewage, bacteria, oxgyen and algae is caused by wind action on the water surface. The liquid overflows into the secondary pond where the production of a non-pollutant fluid is completed and this flows into the ground or a stream. Storm water must not be allowed to run into the ponds in case it causes raw sewage to overflow.

The oxidation pond can also be used in an up-to-date sewage works. Fig 11.31, the flow diagram of Barnsley sewage works, shows oxidation ponds used to take the effluent from secondary humus tanks. The pond has a surface area of 0.19 hectares with an average depth of 0.9 m. Each pond treats $0.77 \times 10^3 m^3$ of effluent per day with a retention time of 60 hours. These ponds produce an average reduction of suspended solids of about 40%. The final effluent discharges into a river, giving a 2.3:1 dilution which is sufficient to enable the river to support coarse fish and a variety of plant life.

iv) If there is no piped sewage system, non-organic liquid refuse from houses must be dealt with separately from human excreta. The waste water from bathroom and kitchen must be taken away by pipes to a *soakage pit.* This is generally about 2 m deep by 1 m wide and 1 m deep. It must be at least 30 m from any well and large enough to take the water likely to be discharged from a house. The pit is filled with coarse stones so

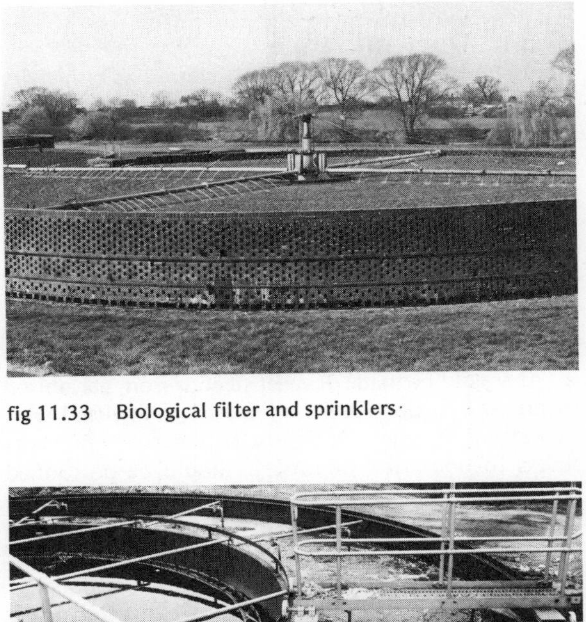

fig 11.33 Biological filter and sprinklers

fig 11.34 Activated sludge tank

fig 11.35 Sewage oxidation lagoons

that the water trickles down through the spaces and seeps into the soil and subsoil. The waste pipe leading to the pit should have a grease trap in order to prevent blockage.

11.34 Domestic refuse

Refuse from occupied houses is classed as either dry or liquid. Dry refuse consists of paper, rags, plastic bags or containers, house sweepings, broken glass and crockery and, in addition, kitchen waste such as vegetable peelings, bones, fruit peelings and other organic waste. It is essential that this kitchen refuse should not be allowed to accumulate since it attracts flies and other disease-transmitting organisms, and furthermore it decays and quickly gives off unpleasant smells.

The refuse is normally collected in a *dust bin* which has the following specifications:

i) It should be made of sheet steel or iron, galvanised to prevent rusting, or heavy plastic, and fitted with two strong handles. A tightly fitting cover prevents access to houseflies or rats. It should be positioned outside the house with ease of access for refuse collectors.

ii) The capacity should be large enough to take at least one week's accumulation of rubbish, although it must not be so large that it cannot be carried when full. Two or more bins may be necessary for houses with large families.

iii) The bin may be lined with a plastic bag to facilitate clearing and prevent fouling of the inside. This is expensive, however, and is not necessary if the inside of the bin is cleaned and disinfected regularly.

In most areas the local authorities generally arrange for *collection of refuse.* This collection is at least once per week, but in large towns it is often collected daily from commercial areas of shops and restaurants. Collection used to be in open carts but this allowed rubbish to blow away and also exposed it to flies. Today it is collected in specially built refuse lorries (dust carts) which are side-loading with sliding doors, or back-loading with a mechanical ram to compress the rubbish and push it to the front of the vehicle.

The refuse lorries deliver the refuse to a refuse station, not too far from the town, where the rubbish is crushed to reduce its bulk. The refuse is passed through sieves or screens to separate bottles and other objects. Metal, mostly tin cans, is extracted by an electromagnet and this is baled to be sold for scrap metal. The remaining material can be burned in a furnace or converted into a low grade fuel which may have a heat output about half that of coal. The refuse station may have to cope with waste from factories and other establishments such as garages. From the latter thousands of gallons of old engine sump oil are collected and sold. It is clear that a great deal of refuse can be recycled by putting it back into the soil cycles or using it again for manufacturing purposes.

Garden refuse and domestic organic waste can be made into a *compost heap.* This should have three sides of corrugated iron or preferably grids to allow air flow. The heap is made of alternate layers of organic waste and soil. The bacteria in the soil help with the breakdown of organic matter and after the heap is completed it should be left to mature. The compost can then be dug out and used as a natural fertiliser. Organic waste can also be composted below ground in a pit or trench dug to a depth of about 50 cm. The bottom of the pit should be lined with stones or clinker to allow air to circulate. The organic waste is layered with top soil or manure to promote decay. In rural areas dry refuse can be burned and the ashes added to garden soil or compost heaps.

fig 11.36 Types of refuse disposal lorry

fig 11.37 Rubbish treatment (a) dumping (b) incineration

11.35 Hygiene of the home

Construction, siting and care of houses, together with proper sanitary services, are most important for the health of the people living in them. If these factors are not given great attention slum areas develop in which all people, but particularly children, will be susceptible to communicable diseases such as tuberculosis, influenza, measles, typhoid and cholera. Poor housing, overcrowding and bad health go together and some of the factors that operate are as follows:

1 *Overcrowding* Houses that are too close together without adequate facilities result in accumulation of rubbish, waste products, sewage and so on. Along with this too many people often occupy these houses and their living and sleeping together can favour the transmission of infectious diseases. Micro-organisms are spread by coughing and spitting, so when people are close together it is very easy for disease organisms to be spread from one to the other. In addition direct transmission by contact can spread body parasites such as ringworm and lice.

2 *Ventilation* The transmission of disease is more certain where there is poor ventilation and air in rooms is not renewed by through draughts or some other type of ventilation. Thus the air becomes more and more impure and micro-organisms accumulate.

3 *Cleanliness and rubbish disposal* As a result of large numbers of people living in houses that are sited close together, and also overcrowded conditions in individual houses, large amounts of refuse and sewage are produced which are often not disposed of properly. The situation is worsened by poor personal habits and lack of cleanliness of the body, clothing, the house and furniture. This aids the breeding of flies, mice, rats, lice and cockroaches. These organisms can also be the vectors of disease.
4 *Lighting* Overcrowding in poor housing conditions also means lack of adequate lighting. Sunlight kills bacteria and if sunlight does not enter the rooms, and they are not properly cleaned, dirt, dust and micro-organisms build up. Furthermore the lack of sunlight resulting from overcrowded housing can mean that the inhabitants' bodies cannot make vitamin D.

Whatever the conditions of housing, the aims of the occupants should be to maintain a high standard of hygiene so that the home is in a clean condition unfavourable to the development and transmission of disease. This can be achieved by the following actions:
1 *Cleanliness* The sweeping and washing of floors, working benches and cupboards ensures that dust and grease do not accumulate. There must be regular washing, with soap and disinfectants, of toilets, kitchen sinks, food preparation areas, wash areas and refuse containers.
2 *Vermin control* If cleanliness is maintained vermin are not attracted by pieces of food and grease. Rodents can be controlled in houses and food stores by (a) rat-proofing entrance and exit points such as ventilators, doors, windows and openings for drain pipes, (b) use of traps and poisons, (c) keeping cats and dogs to catch the rats and mice. Flies are dangerous vectors of disease and so proper storage of food-waste and rubbish is important to prevent breeding. In addition the cleaning of toilets prevents any flies picking up micro-organisms from faecal splashes. Food must be kept covered at all times. Cockroaches can be controlled by insecticides as well as screening of gulley traps with fine wire mesh.
3 *Windows* These should be open whenever possible to allow a free flow of air and to let in sunlight.

11.40 Population

Animal populations are kept in check by three factors:
1 disease
2 predators
3 limitations of the food supply.

Let us consider the theoretical growth of populations of mice, yeast or bacteria, increasing without any natural checks. We would obtain a graph like that shown in fig 11.38. A pair of mice produces on average a litter of six young every three weeks. This litter

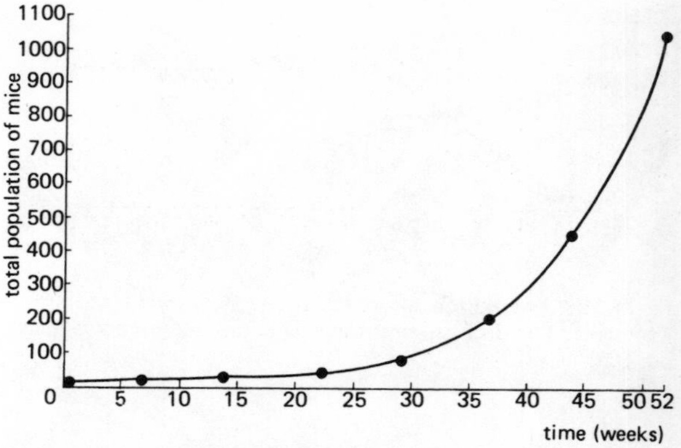

fig 11.38 Theoretical population growth of a colony of mice

could include three males and three females. Death from old age occurs at about three years. On the basis of this information, the theoretical growth rate over a period of one year would be as follows:

Time (weeks)	0	7	14	21	28	35	42	49
Population of mice	2	8	14	38	80	194	434	1 016

A graph of these figures would resemble that in fig 11.38. It is clear from these graphs that unhindered population growth results in a huge and continuous increase in numbers. In real populations, however, the natural checks result in changes in the curve such that it does not go on rising. Figure 11.39 represents the population growth of sheep introduced into the island of Tasmania in 1814. The population increased rapidly over a period of fifty years, but then the curve levelled out as the limiting factors took effect. Figure 11.40 shows a population of deer which increased rapidly when *predators* (lions, coyotes and wolves) were dramatically reduced in the habitat. The population of deer increased very rapidly until it overshot the estimated carrying capacity based on food supply.

Questions
1 What explanation can you give of the rapid increase in the deer population from 1905–1924?
2 If the estimated carrying capacity of the habitat was about 25 000 to 30 000 deer, what happened to the vegetation between 1918 and 1924?
3 What explanation can you give for the sudden downward curve of the population between 1924 and 1930?
4 What would you expect to have happened to the deer population and the predators after 1935?

Figure 11.41 shows this aspect of prey-predator relationship over a long period of time, indicating that these fluctuations in numbers are closely related.

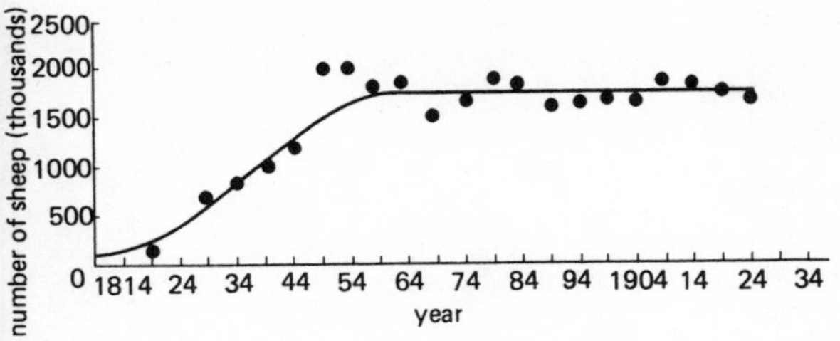

fig 11.39 Population growth of sheep introduced into Tasmania. The dots represent average numbers over five-year periods

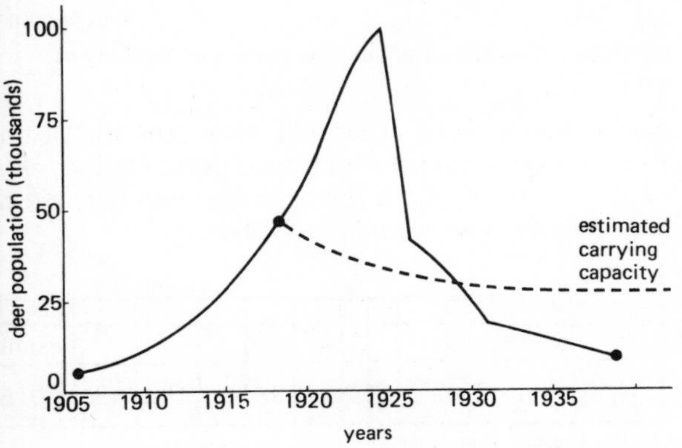

fig 11.40 Population explosion in Kaibab deer following removal of predators in an isolated area in America

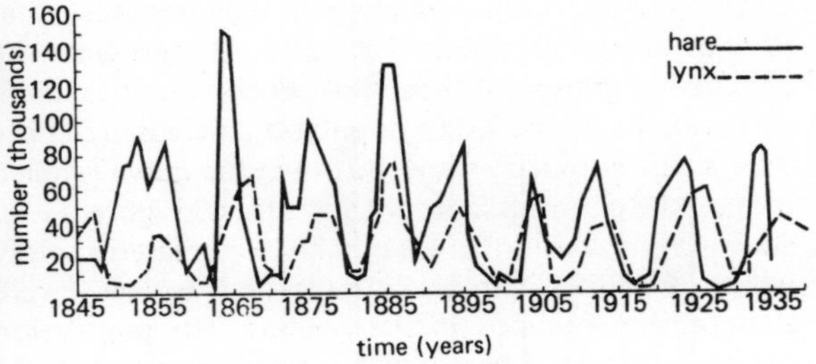

fig 11.41 Changes in the abundance of lynx and snowshoe hare

Examine fig 11.41 and answer the following questions.

Questions

5 Which animal population increased first? Why should this be so?

6 Why is this then followed by a peak in the numbers of the second animal?

7 Why is this second peak always below (less numerous than) the first peak?

Once a population has reached an equilibrium point due to natural checks then the population remains fairly stable (see fig 11.39 Tasmanian sheep). Despite the enormous reproductive potential of plants and animals, the controls acting on a population are quite effective. The parasites and predators feed on those less fitted to survive. Early death results from these factors and thus the population is kept in check.

In addition to the three checks already mentioned, other mechanisms operate in keeping the numbers of each generation below its theoretical reproductive potential. These are the density-dependent factors.

i) *Territorial behaviour* The number of parents are reduced by establishing breeding territories such that each pair occupies an area of sufficient size to supply the needs of itself and its offspring. The males defend the area against other members of the same species. An intruder is attacked and usually withdraws, leaving the occupying pair in sole possession. This territorial behaviour ensures that the food available in the habitat is shared among the breeding pairs, and furthermore it can prevent breeding amongst the surplus members of the population. When food is plentiful for a species the territories become smaller, allowing more breeding pairs.

ii) *Reproduction of offspring produced* Some animals reduce the numbers of their offspring and thus safe-guard against over depletion of food supplies. Fruit flies living under crowded conditions produce fewer eggs. Laboratory rats and mice living in a crowded environment reach a population limit, even though plenty of food is available. Some mammals can reduce the rate of ovulation by a change in hormone output, and in others there is a resorption of embryos in the uterus as a result of stress (e.g. in rabbits, foxes and deer). In the hive-bee, only the queen produces eggs and her rate of egg-laying is adjusted to the needs of the hive. In a cold, wet period (and thus poor flower production) fewer eggs are laid. Although some species show regular fluctuations in numbers over definite time intervals, most maintain a steady population level year after year and thus they seem to be regulated in a homeostatic manner (see Chapter 5).

11.41 Human populations

In human populations, the checks that apply to other animals, namely disease, predators and food supply are not all equally relevant. The problems of infectious and contagious diseases have been greatly reduced in recent years. Man's population growth is not conditioned therefore by this first factor, and furthermore he has no predators. War is often thought to eliminate large numbers of people but up to the second world war (1939–1945) more people died of disease during wars than from injury. Food supply has always been a problem with Man and, as the population has increased rapidly in the twentieth century, more than half of the people in the world suffer from malnutrition. Even so, it can be seen that natural population checks are not really operating to any extent on humans, and there-

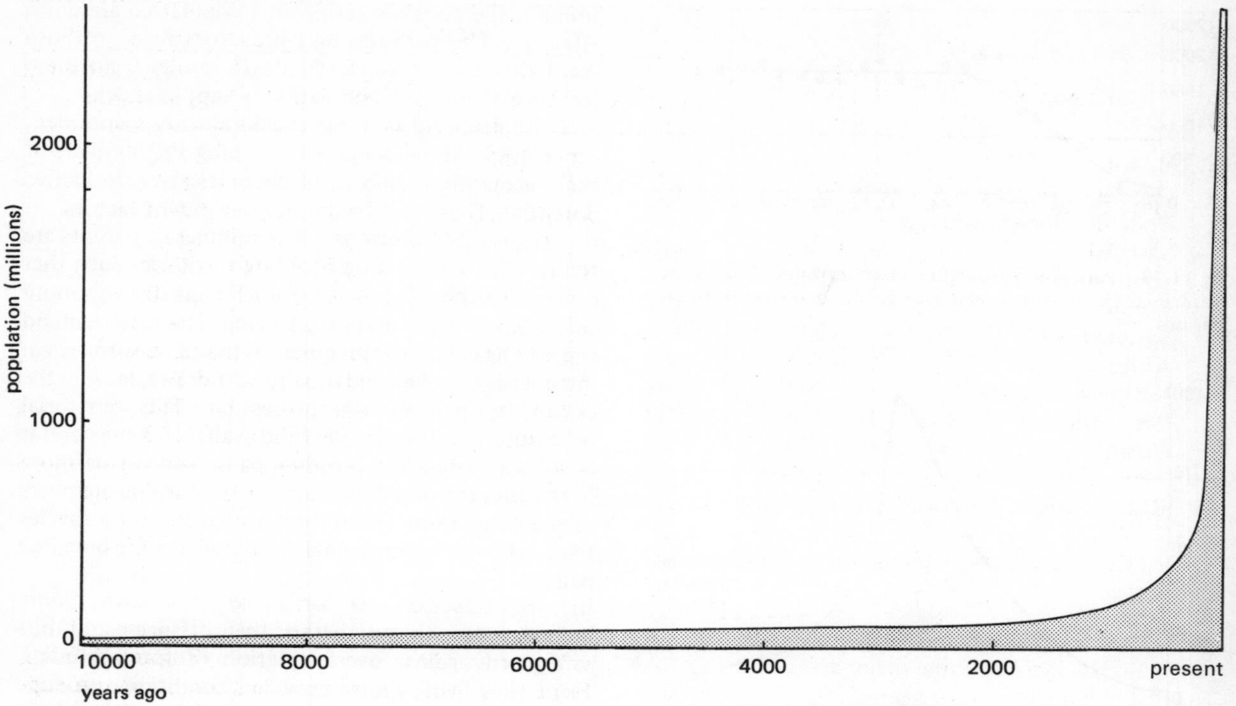

fig 11.42 Growth of the human population from 10 000 years ago to the present time

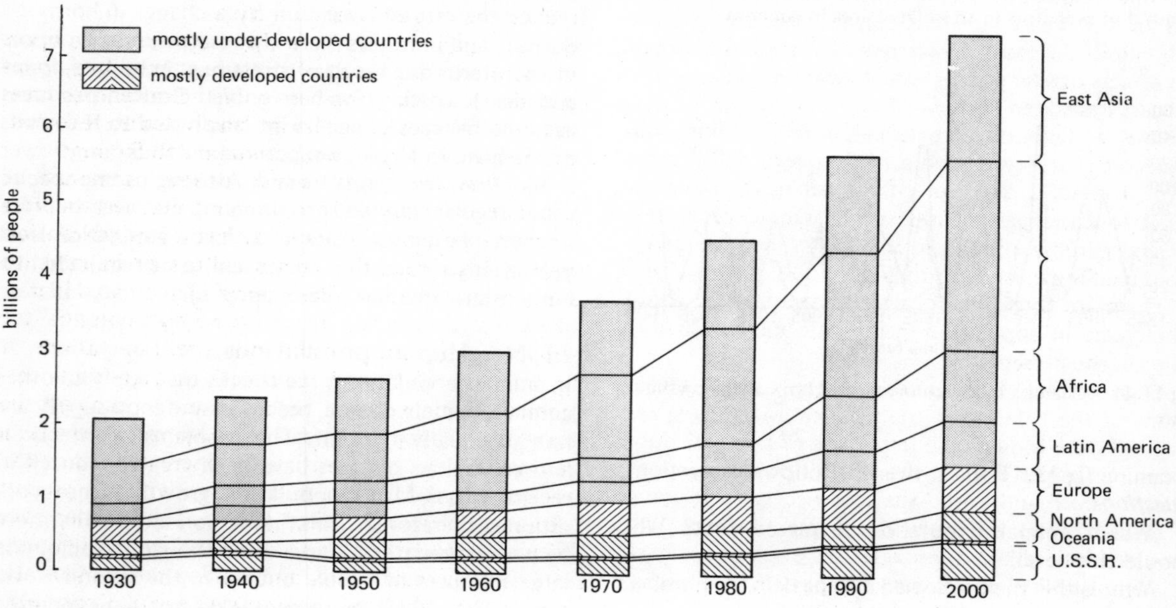

fig 11.43 Projected growth of world population

fore the graph of population increase follows that of a theoretical curve (see fig 11.42), rather than a curve of a natural animal population which levels out or fluctuates as checks come into operation. Table 11.3 shows the projected figures of world population and it is apparent that the world population will double between 1969 and the year 2000. Fig 11.43 shows these figures in the form of a histogram indicating the

Region	1969 population	Projection			
		Low	Medium	High	Constant fertility, no migration
World total	3 551	5 449	6 130	6 994	7 522
Developed regions	1 078	1 293	1 441	1 574	1 580
Underdeveloped regions	2 473	4 155	4 688	5 420	5 942
East Asia	1 182	1 118	1 287	1 623	1 811
South Asia	809	1 984	2 171	2 444	2 702
Europe	456	491	527	563	570
Soviet Union	241	316	353	403	402
Africa	344	684	768	864	860
Northern America	225	294	354	376	388
Latin America	276	532	638	686	756
Oceania	19	28	32	35	33

SOURCE: United Nations, *World Population Prospects as Assessed in 1963.*

Table 11.3 Estimate of population in the year 2000 (in millions)

proportionate increase in the major areas of the world. The first million was reached about 2000 years ago and 500 million was reached in the seventeenth century. The population doubled again in only 200 years by the mid-nineteenth century, and the next doubling was in 100 years, whereas currently doubling takes only about 30 years. Great fears are expressed about the effects of over population. A group of Nobel prize-winners has stated that:

'Unless a favourable balance of population and resources is achieved with a minimum of delay, there is in prospect a dark age of human misery, famine, under-education and unrest which would generate a growing panic, exploding in wars fought to appropriate the dwindling means of survival.'

There are three factors which have brought about an increase in population in the past and continue to do so in the present.

i) Agricultural development – the earliest civilisations of the Nile valley, the great rivers of China and India all grew food on the fertile soil of the river flood plains. Fifty per cent of the world's population now lives in South and East Asia. It is necessary for most of these to work on the land yet only achieve a bare subsistence level. Java provides an extreme example of high density population surpporting more than 1 000 to the square mile (2.29 km²).

ii) Industrial development – the great centres of population in Europe are correlated with areas in which coal is mined. Associated with these areas are large industry-supported populations, from Scotland eastwards through Europe to Poland. Industrial development leads to population build-up such that Holland has 960 to the square mile, England 900 and Belgium 790. These populations are of course supported at a high level of affluence because (a) their technology enables them to exploit their agricultural resources more fully, and (b) they trade manufactured items to food-producing countries.

iii) Potential for trade – this factor depends upon the previous two because trade must have large populations producing food or goods. Great cities have sprung up because they are on land or sea trade routes, e.g. Singapore, Hong Kong, London and Sydney.

We must be careful when looking at a map of population density and pronouncing that certain areas are over-populated. Antarctica has a low population density because it has no agriculture or industry to support a population, whereas highly populated islands like Great Britain and Japan have a high potential for supporting agricultural and industrial populations. It is not easy to classify countries as over- or under-populated for we must always take into account the numbers of people that can be supported by available reserves. Man has continually increased the food productivity of his habitat although with the present explosion of human beings he is unable to keep pace with the amounts of food required. With his high birth rate and low death rate, and lacking the homeostatic mechanisms that regulate animal populations, Man cannot look to any natural process to restrain his growth in numbers. Any reduction must be by Man's deliberately applied efforts. What is the solution? Man could never institute natural checks by allowing disease to take its toll, deliberately starving or killing people. The only solution, therefore, is to prevent, by birth

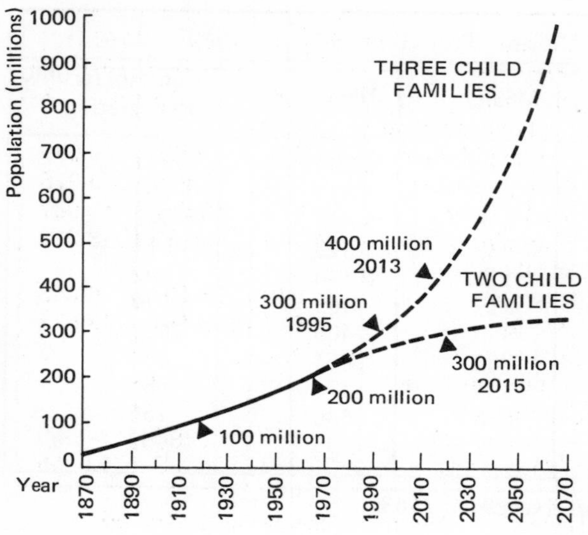

fig 11.44 In order to maintain a population at zero growth families must be limited to a maximum of two children. The graph shows the effect of only one extra child per family in the U.S.A.

control, the birth of too many babies. The technical means of birth control are well-developed and can be mass-produced on a large scale. A massive programme of education of populations in birth control methods has been put into operation in many parts of the world. There are, however, strong emotional and cultural factors working against family limitations in many countries. Efforts must be continued so that all nations can achieve a balance between numbers of people and available food resources. Unless this balance is achieved, even tremendous efforts by governments cannot keep pace with the increasing numbers of people.

11.42 Health and population statistics

All countries are required by law to keep a record of certain vital statistics concerning marriages, births, deaths, fertility and sickness. They are grouped as:

i) population statistics (population, marriage, birth and fertility)
ii) mortality statistics (deaths and causes of death)
iii) morbidity statistics (disease, injuries and disabilities).

This information is used for planning and development of programmes of education, public health, agriculture, business and industry. The collection of such information is carried out by government departments and based upon data from registrars of births and deaths, Health Ministries and so on. The methods used are:

i) *Registration* This is the continuous and permanent recording of births, deaths and marriages. The birth of any child or the death of a person must be reported to the Reigstrar within a stipulated time.

ii) *By census of the population* This is a complete counting of a population that takes place at intervals of ten years. This can give two estimates of a population,

a) the *de facto* census, which is the number of persons counted on the appointed day (the enumerated population),

b) the *de jure* census, which is the number of persons living in each locality of the country (the resident population).

Each household records on a census form the age, sex, occupation, place of residence, marital status and other social information for each occupant. If a population is changing through migration or other causes a government may require information between two major censuses and it then initiates a census of a 10% sample of the population. From this sample trends can be deduced.

iii) *By special notification* These can be certificates of sickness, notifiable diseases, hospital records of outpatients and so on.

In most countries the law requires notification of two types of disease,

a) communicable diseases such as those shown in table 11.4

b) industrial diseases such as silicosis (resulting from dust in mining operations), lead poisoning, mercury poisoning and so on.

The vital information that is obtained in these ways is recorded statistically as a number of different rates.

Crude birth rate

This is the number of births per thousand of the population

$$\frac{\text{Number of births in a year}}{\text{Total population of that year}} \times 1\,000$$

The total population is always taken as the mid-year figure. In the United Kingdom in 1974 the live births were 738 000, with a population of 55 968 000. This gives a crude birth rate of

$$\frac{738\,000}{55\,968\,000} \times 1\,000 = 13.18$$

In India the population is 547 949 000 and increasing at a rate of about 12 000 000 per year. This gives a crude birth rate of 21.89.

Fertility rate

This figure is of great value for indicating population trends. It is expressed as the number of live births per 1 000 women of child-bearing age.

$$\frac{\text{Number of births in a year}}{\text{Number of women between 15 and 45 years of age}} \times 1\,000$$

Infant mortality rate
This is the number of deaths of infants under the age of one year per thousand live births. It is a useful guide to the standards of hygiene in a community.

$$\frac{\text{Number of infant deaths}}{\text{Number of live births}} \times 1\,000$$

In 1970 in the United Kingdom this rate was 18.8 but in 1974 it was 13.6, thus in that short period of time there was a decrease in the rate and therefore an improvement in the care of young children. In other parts of the world we can see similar improvements in the rate, but the definitive rate year by year still shows the need for improved child, post-natal and ante-natal care. For example, in West Malaysia 1966–1970, the rate improved from 47.9 to 40.8. The infant mortality rates for some countries are very high, e.g. Turkey 153.0, Pakistan 142.0 and Indonesia 125.0.

Disease	1964	1969	1974
Diphtheria	20	13	3
Typhoid and para-typhoid fevers	1 041	378	239
Food poisoning	6 530	8 599	6 276
Scarlet fever	22 673	18 662	11 087
Whooping cough	34 243	6 132	18 259
Smallpox	–	–	–
Dysentery	25 640	28 107	9 858
Measles	318 912	163 141	118 638
Acute poliomyelitis	39	14	8
Ophthalmia neonatorum	879	469	350
Tuberculosis	20 972	14 541	12 456

Table 11.4 Notifiable diseases in the United Kingdom

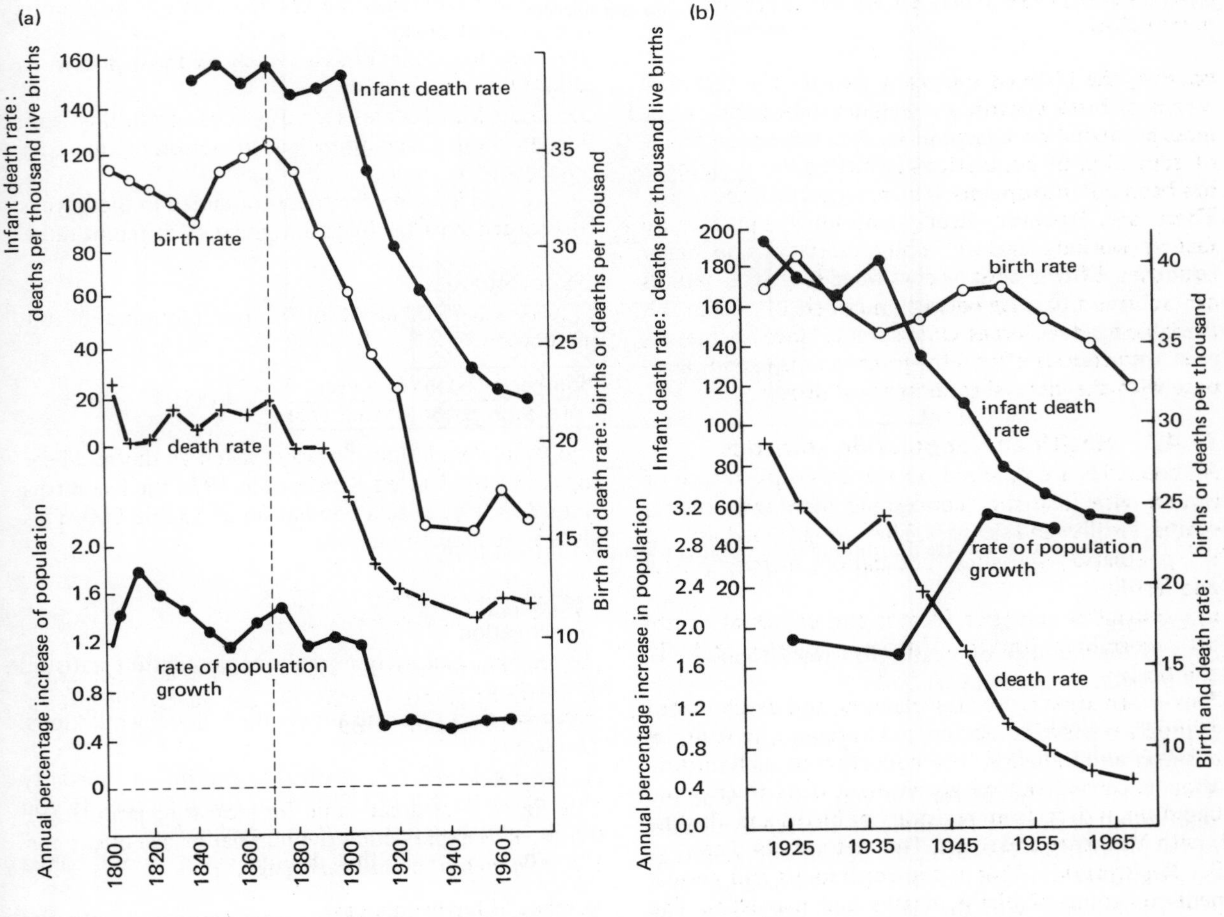

fig 11.45 Birth rate, infant mortality, death rate and rate of population growth (a) England and Wales 1800–1970 (b) Sri Lanka 1922–1967

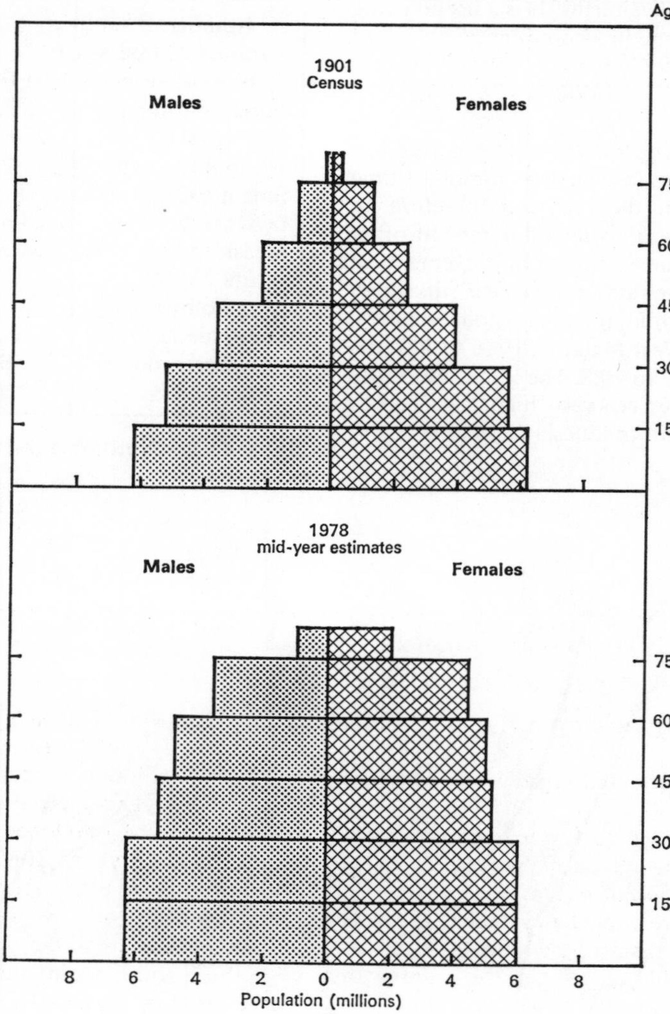

fig 11.46 Population and age distribution in the United Kingdom 1901 and 1978

Death rate
This is the number of deaths for every thousand of the population in each year.

$$\frac{\text{Number of deaths}}{\text{Total population in that year}} \times 1\,000$$

This is a useful figure for any country, or region of that country, since it can indicate the relative health of the population. However, the age grouping must be taken into account and whether it is industrial or rural. When compared with the crude birth rate these two figures can show whether a population is increasing or decreasing.

Investigation
To examine and analyse graphs of population statistics
Procedure
Examine fig 11.45 and answer the following questions.
Questions
1 In fig 11.45 (a), death rate and infant mortality rate began to fall in 1870. What social factors do you think contributed towards this decline?
2 What year did this decline begin in Sri Lanka (graph (b))?
3 Why was it delayed so long compared with the United Kingdom?
4 What was the most important factor influencing

Age	Males		Females	
	Number of survivors from 10 000 births	Expectation of life in years	Number of survivors from 10 000 births	Expectation of life in years
0	10 000	68.8	10 000	75.1
5	9 772	65.4	9 821	71.4
10	9 751	60.5	9 807	66.5
15	9 733	55.6	9 796	61.6
20	9 689	50.9	9 776	56.7
25	9 639	46.1	9 755	51.8
30	9 594	41.3	9 731	47.0
35	9 540	36.6	9 696	42.1
40	9 460	31.8	9 640	37.3
45	9 319	27.3	9 541	32.7
50	9 065	23.0	9 372	28.2
55	8 651	19.0	9 122	24.0
60	7 993	15.3	8 752	19.9
65	7 017	12.1	8 213	16.0
70	5 660	9.4	7 404	12.5
75	4 023	7.2	6 199	9.4
80	2 377	5.5	4 538	7.0
85	1 084	4.0	2 675	5.0

Table 11.5 Life table 1971–73 United Kingdom

the fall in rate of population growth in graph (a) after 1870?

5 Why has the rate of population growth significantly increased in graph (b)?

The death rate applied to varying age groups of the population will show quite different results. These are called *age-specific mortality rates.* From these it is possible to calculate life expectancy tables. In Table 11.5 it can be seen that, starting with 10 000 of the population, it is possible to set out the number of people who are likely to survive at the end of each five years of age. It is clear that there is a sex difference in favour of females throughout the tables, but 70% of men and 80% of women can expect to reach the age of 65 years. This can be seen in a different fashion in fig 11.46, which compares the year 1974 with 1901.

The expectation of life in 1974 and also at the present day is clearly much greater than in 1901 where 65% of the population were dead by the age of 45 years.

The life expectancy in the developing world is much less than in the developed world, e.g. Mali only 27 years, India 45 years. This means that the age distribution of the population is very different for developed parts of the world as shown in the table below: It can be seen that the populations of Africa, Asia (Southern and Eastern) and Latin America are largely made up of young people.

The natural growth of a population can be calculated from the crude birth rate and crude death rate:

Crude rate of increase = crude birth rate – crude death rate

Region	0–14 years	15–64 years	65 years and over
North America	30.0	60.8	9.2
Europe	24.3	64.0	11.7
Africa	43.7	53.5	2.8
Eastern Asia	35.6	60.0	4.3
Southern Asia	43.3	53.8	3.0
Latin America	42.4	53.8	3.8

Table 11.6 Percentage of age groups in the population in 1970

	Birth rate	Death rate	Natural increase	%
India (1970)	40	15	25	2.5
Indonesia (1970)	45	18	27	2.7
Singapore (1971)	22.8	5.4	17.4	1.7
Kenya (1970)	50	17	33	3.3
Nigeria (1970)	57	27	30	3.0
Asia	38	15	23	2.3
Africa	47	21	26	2.6

All figures are rates per 1 000 of the population

Table 11.7 Rate of natural increase

The table above shows the rate of natural increase for five countries and two continents.

The crude birth and death rates do not provide the best indications of natural increase. A very important figure is of course the infant mortality rate, which may be very high. Other factors such as age distribution of the population, migration and so on also have important effects on the population growth rate.

11.43 Results of Man's population growth

Primitive Man lived mainly as a hunter. He took precautions to prevent his numbers increasing by killing unwanted babies and he also made sure that the animals he hunted did not die out. He never killed pregnant females and hunting usually stopped during the breeding season. Primitive hunting peoples had small families because such small groups could survive more easily and the children of these families probably inherited their parent's low fertility. Thus stone-age Man made little impact on his environment and he lived like any other animal. He did not destroy forests, he did not use up the mineral resources, he did not cause pollution and when he died he left little trace of his existence.

Man had always collected plants for food and he also began to cultivate plants. The first of these to be grown could simply be pushed into the ground as a piece of stem or root, e.g. yams and bananas. The most important plants, however, were those that could be grown from seed and such cultivation developed in three main regions of the world: south-west Asia, south-east Asia and North America. Wheat, barley and beans made their appearance first and then oats, rye, maize, peas, grapes, dates and apples.

Animals were domesticated by those familiar with animals through their hunting. Goats first and then sheep, pigs and cattle were herded for their meat and hides. Cattle have been used to draw ploughs since about 3000 B.C.

The production of food by farming resulted in a surplus of food for the first time in the history of Man. This produced a population increase and for the people a division of labour. No longer did one family have to spend all of its time producing food. Now there was time for specialists to appear, notably toolmakers and potters, and at the same time the populations began to group together to form towns. From this time onwards Man began to behave irresponsibly towards nature.

Forests

Man began to cut and burn down forests to provide agricultural land as soon as he became a farmer. The result in hot countries was that soil previously covered by trees was eroded by rain during the wet season. Thus in many places we now have deserts, where once large forests protected the soil. We have continued this ill-treatment of natural resources up to the present day. The great forests of the Amazon are even now being systematically stripped as Brazil continues her rapid industrialisation. These forests are inhabited by primitive Indian hunters who are being wiped out by the prospectors and road builders. The Brazilians have not learned the lessons of the centuries, for already large areas, once forest, now have poor soil, droughts and famines. The forests still remaining in Brazil have a larger area than the British Isles, and some scientists believe that if all of these trees were felled, it would reduce the supply of oxygen on this planet. It must be said, however, that Man has attempted to replace timber by reafforestation but the rate of consumption of wood still outstrips the rate of replacement. The manufacture of newsprint alone consumes vast quantities of woodland every year.

Fossil fuels

The industrial revolution of the nineteenth century was essentially brought about by a change in the main source of energy used. Up to this time the muscles of man, his horses and his oxen had supplied the necessary energy. The Pyramids were built by thousands of slaves over many years. Rich people and land-owners employed vast numbers of servants in their homes and on their farms. It was the increase in the production of coal and the invention of the steam engine that provided a much greater source of energy from 1800 onwards. Industry increased by means of this energy and industrial nations became much richer, although this wealth was not widely shared. In 1750 most people in Britain worked on the land, but by 1900 only one tenth of the population worked on farms, and today only one person in fifty is employed in agriculture. These figures indicate the great change brought about by the industrial revolution.

The spread of industry has been speeded over the last seventy years by the discovery and use of oil as an energy producer. This fossil fuel is being used at a prodigious rate and, although Britain now has her own oil supply, it is not expected to last beyond the year 2000. The supply of coal could last longer but both fuels are being expended rapidly and they can never be replaced. Natural gas is the most recent discovery and in 1977 the whole of England was being supplied with North Sea gas. These natural fuels were formed during a period of 100 million years but, with present consumption, all will have gone by the year 2100.

Other types of mining
Iron ore is used in great quantities and turned into sheet steels. Whether it is made into cars, beer cans, washing machines or bicycles it is eventually abandoned to rust away into iron oxide. This is dispersed and the iron cannot be recovered so that it is a net loss to this planet. Other metals likewise are mined, processed and eventually discarded. Farming relies more and more on artificial fertilisers containing phosphates and nitrates. These occur in natural deposits which have been mined and distributed around the world, spread on the land and finally washed into seas or lakes.

11.44 Pollution

Air pollution
Air is a mixture of gases, 78% nitrogen, 21% oxygen, 0.03% carbon dioxide and minute amounts of argon, neon and helium. The pollution of the atmosphere in industrial countries is causing increasing concern. The following are the main pollutants.

i) *Dust* This consists of small particles produced by industrial processes such as brickworks, cement works and power stations.

ii) *Smoke* Industrial furnaces and domestic fires produce about one thousand million kilograms of smoke every year. In certain atmospheric conditions this smoke combines with fog to cause smog, which at its worst can be lethal. After the great London smog of 1952 that killed 4000 people in five days, the Clean Air Act of 1956 enabled authorities in towns and cities to set up smokeless zones. Since that time the appearance of London in winter has changed completely (see fig 11.47). December sunshine has increased by 70% in the last fifteen years, and the estimated smoke emission is down by 50%. This has been achieved by the use of electricity and smokeless fuels in the clean air zones.

Region	1970/71	1972/73	1973/74
Smoke: average concentration[1]			
England			
North	88	74	55
Yorkshire and Humberside	80	65	44
East Midlands	60	55	44
East Anglia	46	40	35
South East (excluding Greater London)	31	27	27
Greater London	42	36	36
South West	29	26	24
West Midlands	50	48	39
North West	81	65	49
Wales	33	29	30
Scotland	69	49	35
Northern Ireland	72	50	53

[1] Measured in micrograms per cubic metre.

Table 11.8 Air pollution by region of the United Kingdom

iii) *Sulphur dioxide* The combustion of fuels such as coal and oil produces sulphur dioxide. This is an irritant, choking gas. In moist air it forms sulphurous acid and this, falling in rain, can cause corrosion of stone and metal. The use of smokeless fuels has reduced the concentration of this pollutant gas by half.

iv) *Carbon monoxide* This highly poisonous gas is produced by petrol engines. In a concentration of 1000 p.p.m. it kills rapidly, but at a concentration of 100 p.p.m. it causes headaches and stomach cramps. Measurements have been made during London rush hour traffic and these show concentrations as high as 200 p.p.m. The most heavily polluted air containing this gas is found in Tokyo where traffic policemen have to wear gas masks.

v) *Lead* This is a very dangerous atmospheric pollutant and is mainly produced by the internal combustion engines of cars and lorries. Lead is added to

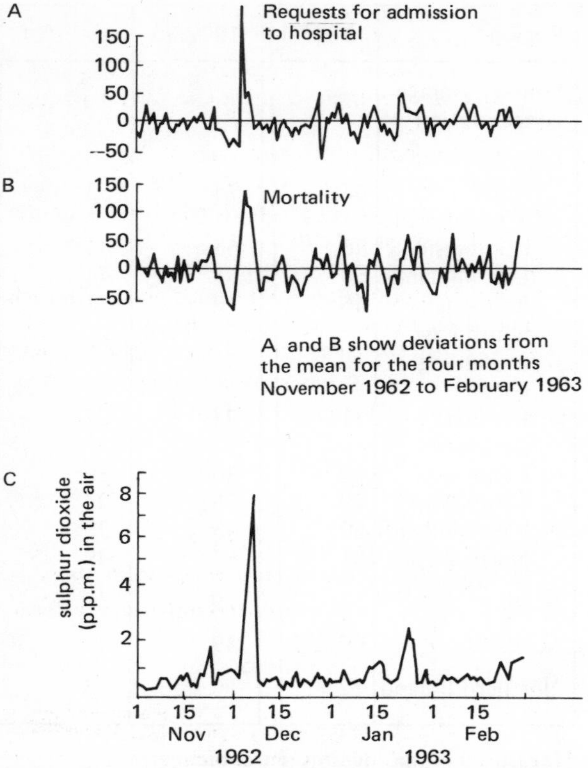

Abnormal smog conditions in early December were correlated with a large increase in hospital admissions and deaths from respiratory disease.

fig 11.47 Changes in air pollution in London, 1962–1963

petrol to make the engines run more smoothly. It is expelled in car exhausts and settles on vegetation at the roadside. Thus blackberries should never be picked from hedgerows near busy roads.

vi) *Smoke from cigarettes* This is far more dangerous than any other form of pollutant. This is true for smokers and non-smokers alike, but in particular inhaling the smoke directly by the smoker is the principal cause of many deaths from lung cancer.

vii) *Nitrogen oxide* This substance is also produced by the motor vehicle and is another possible cause of lung cancer.

viii) *Radiation* The testing of atomic weapons results in a rise of radiation levels due to the dispersal of radioactive particles into the atmosphere. The presence of strontium 90 in the dust of atomic explosions is potentially dangerous since it has a half life of 28 years, that is, one-half of its discharge decays during that time. It follows, for example, that it will still be at one eighth of its strength after 74 years. It can replace calcium in the bones of young children and therefore is potentially dangerous.

Hours of winter sunshine (December to February) at London Weather Centre and Kew Observatory

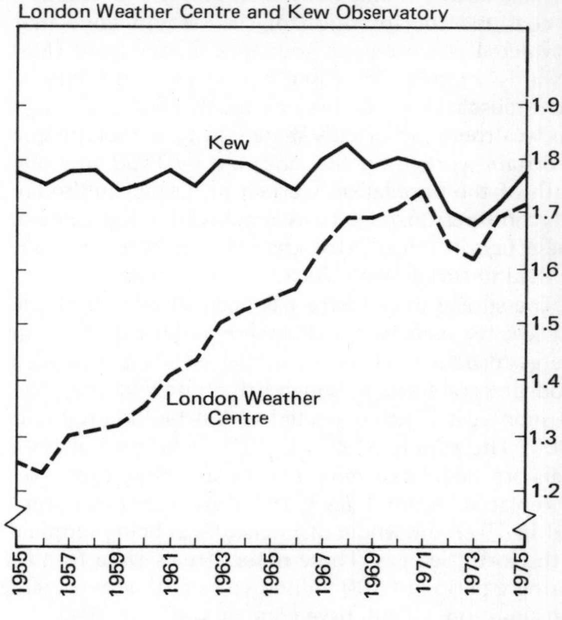

fig 11.48 Improvement in air cleanliness in London as a result of Clean Air Acts

There is considerable controversy today regarding the safety of nuclear power stations. Not only is there fear of an explosion if the station went out of control, but also the disposal of the large amounts of radioactive waste presents an enormous problem. There have been attempts at several different methods, from burial at sea in concrete blocks to storage in disused mines. It is certain that radioactive waste will increase in quantity as more and more power stations come into service and therefore every effort must be made to solve this problem.

It is possible that even more dangerous than these bulk waste products are the number of radioactive isotopes released into the atmosphere. Krypton-85 is one of the radioactive substances released at 'low-levels' by fusion reactors, and there would appear to be no economically efficient way of coping with these emissions at the present day. Similarly atomic power stations using sea water as a cooling fluid have been shown to produce small amounts of radioactivity that are taken up by seaweeds and these can move through food chains. In March 1979 there was a leak of radioactive steam from a power reactor in Harrisburg in the U.S.A., and thousands of people had to be evacuated from the surrounding areas.

Every effort must be made to reduce these hazards, not only because of their dangers to the public, but also because atomic fuel seems to be of vital importance as a replacement for dwindling oil and coal reserves.

We cannot afford a society totally free of pollution,

but we must aim at a cost-effective effort to improve the atmosphere incorporating an acceptable small level of risk.

Experiment

To investigate quantitatively the effects of pollution

Procedure

1 Obtain nine beakers (or jam jars) labelled A to I and two-thirds fill them with tap-water. Fill a tenth beaker (J) with distilled water.

2 Add to the beakers the following substances:

A 5 cm^3 of garden soil
B 5 cm^3 of common salt (sodium chloride)
C 5 cm^3 of detergent
D 5 g of sodium nitrate
E 5 g of superphosphate of lime
F 5 cm^3 of car oil
G 5 cm^3 of concentrated sulphuric acid
H 5 cm^3 of common weed killer (e.g. paraquat, weedol)
I no addition (control) – tap water
J no addition (control) – distilled water

3 Add to each container 10 fronds (lobes) of the common duckweed (*Lemna*).

4 Stand the cultures on a well-lit bench.

5 Count the number of living fronds present in each container after the following intervals:
(i) 2 days (ii) 4 days (iii) 7 days (iv) 2 weeks (v) 3 weeks (vi) 4 weeks. Record also brown or dying fronds (see Chapter 12 section 11.44).

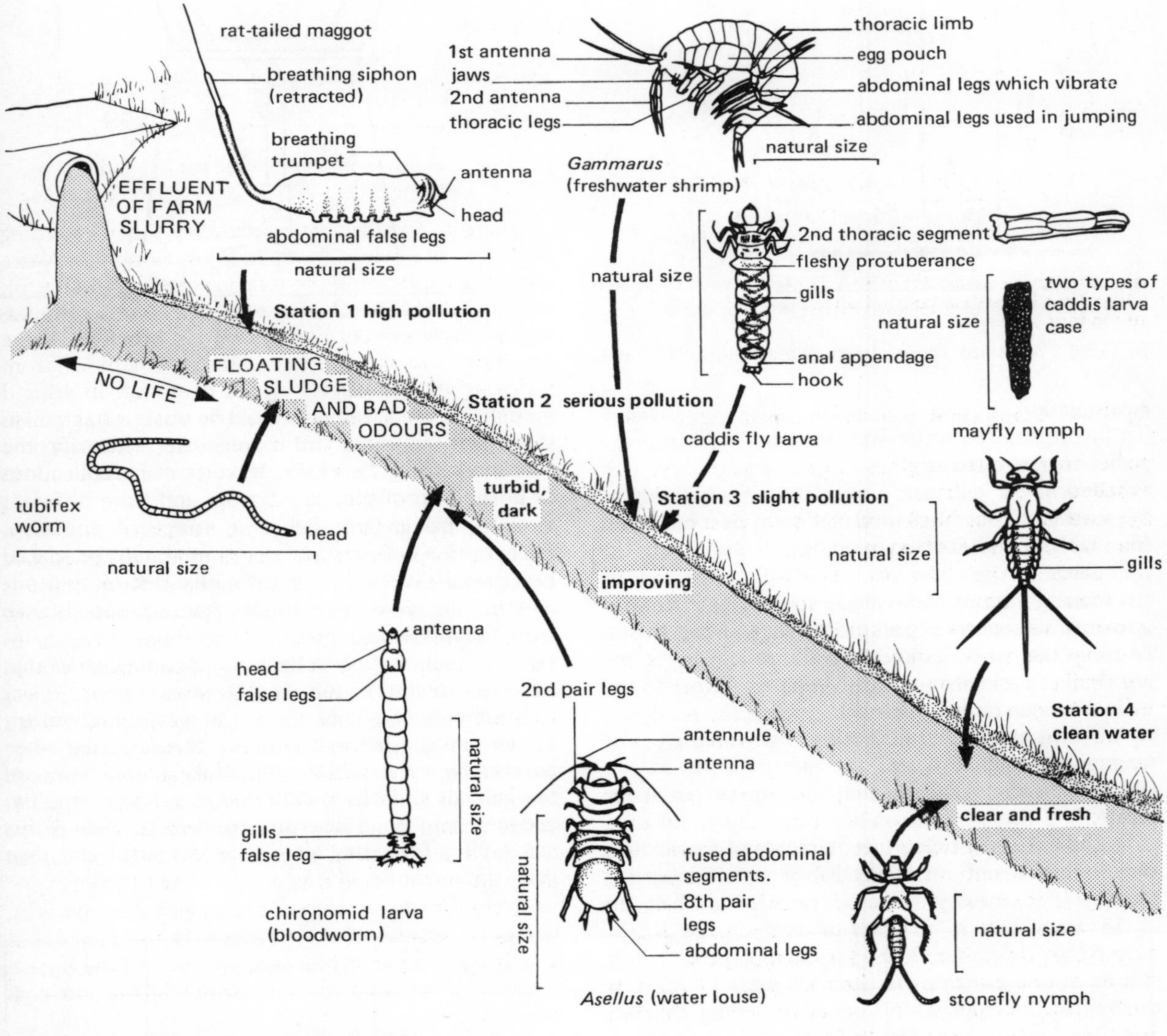

fig 11.49 Organisms related to water pollution

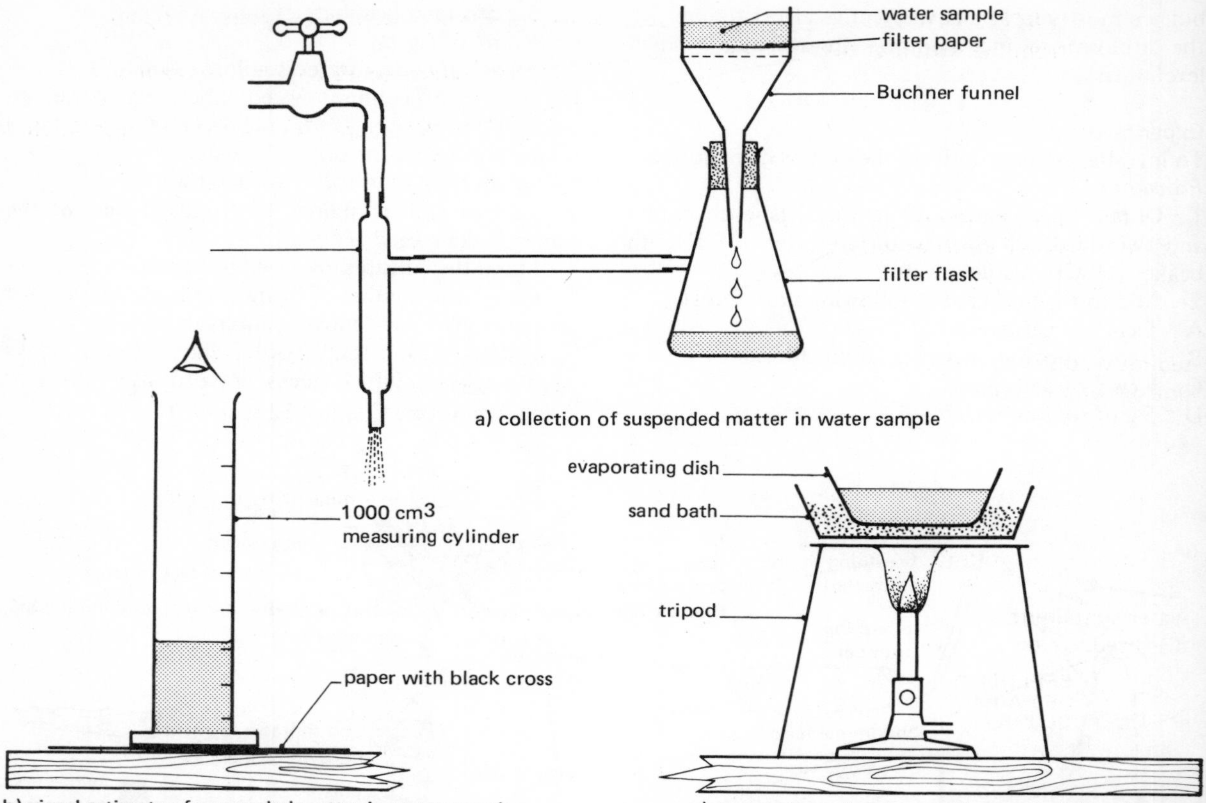

a) collection of suspended matter in water sample

b) visual estimate of suspended matter in water sample

c) evaporation of water sample for obtaining dissolved solutes

fig 11.50 Apparatus for investigating the pollution of a stream

Investigation

To investigate the water and the living organisms in a polluted river or stream

Procedure

1 First gather as much information as possible about the habitat and its surroundings, based upon the following:

i) Make a *map* of the river or stream together with any possible sources of pollution such as sewage works, factories, farms or garages and indicate where there are drains or side-streams providing an inflow to the main water under investigation.

ii) Make regular visits noting *the weather,* and keeping detailed notes of the *rate of flow, colour, presence of debris, foam* and *floating matter* in the water.

iii) Collect *living organisms* by means of a collecting net. A white enamel dissecting dish is useful for sorting specimens. Mud samples should be obtained as well as specimens collected from mid-water. Identify the specimens using fig 11.49. Make as many collections as possible from the same station and from different stations, particularly below a source of pollution. Each station below such an inflow should be spaced at intervals say of 10 metres in order to find out whether the types of organisms change as the distance from the inflow increases.

iv) At each station collect any *plant material* that may be growing in the river or stream bed. Collect also samples of plants above the water line and try to determine whether they have been affected when covered by a rise in the water. Make a table to record the animals and plants collected in parts (iii) and (iv) above. There is no necessity to identify these plants accurately, they may simply be described and then given a number by which to record their presence.

Station on bank	State of water	Animals present	Plants present	Temp. pH

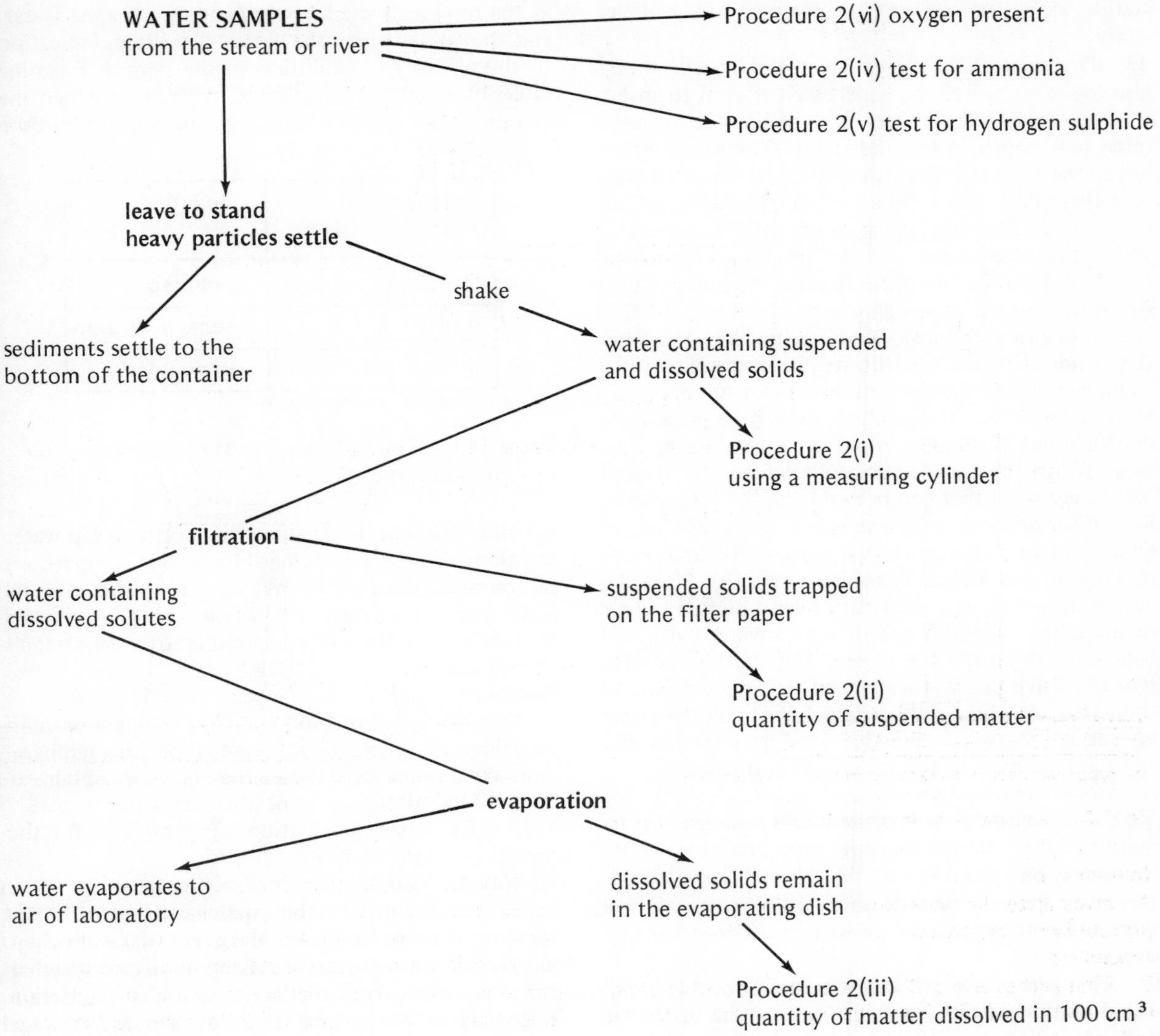

fig 11.51 Sequence of procedures for analysis of water

v) Note that in the table opposite the *temperature* of the water and its *pH value* have also been included. It is important to test these values on each station and at each visit, since they cannot be measured after water samples have been removed to the laboratory.

a) Measure the temperature of the water using an ordinary laboratory thermometer.

b) Measure the pH value by dipping a piece of universal indicator paper into the water and comparing the colour of the wet end with a colour chart (see Introductory Chapter).

2 After the completion of part 1, the next stage is to investigate the *causes of pollution* by an examination in the laboratory of the water from the stream or river. Collect water samples in bottles for each of the stations mapped and investigated in Part 1. Return to the laboratory and carry out the following tests.

i) *How much suspended matter is in the water from the river?*

Obtain a large (1 000 cm^3) measuring cylinder. Stand it on the laboratory bench over a pencilled cross on a piece of paper. Add water to the cylinder slowly and at the same time look down through the water at the cross. When you can no longer see the cross record the volume of water in the cylinder. Make a table to record this volume for the samples taken at each station along the bank of the river or stream under study.

ii) *What quantity of suspended matter is in the water?*
Use the same sample of water, well shaken to ensure that any suspended matter has not settled, to determine how much is suspended in 100 cm^3 of water. Set up the apparatus shown in fig 11.50 which consists of a filter flask and a filter funnel attached to a filter pump. Dry a filter paper in an oven at 105°C overnight. Weigh the filter paper. Fit the filter paper into the funnel and connect the filter flask to the pump. Again shake the sample thoroughly and pour it into a 1 000 cm^3 measuring clyinder. Turn on the tap and filter the water. (Retain the filtrate for procedure (iii).) Remove the filter paper from the funnel and dry again in an oven at 105°C overnight, or if time is available dry for about 30 minutes in the same oven during class time. Weigh the paper again and calculate the mass of suspended solid that had been left on the filter paper.

iii) *What quantity of matter is dissolved in the water?*
Obtain a large, clean, dry evaporating dish and weigh it. Pour in the filtrate from procedure (ii). Place the evaporating dish in a sand bath over a bunsen burner or place the evaporating dish over a water bath, and heat it to drive off the water. This method ensures that the dish is not overheated and thus the evaporated solids are not chemically changed. When all the water is evaporated, weigh the dish and its contents and calculate the mass of the solute left in the dish.

iv) *Does the water contain ammonia?*
Take a piece of turmeric paper and dip the end into a sample of the water. If the paper turns brown ammonia is present. This should be correlated with your investigation on the river bank to show whether there is any farm effluent or sewage passing into the stream or river.

v) *Does the water contain any hydrogen sulphide?*
Put a sample of the water into a test-tube so that it does not quite reach the top of the tube. At the top of the tube place a piece of moist lead acetate paper and secure this with a cork. The paper should not touch the water. If the paper turns black within a few minutes hydrogen sulphide is present.

vi) *Is there enough oxygen for plants and animals living in the water of the stream?*
Take a fresh sample of the water to be tested. Pour the sample into a bottle that has a tight fitting bung or screw cap and then add three drops of methylene blue solution. At this stage the top of the liquid should be just below the bung or screw top. Close the bottle tightly. Shake the bottle well so that the dye is mixed with the water. Label the bottle with the date, the time of testing and the station that it comes from. Put the bottle away on a shelf in the laboratory and place a white sheet of paper behind it so that the colour can be seen clearly. The colour will gradually disappear as the oxygen is used by the organisms in the water, and the rate at which the colour fades is an indication of the amount of pollution in the sample. Examine table 11.9 for an indication of the rate at which the colour disappears related to the amount of pollution.

Colour fades from water (days)	Pollution
0 to 3	polluted
4 to 6	some pollution
more than 6	unpolluted

Table 11.9 Time of colour fade of methylene blue related to pollution

A control should be set up with a bottle of tap water and three drops of methylene blue so that comparisons can be made with the habitat samples throughout the experiment. Complete the whole study of oxygen content with a table of your results from the samples at each station.

Questions

1 Consider your measurements of temperature and pH values. What might have caused increasing temperature? What might have caused the pH values obtained?

2 Looking at the cloudiness of the water in part 2 section (i), what explanation can you give for the cloudiness of the water?

3 Why is it necessary to dry the filter paper?

4 Examine some of the particles from the filter paper by eye and under a microscope. Make drawings and record their appearance. Can you decide whether they have come from the land over which the stream is flowing or have come from a farmyard, sewage works, factory or any other outflow that may be present?

5 What could have caused the turmeric paper to turn brown in your investigation, section (iv)?

6 What could have caused the lead acetate paper to turn black in your investigation, section (v)?

7 If the water does seem to be polluted with ammonia and hydrogen sulphide can you explain why there is a lower oxygen content than would be expected? Has it been caused by respiration by plants and animals or was it due to respiration by bacteria feeding on the organic matter?

Fresh water pollution

Probably 50% of the total water in the rivers of the British Isles is polluted to an unacceptably high level, but River Authorities are taking more and more action to improve this situation. The River Thames has been polluted for many years and as late as 1957 there were no fish present in its lower reaches. By 1968, however, forty-one species had been recorded, this in spite of the fact that the Thames receives 2300 million dm^3 (litres) of treated sewage effluent every day.

The pollution of rivers, streams and lakes can be caused by the following:

i) *Detergents* River weirs and locks used to produce considerable amounts of foam because of unused detergents flushed into rivers in sewage. This problem has been largely overcome by producing detergents that are easily decomposed by bacteria in sewage works. Detergents in household waste that pass into sewage works still contain phosphates, even after treatment, and so these are discharged unchanged into river.

ii) *Sewage* Heavily-polluted rivers may have raw sewage pumped into the water, but local authorities are now endeavouring to treat all sewage before discharge. Unfortunately, the increase in size of populations of towns often outruns the capacities of sewage works. Bacteria in rivers can break down small quantities of sewage, but when it is discharged in large quantities the bacteria increase very rapidly, using up all the oxygen in the water. In these deoxygenated waters all other living organisms will die. Even treated sewage can damage the freshwater environment because the high amounts of phosphate cause an abnormal growth of algae (called a 'bloom'). This effect is most noticeable in lakes and reservoirs and it is called eutrophication. The algae eventually die and their decomposition removes oxygen from the water.

iii) *Farming* The modern practice of factory-farming where stock is kept in covered accommodation all the time also contributes to water pollution. When stock (cattle, sheep, fowl) lives in open fields the manure is spread around and helps to keep the soil fertile. In factory farms, the manure is washed away with large amounts of water and this *slurry*, as it is called, is often discharged into rivers.

Farmers also use chemical fertilisers in their fields instead of, and to supplement, manure. This also causes problems, because rainfall dissolves the unused fertilisers and washes them into rivers and lakes. Much nitrate finds its way into fresh water from this source and can cause eutrophication in the same way as phosphate. Furthermore it can still be present in drinking water that has passed through a filtration plant. A continual check on nitrate levels must be maintained by water authorities and in some areas special bottled water has been supplied for babies where the nitrate content of water is too high.

iv) *Industrial waste* Factory waste can contain many different chemicals which are generally poisonous. These pollutants may include lead, mercury, zinc, cyanide and copper which are cumulative poisons, building up in the living food chains as they pass from organism to organism. Oil, detergents, sulphur compounds and suspended particles are also discharged into fresh water in factory effluent.

	Non-tidal rivers (miles)	Tidal rivers (miles)	Canals (miles)
Total miles	22 317	1 783	1 545
Chemical classification			
Unpolluted	17 279	880	706
Doubtful	3 267	414	614
Poor	939	253	147
Grossly polluted	832	236	78

Table 11.10 River pollution in England and Wales, 1972

Sea pollution

Seventy per cent of the earth's surface is covered by the salt water of the seas. Into these flow the rivers of every country, carrying their own pollutants and adding to pollutants of all types from the towns and factories that border the seas. The following are the major pollutants.

i) *Sewage* Large centres of population bordering the seas have always considered it their right to pump sewage into the sea. It is true that the sewage emerges from large pipes often a mile out to sea but unfortunately it can be swept back by tides onto the beaches. The bacterial content of the sea water off bathing beaches can rise alarmingly and bathers may suffer infections of throat and ear as a result. In 1973 Italy had an outbreak of cholera caused by people eating shellfish contaminated by the disease organisms in sewage. The countries bordering the Mediterranean are largely to blame for the polluted state of its waters. The Spanish government has allocated large amounts of money in an effort to treat sewage and so protect the tourist industry.

ii) *Oil* It is considered that 5 to 10 billion kilograms of oil are released every year into the seas of the world. This includes illegal discharge by tankers as well as accidental losses from damaged tankers. More and more oil drills are in the sea and oil can escape from these in a variety of ways. The tanker *Torrey Canyon* wrecked off the British Isles released 100 million kilograms of crude oil, killing some 100 000 seabirds and necessitating the expenditure of some £1 500 000 in order to clean up the beaches of the West Country. The ultimate catastrophe, the collision of two 300 000 ton super-tankers at sea, occurred off South Africa's southern Cape Coast in December 1977. Only one ship was fully laden with 250 million kilograms of crude oil. It was two weeks before a change of wind blew a large slick of oil onto the beaches of the South Africa coast. In March 1978, the *Amoco Cadiz*, the largest oil tanker ever wrecked, was grounded off the French Atlantic Coast. It lay there for two weeks of continuous gale and eventually lost all of its millions of gallons of crude oil. The fishing and the tourist industries of this part of France were ruined and it cost £10 000 a day for many weeks to clear the oil from miles of beaches.

Oil, being a natural product, can be broken down by bacteria in time but the efforts of man to break down very large quantities of oil spillage can be dangerous to wild life. Detergents used to clear the oil are poisonous to many small sea animals.

iii) *Metals and chemicals* Rivers carry large quantities of these substances into the seas and they become

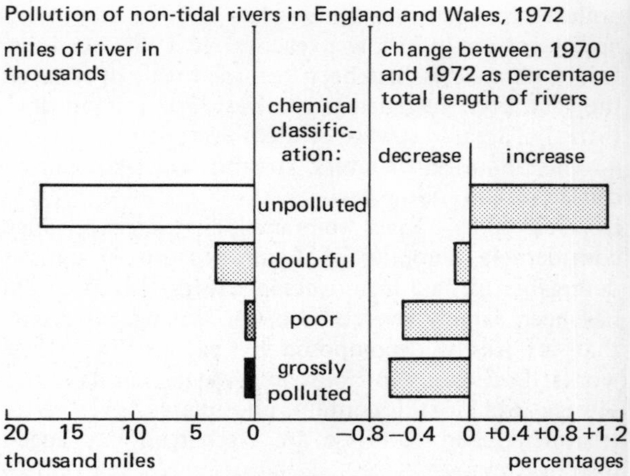

fig 11.52 Reduction of pollution in non-tidal waters in England and Wales, 1970–1972

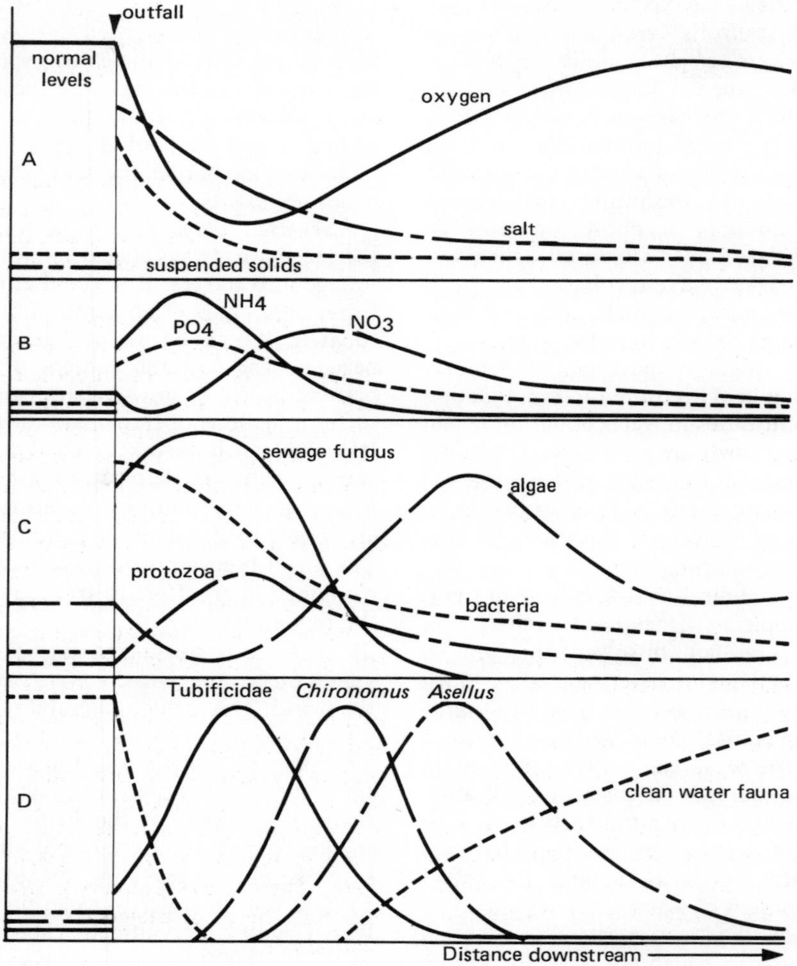

fig 11.53 The effect of sewage effluent on a river. A and B — changes in solids and certain chemicals, C — changes in micro-organisms, D — changes in sensitive small animals

fig 11.54 The pollution of a waterway

concentrated in organisms as they move up through the food chains. Minimata in Japan is a small town in which forty-six people died and many remain permanently injured from eating shellfish containing mercury released into the sea by a plastics company. The dangerous substances dumped in the seas are too numerous to mention but they include canisters of nerve gas, radioactive waste and pesticides.

11.45 Conservation

The wildlife of the world has suffered a severe blow at the hand of Man. Many species have become extinct and many others such as the leopard and certain whales are approaching the stage when they will also disappear. There are many demands to protect wildlife, but should we do so? Let us consider some of the reasons.

1 The useful aspects of wild life and wild places

a) The study of wild plants has provided us with many of our medicines. New crops have been developed by the incorporation of wild stock into breeding programmes. Wild animals originally provided the source of our domestic animals and Man still makes use of wild animals for food by the herding of reindeer, and the domestication of red deer and antelope (eland). Wild plants and animals must not be allowed to disappear or their breeding and genetic potential will be lost for ever.

b) The biological control of pests can be continued only if wild populations of plants and animals are maintained. Only from these sources can we find the control organisms to safeguard our food supplies from pests. This method must be used more and more to avoid the use of chemical pesticides, particularly the persistent poisons that become incorporated into the natural food chains.

c) The countryside must be maintained, not only for agricultural purposes but as a place where the populations of large cities can breathe freely and use it for their recreation and enjoyment. With more freedom from a shorter working week in the future, more people will wish to pursue outdoor activities ranging from mountain-climbing to fishing. It is most important that these varied interests should be satisfied and for this reason National Parks have been set up all over the world. The U.S.A. has very large National Parks attracting millions of visitors every year. Here wildlife is protected with regulations for control of shooting and erection of buildings. The British Isles has ten National Parks with a total area of 13 500 km^2. These often include farms and houses but the government and Parks authorities endeavour to satisfy all the conflicting interests while protecting the wildlife of the Parks. The game parks of Kenya and Tanzania attract vast numbers of tourists from all over the world who come to see the variety of wildlife enclosed within them. These parks contain a small fraction of the enormous herds of game that used to cover East and Central Africa.

2 Aesthetic appreciation of wildlife

By no means the least important aspect of wildlife in its natural surroundings is its beauty. Birds, butterflies, moths, flowers, fungi and fish all provide a source of study and hobby for countless enthusiastic specialist students of their variety of form, lifecycle and behaviour.

Let us therefore consider ways in which we can avoid the future extinction of animal and plant species. The answer lies in conserving what we still have and trying to revitalise old and damaged habitats. In 1977 the President of the U.S.A. exhorted his country to conserve energy, and at the same time an oil drilling platform in the North Sea lost thousands of kilograms of oil through human error in the fitting of a valve. 1977 also saw the tenth anniversary of the Council of Europe's Centre for Nature Conservation. This centre has achieved a great deal in this short space of time and has shown that only Man can preserve Nature in our densely populated countries, and that it is the duty and responsibility of all of us to safeguard our natural resources. In order to do this there must be public understanding and assistance. Each country in the Council of Europe has set up a National Council for Conservation and in 1970 the European Council launched a campaign, known as European Conservation Year, to influence public opinion.

To conserve and revive our environment nationally and in the continent of Europe we must look towards the following solutions:

1 The protection of wildlife.

a) National Parks already established in the British Isles and Europe must be maintained and new areas of outstanding beauty must be added to the list of those already in existence.

b) Hunting animals for their skins (e.g. leopards) and their products (e.g. whales), and other rare animals must be forbidden. More species must be protected by law, whether it be the taking of birds' eggs (Protection of Birds Act 1954) or the importation of skins for making coats.

c) Rare animals must be preserved and bred in zoos and parks.

d) Exploitation of the remaining wild places of the earth for minerals or timber must be stopped or strictly controlled.

2 Prevention of habitat destruction.

a) Pollution must be tackled at all levels. The discharge of chemical waste and sewage into lakes and rivers, the emission of smoke and chemicals into the atmosphere and the fouling of the seas must be stopped by international action.

b) The misuse of land must be halted. The growth of large scale farming, resulting in loss of habitat for countless organisms, must be controlled. The decline of 278 of the rarest British plants has been measured between the years 1963 to 1970 and has been calculated at 30%. This was due to increased agriculture and the elimination of hedges and copses (see fig 11.56). The decline of habitats has not been compensated for by the production of new habitats. Old-fashioned mixed farming resulted in a by-product of varied wildlife and we must try to achieve some compromise between increased food production and efficient wildlife conservation.

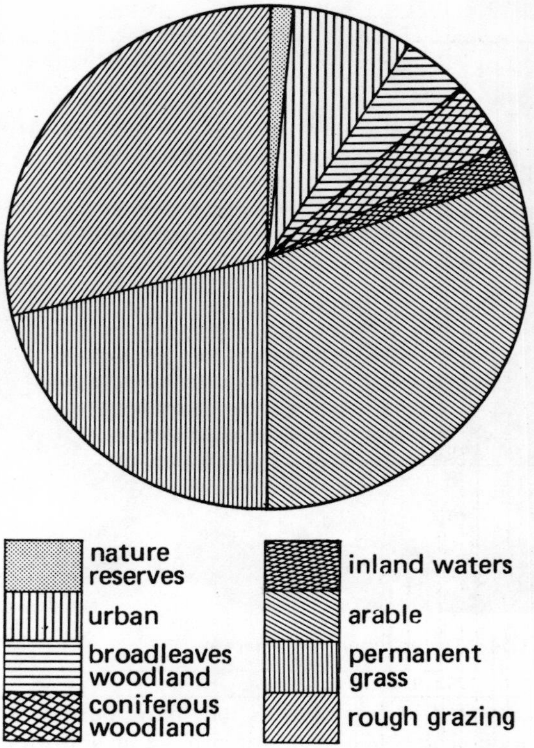

fig 11.55 The pattern of land use in the British Isles — nature reserves take up only 0.8% of the land surface

c) Wherever the waste of deep or open cast mining is left on the surface of the ground this must eventually be landscaped and replanted so that the natural habitat can renew itself.

3 Reduction in the rate of loss of resources

a) The forests of Europe and the world must no longer be cut down to provide enormous quantities of newsprint which has a life of only one day. Very little of this mass of paper is used again although many conservation groups in society do make an effort to use it again by collecting and repulping. Each nation must introduce or extend its reafforestation programme to include not only quick-growing conifers (softwoods) but also the longer maturing hardwoods.

b) The burning of coal, oil and gas must be slowed down so that vital energy resources are not dissipated for ever. All nations with the technology available must continue to research actively for new methods

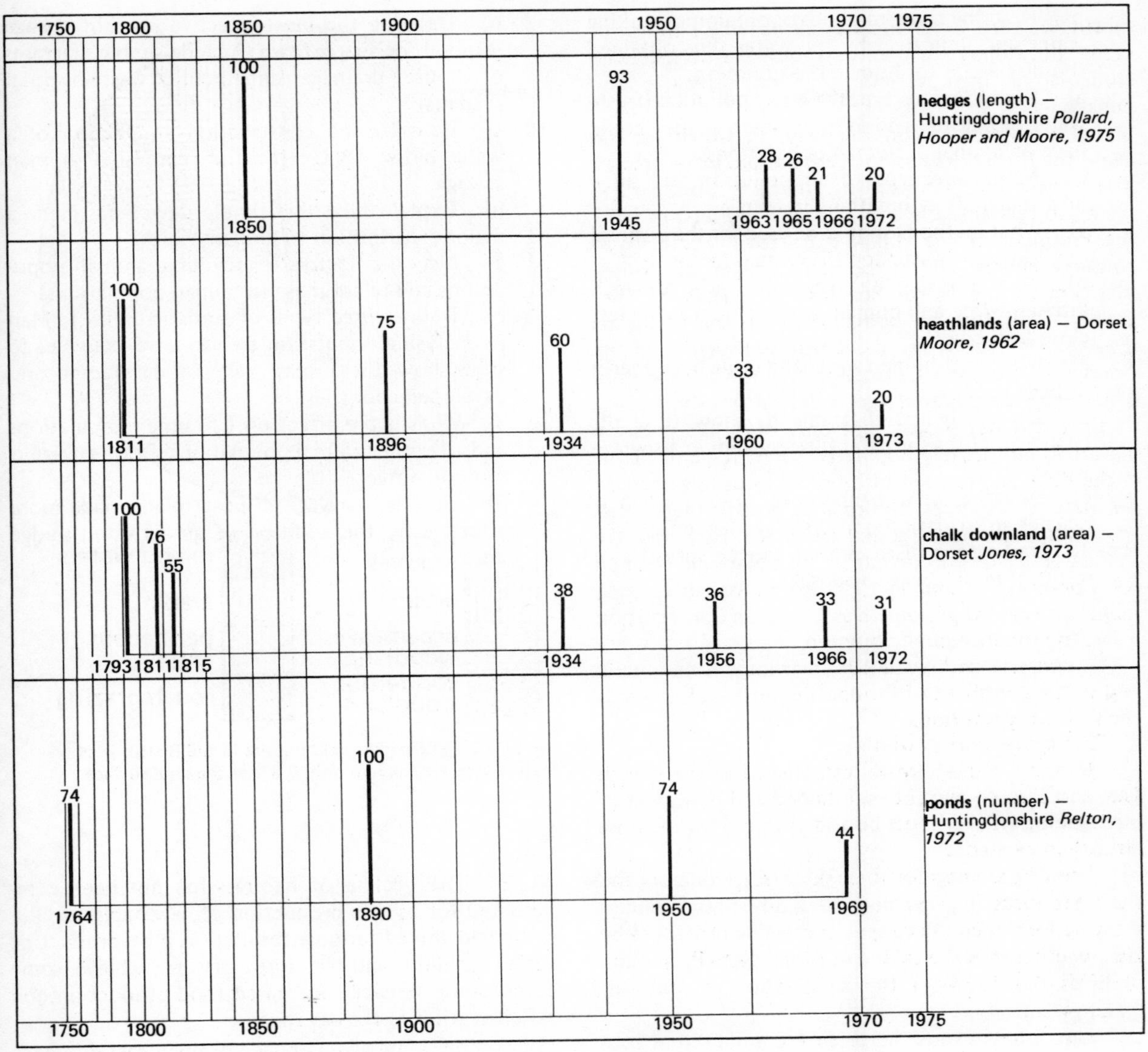

fig 11.56 Four types of land loss in England. (All values are expressed as percentages of the largest value recorded in each case.)

of producing energy from tidal power, tidal barrages, windpower and hydroelectric schemes.

c) There must be recycling where possible of the items used in modern industrial societies. An important first step would be the reuse of the so-called 'disposable' glass bottles.

Questions requiring an extended essay-type answer

1 a) What do you understand by the term *cycles in nature*?
b) Show the importance of natural cycles to (i) Man's agriculture and (ii) his drinking water.

2 a) Give examples of (i) a food chain and (ii) a food web.
b) Describe one example of Man's detrimental effect on the environment.

3 Describe the principles on which you would base the planning and construction of a new town.

4 a) Describe methods of *either* heating *or* ventilating a new house.
b) What principles of (i) construction, (ii) lighting and (iii) sewage disposal contribute towards a healthy environment in a family house?

5 a) State *six* desirable properties of drinking water.

b) Describe *three* methods of obtaining drinking water and indicate why the water thus obtained should be safe for human consumption.

6 a) Describe *three* types of water pollution (freshwater or marine) and show what effect they could have on animal or human populations.

b) Name *one* type of air pollution and show how it can affect breathing in Man.

7 a) What methods should be used for the disposal of (i) kitchen waste and (ii) human faeces?

b) What possible dangers could arise through kitchen waste and human faeces simply being left close to the house?

8 a) Why must human waste and urine be properly disposed of?

b) Draw labelled diagrams to show (i) a pit latrine and (ii) a sprinkler filter system in a sewage works.

c) Describe what happens to sewage when it reaches an oxidation pond (sewage lagoon).

9 a) Name *four* diseases that can be spread as a result of poor sewage disposal.

b) Describe the most effective form of sewage disposal for a large town (include simple diagrams of house drainage and public waste-disposal systems).

10 a) Describe the construction and action of a water closet (W.C.) for the removal of human sewage.

b) Explain how this simple device has made a major contribution to human health.

11 a) List the factors controlling animal populations and relate these to human populations.

b) Define *three* types of statistic useful to Man in analysing population trends. Give examples to show how these may vary in developing and developed countries.

12 a) What is the effect on (i) human populations and (ii) other animal populations if raw sewage is run into a river?

b) How is a sewage disposal plant made more effective by the addition of an activated sludge unit?

12 Answers and discussion

Introductory Chapter

Section I 2.23

1 Fibres of cellulose can be seen at the higher magnification compared with the smoothness of the paper seen by the naked eye alone.
2 Yes.
3 It is now a letter quite different from letter 'p'. It appears to be the letter 'd'.
4 It is upside down and turned from right to left.

Section I 2.24

1 All of the wing can be seen.
2 It is much larger (more magnified) and more detail can be seen.
3 It moves to the left.
4 It moves towards the observer.

Section I 5.40

1 Hydrochloric acid 0.01 M has a pH of 2. Sodium hydroxide 0.000 1 M has a pH of 11.
2 It has a pH of 7 which is between hydrochloric acid (pH 2) and sodium hydroxide (pH 11).
3 It is the same as the given solution of hydrochloric acid.
4 The pH value of the acid increases and the pH value of the alkali decreases. If the hydrochloric acid is diluted exactly 10 times, the pH increases by 1, and each time this is repeated the pH increases similarly.

Thus:	0.001 M	pH 3
	0.000 1 M	pH 4
	0.000 01 M	pH 5

If the sodium hydroxide is diluted 10 times the pH decreases by 1, and each time this is repeated the pH decreases similarly.

Thus:	0.000 1 M	pH 10
	0.000 01 M	pH 9
	0.000 001 M	pH 8

Section I 5.50

1 Measure the difference between the new height of the liquid in the capillary tube and its initial level. This measurement will vary according to the diameter of the capillary tube and of the visking tubing.
2 Water molecules must have entered the visking tubing from the water in the beaker.
3 The liquid in the capillary tube must be exerting a hydrostatic pressure on the liquid in the visking tubing. As the liquid continues to rise in the capillary tube, the inflow of water must be overcoming this pressure.
4 It looks distended or swollen. It feels firm; it is no longer as flabby as it was at the beginning of the experiment.
5 Water has passed into the visking tubing, and since the screw clip is closed the bag is distended.
6 The liquid inside the bag and tube shoots out in a fine stream from the glass nozzle. The water that has entered the visking tubing has filled it and built up a pressure, possibly stretching it. When the clip is opened the pressure is released and the liquid spurts out.
7 It is still blue in colour, but a much paler blue.
8 The copper sulphate solution has diffused throughout the water and is therefore diluted. The dilution (mixing of water with the solution) has produced a paler coloured liquid. The same amount of copper sulphate is present but it is dispersed over a greater volume.
9 The cotton wool pieces soaked in phenolphthalein change colour to pink in succession along the glass rod. The cotton wool nearest to the ammonium hydroxide changes first.
10 The ammonia fumes cannot be transmitted in any way to the phenolphthalein except by diffusion through the air in the tube.
11 It is much faster in gas than in a liquid.
12 Blood cells could be seen under low and high power in tubes (ii) and (iii). No trace of cells could be seen in tube (i).
13 In tube (ii) the blood cells appeared as biconcave discs with a smooth outline. In tube (iii) the blood cells had a wrinkled appearance and the outline was no longer smooth.
14 In tube (ii) the blood cells were unchanged because their surroundings have the same salt concentration as normal blood. In tube (iii) the cells had lost water to the strong salt solution.
15 In tube (i) the liquid was a clear red colour. (ii) the liquid was red in colour but slightly opaque. (iii) the liquid was red in colour but slightly opaque.

16 i) The blood cells in this tube took up water from their surroundings by osmosis. This resulted in a stretching of the cell membranes until they burst, releasing the red pigment haemoglobin into the water.
ii) The liquid appears red because of the presence of blood cells, but it is not a clear red colour as in (i).
iii) The blood cells in this tube lose water to the surroundings by osmosis, so that the cell membrane shrinks. The cells are still present and give the liquid its red colour, but again it is not as clear a red as in (i).
In both (ii) and (iii) the cells will sink to the bottom and in time the liquid will become clear.

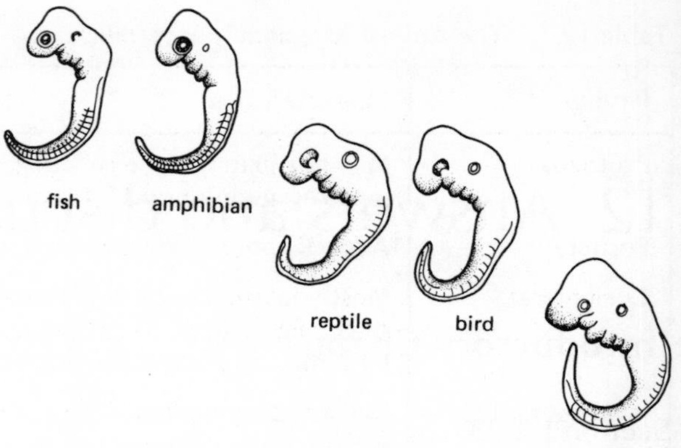

fig A Early embryos of vertebrates

1 A brief history of Man

Section 1.01

1 (a) and (b) Using the green pigment chlorophyll they harness the radiant energy of the sun to make sugars from carbon dioxide and water. This process is known as *photosynthesis.* Organisms (c), (d) and (e) can only obtain their food by eating other organisms.
2 (c), (d) and (e) This *locomotion* is associated with the need to find other organisms to feed upon.
3 (a) and (b).

Section 1.02

1 (d) and (e) In these animals, the bony skeleton is internal (an endoskeleton). The insect, on the other hand, is supported by an exoskeleton made of a horny substance called chitin, and the earthworm has a 'hydrostatic skeleton' produced by contraction of the body wall muscles against the fluid in the body cavity. The shape of the single-celled *Paramecium* is maintained by a tough outer layer of cytoplasm called the pellicle.
2 (d) and (e) In these animals, the dorsal nerve cord is protected by the backbone. The main nerve cords of the insect and earthworm are situated ventrally, and *Paramecium* has no specialised nervous system as such.
3 (d) only However, if we look carefully at fig A, which shows an early human embryo, we can see structures which correspond to the gill slits of fish. Most of these disappear during the course of development, but one of them stays to form the opening of the ear.

Section 1.03

1 (d) and (e) Reptiles and fish have scaly skins, birds have feather-covered bodies and scaly legs, and amphibian skins are thin, smooth and moist.
2 (d) and (e) Birds and amphibians lay eggs, as do many fish and reptiles. Other fish and reptiles may produce live young, but these develop from yolky eggs, which are held within the maternal body as a means of protection only. Their nourishment comes from the yolk and not from the maternal blood stream via a placenta.
3 (d) and (e) The females of both species possess mammary glands, which produce milk to nourish their young offspring.

Section 1.04

1 (d) and (e) In apes, this applies to both forelimbs and hind limbs; in man, only the hands are so modified. The thumb is said to be *opposable* to the four fingers.
2 (d) and (e)
3 (d) and (e) With the reduction in number of the turbinal bones in the nose, which are essential to a strong sense of smell, the snout has become shorter during the course of evolution. Thus the eyes have moved forward so that their fields of vision overlap and accurate judgement of distances by stereoscopic vision has become possible. At the same time, the width of the visual field has narrowed (see figs 7.2 and 7.3).

2 Nutrition

Section 2.11

1 The colour changes from the brown colour of iodine solution to an intense blue-black. In testing foods for the presence of starch many different shades of blue may be observed, but all indicate a positive result.
2 The copper sulphate in the Benedict's or Fehling's solution is reduced to copper oxide, which forms a precipitate. This goes through a series of colour changes finalising at red or brown. If the tube is left to stand, the precipitate will sink to the bottom of the tube.

Table 12.1 The Animal Kingdom – Invertebrata

Phylum	Characteristics	Examples
Protozoa	Mainly aquatic; single celled; generally one nucleus; move by pseudopodia, cilia or flagella; some parasitic	*Amoeba, Paramecium, Euglena, Plasmodium*
Porifera	Aquatic; pores through which water circulates	Sponges
Coelenterata	Mostly marine; body composed of two layers of cells; single digestive cavity with one opening; mouth surrounded by tentacles; possess stinging cells (nematocysts)	*Hydra*, sea-anemones, corals, jellyfish
Platyhelminthes	Flattened segmented worms; alimentary canal but no mouth; body composed of three layers of cells; no body cavity; two classes are parasitic	Planarians, flukes, tapeworms
Nematoda	Unsegmented worms, pointed both ends; gut with mouth and anus; three layers of cells; many parasitic	Hookworms, threadworms, roundworms
Annelida	Segmented worms; body composed of three layers of cells; gut with mouth and anus; bristles (chaetae) in two classes; some are parasitic	Earthworms, lugworms, ragworms, leeches
Arthropoda	Segmented, chitinous exoskeleton, jointed limbs	Insects, spiders, crabs
Mollusca	Unsegmented soft body often with a shell; large single muscular 'foot'	Snails, slugs, mussels, octopus
Echinodermata	Unsegmented, radially symmetrical; five arms; spiny skin with hard plates; possess tube feet; all marine	Starfish, sea urchin, bristle star, sea cucumber

Table 12.2 The Animal Kingdom – Chordata

Phylum	Characteristics	Examples
Chordata	Notochord present in adult; tubular, dorsal, hollow nerve cord; closed blood system; post-anal tail	
Sub-phylum		
Acrania	No true brain or skull, heart or kidneys	*Amphioxus*
Craniata or vertebrata	Well developed head and brain; muscular ventral heart; kidneys, organs of excretion; two pairs of limbs; endoskeleton	
Class		
Pisces (fish)	Paired fins; gills for gaseous exchange; external scales; lateral line system	Cod, plaice, carp
Amphibia	Paired pentadactyl limbs; gills present in the tadpole larva; lungs in the adult; soft skin, no scales; middle and inner ear, no outer ear	Frog, newt, toad, salamander
Reptilia	Paired pentadactyl limbs; lungs for gaseous exchange; dry scaly skin; large heavily yolked eggs with shell are laid; no larval stage	Grass snake, turtle, lizard, crocodile

Phylum	Characteristics	Examples
Aves (birds)	Paired pentadactyl limbs, forelimbs are wings; lungs for gaseous exchange; feathers over the body, legs have scales; heavily yolked eggs with a calcareous shell are laid, no larval stage, warm-blooded;	Chaffinch, eagle, heron, stork
Mammalia	Paired pentadactyl limbs, lungs for gaseous exchange; skin bears hairs and two types of gland, sebaceous and sweat; primitive mammals lay eggs (oviparous) but the majority are viviparous; warm-blooded; possess milk glands to feed the young; external, middle and inner ear; diaphragm separates the thorax from the abdomen;	Mole, mouse, dog, cow, elephant, whale, bat, monkey, ape, man

3 The colour changes in the following sequence: green, yellow, orange, red and brown. In the test for reducing sugars, the colour green indicates a slight amount of the sugar and brown means a substantial amount.

4 No change. The blue colour of the copper sulphate in the test reagents remains.

5 No change in the experimental or the control tube.

6 Positive reaction through green to red or brown.

7 The sucrose has been hydrolysed to monosaccharide molecules which can now reduce the test reagents and give a positive result. The test is thus positive for non-reducing sugar, but **not** reducing sugar.

8 The oil breaks up and dissolves in the ethanol.

9 A milky white emulsion is formed. This is the positive indication of the presence of oil or fat. When foods are being tested, particles of food should be allowed to settle before the ethanol and its dissolved fat is poured into the water.

10 The water spot has dried out, but the oil spot remains and the light shines through. When foods are being tested they can be rubbed on the paper to see if they make an oily mark.

11 A violet colouration is a positive indication of peptides and hence all proteins. This is a general test for proteins since it does not test for any specific amino acid. It is therefore widely applicable.

12 Suppose that 10 drops of orange juice just decolourised 1 cm^3 of DCPIP solution, and that 8 drops of the ascorbic acid solution did the same.

Then the ascorbic acid solution contains 1 mg of ascorbic acid per cm^3, and the orange juice must have contained $1 \times \frac{8}{10} = 0.8$ mg of ascorbic acid per cm^3.

Repeat for all juices tested.

13 Exposure to the oxygen of the air destroys vitamin C in fruit and vegetables, therefore the values of vitamin C (ascorbic acid) should be less after exposure for several days. Cooking also destroys vitamin C.

Section 2.21

1 It is green in colour. The chlorophyll has been extracted by the ethanol.

2 The leaf or leaf discs show a blue-black colour *between* the veins of the leaf. The colour of the veins is not blue-black. Starch must be present in the leaf, but only in the cells of the blade, not in the veins.

3 The leaf is the colour of the iodine solution, brown or orange-brown. No starch is formed in the absence of light.

4 Depending on the size of the leaf and the amount of sugar present, the colour could vary from green to brown. The test is positive for the presence of reducing sugar.

5 The green areas of the leaf show a blue-black colour indicating the presence of starch. The white areas of the leaf show the brown colour of the iodine solution indicating that no starch has been formed.

6 Starch is only formed in those areas of the leaf which are green in colour, that is, where chlorophyll is present.

Section 2.22

1 No. The concentration did not drop below 0.032%.

2 After twelve midnight the carbon dioxide concentration rose to a peak at 7 a.m. and then fell steadily to reach its lowest concentration during the later afternoon. It then began to rise again through the hours of darkness.

3(a) 7 a.m. 0.054% carbon dioxide

(b) 4 p.m. 0.032% carbon dioxide

4 The carbon dioxide concentrations in the forest rose during the hours of darkness and fell during the hours of daylight. It could be that the vegetation is giving out carbon dioxide at night and taking it in during daylight.

5 *Tube A* changes to yellow

Tube B stays unchanged (or slightly yellow or slightly purple, depending on the density of the muslin screen).
Tube C changes to purple.
Tube D stays unchanged (orange-red).
6 *Tube C* The leaf in light has withdrawn carbon dioxide from the air above the indicator, and carbon dioxide from the indicator has diffused out decreasing the acidity thus changing the colour to purple.
Tube A This colour change to yellow represents an increase in acidity, indicating that carbon dioxide has been added to the indicator. This has occurred in the dark and is in accordance with the graph of carbon dioxide in the forest.
Tube B This tube has the leaf in half light or twilight and the uptake and output of carbon dioxide could be completely balanced. It depends on the thickness of the muslin screen; a thinner screen could tip the indicator towards a purple change, a thicker one could tip it towards a yellow change.
Tube D Has no leaf and the indicator is unchanged, showing that any change in the other tubes must be due to the enclosed leaf.
7 Control.
8 Yes. The indicator shows that carbon dioxide is given out in the dark and taken into the leaf in the light. In half light or twilight either there is a balance or it could be that there is none given out and none taken in. This can only be decided by further experimentation.
9 Emission of light is always accompanied by heat. The changes that take place in the indicator could be due to the action of heat emitted from the light bulb or the sun. To eliminate this variable the tubes are placed in water, so that the glass and the water act as a heat shield.
10 *Leaf A* (deprived of carbon dioxide) This shows no starch in the leaf, which remains the orange-brown colour of iodine solution after testing.
Leaf B (in the flask but carbon dioxide available) This gives a positive result for starch, but the blue-black colour is not strong.
Leaf C (in the air with more carbon dioxide available) This gives a much stronger positive result for starch with the colour a deep blue-black.
11 The carbon dioxide contains two of the elements needed for the formation of carbohydrate. The third element could come from water which is always present in living organisms. It would appear that the carbon dioxide contributes towards the manufacture of carbohydrate in the leaf.
12 Since the experiment investigates the supply of carbon dioxide to the leaf, it would probably be better to have flowing over the plant a supply of air from which the carbon dioxide has been removed. Enclosing the leaf in a flask limits the supply of other gases present in the air, and therefore a better experimental set-up would be shown in fig B. What would be the control apparatus for this experiment?

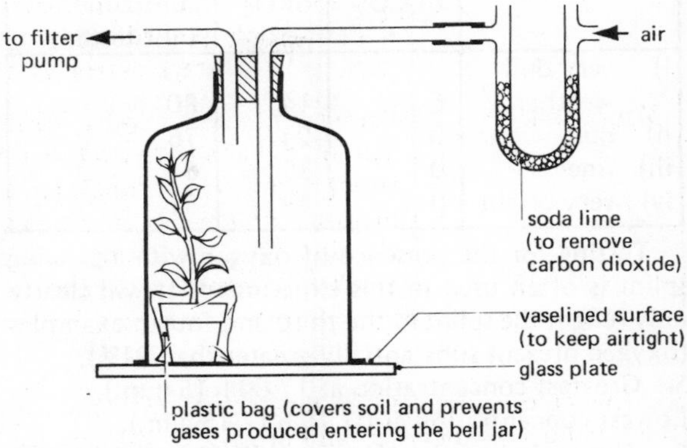

fig B Alternative apparatus to investigate the use of CO_2 by green leaves

13 *Leaf A* (deprived of light) This shows no starch in the leaf, that is orange-brown colour of iodine solution after testing.
Leaf B (in the flask but light available) This gives a positive result for starch with a blue-black colour.
Leaf C (in the air with light available) This gives a positive result for starch with a blue-black colour.
14 Light is essential for starch formation in the leaf.
15 Carbon dioxide might be used up in the flask and become a limiting factor. Another method is to shut off light from part of the leaf only, so that the remaining parts receive all conditions and therefore give a positive result for the starch test. The black cardboard must be raised slightly from the leaf in order to allow a free flow of air containing carbon dioxide.

Section 2.23

1 In order to stabilise the temperature of the gas bubble. Gas expands considerably with a small rise in temperature, thus when the apparatus is handled the body heat may cause expansion. The temperature and therefore the volume, is restored by returning the J-tube to the water trough.
2 In order to present the surface of the bubble with the potassium hydroxide and the potassium pyrogallate. The chemicals absorb the carbon dioxide molecules and the oxygen molecules respectively.
3 The results will vary according to the light intensity and the length of time the weed is exposed to light during each twenty-four hour period. The oxygen content can increase by up to 50% of the gas, and carbon dioxide by up to 10% compared with normal air.

4 Light intensity will alter the rate of photosynthesis: below are some figures which indicate the relationship:

	% of CO_2	% of O_2	% of N_2 by subtraction (not tested)
i) very dull weather	6	14	80
ii) dull	0	24	76
iii) fine	0	39	61
iv) very bright	0	49	51

Testing for the presence of oxygen with a glowing splint is often used in this experiment. It will clearly only relight the splint in the third and fourth examples (oxygen present substantially greater than 21%).

5 Greatest concentration at 17 00 h (5 p.m.).
Lowest concentration at 07 15 h (7 15 a.m.).

6 The period of time between 07 00 h and 14 00 h shows the most rapid increase, reaching a maximum at 17 00 h, broadly speaking from early morning to late afternoon.

7 8.05 – this value does agree with 'less carbon dioxide' since this pH is more alkaline or less acid indicating that there is less carbonic acid present. Carbonic acid is formed from carbon dioxide.

8 7.5 – this value does agree with 'more carbon dioxide' since this pH is more acid or less alkaline than the previous value for pH. Thus there is more carbonic acid and pH falls in value, but notice that it does not become acid since the pH7 is neutral (see Introductory Chapter section 5.40).

9 Fig 2.13, graph of the concentration of carbon dioxide in a forest.

Section 2.24

1 The gas is produced in the leaves during photosynthesis. Assuming that there is a connection through the stem to the leaves then the gas will pass through the stem and bubble from the cut shoot.

2 The red filter between the lamp and the beaker causes the most rapid rate of bubbling.

3 The green light is reflected or passed through the leaf, whereas the red light is absorbed by the chloroplasts and used in photosynthesis.

Section 2.26

1 Cotton wool is largely composed of cellulose and if wet could be a medium for growth of fungi and bacteria which would damage the seedling.

2 No air is available. Roots in soil must have access to soil air containing oxygen. Bubble air through the solution each day by means of a glass tube.

3 a) *Tube 1* containing complete solution.
b) *Tube 4* containing distilled water.

4 *Tube 3* Calcium ions are required by the cell for middle lamella development between the cell walls. Deficiency of the element results in poor root growth.

5 Because poor root growth would limit the passage of water and salts to the leaves and hence there would be poor leaf growth. Conversely poor leaf growth would provide little food from photosynthesis for root growth.

6 The seedlings extract water and salts at different rates. Water is removed more quickly and therefore the addition of more culture solution would increase the concentration of salts. This could reach too high a level and result in water passing out of the roots.

Section 2.31

1 Recent rainfall can make a considerable difference, so that soil examined in the rainy season will show a much greater amount of water than that examined in the dry season. Nevertheless at any one season, the amount of water depends on the drainage characteristics of the soil.

2 Forest soils will have the greater amount of humus compared with garden soil. Swampy ground, where humus cannot decay, will show considerable amounts of humus. Sandy soils lack humus (see table 12.6).

3 Humus is missing since plants and animals cannot live easily in the hard-packed soil. Water also shows a smaller percentage.

4 Topsoil has more humus and less mineral matter than subsoil. If the soil contains much chalk (calcium carbonate) some of the weight loss could be due to the decomposition of chalk on heating.

5 An identical sample of soil could be dried to eliminate water content and the amount should be subtracted from the calculated air.

6 The soil could be compressed into the tin beyond its natural state. The soil in the cylinder was not stirred properly and thus all the air was not released.

7 It provides oxygen for the respiration of soil organisms and plant roots.

8 Limewater turns milky or bicarbonate indicator turns yellow in tube A but not tube B.

9 Carbon dioxide has been produced in tube A, indicating the presence of living organisms. The organisms in the soil of tube B were killed by heating.

10 The light, heat and dryness produced by the bulb are all factors causing organisms to move away from the upper parts of the soil.

11 To reflect the light and heat into the soil.

12 There are more different types of organism and they are more numerous in the topsoil. This is due to the presence of decaying plant and animal matter on which they feed, while some of the carnivorous forms feed on the other living organisms.

Section 2.32

1 Sand.
2 The larger air spaces enable the water to rise more rapidly in the first hours (see fig 13.5).
3 Clay.
4 The smaller air spaces, although slowing the early rise of water, enable the water finally to rise higher than in the sand (see fig C).

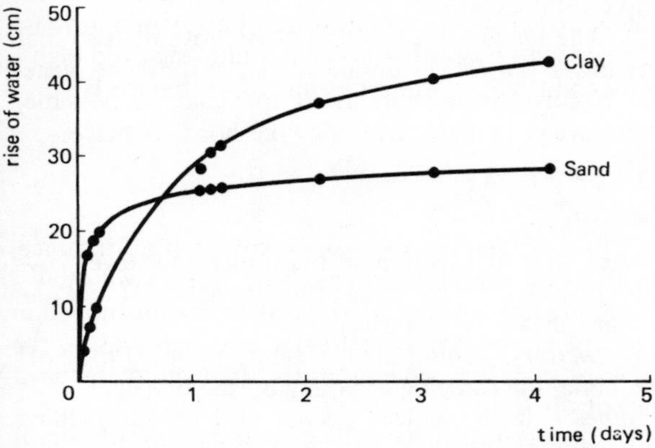

fig.C A graph to compare capillarity in clay and sand soils

5 Sand.
6 The larger air spaces enable the water to drain through more rapidly.
7 Clay.
8 Clay can hold more water around its small particles.
9 In beaker A more clay has settled out. There is a greater depth of clear (or slightly milky) water at the top of the beaker and a deeper layer of clay at the bottom. The clay particles in A have clumped together, whereas those in B have not. The clumping together (flocculation) of clay particles into larger lumps has caused them to settle out.
10 Lime causes flocculation of clay into larger particles. This improves the porosity of the soil and allows greater penetration of air and water to root systems.

Section 2.40

1 *The box of soil inclined at 40°*. No plant material was present to bind the soil. The 40° slope allowed the water to travel down quickly taking soil particles with it into the container.
2 *The box containing the cut turf.* The leaves deflected the water drops thus protecting the soil, the roots bound together the mineral fraction and the humus absorbed the water.
3 The process is called erosion.

Section 2.52

1 Yellow The colour is not evenly dispersed, but is restricted to the vascular bundles.
2 Most of the sections between the vascular bundles are stained blue-black, showing that starch is present.

Section 2.70

1 No.
2 No starch was present, but the glucose test was positive.
3 Glucose passed through the tubing. The glucose molecules are much smaller than the starch molecules and thus able to pass through the minute pores in the visking tubing. The pores are not large enough for the starch molecules to pass through.
4 Two controls would be appropriate:
i) The same experimental set-up but with starch only in the visking tubing.
ii) The same experimental set-up but with glucose only in the visking tubing.
5 The starch had disappeared from tube A.
6 Something in the saliva had caused it to disappear, because it had not disappeared from tube B where saliva is absent.
7 The starch had been hydrolysed to reducing sugar because the Benedict's test on tube A was positive. The Benedict's test on tubes B and C was negative.
8a) It shows that something in the saliva breaks down the large, insoluble molecules of starch to smaller, soluble reducing sugar molecules. This is the essential function of digestion.

Dried soil	Colour before burning	% loss of weight on burning	Colour of remaining mineral matter
Forest (topsoil)	Black	30	grey
Garden (topsoil)	Dark brown	15	grey
Soil from a school field	Light brown	5	red
Subsoil	Red	3	red

Table 12.3 Soil data

b) Test saliva only with Benedict's solution. The result should be negative.
9 Tube C. The liquid in the tube was clear, indicating that the globules of albumen must have disappeared.
10 Enzyme activity must have been stopped by boiling. The enzyme is denatured above a temperature of about 45°C. Note that it is **not** killed since enzymes are not living material.
11 It could be that hydrochloric acid changes the egg albumen **only** in the presence of pepsin. What other control would be needed to test this hypothesis?
12 Since the pepsin works best in acid conditions, the pH must be less than pH 7 (neutral). Pepsin in the stomach works at about pH 1.5 to 2.0.

Section 2.71

Investigation 1
4 i) About 1 year of age ± three months.
ii) 4 in the upper jaw, 4 in the lower jaw.
iii) Incisors.
iv) Third molar in each half jaw, often called the 'wisdom teeth' since they appear at about the age of 19–21 years, when 'wisdom' has arrived!

Investigation 2
1 Incisors 0, canines 0, cheek teeth 6 (premolars 3, molars 3) in half jaw.
2 Incisors 3, canines 1, cheek teeth 6 (premolars 3, molars 3) in half jaw.
3 Front teeth (chisel shaped incisors and canines) on the lower jaw act against a horny pad at the front of the upper jaw. Sometimes food is chopped off and at other times it is pulled off the plant or out of the ground.
4 A large gap in the lower jaw, between the canines and the premolars, enables newly taken vegetable matter to be stored. The food is kept apart from that being chewed by the cheek teeth.
5 $i\frac{0}{3}\,c\frac{0}{1}\,p\frac{3}{3}\,m\frac{3}{3}$

6 The lower jaw performs a circular type of motion from outside to centre, on one side and then the other.
7 *Incisors* – chisel shaped, flattened surface not very sharp; chopping and pulling.
Canines – similar to incisors.
Premolars – ridged across the jaw so that the W-shaped ridges on the upper jaw fit into the M-shaped ridges of the lower jaw; chewing and grinding.
Molars – similar to premolars.
8 The root is open. Growth is continuous so that as material is worn away from the surface it is replaced by new material transported by the blood vessels through the root and into the pulp cavity.

Investigation 3
9 Incisors 3, canines 1, cheek teeth 6 (premolars 4, molars 2) in half jaw.
10 Incisors 3, canines 1, cheek teeth 7 (premolars 4, molars 3) in half jaw.
11 The wild dog uses its canines for killing its prey. Premolars and molars are used for cutting up and crushing the corpse. The domestic dog does not have to catch its prey, but still uses its teeth for cutting and crushing its food.
12 The hinge joint has no lateral movement so that the lower jaw moves up and down in the same place. No circular movement, as in the goat, is possible, therefore chewing or grinding does not take place.

13 $i\frac{3}{3}\,c\frac{1}{1}\,p\frac{4}{4}\,m\frac{2}{3}$

Notice that in this animal the upper and lower jaws are not exactly similar, but the two halves of any one jaw are always symmetrical.
14 *Incisors* – chisel shaped, blunt; scraping food off the bone, the carrying of young by the female.
Canines – large, conical, curved, very pointed; killing prey, ripping food.
Premolars and molars – ridged along the jaw with sharp points, lower jaw fits inside upper jaw. Teeth work like a pair of scissors, particularly the *carnassial* teeth (i.e. upper premolar 4 and lower molar 1). Cutting and crushing (**not** chewing).
15 The tooth has been dissolved away on its surface. It has a pitted appearance.
16 The tooth is much softer.
17 The dentine.
18 Yes. The increase of the fluoride up to 2.0 ppm gives a rapid decrease in the incidence of cavities per child.
19 10.
20 3.
21 1.5 ppm (from 10 per child to 2.5 per child).

Section 2.80

1 0.210 kJ.
2 1.050 kJ.
3a) Some heat is lost around the edge of the tube.
b) Some heat is used up in warming up the glass of the tube.
c) The cashew nut or ground nut does not burn completely.
d) Heat could be lost from the water through the glass, or through the cotton wool.
e) The cashew nut is not a pure food, it contains oil, protein, carbohydrate and water.
Improvements:
i) Use a flat bottomed container so that heat is kept below the container and the water.
ii) Insulate the vessel in some way to stop heat loss.

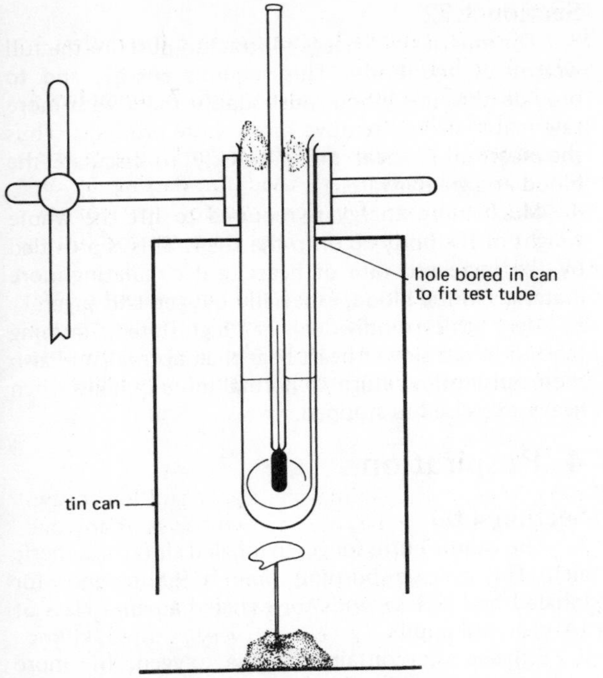

fig D Improved method for experiment and apparatus shown in Chapter 2, fig 2.70

iii) Use a jet of oxygen from an oxygen cylinder to make the nut burn freely and completely.
iv) Use a pure food such as sugar or oil.

Considerable heat loss can be prevented by simply surrounding the test-tube with a tin can (see fig D). A much more complex apparatus, satisfying many of the above criticisms, is shown in fig E.

4 Most of the shortcomings of the apparatus will tend to reduce the amount of heat entering the water in the test-tube and this will make the resultant energy value appear artificially small. Therefore, if we assume complete combustion externally is equivalent to the processes of oxidative decomposition of food in the body, more energy will be produced by the food in the body than would be indicated by this experiment because of the external losses in combustion.

Section 2.81

1 *Protein requirement*
Total weight of protein required = 1 x 60 = 60 g
Weight of animal protein = 1/5 of 60 g = 12 g
Weight of plant protein = 4/5 of 60 g = 48 g
Fat and carbohydrate requirement
Energy to be provided by fat = 1/6 of 13 500 kJ
= 2 250 kJ

Weight of fat to provide this energy $= \frac{2\,250}{38}$ g $= 59.2$ g

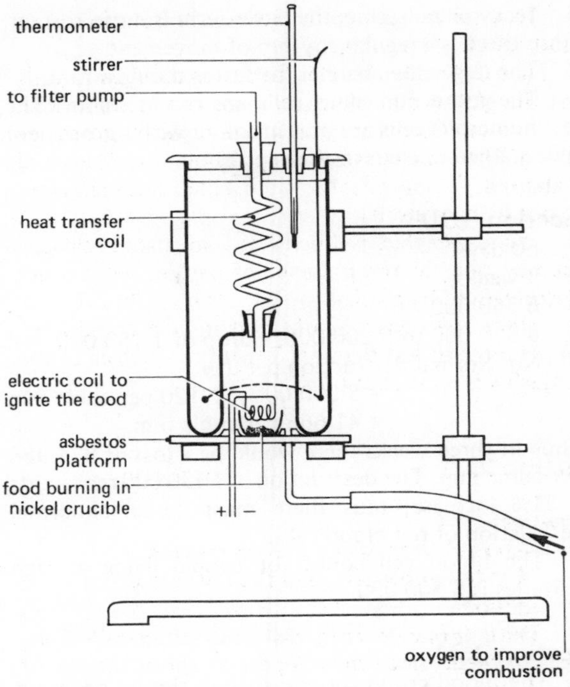

fig E A calorimeter for estimating the calorific value of foodstuffs

Energy to be provided by carbohydrate
= 13 500 − 2 250 kJ
= 11 250 kJ

Weight of carbohydrate to provide this energy
$= \frac{11\,250}{16} = 703.13$ g

2 The abdomen is distended. Skin is peeling and blotchy.

3 There has been an acute lack of protein in the child's diet.

3 Circulation

Section 3.10

1 The colour penetration is much faster in the 0.5 cm cube than in the 1 cm cube. When the smaller cube is uniformly red in colour the 1 cm cube still has a clear area in the centre.

2 i) The colour would penetrate at the same rate.
ii) The colour would penetrate at a much faster rate because there are a larger number of molecules to diffuse into the block.

Section 3.20

1 Yes.

2 No. In some channels the particles flow outwards towards the edge of the fin, in some they flow across the fin, and in some they flow back towards the centre.

3 Jerky or pulsating; the latter term is more correct, since there is a regular rhythm of movement.
4 The larger the channel the faster the flow rate.
5 The more numerous cells are red in colour. The less numerous cells are blue in colour with a prominent blue-stained nucleus in the centre.

Section 3.30

1 60 days.
2 9 weeks.
3 6 350 000.
4 6 350 000 to 5 200 000, a drop of 1 150 000.
5 No. Normal destruction per day

= 5 000 000 x 1/120 per cubic mm
= 41 666 per cubic mm

Thus in three weeks there would be a loss of 874 986 per cubic mm. The destruction of 1 150 000 represents a 31% increase; thus there must be an abnormal destruction of red blood cells.
6 The blood cell count for people living at high altitudes is 6 500 000.

Section 3.31

1 The blood group for transfusion should be group O or A and Rh–. Groups O and A are the only two groups compatible with her blood (O is the Universal Donor). It is also advisable to have Rh– blood, since Rh+ antigens would produce antibodies in her blood. These could affect the unborn child in any future pregnancy, since the foetus could be Rh+. This situation is common when the father is Rh+.

Section 3.41 The mammalian heart – external examination

Experimental procedure 2. The auricles have wrinkled walls. They are smaller and darker in appearance than the ventricles. The ventricles are larger, smoother and pink.
Procedure 4. Arteries have semilunar valves at their bases.
Procedure 5. To supply oxygen and dissolved food materials to the heart muscles. To remove the waste products produced by the working of the muscle.
Questions
1 No. There is no connection between right and left sides of the heart.
2 The left ventricle has the larger cavity and the thicker muscular walls.
3 The right auricle has the larger cavity.
4 A and E.
5 To stop the backflow of blood into the ventricles when the heart rests between beats. Also, the elastic walls of the arteries press on the blood.
6 To prevent the valve turning inside out when the muscles of the ventricle walls contract.

Section 3.52

3 The muscles of the legs contract to push up the full weight of the body. This requires energy, and to provide this the blood must supply them with more raw materials and remove more waste products. Thus the heart must beat more quickly to circulate the blood at a greater rate.
4 Much more energy is required to lift the whole weight of the body up onto the chair. This is provided by the increased rate of heart beat circulating more materials in the blood, especially oxygen and sugar.
5 Very athletic individuals in a high state of training tend to have a slower heart beat than normal, and also their pulse rates return to normal more quickly when heavy exercise has stopped.

4 Respiration

Section 4.00

1 The candle burns longer in inhaled air (atmospheric air). The average burning time is 9.5 seconds for inhaled and 5.9 seconds for exhaled air in a class of 14-year-old pupils.
2 Exhaled air contains (i) less oxygen, (ii) more carbon dioxide, (iii) more water vapour.
3 Condensation of vapour to form liquid drops. It is a colourless liquid. (You must not conclude that it is water yet – there is no evidence.)
4 The blue cobalt chloride paper turns pink or the anhydrous copper sulphate turns from white to blue. This test shows that the colourless liquid contains water.
5 The exhaled breath contains a great deal of water vapour which condenses on the glass.
6 When you are breathing in, bubbles stream out of the long glass tube in boiling tube B into the indicator. When you are breathing out, bubbles stream out from the long glass tube A.
7 When you are breathing in, the pressure decreases in boiling tube B, owing to the withdrawal of air, with the result that external air pressure forces air through the liquid. When you are breathing out, the increased pressure in the long tube in boiling tube A, forces exhaled air out through the liquid.
8

Tube A	Tube B
Bicarbonate indicator turns yellow. Lime water turns milky (white).	Indicator remains unchanged. Lime water remains unchanged.

9 The exhaled air passing through tube A contains much more carbon dioxide than the inhaled air passing through B.
10 This tube is the control.

11 To stabilise the temperature to a constant value, so that the volume of the gas bubble is always measured at the same temperature.

12 Mean values are of the following order:

Inhaled air		Exhaled air	
$\%CO_2$	$\% O_2$	$\%CO_2$	$\% O_2$
1	20	4	16

These results generally show clearly that the carbon dioxide concentration increases to 4–5% in exhaled air. This provides the cause of the change in the indicator in the previous experiment. It also indicates that about the same amount of oxygen is absorbed, so that its percentage in air reduces from 21–20% to 16%.

The proportion of carbon dioxide in atmosphere air is very small (0.03%) and the apparatus is not accurate enough to show this value. It generally indicates around 1–2%.

13 Exercise will increase the rate of breathing and also its depth, such that the air will be rapidly exchanged. The analysis of air from deep in the lungs generally gives the following, even for a resting subject:

Carbon dioxide	5.55%
Oxygen	14.08%

Arrange for a fellow student (or yourself) to perform vigorous exercises such as running up and down stairs, or stepping on and off a chair. Immediately the exercise is completed and the subject is breathing heavily, take a sample of exhaled air and analyse it by means of a J-tube.

Section 4.02

1 To ensure that no acid or alkaline deposit will be present on the apparatus to change the pH of the indicator.

2 Green plant material contains chlorophyll and under normal laboratory conditions of light and temperature, it would be photosynthesising in the daytime. This process results in a gaseous exchange opposite to that under investigation. For this reason the plant material used must not contain chlorophyll. There is of course another alternative, that all plant material could be used but it must be kept in the dark. The tubes could be surrounded by black paper or black polythene. In this way the gas exchange of photosynthesis is prevented.

3 The tubes containing animals turn yellow quite quickly, dependent on the activity of the animal. The tubes containing plant material do turn yellow eventually, but it takes longer.

4 The process which uses up oxygen and gives out carbon dioxide must proceed much more quickly in animals than in plants. It is clear from the rates of change of the indicator in the animal tubes that this depends to some extent on the activity of the animal concerned. For example, a cockroach will cause a more rapid change than a snail or slug.

5 To absorb carbon dioxide.

6 The coloured water moves up on the left side and down on the right side.

7 No. The oxygen is removed from the air by the gas exchange organs of the animal within the flask and an equal quantity of carbon dioxide is produced. The carbon dioxide is absorbed by the potassium hydroxide and thus the volume of gas in the flask decreases by a volume equal to that of the oxygen produced.

Section 4.03

1 The following are examples of results with pupils of 15 years of age. Vital capacities shown are mean values for the groups indicated.

Age 15 years and

1 to 4 months. . .	average vital capacity 3.25 litres
5 to 8 months. . .	average vital capacity 3.40 l
9 to 12 months. .	average vital capacity 3.90 l

Height

150 to 157 cm	average V. C. 2.40 litres
158 to 165 cm	average V. C. 2.75 l
166 to 173 cm	average V. C. 3.80 l
174 to 181 cm	average V. C. 3.90 l

Weight

45.0 to 51.3 kg.	average V. C. 2.75 litres
51.4 to 57.7 kg.	average V. C. 2.90 l
57.8 to 64.1 kg.	average V. C. 3.30 l
64.2 to 70.5 kg.	average V. C. 3.80 l
Over 70.5 kg.	average V. C. 3.90 l

Activities

Very active.	average V. C. 3.60 litres
Active	average V. C. 3.20 l
Some activity	average V. C. 2.75 l
No activity.	average V. C. 2.70 l

If time permits the vital capacity of students over the whole age range of a school can be measured. The following results were obtained measuring about 100 pupils:

Age

12 and 13 years.	average V. C. 2.6 litres
14 years.	average V. C. 3.3 l
15 years.	average V. C. 3.8 l
16 years.	average V. C. 4.6 l
17 years.	average V. C. 4.7 l
18 years.	average V. C. 5.0 l

Weight

45 kg	average V. C. 2.6 litres
51 kg	average V. C. 3.7 l
57 kg	average V. C. 4.2 l
63 kg	average V. C. 4.9 l
69 kg	average V. C. 4.8 l
69 kg and over	average V. C. 4.9 l

Activity

Very active.	average V. C. 4.9 litres
Active	average V. C. 4.8 l
Some activity	average V. C. 4.5 l
No activity.	average V. C. 3.9 l

2 From the figures quoted above there is a clear positive correlation of vital capacity with age, height, weight and degree of physical activity.

Section 4.10

1 Shiny and moist. This is an indication of the presence of fluid produced by the pleural membranes, which line the thoracic cavity and cover the lungs, so that they slide smoothly against each other.
2 Large blood vessels can be seen. The pulmonary artery from the right ventricle of the heart divides into two main vessels, one for each lung. Pulmonary veins discharge from each lung into the left auricle.
3 It is domed, with a tendinous centre portion attached to muscles at the edge.

Section 4.21

1 The movement of smoke through the indicator causes it to change to a yellow colour from red.
2 The colour of the indicator changes back to red.
3 The smoke contains acid gases and therefore the indicator becomes more acid as shown by the yellow colour. When the finger is removed more air flows through the indicator re-establishing the red colour as the acidity is decreased. This acidity is probably due to the carbon dioxide and other gases in the smoke.
4 The temperature of the smoke is 2–3°C higher.
5 The glass wool is brown in colour and has a tarry smell. These are due to the products of combustion of the cigarette which are retained by the glass wool.
6 They both show the diseases related to areas of high population density, particularly in the South-East and London, the industrial North and South Wales.
7 Whereas bronchitis is restricted to the very large conurbations such as London, Liverpool and Cardiff, lung cancer is more widely spread. It appears wherever there are towns whether they are large or small.
8 Bronchitis is clearly a disease of the larger industrial areas and therefore is probably related to air pollution.
9 Lung cancer is also associated with the major industrial areas but it is also shown as present in the smaller towns. It could be, therefore, that there are non-industrial factors involved such as cigarette smoking.

An important enquiry was carried out by two doctors, Stock and Campbell, in 1955, when they studied the death rate from cancer of the lung among men between the ages of 45 and 74 years living in two different areas, Liverpool (urban industrialised) and North Wales (rural and non-industrial). Their findings are summarised in the table below:

	Death rate per 100 000 per year	
	North Wales	Liverpool
Non-smokers	0.14	1.13
Pipe smokers	0.41	1.42
Cigarette smokers –		
light	1.53	2.87
medium	2.13	2.97
heavy	3.03	3.94

These results are in agreement with the analysis of the maps in fig 4.12 and 4.13, i.e. that although living in industrial polluted air can contribute to lung cancer, the benefits of living in the countryside are vastly outweighed by the adverse effects of smoking.

Section 4.31

1 The energy comes from the heat of the flame which is released by the burning of the fuel (gas or paraffin).
2 The *sun's energy* was harnessed in the fuel. The burning fuel produced *heat energy* and *light energy.* The heat energy provided the latent heat of vaporisation of water. The steam expanded and passed out

with *kinetic energy*. The steam moves the vanes of the wheel producing *mechanical energy*.
3 The level of liquid in the arm of the manometer nearest the flask falls, while the other level rises.
4 The air in the test-tube and the connection to the manometer must have increased in volume in order to depress the liquid in the manometer. Air expands considerably for a small rise in temperature. This rise in temperature must have been due to heat generated by the living organisms in the flask.
5 Use exactly similar apparatus, but with no living organisms in the flask. An equal mass of non-living material (e.g. small stones) could be substituted.

Section 4.40
1 The amount of heat produced is very small and the vacuum flask prevents its loss to the air.
2 Yes. The temperature rises showing heat production in the yeast-glucose solution.
3 To prevent atmospheric oxygen from diffusing into the yeast-glucose solution.
4 Bubbles of gas appear, rising slowly through the oil layer and bubbling out of the end of the delivery tube.
5 The indicator has turned yellow, showing that the gas is carbon dioxide.
6 Alcohol (ethanol).

Section 4.41
1 58 mg/100 cm^3 of blood.
2 40 mg/100 cm^3 of blood.
3 Assuming that the rate of loss of lactic acid continues to be the same as that shown by the last line on the graph, extend the graph until it cuts the 20 mg/100 cm^3 blood line. It cuts the line at about 73 minutes. Thus from the end of the exercise it takes: 73–9 = 64 minutes.

5 Excretion and regulation — homeostasis

Section 5.10
1 The body temperature of the frog stays at the same level as the external temperature.
2 The body temperature of the students stays at a constant level, independent of the external temperature.
3 The students can remain active at night when temperatures fall, whereas the frog becomes inactive as the body temperature drops with the external temperature. Thus the students can study throughout the 24 hour period if necessary!
4 The constant body temperature enables Man to be independent of the external temperature and so live anywhere in the world, from the Poles to the Tropics.
5 The body temperatures of the insect larvae increase as the external temperature increases.

Section 5.12
1 To remove natural oils from the surface of the skin, and thus to prevent the liquid from running off.
2 The skin feels cool at this point.
3 The movement of air from the mouth, across the water drop, causes the skin to feel even cooler.
4 The ether feels much colder than the water.
5 The ether evaporates more quickly and thus draws more heat from the surface.
6 The movement of air from the fan cools the body because the sweat produced by the body at high temperatures is evaporated.
7 Figures for this experiment will depend upon the environmental temperature. For an environmental temperature of 20°C the following were calculated:
a) 0.8°C per minute
b) 1.26°C per minute
c) 0.89°C per minute (table 12.4 and fig F).

Time (mins)	Fall in temp. first 10 minutes (°C)	Fall in temp. when wiped with ethanol (°C)	Fall in temp. second 10 minutes (°C)
1	94.0	84.5	77.4
2	93.2	82.2	76.5
3	92.4	80.9	75.5
4	91.5	79.5	74.6
5	90.6	78.2	73.6
6	89.8		72.6
7	89.0		71.5
8	88.0		70.5
9	87.0		69.4
10	86.0		68.5

Table 12.4 Figures showing effect of ethanol on the rate of cooling
(refers to 5.12 no. 7)

8 The latent heat of evaporation of the ethanol draws heat more quickly from the flask, and the rate of fall of temperature increases.
9 Prevent heat loss to the air by surrounding the flask with insulating material. This slows down convection and radiation of heat.

Section 5.13
1 As in section 5.12 the figures for this experiment will depend upon the environmental temperature. For an environmental temperature of 20°C the following were calculated: Flask A 0.79°C per min., flask B 0.44°C per min. (See table 12.5 and fig G).
2 The slower rate of fall of temperature for flask B shows that the insulating material reduces heat loss.

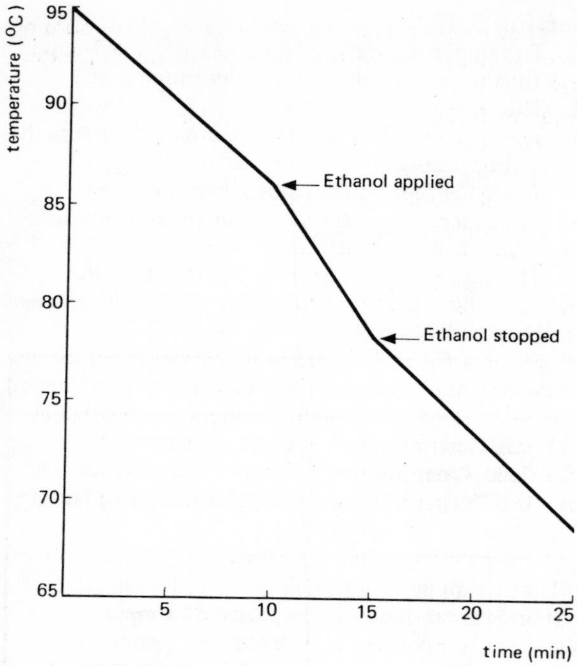

fig F A graph to show the cooling effect of ethanol

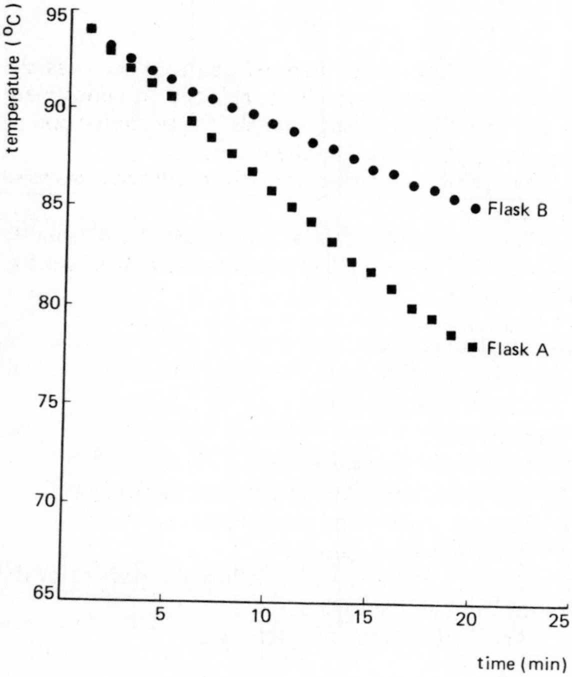

fig G A graph to compare cooling curves in insulated and uninsulated flasks

Time (mins.)	Flask A (°C)	Flask B (°C)
1	94.0	94.0
2	93.0	93.2
3	92.0	92.5
4	91.4	91.8
5	90.5	91.5
6	89.2	90.8
7	88.4	90.5
8	87.6	90.0
9	86.7	89.7
10	85.8	89.3
11	85.0	88.8
12	84.2	88.3
13	83.2	88.0
14	82.3	87.6
15	81.8	87.0
16	81.0	86.8
17	80.0	86.2
18	79.5	86.0
19	78.8	85.6
20	78.2	85.2

Table 12.5 Information for the plotting of cooling curves
(refers to 5.13 no.1)

Section 5.14

1 The ears of the foxes show adaptation to different environmental temperatures. The African fox has large ears, as does the elephant. They enable a quicker heat flow from the blood to the surrounding air.
2 The Arctic fox has the greatest amount of fur, and this slows down heat flow to the cold air surrounding the animal.
3 The African fox has the smallest head and the Arctic fox the largest. Thus the head of the African fox has a higher S.A./V. ratio to enable greater heat loss to take place.

Section 5.20

1 We have to drink more water, although under these conditions loss of water in urine is reduced.
2 If the external temperature is low, the rate of sweating is reduced and urine output increases.
3 Diarrhoea is a condition in which water is not reabsorbed from the gut, and so more than 8 litres a day could be discharged. In conditions other than intensive hospital care this water cannot be replaced.

Section 5.21

1 Renal artery; renal vein; ureter.
2 Blood vessels are distinguishable from the ureter

since they contain blood. The ureter is much paler in colour. The artery has thicker walls than the vein.
3 It is divided into an inner and outer portion (the medulla and cortex).
4 It receives the urine from the pelvis.

Section 5.22
1 No protein is filtered through the Bowman's capsule. The molecules are too large.
2 It must be reabsorbed during its passage down the nephron.
3 They become very concentrated in the urine. Urea is 60 times more concentrated in the urine, ammonia 40 times and creatinine 75 times.
4 These are two possible reasons:
i) these three substances have been actively secreted into the tubules along their length;
ii) water has been removed from the fluid of the tubules thus concentrating the solution.

The latter is the correct explanation. From the text previous to the table, it can be seen that the production of filtrate is of the order of 100 litres a day whereas urine production is 1 000 to 1 500 cm³ per day.

Section 5.23
1 Two hours.
2 This strength of salt solution is near to that of the blood and thus it does not alter the osmotic pressure of the blood. Urine output is not increased, but body weight increases by the weight of 1 litre of salt solution.

6 The skeleton and locomotion

Section 6.00
1 It has become shorter and thicker as it pulls on the forearm to bend it upward. In other words, it has contracted.
2 It has become longer and thinner as it allows the elbow to bend. In other words, it has relaxed. These two muscles act against each other antagonistically to produce bending and straightening of the arm.

Section 6.12
1 The tube.
2 A tubular structure is stronger than a solid structure of the same mass.
3 Having the heavier compact bone tissue arranged in the form of a tube around the lighter marrow enables the limb to be relatively light and strong.
4 The relative strength of the structure decreases with increased length.
5 The longer structure has to support its own increased weight as well as the weight suspended from it. Also, part of this weight is at a greater distance from the fulcrum (i.e. the chair) and thus a greater movement is produced.
6 They must be wider, as well as larger, than those of short animals, even if the trunk is not much greater.

Section 6.21
1 The atlas.
2 Nodding only.
3 Partial rotation or shaking.
4 Attachment of muscles.
5 Articulation with the ribs.
6 See table 12.6.

Thoracic	Lumbar
a) Small centrum	Large centrum
b) Longer neural spine	Smaller neural spine
c) Smaller articulating facets	Larger articulating facets
d) Facets present for articulation with ribs	Absent
e) Shorter transverse processes	Longer transverse processes
f) Absent	Additional processes present for muscle attachment

Table 12.6 Comparision of Thoracic and Lumbar Vertebrae
(refers to 6.21 no. 6)

7 In an adult rabbit there will be four (or possibly three) fused sacral vertebrae, referred to collectively as the sacrum. In young animals, the posterior one or two may not be fused to the others.
8 At the first sacral vertebra (the modified transverse processes).
9 The posterior caudal vertebrae have smaller projections. The very last ones consist only of centra.
10 Twelve.
11 Seven pairs.
12 Two pairs.
13 Three pairs.

Section 6.31
1 Lateral (or dorso-lateral).
2 The breast-bone or sternum (see fig 6.17).
3 It is cup-shaped and smooth.
4 Attachment of muscles.
5 During birth it 'gives' to allow the passage of the foetus through the vagina.
6 Articulation with the sacrum.

Section 6.32
1 Attachment of muscles.
2 Lateral.

3 Facing downwards.
4 No.
5 Prone.
6 Because Man walks on two legs instead of four, the forelimbs can be used for a variety of functions, such as feeding, and carrying and manipulation of tools. Thus, during the course of evolution, specialisation and strength has been sacrificed for versatility. The act of turning the hand from a prone to a supine position is an essential feature in such complex activities as using a screwdriver.
7 There is a ridge (the *deltoid tuberosity*) on the anterior face, for the attachment of the deltoid muscle.
8 Seven: arranged as one row of two and a more distal row of five.
9 Seven: arranged as one row of three and one row of four.
10 Its weight is carried on the bases of the digits of its hind feet: i.e. it walks on the balls of its feet. This means that the elongated metatarsals are free to increase the effective length of the leg and thus increase the potential speed. The same is true of humans when they run on the balls of their feet. Almost all fast-running mammals have elongated metatarsals that function in this way. The next time that you see a dog or a cat, note that the backward-pointing joint halfway up the animal's back leg is in fact the heel.

In ungulates, such as horses or antelope, the walking or running surface is the tip of the toe: this allows an even greater effective length of leg.

Section 6.50

1 Greyish-pink with faint lines running along the main axis of the limb.
2 Near the heel bone, the tissue takes on an opaque, white appearance. This is in fact the tendon that joins muscle to bone.

Section 6.52

1 The arm cannot maintain the rate of lift it achieved in the first minute. In subsequent minutes the rate decreases.
2 The rest periods help as shown by the rate of lift in the second and third minutes compared with the result with no rest.
3 The muscle becomes fatigued due to the build up of lactic acid. Rest periods enable some recovery to take place.

Section 6.60

1 With the upper arm vertical.
2 The tendon attaching the muscle to the bone is at right angles to the forearm so when the muscle contracts it exerts its force at a similar angle. Less force is applied if the tendon is inclined at an angle greater than 90°.

7 Co-ordination

Section 7.11

1 The projecting eyebrow ridges above the eye socket together with the bridge of the nose protect the eyes from damage if the head hits a blunt object. The eyebrows and eyelashes also form a protective screen. The eyelids form some protection but their principal purpose is to wash dust from the front of the eyeball when blinking. The eyeball is lubricated by the watery fluid discharged from the tear gland.

Section 7.12

How do the eyes react to light and dark? Responses to questions posed in procedures 1 and 2.
1 Owing to the lack of light in the left eye the right eye adjusts to allow in more light. The pupil of the right eye can be seen to enlarge.
2 When the left eye is covered the pupil enlarges to its maximum in the absence of light. Immediately the eye is uncovered the light causes the eye to react, and the pupil closes due to the expansion of the iris. Thus the eye is now adjusted to the external light intensity.

Section 7.13

1 The dot disappears when the left eye reaches the region of numbers 5, 6 and 7. Examine fig 7.5 in Chapter 7. Can you give a possible explanation why this should occur?

Section 7.14

1 The discs have been cut to different sizes, but one eye is unable to distinguish this fact, or which disc is nearer to the eye.
2 The discs appear to be their normal size, that is when the right eye is opened the two eyes can see the discs at their correct distance and correct size.
3 Two eyes set in the front of the head can distinguish clearly the different distances of objects from the head, whereas only one eye finds difficulty in interpretation of distance.

Section 7.15

1 This is muscle tissue and the strips are concerned with the movement of the eyeball.
2 There are six strips.
3 The other end of each strip is attached to the socket of the skull so that the contraction of the muscles moves the eyeball within the socket.
4 This is the nerve connecting the receptor cells of the eye with the brain. It conveys nerve impulses.
5 The black choroid prevents the reflection of light rays inside the eyeball. In a similar fashion, the camera is black inside to prevent internal reflection.
6 The humor helps to keep the eyeball spherical. It carries nutrients to the non-vascular portions of the

eye. For example, it contains large amounts of vitamin C.

7 The pupil shape is elongated compared to our own eye when seen in the mirror. This is typical of the ungulate eye. (See page 172.)

Section 7.16

1 The comparison is very close except for the focusing mechanism. In the camera, the lens moves forward or backwards, whereas in the eye the lens stays in the same position, but alters its shape to increase or decrease its power.

2 At a certain position of the flask an image of the window and the outside scene appears on the white paper screen. It is obvious from inspection of the image that it is upside down (inverted).

Section 7.17

1 No. Although one can see the pencil because an image must be present on the retina, this must be formed by rods only. No cones are present which permit colour perception. The image must have just come on to the edge of the retina so that in this region the sense cells are rods only.

Section 7.20

1 To equalise air pressure on either side of the tympanum.

Section 7.21

1 Perforation of the tympanic membrane would stop it reacting to the small changes in pressure caused by sound waves, so these could not be transferred to the ear ossicles and hence to the inner ear.

2 The vibrations of the tympanum are transferred, by the ear ossicles, to the oval window which also vibrates at the same frequency. Since the liquid of the inner ear is incompressible there must be some mechanisms by which pressure can be transferred. The round window is a membrane which acts in unison with the oval window, except that as one moves inwards, the other moves outwards and vice versa.

Section 7.31

1 Yes. The four taste areas are shown in fig H.

2 There is overlap of each area.

3 The basic pattern is similar but the overlapping areas may vary considerably. Some people are unable to separate sourness and bitterness, while others are able to differentiate types of sweetness.

Section 7.40

1 Yes, the two fingers feel at the same temperature. To ensure that they were initially at the same temperature before commencing the experiment.

2 The right index finger coming from the ice water indicates a much warmer liquid than does the left index finger which as come from the hot water. The water feels warm to the right finger, and feels cold to the left.

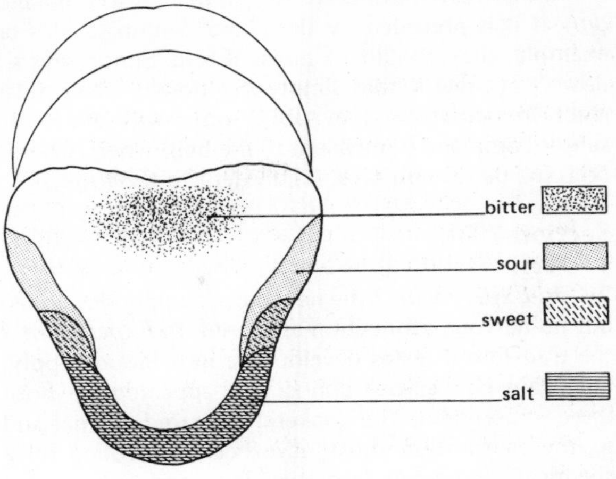

fig H Taste areas of the tongue

3 The temperature receptors cannot indicate absolute temperatures although there are detectors of cold and heat. It can detect changes in temperature and in fact is sensitive to quite small temperature differences, such as 0.2°C on the arms and 0.5°C to 1.0°C by the fingers.

4 In some parts of the skin the receptors must be further apart than 2 mm since the pins cannot be detected as two separate points.

5 The areas of the skin vary considerably. The most sensitive are tongue, nose, lips and fingertips. The least sensitive are middle of the back, shin and thick parts of the sole.

Section 7.53

1 The lower part of the leg and the foot jerk forwards.

2 The upper muscle of the thigh can be felt to bunch up as it contracts.

3 The tap below the knee is a stimulus which initiates a nerve impulse that travels by way of sensory neurons to the spinal cord. The return impulse travelling along the motor neurons causes the leg muscle to contract.

4 The vapour from the cut onion causes the eyes to discharge a liquid (tears) which can spill over onto the cheek or run down into nasal ducts.

Section 7.60

1 Less than 1, say 11/13 or 0.85.

2 Greater than 1, say 5/3 or 1.7.

3 The word ‘a’.

4 Making a tick.

5 Tapping the pencil.

6 'Simon says'. Children are told to obey a command *only* if it is preceded by the phase 'Simon says'. For example, they should sit down if told 'Simon says sit down', but not if told simply 'Sit down'. They must react *immediately*. Inevitably some will obey the simple command (equivalent to the bell) when it is *not* prefixed by 'Simon says' (equivalent to the food).

Section 7.71

1 Yes.

2 In *Experiment 2* the new blood supply developed but no nervous connection appeared. In *Experiment 3* the transferred testes developed a new blood supply, but again no nervous connection appeared. In both these experiments the cockerel appeared normal and so the maturation must have been controlled by a chemical discharged into the blood and not nervous control.

3 Kamimura showed that even with the pancreatic duct tied off no sugar was discharged into the urine, but if a pancreas was removed (Mering and Mindowski) the animal died. Thus the hormone controlling blood sugar must have been discharged into the bloodstream.

4 The extract only lowers the glucose circulating in the blood, and therefore has to be repeated again when the glucose builds up. It is not a cure for the disease.

8 Reproduction, development and growth

Section 8.11

1 It bears an artery and a vein.

2 It is the vas deferens, which carries the sperm from the testis to the urethra.

3 Six or possibly seven: the bladder; the two vasa deferentia; the two seminal vesicles (which are very difficult to seperate from the coagulating glands which lie anterior to them), and the prostate gland (which may or may not appear paired).

Section 8.12

1 At the base of the bladder, near the urethra.

2 No. The right kidney is more anterior than the left.

3 Conduction of urine only to the exterior. It is separate from the reproductive tract.

4 The following structures are absent: penis, testes, vasa deferentia, prostate, Cowper's glands and seminal vesicles.

The following is present in the female but not in the male: a large Y-shaped structure with 'knobs' of tightly-coiled tubules at the tips of the arms.

Section 8.30

1 i) Accumulation of alcohol (ethanol) from anaerobic respiration will produce a toxic effect.

ii) Alternately, the substrate may be exhausted.

Section 8.31

1 S-shaped or sigmoid. This shows a slow increase in weight at first, followed by a greatly increased rate of weight increase, followed in turn by a gradual levelling off of the rate of weight increase at maturity. Compare this curve with that of the population of yeast in fig

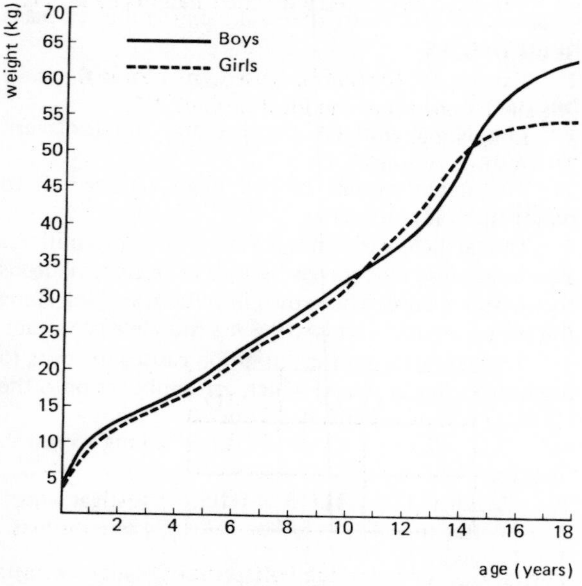

fig I A comparison of weight in boys and girls

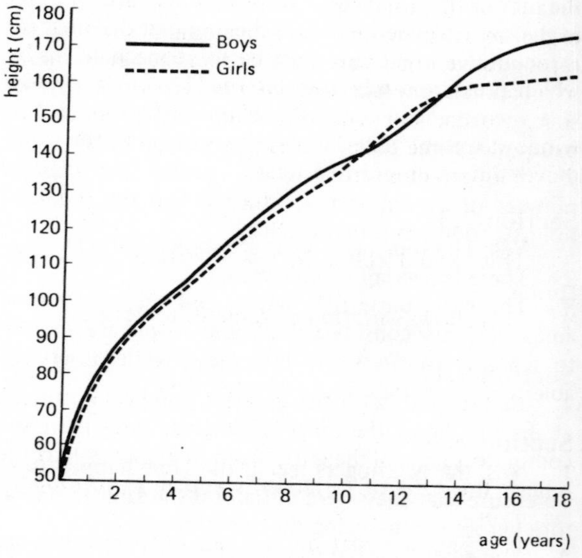

fig J A comparision of height increases in boys and girls

8.23. You will note that colonies of micro-organisms, where the individuals are separated from each other, conform to the general growth pattern of a multicellular organism when treated as one mass of cells.
2 Sigmoid.
3 The curve for males. In the curve for females, the increase in rate of growth starts earlier and flattens out both earlier and at a lower weight. This reflects the facts that girls tend to mature earlier than boys and that men tend to be heavier than women (see fig I).
4 The 'height spurt' appears earlier in girls (12–13 years) than in boys (13–15 years). This indicates earlier maturity. Boys on average are 12 cm taller than girls when the growing period has been completed (see fig J).

9 Heredity

Section 9.00

1 i) Both contain a nucleus.
ii) Both contain cytoplasm.
iii) Both are surrounded by a cell membrane.
2 i) It has chloroplasts.
ii) It has a large vacuole.
iii) It has a cell wall outside the membrane.

Section 9.02

1 i) No asters are present.
ii) The parent cell does not constrict at telophase: instead a new cell wall grows inwards at the equator.
2 They are a part of the plant that is growing rapidly, and therefore, the cells are continually dividing.

Section 9.04

1 Theoretically the answer is four. However, the details of meiosis vary according to whether the organism is an animal or a plant, and whether the reproductive organs are male or female. For example, in mammals one spermatogonium gives rise to four spermatozoa, but one oogonium will produce one ovum plus three polar bodies (see section 8.21).
2 The gametes are said to contain the *haploid* number of chromosomes, which is half the (*diploid*) number found in somatic cells.
3 This would result in less variation in the offspring.
4 This allows *variation* of gametes to occur.

Section 9.11

1 Bell-shaped, with the greatest numbers occurring at the middle of the range of heights. Since there is a complete range of heights, from the shortest to the tallest, this type of variation is known as *continuous variation.*
2 No. The fact that no two sets of fingerprints are exactly the same has made fingerprinting an extremely useful technique in crime detection. Since there is a complete range of the different types of fingerprint, this is another example of continuous variation.
3 The approximate proportions in a large population are: loops 70%, whorls 13%, arches and other types (including combinations) 17%.
4 No. Either the tongue can be rolled or it cannot be rolled. This is an example of *discontinuous variation.*
5 This is clearly different from the type of inheritance for height and fingerprints, which show a wide range of variation because a number of different factors are involved. Inheritance of tongue rolling is controlled in a relatively simple way: the following is a full explanation, which you will find easier to understand after you have worked through sections 9.12 to 9.16.

The ability to roll your tongue is inherited through a single dominant gene. Thus the homozygote for this gene (RR) and the heterozygote (Rr) both have the ability to roll their tongues. Only the homozygote for the recessive gene (rr) cannot roll his tongue.
6 i) Small numbers of offspring.
ii) Long generation time.
iii) Unsuitable for experimental crosses.
iv) Unsuitable for breeding of close relatives.

Section 9.12

1 25% (approximately).
2 i) The time between generations is very short (10–14 days).
ii) It has large numbers of offspring.
iii) It is easily bred in the laboratory.
iv) It has a number of characters which show discontinuous variation.
v) These characters are easily visible.
vi) Flies can be crossed with their parents or *siblings.*

Section 9.15

1 The greater the numbers used, the nearer the ratio obtained approaches the ideal ratio of 3:1.
2 Long; long; vestigial.
3 Approximately 3:1.

Section 9.16

1 Ll and ll.
2 L and l.
3 l only.
4 P_2 Ll x ll

ll

Ll

gametes	l	l	
L	Ll	Ll	2 long wings
l	ll	ll	2 vestigal wings

F_2 phenotypes

The ratio of longed winged offspring of vestigial winged offspring is 1:1.

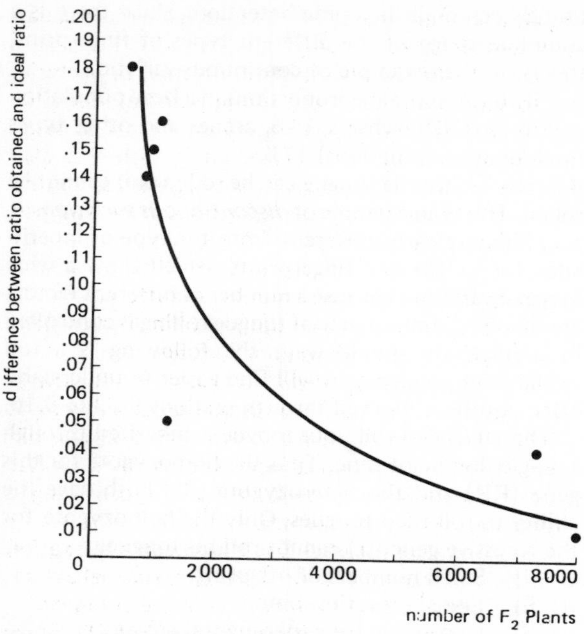

fig K The difference between the ratio obtained and the ideal ratio

Section 9.17

1 P_1 TT x tt

tall short

F_1 Tt x TT

TT

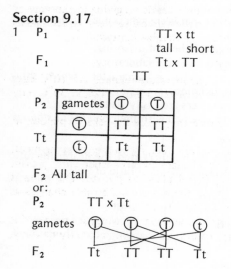

P_2 gametes	Ⓣ	Ⓣ
Ⓣ	TT	TT
Tt ⓣ	Tt	Tt

F_2 All tall

or:

P_2 TT x Tt

gametes Ⓣ Ⓣ Ⓣ ⓣ

F_2 Tt TT TT Tt

2 50%. The other 50% are heterozygous: that is to say they (like the F_1 plants) possess genes for tallness and shortness.

3 i) They can be self-pollinated.

ii) Since they are naturally self-pollinating, pure-breeding strains are easy to obtain.

iii) The numbers of offspring can be even greater.

4 50% smooth seeds and 50% shrunken seeds.

5 The time between pollination and appearance of the characters is very short. Experimenters do not have to wait until a plant has grown and matured.

Section 9.20

1 P $G^AG^O x G^BG^O$

AO x BO

BO

gametes	Ⓑ	Ⓞ
Ⓐ	AB	AO
AO Ⓞ	BO	OO

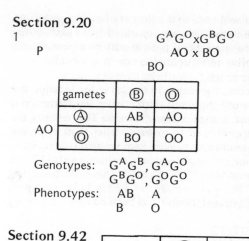

Genotypes: G^AG^B, G^AG^O, G^BG^O, G^OG^O

Phenotypes: AB A B O

Section 9.42

1

gametes	X^R	X^r
X^r	X^RX^r	X^rX^r
Y	X^RY	X^rY

F_1 genotypes X^RX^r X^rX^r X^RY X^rY

phenotypes daughters sons

normal sighted, colour blind, normal sighted, colour blind

2 Married to a normal sighted male, all of her sons would be colour blind but her daughters would be normal sighted, although 'carriers'.

Section 9.53

1 Malaria.

2 People possessing the sickle cell trait also have a high resistance to a certain type of malaria. Thus, people living in regions where malaria is common are less likely to die before producing children if they possess a single S gene than if they possess none at all: the S genes in these areas are therefore selectively advantageous.

3 Mutations of *Staphylococcus* are produced which are resistant to Penicillin. These mutated forms gradually increase.

4 Vary the type of antibiotic used so that mutated forms do not become dominant.

5 46 (23 pairs).

6 47 – one extra chromosome.

Section 9.55

1 Light moths are more vulnerable in Birmingham and dark moths in Dorset, because they stand out against the tree backgrounds.

2 With the advent of the industrial revolution, smoke and soot from the factory chimneys blackened the bark of trees in industrial areas. The prevailing winds carried this smoke and soot to the east.

8 The overall colour of tree bark would lighten and dark moths would become more obvious to predators than light moths. More light moths than dark moths would survive to reproductive age and would pass on the genetic character for light wings to their offspring. Thus the next generation would contain a greater proportion of light moths. This process would continue until almost all peppered moths in the area would have light wings. This is an example of evolution taking place by natural selection responding to changes in the environment.

10 Organisms and disease

Section 10.12

1 *Tubes* 1, 2 and 3 have changed in appearance.
2 The broth has become cloudy (turbid).
3 The broth has gone bad. It has an unpleasant smell, and it is probable that this has been caused by micro-organisms.
4 *Tube 1* goes cloudy even though heated, but it is open to the air.
Tube 2 although protected from the air is not heated and goes cloudy, but rather more slowly.
Tube 3 is open to the air through the glass tube and although heated turns cloudy.
Tube 4 is the only tube to remain clear. It has been heat treated but is open to the air. Thus air itself does not cause change in the broth but something present in the air. The micro-organisms present in the air must be trapped in the S-shaped glass tube.
5 Water condenses inside the Petri dish and if it were the correct way up the drops on the lid could fall onto the culture. Therefore the cultures are placed upside down and marking them on the base enables the markings to be seen clearly. Furthermore since the lid can be twisted around markings on the lid could be moved out of their correct position.
6 To ensure that they are not taken off accidentally or fall off as the dishes are picked up.
7 Growth of micro-organisms shows up as patches on the jelly.
8 This could be regarded as a control to make sure that the broth is sterile, but it could be contaminated by the hairs when placed on the jelly. It is better to have a separate dish with jelly only as a control.
9 There are fewer colonies on the jelly in 1B than in 1A, showing that soap and water can remove many bacteria, but not all.
10 i) Hands and other parts of the skin should be kept very clean.
ii) Hair should be contained under a hat in order that stray hairs should not fall on the food.
iii) Mouth masks would prevent the transfer of micro-organisms by coughing or sneezing.
11 At a temperature of 37°C which is the mammalian body temperature.
12 Dish 3 showed fewest or no colonies of bacteria, and so food stored at the freezing point of water (or slightly below) would be prevented from decaying. Heat treatment of food as shown in dish 5 is also effective in that it destroys micro-organisms, and no decay will occur if micro-organisms can be stopped from reinfecting the food.

Section 10.25

1 a) The gonococci have developed strains resistant to certain antibiotics and sulphonamides.
b) Many women show no symptoms of gonorrhoea and thus may pass the disease on unknowingly.
c) The development of the contraceptive pill means that fewer couples use sheaths, which offer some protection from the disease by providing a physical barrier to the bacteria.
d) In many societies, there has been a relaxation of traditional standards with regard to sexual behaviour, and thus casual affairs have become more common.
e) Embarrassment at contracting an STD causes people to postpone seeking medical attention until the disease has progressed considerably. By this time they may have infected other people. Embarrassment and shame also result in patients giving incomplete or false information to the medical services, which makes it difficult to trace all possible contacts.

Section 10.26

1 The time is temperature-related. At 10°C eggs may take 4 weeks to develop into larvae, at 30°C only 3 days.
2 At 10°C larvae may take 8 weeks to pupate, at 30°C, 3 days.
3 Larvae are actively feeding on the decaying meat. They moult and increase in size. When fully grown, they seek a sheltered place to pupate. The pupa is inactive but inside there are considerable changes as the larval form changes to the adult.
4 It crawls out of the pupa. Its wings are limp and it waits for these to dry and harden before flying away.

Section 10.44

1 The patient forms antibodies to the 'foreign' proteins of the donated heart tissue.
2 Naturally immune people may act as unsuspecting carriers of the disease.
3 When a patient has already been infected with pathogenic micro-organisms. In severe cases, the patient's own lymphocytes may not have time to develop the ability to produce antibodies before the disease proves fatal. In other cases, passive immunisation simply speeds recovery. People who have not previously been actively immunised against tetanus

are often immunised passively if they receive a deep wound.

Section 10.45

1 The dose taken was not sufficient to kill all the bacteria, with the result that the more resistant forms survived. The antibiotic also killed some harmless gut micro-organisms which had previously competed with the pathogens. Thus the resistant, pathogenic bacteria multiplied, causing the disease to become more acute.
2 Antibodies are continuously being broken down in the blood system, and the white blood cells of the patient cannot replace them.

Section 10.70

1 Old people are particularly vulnerable to falls and hypothermia. Care should be taken to ensure that stairways are well lit and not too steep. Stair coverings and rugs should be firmly anchored and highly polished floors should be avoided. Baths should have handrails and non-slip mats. Footwear should be well-fitting.

Since many old people live alone and are not very mobile, falls resulting from slips, faints, heart attacks or strokes can mean that they are left lying on the ground for hours or even days before help arrives. Their ability to produce body heat is also limited and they may well die from hypothermia in cold weather. Added to this is the problem of poverty in old age, and the expense of providing adequate heating. Therefore, wherever possible the neighbours and families of old people should visit them regularly (if only briefly) and should look out for signs such as uncollected milk or newspapers, which might indicate that all is not well. Where possible, flats and bungalows should be purpose-built and fitted with telephones and alarm bells.
2 Young children are particularly vulnerable to burns and scalds, electrocution, suffocation and poisoning. Older children are also the main victims of drowning accidents. Care should be taken to keep domestic hot water supplies below 65°C, to guard all fires and to turn the handles of all cooking pots inwards from the edge of the cooker. Electric sockets should be kept out of the reach of young children and fireworks should be restricted to organised displays. Plastic bags should be removed from all new toys and clothes, since these can cause death by suffocation if placed over the head. Like old people, babies are vulnerable to hypothermia in cold weather. This is because they lose heat quickly through a surface area which is large relative to body volume, and they should be kept well covered during the night. Medicines should be kept in locked bathroom cabinets and all poisonous cleaning fluids etc. should be kept in cupboards well out of reach of young children. Weedkillers and other poisons should be kept where children cannot reach them and should be labelled clearly and not kept in old sof drink bottles.

11 Man and his environment

Section 11.11

1 to 6 Answers specific to the community studied.

Section 11.40

1 This must be partly due to the removal of predators but also due to the large amounts of vegetation available for browsing deer. In 1905 only about 5 000 were feeding in an area capable of supporting 30 000. This probably meant that not only would the birth rate rise, but many more young deer would survive if there was plenty of food and few predators.
2 As the population rose above 30 000 the vegetation was unable to support the population. Finally the whole area was considerably overcropped.
3 The overcropping meant lack of food, lower birth rate, fewer surviving young and furthermore an increased birth rate and survival for predators with plenty of available prey (the deer). There would be intraspecific competition for food between the deer, and interspecific competition between deer and predators. Disease would also play a part in reducing the heavy population of deer.
4 The population of predators would fall back as its food disappeared (10 000 deer in 1939), then the deer population would recover, followed again by that of the predators. The populations would again reach a peak but they would fluctuate around the 30 000 mark for deer. See fig 11.41 for the lynx and snowshoe hare population cycles.
5 The hare population. They are able to take advantage of the greater space and food available. The carrying capacity is greater than 20 000.
6 As the number of hares increases greater amounts of food are available for the predators and thus more young lynx survive in each litter. The lynx population therefore increases with a lag behind the hares.
7 Because the secondary consumer of any food chain must be sustained by larger numbers of primary consumers (in this case hares).

Section 11.42

1 Better hygiene, proper disposal of sewage, the understanding of the link between disease and causal agents (such as bacteria), antiseptic and disinfectant techniques in hospitals and houses.
2 About the year 1938.
3 Lack of investment in public health. As a result public health measures in towns and villages such as piped water, sewage works, rubbish disposal, adequate medical care and so on were not available.

4 Family size decreased in the late nineteenth century and early twentieth century due to increased wealth and the beginning of family planning through birth control.

5 Improved social hygiene and medical care reduced the infant mortality rate and the death rate. More children survived and families became larger.

Section 11.44

Experiment page 310

Experiment beakers	Number of days after beginning experiment					
	2	4	7	14	21	28
A	10	10	10	13	18	24
B	10	losing green colour	decolourised	—	—	—
C	10	10	10	10	10	dead – fungal attack
D	10	10	10	10	10	dead
E	10	10	12	13	19	36
F	6 sunk	9 sunk	all sunk	—	—	—
G	10 yellow	brown	—	—	—	—
H	10	pale	pale	decolourised	—	—
I	10	10	13	20	29	48
J	10	10	11	12	16	19

Table 12.7 **Results of experiment: numbers and condition of fronds**

Investigation of water pollution of a stream

1 to 7 Answers will be specific to the stream or river under investigation and the type of pollution involved.

Index